ANCIENT GLASS

This book is an interdisciplinary exploration of archaeological glass in which technological, historical, geological, chemical and cultural aspects of the study of ancient glass are combined. The book examines why and how this unique material was invented some 4,500 years ago and considers the ritual, social, economic and political contexts of its development. The book also provides an in-depth consideration of glass as a material, the raw materials used to make it and its wide range of chemical compositions in both the East and the West from its invention to the seventeenth century A.D. Julian Henderson focuses on three contrasting archaeological and scientific case studies: late Bronze Age glass, late Hellenistic–early Roman glass and Islamic glass in the Middle East. He considers in detail the provenances of ancient glass using scientific techniques and discusses a range of vessels and their uses in ancient societies.

Julian Henderson is Chair of Archaeological Science in the Department of Archaeology at the University of Nottingham. The author of *The Science and Archaeology of Materials*, he has published more than 200 contributions to books and journals, including in *Antiquity, Journal of Archaeological Science, Journal of Glass Studies* and *Journal of Analytical Atomic Spectroscopy*.

∽ ∽ ∽ ∽

ANCIENT GLASS

An Interdisciplinary Exploration

Julian Henderson

University of Nottingham

CAMBRIDGE
UNIVERSITY PRESS

University Printing House, Cambridge CB2 8BS, United Kingdom

One Liberty Plaza, 20th Floor, New York, NY 10006, USA

477 Williamstown Road, Port Melbourne, VIC 3207, Australia

314-321, 3rd Floor, Plot 3, Splendor Forum, Jasola District Centre, New Delhi - 110025, India

79 Anson Road, #06-04/06, Singapore 079906

Cambridge University Press is part of the University of Cambridge.

It furthers the University's mission by disseminating knowledge in the pursuit of education, learning and research at the highest international levels of excellence.

www.cambridge.org
Information on this title: www.cambridge.org/9781107551909

© Julian Henderson 2013

This publication is in copyright. Subject to statutory exception and to the provisions of relevant collective licensing agreements, no reproduction of any part may take place without the written permission of Cambridge University Press.

First published 2013
Paperback edition first published 2016

A catalogue record for this publication is available from the British Library

Library of Congress Cataloging in Publication data
Henderson, Julian, 1953–
Ancient glass / Julian Henderson.
p. cm.
Includes bibliographical references and index.
ISBN 978-1-107-00673-7 (hardback)
1. Glassware, Ancient.   2. Glassware, Classical.   3. Glass manufacture – History.   I. Title.
TP850.H46   2012
748.2009′01–dc23   2012017636

ISBN 978-1-107-00673-7 Hardback
ISBN 978-1-107-55190-9 Paperback

Cambridge University Press has no responsibility for the persistence or accuracy of URLs for external or third-party internet websites referred to in this publication, and does not guarantee that any content on such websites is, or will remain, accurate or appropriate.

For Rosette and Yvette
This book is also dedicated to the memory of my parents.

CONTENTS

Figures		page xiii
Tables		xvii
Preface		xix

1	GLASS AS A MATERIAL: A TECHNOLOGICAL BACKGROUND IN FAIENCE, POTTERY AND METAL?	1
	1.1 Glass as a Material	1
	1.2 The Formation of Glass: Of Volcanic Glass, Asteroids, Slags and Scums	5
	1.3 Production of the First Glasses	8
	1.4 The First Glass: A Paradigm Shift?	12
	1.4.1 Glazed Steatite, Egyptian Blue and Faience	13
	1.5 Evidence of Production Sites	18
	1.6 Conclusions	19
2	WAYS TO FLUX SILICA: ASHES AND MINERALS	22
	2.1 Glass Raw Materials	22
	2.2 Halophytic Plant Ashes	23
	2.2.1 Occurrence and Range of Types	23
	2.2.2 Technological Considerations	24
	2.2.3 Chemical Analyses of Plant Ashes	26
	2.2.3.1 The Choice of Scientific Techniques for the Analysis of Plant Ashes	26
	2.2.3.2 Turner's Research	26
	2.2.3.3 Brill's Work	28
	2.2.3.4 Other and More Recent Research	29

Contents

	2.3 Major Elemental Compositions of Syrian Halophytic Plants and the Implications for Ancient Glass Production	34
	2.3.1 A Relationship between Halophytic Plant Ash Compositions and Plant Species?	36
	2.3.2 Glass Coloration and Halophytic Plant Ashes	42
	2.3.3 Alkali Levels in Halophytic Plants	44
	2.3.4 Calcium Levels in Halophytic Plants	46
	2.4 Forest Plant Ash Compositions	48
	2.5 Natron	51
	2.6 Reh and Oos	53
	2.7 Nephaline	54
	2.8 Conclusions	54
3	**Silica, Lime and Glass Colourants**	**56**
	3.1 Silica	56
	3.2 Calcium and Aluminium	64
	3.3 Colourants and Opacifiers	65
	3.3.1 What Causes Glass Colour?	65
	3.3.2 The Use of Colourant Materials	68
	3.3.2.1 Cobalt Coloration in Glass	69
	3.3.2.2 Manganese-Coloured Glass	75
	3.3.2.3 Copper- and Gold-Coloured Translucent Glasses	75
	3.3.3 Decolorised Glass	76
	3.3.4 Glass Opacification	77
	3.4 Conclusions	81
4	**Glass Chemical Compositions**	**83**
	4.1 The Middle East and Europe c. 2500 B.C.–A.D. 1700	85
	4.1.1 Plant Ash Glasses c. 2500–c. 800 B.C.; A.D. 800–1700	85
	4.1.2 Mixed-Alkali Glass and Its Compositional Variations (c. 1100–c. 750 B.C.)	90
	4.1.3 The Transition from Plant Ash Glass to Natron Glass, c. A.D. 800	91
	4.1.4 Natron Glasses c. 800 B.C.–c. A.D. 800	92
	4.1.5 The Demise of Natron Glass Technology: The 'Reintroduction' of Plant Ash Glass Technology from c. A.D. 800 in the Middle East	97
	4.1.6 The Demise of Natron Glass Technology: From Natron Glass to Wood Ash Glass North of the Alps	102
	4.1.7 Wood Ash Glasses and the Use of Other Alkali-Rich Plants (c. A.D. 800–1700)	104

		4.1.8 Lead Oxide-Silica Glass in the West	108
	4.2	*India, Pakistan, and Sri Lanka*	111
		4.2.1 The Compositional Types	111
		4.2.2 Soda-Alumina Glass: Technology and Trade (Beads and Bangles)	113
	4.3	Africa and Madagascar	116
		4.3.1 The Compositional Types	116
		4.3.2 High Lime–High Alumina Glass	117
	4.4	China and Southeast Asia	118
		4.4.1 The Compositional Types	118
		4.4.2 Lead–Barium Oxide Glasses	123
		4.4.3 Potassium-Silica Glasses	123
		4.4.4 Potassium-Lime Glasses	124
		4.4.5 Lead-Potassium and Lead-Soda Glasses	125
		4.4.6 Mixed-Alkali Glasses in Southeast Asia	126
		4.4.7 Lead-Silica Glasses	126
5	EARLY GLASS IN THE MIDDLE EAST AND EUROPE: INNOVATION, ARCHAEOLOGY AND THE CONTEXTS FOR PRODUCTION AND USE		127
	5.1	The Social and Political Contexts of Early Glass Production in the Middle East	127
	5.2	The Occurrence of Early Glasses in Mesopotamia	132
		5.2.1 Alalakh, Tell Atchana (Plain of Antioch), Turkey	136
		5.2.2 Ugarit-Ras Shamra, Syria	138
		5.2.3 Tell Brak, Syria, Northern Mesopotamia	139
		5.2.4 Nuzi, Near Kirkuk, Iraq, Northern Mesopotamia	140
		5.2.5 Assur, Iraq, Northern Mesopotamia	141
		5.2.6 Tell al-Rimah, Iraq, Northern Mesopotamia	142
		5.2.7 Ur, Iraq, Babylonia	142
		5.2.8 Nippur, Iraq, Babylonia	143
		5.2.9 Tchoga Zanbil, Iran, Babylonia	143
	5.3	Innovation and Glass Technology in Mesopotamia	144
	5.4	The Social and Political Contexts for the Earliest Evidence of Glass Production in Europe	148
		5.4.1 Frattesina, Provincia Rovigo, a Key Site for Early Glass in Europe	152
		5.4.1.1 The Chronology of Frattesina (Twelfth–Ninth Centuries B.C.)	154
		5.4.1.2 Evidence of the Glass Industry at Frattesina	154
	5.5	Conclusions	156

Contents

6		EARLY GLASS IN THE MIDDLE EAST AND EUROPE: SCIENTIFIC ANALYSIS	158
	6.1	Scientific Analysis of Mesopotamian and Syro-Palestinian Glasses (c. 1500–1000 B.C.)	160
	6.2	Scientific Analyses of Egyptian Glasses	164
	6.3	Scientific Analyses of Mycenaean Glasses	165
	6.4	Did Primary Glass Production Occur in Different Parts of the Bronze Age Mediterranean?	167
		6.4.1 An Isotopic Approach	175
		6.4.2 The Way Forward	181
	6.5	Scientific Analyses of Mixed-Alkali and Potassium Glasses, and Mixed-Alkali Glassy Faience and Faience from Europe: Evidence of Production Zones?	183
		6.5.1 Potassium Glasses	196
	6.6	Local Production Centres in Bronze Age Europe	197
	6.7	Conclusions	199
7		HELLENISTIC TO EARLY ROMAN GLASS: A CHANGE FROM SMALL- TO LARGE-SCALE PRODUCTION?	203
	7.1	The Hellenistic and Roman Middle East	205
	7.2	Hellenistic Glass Production	209
		7.2.1 Primary and Secondary Production of Hellenistic Glass	214
		7.2.1.1 Rhodes	214
		7.2.1.2 Beirut	215
		7.2.1.3 Syria, Greece and North Africa	222
	7.3	Some Hellenistic Vessel Forms	223
		7.3.1 Hellenistic Core-Formed Vessels	223
		7.3.2 Hellenistic Glass on Rhodes	226
	7.4	Early Roman Glass Production	227
	7.5	The Use of Hellenistic and Roman Glass	232
8		LATE HELLENISTIC AND EARLY ROMAN GLASS: SCIENTIFIC STUDIES	235
	8.1	Levantine Glass and a Comparison with Other Glasses	236
	8.2	The Relevance of Furnace Glass	242
	8.3	Greece Rhodes and Transalpine Europe	246
	8.4	Links with Roman Glass	247
	8.5	Glass and Society	250

Contents

9		ISLAMIC GLASS: TECHNOLOGICAL CONTINUITY AND INNOVATION	252
	9.1	The Social, Political and Religious Contexts of Islamic Glass Production	252
	9.2	Islamic Glass Technology	257
		9.2.1 Raw Materials and Historical References to Their Use	260
		9.2.2 The Change from Natron to Plant Ashes	265
	9.3	Production Centres	266
	9.4	Archaeological Evidence for Production Sites and Furnaces	270
	9.5	Archaeological Considerations for Scientific Analysis	276
10		ISLAMIC GLASS: SCIENTIFIC RESEARCH	279
	10.1	Umayyad Glass	280
	10.2	'Abbāsid Glass	282
	10.3	Glass Compositions and Vessel Types from the Middle East	290
	10.4	Other Islamic Glass Compositions	296
		10.4.1 Islamic Glass from North Africa and the Persian Gulf	300
	10.5	Conclusions	303
11		THE PROVENANCE OF ANCIENT GLASS	306
	11.1	Models for Glass Production and Glass Provenance	307
	11.2	Glass Provenance and the Chemical Composition of Raw Materials	309
		11.2.1 Silica	309
		11.2.2 Natron, Trona, Nepheline and Reh	310
		11.2.3 Plant Ashes	310
	11.3	Glass Provenance and Batch Formation	314
		11.3.1 Alkali Sources	315
		11.3.2 Ashing and Fritting	316
		11.3.3 Bronze Age Glass	317
	11.4	Glass Provenance and Chemical Compositions	319
		11.4.1 Technological and Cultural Factors	319
		11.4.2 The Chemical Compositions of Plant Ashes and Plant Ash Glasses	320
		11.4.3 Summary	325
	11.5	The Use of Isotopes to Provenance Glass: The Importance of an Environmental Approach	326
		11.5.1 Calcium Sources, Strontium Isotopes and Glass Provenance	328

	11.5.2	Silica Sources, Neodymium and Oxygen Isotopes and Glass Provenance	330
	11.5.3	Lead Isotopes in Glass	333
	11.5.4	Mixing of Raw Glass and Glass Raw Materials	334
11.6	A Case Study: Strontium, Neodymium, Oxygen and Lead Isotopes of Glass Found at the Glassmaking Site of Al-Raqqa, Syria		335
	11.6.1	New Information about the Use of Glass Raw Materials	339
	11.6.2	Evidence for Mixing of Raw Materials	340
	11.6.3	Local Production and Imports	342
11.7	Lead Isotope Analyses of Bronze Age and Islamic Plant Ash Glasses		345
11.8	The Future Potential of Isotope Studies		347
	11.8.1	Local Manufacture	348
		11.8.1.1 Bronze Age and Islamic Plant Ash Glasses	348
		11.8.1.2 Phoenician, Hellenistic, Iron Age and Roman Natron Glasses	351
	11.8.2	The Supply System	355
		11.8.2.1 Bronze Age and Islamic Plant Ash Glass	355
		11.8.2.2 Phoenician, Hellenistic, Iron Age and Roman Natron Glasses	356
	11.8.3	Trade	357
		11.8.3.1 Trade in Bronze Age and Islamic Plant Ash Glasses	357
		11.8.3.2 Trade in Phoenician, Hellenistic, Iron Age and Early Roman Natron Glasses	359
11.9	Conclusions		360
12 CONCLUSIONS			363
Appendix			373
Bibliography			381
Index			425

FIGURES

1.1 Australian aboriginal arrowheads knapped from bottle glass, 3
1.2 Two faience ushabti figures, 7
1.3a A magnified (backscattered electron) image of a faience specimen. 1.3b An annotated diagram of 1.3a, 9
1.4 A large droplet of copper sulphide (Cu_2S) surrounded by crystals of copper in a glass matrix of an eleventh-century opaque orange glass tessera, 11
1.5 Two small pieces of overheated frit found at the ninth-century glassmaking site of al-Raqqa, Syria, 21
2.1 *Anabasis syriaca*, 31
2.2 *Chenopodium murale*, 32
2.3 *Girgensonia* sp., 33
2.4 *Halopeplis* sp., 33
2.5 *Haloxylon articulatum*, 34
2.6 *Salsola jordanicola*, 34
2.7a *Salsola kali*. 2.7b *Salsola kali* (detail), 35
2.8 Map of the sampling locations in Syria and Lebanon from which halophytic plants were collected, 37
2.9 A bi-plot of weight % magnesia versus calcium oxide in ashes of *Salsola* and non-*Salsola* plants, 38
2.10 A bi-plot of weight % magnesia versus calcium oxide in ashes of various species of *Salsola*, *Anabasis*, *Haloxylon*, *Halopeplis* and *Arthrocnemum*, 39
2.11 A bi-plot of weight % magnesia versus potassium oxide in sodium-rich plant ashes of the genera *Salsola*, *Anabasis*, *Haloxylon*, *Halopeplis* and *Arthrocnemum*, 40
2.12 A bi-plot of weight % magnesia versus calcium oxide in ashes of plants of the genus *Salsola* compared with ashes of non-*Salsola* plants, 41
2.13 A bi-plot of weight % ferric oxide versus aluminium oxide in ashes of *Salsola* and non-*Salsola* plants, 43
2.14 An X-ray diffraction pattern derived from the analysis of ashes of *Salsola jordanicola*, 45
2.15 A bi-plot of weight % calcium oxide versus aluminium oxide in ninth- and eleventh-century plant ash glass, 47
3.1 A backscattered electron micrograph of a high iron and alumina mineral inclusion, 58
3.2 A backscattered electron micrograph of an iron oxide inclusion in an eleventh-century intensely dark green glass mosaic tessera, 58
3.3 Five-sided zircon crystals in the glassy matrix of ninth-century faience, 59
3.4 A backscattered electron micrograph of a section of a shell found in a sand sample, 61
3.5 A bi-plot of weight % lead oxide versus zinc oxide in a variety of glasses, 74
3.6 A backscattered electron micrograph of a sample of glass consisting of opaque white calcium antimonate crystals, 78

3.7 A bi-plot of calcium oxide against sulphur trioxide in second- and third-century Roman glass tesserae, 78
3.8 A backscattered electron micrograph of second-century B.C. chunk of opaque yellow raw glass, 79
3.9 A backscattered electron micrograph of calcium fluoride crystals in a sixteenth-century A.D. Chinese cloisonné enamel, 79
3.10 A backscattered electron micrograph of metallic Cu drops in ninth-century opaque red glass, 81
4.1 A bi-plot of weight % magnesia versus potassium oxide in Bronze Age glasses, 89
4.2 A bi-plot of weight % soda versus soda+potassium oxide in Middle Eastern furnace glasses, 97
4.3 A bi-plot of weight % magnesia versus potassium oxide in ninth-century glass, 99
4.4 A bi-plot of weight % magnesia versus potassium oxide in eleventh- and twelfth-century glass mosaic tesserae, 101
4.5 A bi-plot of weight % potassium oxide versus calcium oxide in early medieval (ninth–tenth century) and high medieval (eleventh–twelfth century) glass, 103
4.6 A bi-plot of weight % soda versus potassium oxide in postmedieval (sixteenth–seventeenth century) glass, 107
4.7 A bi-plot of weight % calcium oxide versus potassium oxide in postmedieval (sixteenth–seventeenth century) glass, 108
4.8 A bi-plot of weight % alumina versus magnesia in postmedieval (sixteenth–seventeenth century) glass, 109
4.9 A bi-plot of weight % ferric oxide versus alumina in postmedieval (sixteenth–seventeenth century) glass, 111
4.10 A bi-plot of aluminium versus calcium oxides in glasses from India, Sumatra and northern Syria (al-Raqqa), 119
4.11 A bi-plot of calcium oxide versus potassium oxide in high potassium and mixed-alkali glasses, 125
5.1 Locations of principal sites mentioned in the text, 129
5.2 A fourteenth-century opaque turquoise bottle from Tell Brak, Syria, decorated with combed opaque white and yellow glass strands vessel, 135
5.3 A fourteenth-century B.C. turquoise blue glass ingot, 137
5.4 Fragments of fourteenth-century B.C. raw glass, 140
5.5 Map of Italy showing the main sites mentioned in the text, 148
6.1 A bi-plot of weight % soda versus calcium oxide in late Bronze Age cobalt blue and turquoise blue late Bronze Age glass, 161
6.2 A bi-plot of weight % magnesia versus potassium oxide in late Bronze Age glass, 171
6.3 A bi-plot of weight % alumina versus manganese oxide in early Iron Age glass, 173
6.4 A bi-plot of weight % alumina versus manganese oxide in Syrian halophytic plant ashes, 174
6.5 A bi-plot of weight % nickel oxide versus manganese oxide in glasses, 175
6.6 A bi-plot of weight % alumina versus ferric oxide in mixed-alkali glasses, 176
6.7 Bi-plot of $^{87}Sr/^{86}Sr$ versus the concentration of Sr (ppm) in glasses, 177
6.8 Bi-plot of $^{143}Nd/^{144}Nd$ versus the concentration of Nd (ppm) in glasses, 178
6.9 Bi-plot of the concentration of Sr (ppm) versus the concentration of Nd (ppm) in glasses, 179
6.10 A bi-plot of chromium/lanthanum versus 1,000× zirconium/titanium in late Bronze Age glasses, 180
6.11 Bi-plot of $^{87}Sr/^{86}Sr$ versus $^{143}Nd/^{144}Nd$ in glasses, 181
6.12 A crystalline silica inclusion in a mixed-alkali glass from eleventh-century B.C. Frattesina, 185
6.13 A bi-plot of weight % soda versus potassium oxide in eleventh-century B.C. glasses, 189
6.14 A bi-plot of weight % magnesia versus calcium oxide in eleventh-century B.C. glasses, 190
6.15 A bi-plot of weight % phosphorus pentoxide versus calcium oxide in eleventh-century B.C. glasses, 191

Figures

6.16 A bi-plot of weight % calcium oxide versus magnesia in eleventh- to eighth-century B.C. glasses, 192
6.17 A bi-plot of weight % soda versus potassium oxide in eleventh- to eighth-century B.C. glasses, 193
6.18 A bi-plot of weight % soda versus potassium oxide in French and Italian middle Bronze Age mixed-alkali glasses, 194
6.19 A bi-plot of weight % calcium oxide versus magnesia in middle Bronze Age French and Italian glass and (vitreous) faience, 195
6.20 A bi-plot of weight % soda versus potassium oxide in early Bronze Age French glass and (vitreous) faience, 197
7.1 Map of main locations mentioned in the text, 206
7.2 Conical grooved bowl, 213
7.3 Glass vessel fragments associated with the Beirut 015 tank furnaces dating to before A.D. 50, 217
7.4 Examples of some of the massive amount of raw glass found attached to the tank furnace floors in Beirut 015 dating to before A.D. 50, 217
7.5 General view of excavated glass tank furnaces showing TFC 2 in the background, 218
7.6 N-S section tank 2 Section (N-S) through Tank 2 (TFC 2) with lower floor remains and still unexposed earlier wall, 218
7.7 Plan of Tank Furnaces Complex 2 remains within the context of the excavated remains, 219
7.8 Schematic plan of Tank Furnaces Complex 2 remains within the context of the excavated remains, 220
7.9 Schematic plan of Tank Furnaces Complex 3 remains within the context of the excavated remains, 221
7.10 Chronological development of core-formed Hellenistic vessels, 224
7.11 An example of a vessel from the third-century B.C. 'Canosa group' found in Italy, 225
7.12 A large raw piece of glass found on the Sanguinaire shipwreck on the coast of Corsica, 227
8.1 A bi-plot of the weight % calcium oxide versus alumina in Italian, French and Levantine glasses, 237
8.2 A bi-plot of the concentrations of zirconium versus strontium in Italian, French and Levantine glasses, 239
8.3 A bi-plot of the concentrations of barium versus zirconium in Italian, French and Levantine glasses, 240
8.4 A bi-plot of the weight % calcium oxide versus alumina in Beirut glass tank furnaces and the eighth-century Bet Eli'ezer glass furnace, 241
8.5 A bi-plot of the weight % calcium oxide versus alumina in Beirut glass tank furnaces compared with late Hellenistic–early Roman vessels, 244
8.6 A bi-plot of the weight % soda versus alumina in contemporary blown and moulded glass vessels found in Beirut, 245
9.1 The principal sites mentioned in the text, 253
9.2 Mosaics composed of glass tesserae in the Umayyad mosque in Damascus, Syria, constructed in the early eighth century, 255
9.3 Cut glass bottle, 258
9.4 Lustre painted, 258
9.5 Enamelled horseman on a twelfth- to fourteenth-century colourless bottle, 259
9.6 Millefiori wall plaque from Samārrā no. 200, 261
9.7 The southern wall of the al-Raqqa tank furnace showing a rill of vitrified cement that divided the slabs attached to the wall, 272
9.8 The two-phase floor of the glass furnace, 272
9.9 A section through the furnace showing the heating chamber located below the furnace floor, 273
9.10 A view of the furnaces from above facing east showing the sectioned tank furnace, a sectioned circular furnace and the mud-brick structure built against the western wall of the tank furnace, 273
9.11 The rills of glass and cement that divided the slabs attached to the "circular" furnace floor, 275

Figures

10.1 A bi-plot of weight % alumina versus calcium oxide in specific types of mainly ninth- and tenth-century Islamic vessel decorative types, 281

10.2 A bi-plot of weight % magnesia versus alumina in glass from ninth- and eleventh-century contexts in al-Raqqa and probably seventh- to eighth-century Qasr al-Hayr al Sharqi, 283

10.3 A bi-plot of weight % magnesia versus alumina in ninth- and eleventh-century glass from al-Raqqa according to glass type and artefact type, 285

10.4 A bi-plot of weight % alumina versus calcium oxide in raw ninth-century (TZ) and twelfth-century (TB) furnace glass, 285

10.5 A bi-plot of weight % calcium oxide versus phosphorus pentoxide versus calcium oxide in ninth- and eleventh-century al-Raqqa glasses, 286

10.6 A bi-plot of weight % alumina versus ferric oxide in ninth- and eleventh-century al-Raqqa glasses, 287

10.7 A bi-plot of weight % manganese oxide versus ferric oxide in al-Raqqa glasses by compositional type, 288

10.8 A bi-plot of weight % manganese oxide versus ferric oxide in specific glass colours of al-Raqqa glasses, 289

10.9 A bi-plot of weight % alumina versus calcium oxide in tenth- to fourteenth-century glasses from al-Raqqa, Fustat, enamelled vessels and enamelled mosque lamps, 293

10.10 A bi-plot of weight % magnesia versus calcium oxide in Sasanian, ninth- to twelfth-century raw (furnace) glass from al-Raqqa and glass from twelfth- to fifteenth-century Nishapur, the Serçe Limani shipwreck, Damascus, Beirut and Bahrain, 295

10.11 A bi-plot of weight % magnesia versus calcium oxide in ninth- to tenth-century glass vessels of specific types, 297

10.12 A bi-plot of weight % alumina versus potassium oxide in twelfth-century and Ottoman glass armlets, 298

10.13 A bivariate plot of weight % alumina versus potassium oxide in four groups of Islamic glass, 299

11.1 Log % oxides of aluminium (Al), iron (Fe), magnesium (Mg), calcium (Ca), sodium (Na), potassium (K) and phosphorus (P) in four ashed *Salsola* plant samples, 311

11.2 Log % oxides of barium (Ba), copper (Cu) and strontium (Sr) in four ashed *Salsola* plant samples, 313

11.3 Weight % magnesium versus potassium oxide (divided by 10) in raw furnace glass, 321

11.4 Weight % magnesium versus calcium oxide in raw furnace glass, 322

11.5 Weight % magnesium versus calcium oxide in plant ash glasses, 323

11.6 Weight % alumina versus iron oxide in plant ash glasses, 325

11.7 Weight % alumina versus iron oxide in raw furnace glass, 327

11.8 Strontium isotope variations across the Syrian landscape showing a clear dip in the values for the area, 337

11.9 The concentration of strontium versus strontium isotope ratios in glass, 339

11.10 The concentration of neodymium versus neodymium isotope ratios in glass, 341

11.11 The concentration of neodymium versus neodymium isotope ratios in glass, 341

11.12 $^{143}Nd/^{144}Nd$ versus $\delta^{18}O$ vsmow in glasses, 343

11.13 $^{143}Nd/^{144}Nd$ versus $^{87}Sr/^{86}Sr$ in al-Raqqa, Banias and Tyre glasses by artefact type, 344

11.14 $^{143}Nd/^{144}Nd$ versus $^{87}Sr/^{86}Sr$ in al-Raqqa glasses by compositional type, 345

11.15 $^{208}Pb/^{206}Pb$ versus $^{207}Pb/^{206}Pb$ in plant ash and natron glasses, 347

TABLES

2.1 The average composition of 50 samples of Syrian keli in parts per 1,000, 27
2.2 Chemical analyses of the ashes of three halophytic plant species from Syria, 35
2.3 Major elemental compositions of the ashes of three species of forest plants from the UK, 48
2.4 The chemical composition (weight % oxide) of glasses made from beech, oak and bracken ashes, analysed using wavelength-dispersive X-ray fluorescence, 50
4.1 Examples of plant ash glass compositions, 86
4.2 Examples of mixed-alkali and high potassium European glass compositions, 90
4.3 Examples of natron glass compositions, 93
4.4 Examples of European medieval/postmedieval compositions, 105
4.5 Examples of western lead oxide–silica glasses, 110
4.6 The composition of high lime–high alumina glass, 118
4.7 Examples of glass compositional types found in China, 120
4.8 Examples of glass compositional types found in Southeast Asia, 121
5.1 The chronology of later prehistoric Italy, 149
6.1 The chemical composition of blue glass from Eridu, Iraq, c. 2300 B.C., 159
11.1 The strontium and neodymium isotope results for twenty-five samples of glass from al-Raqqa, Syria, 336

PREFACE

This book provides an integrated interdisciplinary approach to the study of a complex and fascinating ancient material. A variety of aspects of ancient glass are discussed, including principally archaeology, history, chemical analysis, materials science, geology and botany. The aim of this book is to explore these aspects by using a combination of focused studies and case studies in various ancient and historical periods. Each case study – in Bronze Age Mesopotamia, the late Hellenistic–early Roman Middle East, and the Islamic world – has been selected to incorporate contrasting social, political, economic and ritual contexts in which glass was manufactured, traded and used. The contrasting characteristics of these societies therefore influenced the ways in which glass was manufactured and used by them. There are complex relationships between the production, trade and use of ancient materials, including glass. The scale of production, involving a range of facilities and critical combinations of raw materials from a variety of sources, was characteristic of specific societies and their ideologies. Each step in the *chaîne opératoire* involved decisions, each with a social impact and significance leading to the manufacture of glass artefacts characteristic of that society. The control over each aspect of production was a reflection of the degree of social hierarchy (perhaps involving social elites) and complexity at the time.

The use of a combination of established and new scientific techniques to examine glass is increasingly providing more precise answers to questions about glass technology, provenance and trade and therefore its place in ancient societies. The results of such scientific projects are described here.

The first four chapters of this book provide an essential background for understanding the case studies. These are a definition of glass as a material (Chapter 1), the definition and sources of glass raw materials (Chapters 2 and 3)

Preface

and a review of ancient glass chemical compositions found in a range of countries in the Eastern and Western Hemispheres. The three case studies (Chapters 5–10) have two chapters, each consisting of the archaeological background and scientific investigations. Another chapter (Chapter 11) brings together many of the issues dealt with in previous chapters. It focuses on the question of whether glass can be provenanced using a range of techniques; the provenance of glass impinges on a range of research questions. The issue of glass provenance considered here was partly inspired by archaeological and scientific investigations at al-Raqqa, Syria, which was funded with a substantial grant from the UK Arts and Humanities Research Council: the Raqqa (Syria) ancient industry project. The discovery and excavation of glass and pottery production on a large scale in an urban context led to a variety of archaeological, cultural and scientific questions, including those relating to provenance. I am grateful to the Director General of Antiquities of Syria and to his staff in Damascus and al-Raqqa, especially Dr Michel Maqdissi and Murhaf al-Halaf.

The writing of this book occurred over several years during which a range of important studies have been published. I have tried to make it as up to date as possible, but inevitably I have not been able to reference all of the most recent publications. The bulk of the time during which it was written was funded by the British Academy in the form of a Research Readership. I am extremely grateful to the British Academy for this funding and for other associated grants from them.

Overall, it is hoped that this book will fill a niche in the literature by considering ancient glass in a fully integrated interdisciplinary fashion, producing an enriched and enriching result.   Nottingham, November 2011

Further Acknowledgments

I thank the following colleagues for discussions about ancient glass, some of whom I have collaborated with: Dr Irina Andreescu-Treadgold, Professor Youssef Barkoudah, Dr Paolo Bellintani, Dr Sophie Bertier, Dr Anna-Maria Bietti Sestierri, Dr Isabelle Biron, Dr Cristina Boschetti, Dr Stephanie Boulogne, Dr Robert Brill, Keith Challis, Dr Simon Chenery, Dr Hans Curvers, Professor Dr Patrick Degryse, Professor Jane Evans, Dr Hamish Forbes, Professor Ian Freestone, Dr Sylvia Funfschilling, Dr Caroline Jackson, Dr Jens Kroeger, Dr Doug Lee, Professor Cristina Leonelli, Dr Fiet van Lith, Dr Kimioshi Matsumura, Stephen Minnitt, Professor Nakai, Dr Marie-Do Nenna, Dr Kalliopi Nikita, Dr Joan Oates, Dr Sachihiro Omura, Professor Sarah O'Hara, Dr Yvette Sablerolles, Dr Marco Verità, Dr David Whitehouse and Dr Nikos Zacharias. I am grateful to Kay Smith for reading through a draft of this book in its entirety and to my wife, Yvette Sablerolles, for making a range of constructive comments about Chapter 7.

ONE

GLASS AS A MATERIAL

A TECHNOLOGICAL BACKGROUND IN FAIENCE, POTTERY AND METAL?

Glass is a thing in disguise, an actor, is not solid at all, but a liquid, that an old sheet of glass will not only take on a royal and purplish tinge but will reveal its true liquid nature by having grown fatter at the bottom and thinner at the top, and that even that it is frail as the ice in a Parmatta puddle, it is stronger under compression than Sydney sandstone, that it is invisible, solid, in short, a joyous and paradoxical thing, as good a material as any to build a life from.
   Peter Carey, *Oscar and Lucinda*, 111 chapter 32, 'Prince Rupert's Drops'

1.1   GLASS AS A MATERIAL

Glass was the first man-made translucent 'solid'. Those who first created it must have been impressed and greatly mystified by the way the glowing red-hot liquid cooled and appeared to 'freeze' into a block of 'solid' glossy material that reflected and refracted light. A quintessential characteristic of early glass was its colour, which could be used to imitate semi-precious stones such as lapis lazuli and turquoise, and there was even the potential to modify it. Indeed, the first appearance of glass is likely to have been unexpected in a high-temperature environment, leading inquisitive minds to question how it formed. Even if produced adventitiously, it may have been highly coloured and is likely to have been attributed a ritual significance. Some of the earliest glass, made from plant ash and silica, was certainly intended to imitate semi-precious stones and was attributed apotropaic properties. As the scale of glass production increased and the roles that glass played in society changed over time, the processes of its production became less mysterious and less enveloped in ritual. Nevertheless, even today in Murano, the famous centre for glass production in the Venetian lagoon in Italy, glassmaking families retain closely guarded technical secrets.

Clearly, glass production involved a series of inter-related aspects. The more practical aspects included the selection and use of raw materials; the collection of enough of the appropriate fuel types; the production of bricks and construction and use of particular furnace types; observation of the glass being melted; the production of a range of vessel and bead forms by blowing, moulding and winding; the decoration of vessels and beads and the relationship between glass technology and other industries, with the potential for sharing knowledge and raw materials. The ethnicity of all of the groups of people connected to various aspects of production, whether they were responsible for gathering glass raw materials, preparing them, building furnaces and ancillary structures, making crucibles and moulds and blowing and shaping glass, for example, would have had an impact on the object forms and decorative styles produced and on how they were used. The forms and colours of glass vessels and beads produced were a reflection of the period in which they were made. Depending on the social contexts in which they were used, glass colours may have been significant in a variety of ways. For example, during the sixteenth and seventeenth centuries, native Indians in North America believed that glass properties could be symbolic of the mind, knowledge and life with white glass equating with good things, including peace and a desire for understanding and red glass with war, intense experience, animation and fire heat (Turgeon 2004, 34–5).

Well before the first glass was made, a naturally occurring glass, obsidian, was worked into tools by percussive flaking. Obsidian can be grey, dark green, or apparently black, but the colours do not even come close to the beautiful range that can be achieved in man-made glass, and obsidian could not be worked at the temperatures achieved by ancient man. Man-made glass could also be flaked, but evidence of this is relatively rare (Fig. 1.1). The reason both glass and obsidian can be flaked as if they were stone is that both are amorphous materials. The word *amorphous* has a specific meaning to material scientists; it is a characteristic that helps define a material that lacks long-range structural order and can be described as a state of matter (Brill 1962). Crystalline silica, as opposed to glass, is composed of silicon and oxygen atoms arranged in a regular way; glass is arranged in a far less regular way, with the bridges between atoms being broken and the other components, such as sodium and calcium atoms, distributed in a relatively random way. Most glasses are composed of a network of silica, SiO_2 (the network former) and other metallic oxides. In pure crystalline silica, each silicon atom is bonded with four oxygen atoms, forming what is known as a tetrahedron $(SiO_4)^{4+}$. When arranged into a three-dimensional network, with the adjacent tetrahendra sharing one oxygen atom, it forms pure silica glass (with a melting point of c. 1700°C, unattainable by ancient man). A typical network modifier, such as soda (Na_2O), bonds ionically

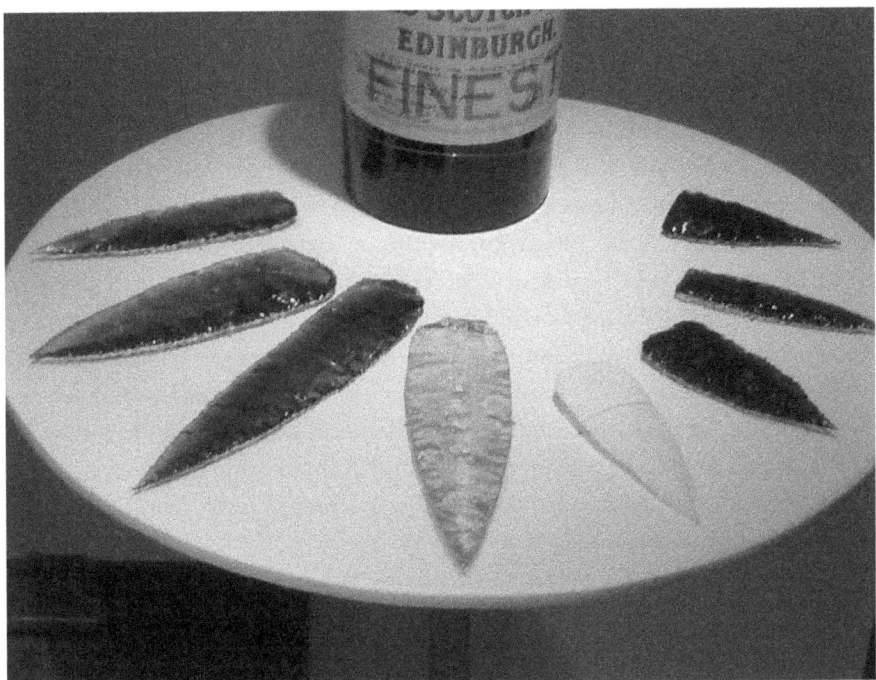

1.1. Australian aboriginal arrowheads knapped from bottle glass (photo: J. Henderson; reproduced with kind permission of Håken Wahlquist and the Ethnographic Museum, Stockholm).

with oxygen within the network and disrupts bridging atoms within the silica tetrahedral network. Network stabilisers (such as calcium oxide, CaO), another type of network modifier, increase the durability of the glass: the bond strength of Ca^+ is twice that of sodium (Na^+) and strengthens the structure. Trivalent alumina (Al_2O_3) and iron oxide (Fe_2O_3) can act as either stabilisers or fluxes.

By using techniques such as X-ray diffraction spectroscopy and neutron diffraction spectroscopy a 'structure' of glass can be revealed (e.g. Hannon and Parker 2000), although this in no way approaches the tightly ordered lattice of metal. Greaves *et al.* (1991) have shown that alkalis and nonbridging oxygen atoms are not arranged in a random way but tend to be concentrated in channels, giving rise to the term 'modified random network model'. This model therefore does not quite fit that for a 'liquid'. Indeed, the glassy state does not fit any of the three classical states of matter – solids, liquids and gases (Brill 1962, 129). This has implications for the durability of ancient glass, as has the distribution of pores through the glass (Lester *et al.* 2004).

Because glass is amorphous and assuming that there are no inclusions in it when it is struck, the 'shock waves' will pass through it in an unhindered way. In contrast, when a highly structured material like metal is struck, the 'shock waves' are prevented from being transmitted through it by the presence of ordered crystalline lattices consisting of repeated structural units; this is known as short-range order. When struck, the amorphous property of glass, obsidian and flint produces conchoidal (shell-like) fractured facets (see Fig. 1.1). However, the properties of glass that we are mainly concerned with here are not those that link it to other siliceous materials, like obsidian and flint, but those that come about when it is fused from raw materials, moulded, drawn, blown and wound to create objects and then cooled to a 'solid' state.

The apparently 'solid' property of glass is somewhat deceptive. Material scientists actually refer to glass as a 'super-cooled liquid' (Shelby 2005, 4–5). In 1956, Jones provided the following definition of glass: 'an inorganic product of fusion which has been cooled to a rigid condition without crystallisation' (Jones 1956, 1). Transparency and translucency must have been considered important properties in the past, so it would have been necessary to avoid the formation of crystals as a result of cooling the glass at the appropriate rate. In relation to the clarity of glass, Al-Buhturi (820–897), a celebrated Arab poet, described a glass containing wine in the following way:

Its colour hides the glass as if it is standing in it without a container.

Another property that has been commented on is the brittleness and transitory nature of (some) glass. In his *Pirotechnia*, published in the sixteenth century, Vannoccio Biringuccio used these properties as a metaphor for man's own transitory existence:

Considering its brief and short life, it cannot and should not be given too much love, and it must be used and kept in mind as an example of the life of man and of the things of the world which, though beautiful, are transitory and frail.

However, the numbers of complete Hellenistic and Roman glass vessels that have survived intact are a testament to the durability of glass rather than its brittleness. This piece seems to be more of a commentary on the fragility of human life than on (most) glass! Glass came late to China. In the medieval period the Chinese were fascinated by the transparency of glass and at the same time they were puzzled by the fact that glass could be both hard and fluid. They compared it with ceramics, metal, precious stones (especially jade) and even water 'but were unable to find a satisfactory analogy to further people's understanding of the material'. Moreover glass did not fit their 5 element system (metal, wood, water, fire and earth) (Hsueh-Man 2002a, 72–3).

The appearance of ancient glass, including some of the earliest, was deliberately altered to make it appear more like semi-precious stones, such as the rare mineral lapis lazuli, with a principal source in Afghanistan. This was achieved by rendering the glass *opaque* by adding crystalline materials or developing crystals out of the glass by reheating (striking) the glass.

When glass is cooling, one particular temperature, the *transition temperature* (T_g), is critical to the formation of a glass rather than a crystalline silicate (Henderson 2000, 24, fig. 3.1). At this temperature, the properties of glass pass from those like liquids to those like solids, although 'solidification', as such, does not occur. The crystalline silicate has a lower volume than the equivalent glass, and it is at the transition temperature that an abrupt change of volume occurs in glass and a slowing down of the rearrangements in its structure. When glass cools, its *viscosity* increases, so much so that for a Roman soda-lime natron glass of a typical chemical composition (see Chapters 4 and 8), Brill (1988) showed that the glass could be gathered and marvered (rolled across a flat [metal] surface to regularise the shape of a gather) at between *c.* 1100°C and 1000°C and softened sufficiently at *c.* 1000°C to blow it. Most ancient glass is composed of three major components. The first is silica (SiO_2), which is generally present at between *c.* 65% to 70%; the second is a flux, soda (Na_2O), which reduces the melting temperature of silica from between *c.* 1710°C. and *c.* 1730°C. to *c.* 1150°C; the third is 'lime' calcium oxide (CaO), which provides durability to the glass. Without calcium oxide the glass would dissolve in water. An alternative alkali, potassium oxide (K_2O), was also used. It has been found in glasses dating to as early as the tenth century B.C. in the Mediterranean, western Han Chinese glass and glass dating to the medieval period in northwestern Europe (see Chapter 4).

Wood or plant ashes and alkali-rich minerals are sources of glass fluxing materials (see Chapter 2). By changing the proportions of soda, lime and silica, the melting and working properties of glass also change. Therefore, the chemical composition of glass has a direct relationship to the ways in which it can be worked. Soda-lime-silica glass has a minimum *liquidus temperature* (the absolute melting temperature of the glass above which nuclei and crystals cannot form) at the ternary *eutectic* of 725°C for a composition of 21.9% soda, 5.1% calcium oxide and 73.1% silica (Morey 1964, fig. 20, tables 13 and 33). A eutectic mixture of compounds is one that has the lowest freezing point of all possible mixtures of sodium, calcium and silicon oxides. The wide range of ancient glass chemical compositions that has been found is discussed in Chapter 4. Variations in the balance of each major (and some minor) component in the glass would have had a direct affect on its working properties.

1.2 THE FORMATION OF GLASS: OF VOLCANIC GLASS, ASTEROIDS, SLAGS AND SCUMS

When magma is spewed from volcano vents and then chilled, obsidian is formed. Nuclear explosions, such as the first atomic bomb test at the Trinity site in New Mexico in 1945, can lead to glass formation. More recently, the use of another natural glass has been suggested as the material used for the carved scarab that forms the decoration of a breastplate of King Tutankhamen: an asteroid. Asteroid impacts (Mayell 2005) can leave a 'carpet' of glass. One exploded 29 million years ago above the Sahara Desert, turning the sand into glass, with a heat that was equivalent to a 110-megaton bomb (see http://www.sandia.gov/news/publications/technology/2006/0804/glass.html). Impact metamorphism can produce 'diaplectic glasses' from quartz and feldspars (Heide and Heide 2011, 28).

In considerably less dramatic contexts, glassy slags can be produced in virtually any high-temperature environment and it is these that probably constituted the first glassy material seen by ancient peoples. Glassy slags can be produced by burning cereals rich in silica-rich opal phytoliths, which provide rigidity to the plant structure (Dimbleby 1978, 129–30), and the combustion of haystacks (Baker 1968). Folk and Hoops (1982) found 'attractive' twelfth-century B.C. blue-green glass at Tel Yin'am in Israel, which they interpreted as being the adventitious fusion of silica-rich plant ash and silica. Even though vesicular (*ibid.*, 460, fig. 14), the chemical composition and colour suggest that it was possibly man-made. Youngblood *et al.* (1978) published scientific analyses of glasses that were formed when the ramparts of Scottish Iron Age forts were ignited, the result of silica in the soil fusing with alkaline-rich materials in the fort ramparts. This actually led to a strengthening of the defences. 'Bone-ash' slags can be produced in cremations, and these also typically have a vesicular appearance (Henderson, Janaway and Richards 1987); Photos-Jones *et al.* (2007) have shown that seaweed known as cramp fused to bone in Bronze Age burials also produced a kind of vitreous slag. Glassy (fuel ash) slags are also often produced in hot furnaces and kilns in which metals are smelted or pots fired; the ashes from the fuel interact with silica present in both the bricks used to build the furnaces and kilns and in crucibles containing hot metals (Biek and Bayley, 1979; Henderson 2000, 53). Indeed *vitrification* in pottery, which results from alkali-bearing minerals interacting with silica in the clay (Kingery *et al.* 1976, 490), can produce hard ceramics such as stoneware and porcelain (Henderson 2000, 133). Pottery wasters from kiln sites often resulted from pots being heated to such high temperatures that the clays became glassy, having started to bloat and the pots then lost their shapes (Henderson 2000, 133). Even glass production generates a range of glassy slags resulting from the interaction of the fuel ashes with the silica-rich bricks used to construct the glass furnace. Glass production can also generate vitreous 'scums' (the nonreactive ingredients of the glass batch) on the surface of glass melts, or they are deposited on the sides and lips of crucibles as the raw materials in the glass batch fuse and the whole melt contracts. Thus these very different formation environments led (and still lead) to the formation of vitreous slags of a range of distinctive chemical compositions, some of which are highly coloured.

This adventitious formation of glass may be regarded as somewhat prosaic in the context of ancient technologies. However, the brilliant red glassy material

1.2. Two faience ushabti figures from Deir el Bahr (c. 1000–900 B.C.; photo: J. Henderson; produced with kind permission of the Museum of Mediterranean and Near Eastern Antiquities, Stockholm).

produced by the presence of reduced copper (Cu I) in metal smelting (Henderson 2000, 54) must have been impressive. Its blood-red colour would undoubtedly have had a ritual significance. Moreover, it is striking that one of the commonest early glass colours in Bronze Age Mesopotamia, Mycenaean Greece and in parts of Europe is the oxidised form of copper (Cu II), which is a turquoise green colour. So although the occurrence of copper in early glasses shows that a copper-rich colourant was available, it does not prove that metallurgy was the driving force behind the emergence of the first glass. As one of the last primary ancient inorganic materials to have been manufactured, it would be obtuse to suggest that only pottery or metal or faience technology led to the development of glass (see Figs. 1.2 and 1.3). Peltenburg (1987, 20–2) has stressed the paucity of solid evidence for the link between glass and metal technologies. Nevertheless, copper-rich minerals did provide the critical colourant that allowed early glassmakers to imitate turquoise, a stone considered to have health-giving properties with ritual playing a dominant part in every aspect of ancient society. Moreover, early stone glazing was achieved by heating copper ore on the powdered surface of talc (Hodges 1970, 62). So the availability of copper can be regarded as one of a number of parameters that played a part in the emergence of early vitreous materials, including glass.

Thus, the adventitious production of glass can be regarded as especially significant in two ways: (1) if brilliantly coloured, it would have made a great impression on those who first observed it, and (2) its very formation would have been striking and almost certainly would have motivated those who saw it to manufacture glass deliberately.

Evidence for an apparent value for early vitreous slag is provided by the discovery of (unpublished) greenish glassy slag placed in an Akkadian *inhumation* burial (c. 2300 B.C.) at Tell Brak excavated by David and Joan Oates. This was characterised by elevated levels of phosphorus and calcium and is therefore probably a bone-ash slag. Although to our eyes it may appear insignificant, given that it was found in an area where the first glasses in the world are likely to have been made and dates to around the time when they were first made, it suggests that such materials may have played a part in the experiments involved in the production of the first glasses. It is, however, difficult to agree with J. B. Lambert's statement (1997, 105) that 'Refinement of slag could have eventually led to glass manufacture' simply because it is difficult to envisage how this could be achieved. Some of the earliest glass known, such as that from Eridu, Iraq (Garner 1956a), dating to c. 2300 B.C., has a chemical composition that is similar to glasses found in archaeological contexts that date to a period covering a further 1,300 years (see Chapter 6). The chemical composition of the block of cobalt blue glass from Eridu indicates that it was made deliberately from a combination of plant ash and silica and coloured with a cobalt-rich material. There is no hint of a compositional link to 'refined' slag, and it is unlikely to be a by-product from an experiment. The technology of ancient glass production may therefore have been 'fully formed' by this time.

1.3 Production of the First Glasses

The first glass appeared c. 2500 B.C. in modern-day northern Syria and Iraq (Moorey 1994). Initially the glass made from plant ash and silica would have been fused in a crucible in relatively small volumes. Most of this early raw glass was then made into beads, and it was not for another 1,000 years or so that larger volumes of raw glass and greater quantities of glass objects, including the first (late Bronze Age) core-formed vessels, were produced. Therefore, there was a period of *c.* 1000 years during which only small quantities of glass were made (see Chapter 5).

The first appearance of glass, and subsequent technological developments, seem to fit Cyril Stanley Smith's contention that 'the discovery of the materials, processes, and structures that comprise technology almost always arose out of aesthetic curiosity, out of the desire for decorative objects and not, as the popular phrase would have it, out of preconceived necessity' (Smith 1981, 347). When seen for the first time, the shiny, coloured, translucent, refractive and smooth properties of glass must have been both exciting and inspiring. As Smith went on to say, 'discovery *is* art, not logic, and new discoveries have to be cherished for reasons that are far more like love than purpose' (Smith 1981, 347). By stating this, Smith, who had a worldwide reputation as an Massachusetts Institute of Technology–trained material scientist, was removing

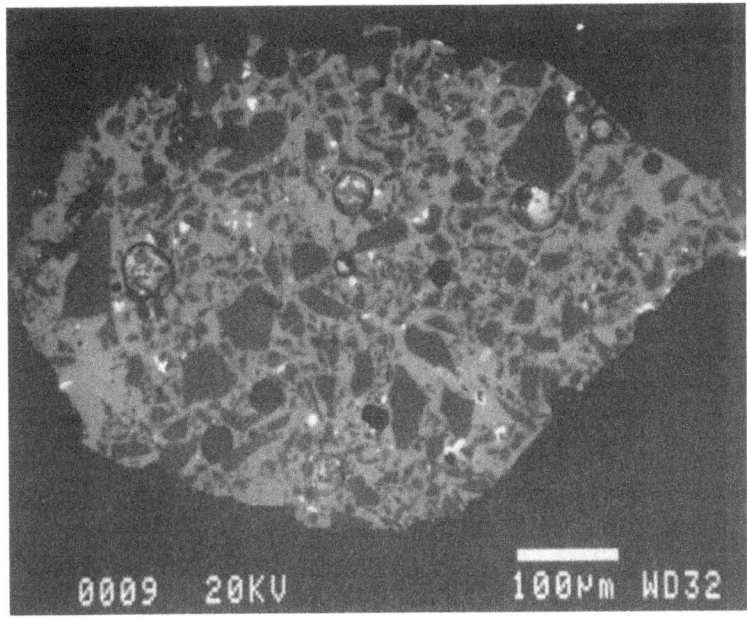

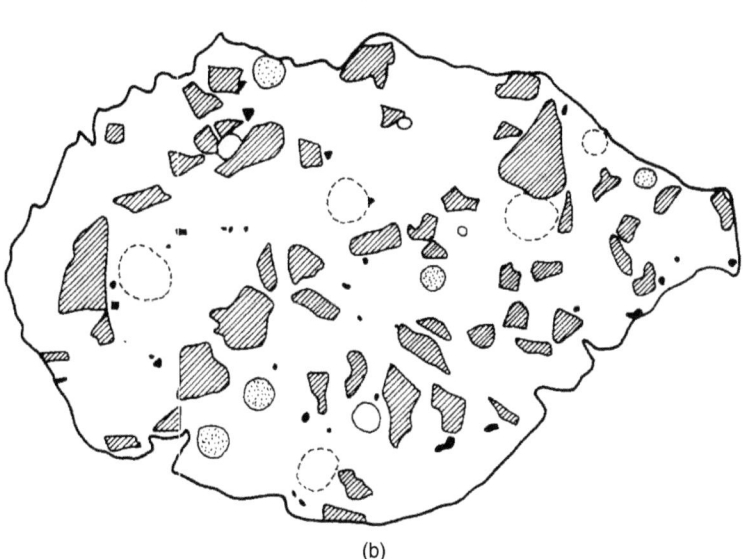

1.3a. A magnified (backscattered electron) image of a faience specimen from Hauterive Champrevèyres, Lake Village, Switzerland c. 1100 B.C. section (pale grey = glassy copper-rich glass matrix, dark grey crystals = silica; white crystals = tin oxide). 1.3b Diagram of 1.3a: hatched crystals = silica, broken lines and stipled areas = pores in sample.

the predictable aspects that characterise much contemporary positivist material science research, consisting of a mechanical and inevitable series of established procedures in the primary manufacture, formation and shaping of materials. For example, the subtle and progressive changes in the colour of glowing iron as a blacksmith works it is a reflection of its changing crystalline structure and something that has little relevance to modern industry. Such characteristics are therefore more connected to the art of creation (and in ancient contexts also to ritual and religion) than to logic.

Although the first glass may not have been deliberately coloured, the process of creating and then modifying the glass eventually led its creators to produce glass opacity, involving the incorporation of masses of tiny crystals (Henderson 2000, 35–8) in imitation of semi-precious stones (Fig. 1.4). Pale blue glass was produced in imitation of turquoise; a deep cobalt blue colour was produced in imitation of lapis lazuli; opaque yellow perhaps in imitation of gold and opaque red in imitation of blood. Most of these colours first started to be made from about the mid-fifteenth century B.C. when the first glass vessels were produced in Mesopotamia (Nolte 1968; Moorey 1994, 193). The Mesopotamian concept of raw material sources was that most had to be imported from specific 'mountains', some more mythical than others (Oppenheim 1970, 9). For example there was a 'cedar mountain', a 'gold mountain', a 'silver mountain' and a 'lapis lazuli mountain'. The wide range of terms used to describe the wide range of shades of lapis lazuli, such as 'beet coloured', 'wild donkey coloured' and 'star-like, starry' [the latter relating to flecks of pyrite in lapis lazuli] is a reflection of how lapis lazuli colours were the most cherished.

There are important questions concerning the manufacture of the first glass that we may never be able to answer adequately. Any evidence for the earliest phases of primary glass production is likely to be on a small scale. It is possible that the first glasses were made deliberately using a range of raw material proportions so as eventually to optimise the process. Perhaps the relatively common production of vitreous slags prompted early glassmakers to somehow establish that plant ash and silica were the primary raw materials. The successful production of glass or glassy materials would, however, be limited by the proportions that could successfully lead to glass production, a ratio of 2:1 by weight of plant ash to silica. Rehren (2000, 15; 2008, 1353) has argued that glass melts follow minimum eutectic troughs, leading to relatively narrow glass compositions. Evidence for experimentation could potentially be shown scientifically if the chemical compositions of contemporary vitreous materials and glasses found on a production site occupy a range or a continuum of *connected* compositions. The compositional continuum would reflect the use of different proportions of raw materials in the glass batch (ideally mixed in a ratio of 2:1 by weight of alkali to sand) and reveal which were unsuccessful. Experiments with different proportions of beech ash and sand have revealed how the all-important melting behaviour of the resulting glasses is affected (Smedley and Jackson 2002). By plotting proportions of one component against another, a mixing line – the 'dilution' of one major component by another – would be produced (see Chapters 10 and 11). The caveats to the possible

1.4. A large droplet of copper sulphide (Cu_2S) surrounded by crystals of copper in a glass matrix in an eleventh-century opaque orange glass tessera from the west wall of basilica of St. Maria Assunta on the island of Torcello in the Venetian lagoon (photo: E. Faber).

discovery of early experimental glasses are, first, that the chemical compositions of some partially reacted raw materials would be altered by weathering and, second, that the chemical compositions of the glassy materials may not be a direct reflection of the raw materials used to make them (Turner 1956b; Rehren 2008). Some 'unsuccessful' glass *batches* (the combination of raw materials in the crucible) are likely to have been vitreous materials containing unfused or partially fused silica. Returning to Cyril Stanley Smith's contention, we should remind ourselves that the actual driving force (rather than the physical evidence) for the earliest glass is likely to have been the excitement of creating it!

During the incipient stages of glass production, for whatever reason, artisans may have found that small changes in the type or proportion of raw materials used in the glass batch would have caused the glass to be difficult to melt and work. For example, an increase in the proportion of silica in the batch could produce a glass that would melt at higher temperatures and, moreover, was workable for a shorter time (a so-called 'short' glass). Alternatively, a smaller proportion of *reactive* plant ash would have produced the same result (also potentially leading to a relatively high proportion of silica). In contrast, a drastic change in the batch recipe would probably have caused a nightmare of unpredictability in the ways in which the glass behaved when worked. Indeed, this was the reaction that a contemporary glass artist gave to the prospect of a new batch composition. Therefore, it is possible that early glass batch compositions may have been modified in a gradual way to be able to learn from, and react to, any changes. Constantly changing the conditions in which glass was produced would obviously have caused frequent and unpredictable results. Indeed, predictability is clearly important during each stage in the *chaîne opératoire*, leading to the control needed to produce the fully formed plant ash glass that was in use from *c.* 2500 to 1000 B.C. in the Middle East and used to make the first glass vessels. This is not to underestimate the excitement and perhaps the reverence that those involved had for the act of creating and experimenting with a 'new' material. As Vandiver and Kingery (1986, 32) have noted in the context of faience production, a period of innovation was followed by a phase when the technology became 'a traditional method, difficult to change and suffering new innovations slowly and in small increments'. In other words, why innovate further if what you already have is satisfactory? Once the behaviour of hot liquid glass of a particular composition had been monitored, understood and 'tamed' to the extent that the required glass objects could be manufactured, there would have been little incentive to change raw materials or other aspects of the technology. This may well have been the case, but we should not exclude the possibility that the acts of experiment and innovation were viewed as essential parts of the ritual and economic fabric of society, and many products of such activities may not have survived.

The experimental phase would have been followed by a developmental phase. Although further experimentation with, for example, colourant materials, may well have occurred, the scientific proof for this phase could be provided by evidence of repeatedly combining raw materials to produce *consistently* similar results. In the experimental phase, the variation in the proportions of raw materials used might have been relatively wide; by the time the *chaîne opératoire* had become embedded, fewer decisions needed to be made and the procedures in glass production would have become more restricted and constrained in terms of choice. Therefore, the glass chemical compositions produced would also have become more tightly constrained. It is notable that during times of transition from one major glass technology to another, glass chemical compositions can become far more variable. This can be seen in, for example, the change from natron glass to wood ash glass in medieval Europe north of the Alps (see Chapter 4) when these two types of glass were mixed.

1.4 The First Glass: A Paradigm Shift?

Glass was the latest of the three main (inorganic) technologies – pottery, metal and glass – to appear. To assess whether the manufacture of the first glass can be regarded as a paradigm shift in the development of ancient technology, it is necessary to consider whether the processes involved in its production had a distinctiveness that sets the material apart from the technologies that already existed. The materials involved are those made from a vitreous component. These preceded glass and include glazed steatite and faience; these are focused on here. Like glass production, pottery and metal technologies frequently involved the construction of kilns and furnaces respectively. These structures are necessary so as to produce high temperatures and to be able to control heat and the gaseous atmosphere. Therefore, access to and preparation of ceramic materials or stone for constructing kilns and furnaces creates a link between all three technologies. Similarly, access to appropriate types of fuel would be necessary to achieve the temperatures and atmospheres.

1.4.1 Glazed Steatite, Egyptian Blue and Faience

Glazed steatite (or soapstone) was produced before the first faience. Steatite is mainly composed of talc with minor amounts of other minerals, such as chlorite; it is a hydrated magnesium silicate. Examples have been scientifically analysed by Vandiver (1983a) and Tite and Bimson (1989). It has been found that the glazed surface of steatite contains high magnesia levels (c. 27%), indicating that the alkali forming the glaze and the copper must have reacted with the steatite. The key factor here is that the material on which glaze is formed is silica-rich.

Egyptian blue was produced at high temperatures perhaps using repeated firings. It was made by heating together silica, copper alloy filings or crushed copper ore (such as malachite), calcium oxide, and a fluxing material, such as natron or trona (see Chapter 2) or, according to Lee and Quirk (2000, 109), potash. It is a crystalline material and contains rectangular blue crystals of cuprorivaite, $(Ca,Cu)Si_4O_{10}$), unreacted quartz, and alkali-rich glass. Copper-bearing wollastonite is often present, and minor components include pyrite (FeS) and titanomagnetite (Fe_3O_4-Fe_2TiO_4), considered to have been introduced in desert sand (Jaksch 1983, 533–5) and cassiterite (SnO_2). Tite, Bimson and Cowell (1987) recognised three shades, dark blue, light blue and diluted light blue, the colour variation mainly being due to the size of cuprorivaite crystals and the proportion of alkaline glass. The shaping of objects made from it involves moulding, carving, cutting and grinding.

The earliest occurrence of Egyptian blue is the Fourth Dynasty (2575–2465 B.C.; Lucas 1962/1989, 342) with other examples dating to the Fifth Dynasty (2465–2323 B.C.; Blom-Böer 1994). It was even used to decorate the Eighteenth Dynasty head of Queen Nefertiti (Wiedemann and Bayer 1982). Studies of later Egyptian blue include that from Ugarit, Syria. The occurrence of it there, in late Bronze Age contexts, was associated with suitable raw materials and its scientific analysis using proton-induced X-ray emission and scanning electron microscopy suggest that it was made locally (Matoïan and Bouquillon 2000,

990–3, fig. 8). Trace occurrences of strontium, chlorine, alumina and magnesia were found to be particularly discriminative.

The sight of another vitreous material, *faience*, would clearly have reminded its makers that the *vitreous state* existed: it was a hard, highly coloured, glazed material that often reflected the light. Before the first glass appeared, a vitreous layer would have been seen as *decorative* and/or as something *functional* that held together the sandy core of the object and also 'held' the colourant in place. In the production of faience, there must have been a deliberate control over the maximum temperatures achieved because over-firing would have produced greater fusion of silica and alkali and, at sufficiently high temperatures, the production of glass. Over-firing was therefore apparently to be avoided, although the kilns used (e.g. Nicholson 2003) would have minimised the possibilities. Hodges (1992, 125) suggested that overheated faience might have provided the first sight of true glass. Tite *et al.* (2002) also have suggested that this was the case and that, in addition, poor compositional control might have played a part.

Hodges (1992, 62) has suggested that the production of faience from c. 4000 B.C. 'can be looked upon as man's first real move into the world of synthesising the material he required'. Essentially the same raw materials, silica and plant ash, were used for making both faience and the first true glass, so the relationship between the technologies cannot be ignored. The Qom process of faience manufacture (discussed below), observed in Iran in the 1960s, involved heating these raw materials together at temperatures up to *c.* 950°C. However, to produce properly fused glass, at least 200°C. higher temperatures are necessary, involving more and/or different kinds of fuels. Nevertheless, by using the same principle of synthesising raw materials to produce a different material, the first glass was produced. Although the *chaîne opératoire* would have involved similar initial preparatory steps as in faience production, the crushing of quartz pebbles and ashing of plants, there would have been a number of differences. The higher temperatures involved in making glass would have necessitated a furnace rather than a small kiln. A second difference is that faience objects resulted directly from the single process of firing, whereas the production of raw glass was a starting point: to make objects, the raw glass needed to be reheated and worked.

The first (sixteenth–fifteenth century B.C.) glass vessels echo the shape of faience vessels (Peltenburg 1987). It is therefore likely that certain aspects of the production processes of faience contributed to a *chaîne opératoire*, leading to the production of the first deliberately made raw glass and the first glass vessels. Some of the earliest glass vessels have been found at Alalakh, in modern-day Turkey (see Chapter 5). They are stratified with faience vessels, so in theory it ought to be possible to see whether the faience industry there reacted to the introduction of glass (Peltenburg 1987, 14). Although the number of artefacts is small there is no clear technological 'anticipation' in 16th century B.C. faience technology and the emergence of the first glass vessels. As Peltenburg (1987, 18) has noted, 'Faience-working therefore provided but a general background rather than a direct impetus for glass production, which, as has often been stated, appears suddenly.' If anything, the closest link, in terms of vessel shape, is between glass and glazed pottery (Peltenburg 1987, fig. 3). The raw materials for faience production are crushed quartz, plant ash and copper (oxide). Using laboratory replication and microstructural examination of polished sections, Tite and Bimson (1986) defined three techniques of faience production. Moreover, Vandiver (1983, 1998) emphasised the importance of observing drips of glaze, drying or firing marks and variations in glaze thickness as ways of suggesting the methods of glazing used. As with the production of glazed steatite, fluxing of the silica by an alkali-rich plant ash created the lustrous glassy surface, and the vitreous network in some cases also provided strength to the faience object. The *efflorescence* technique involved mixing the raw materials with water to create a mouldable paste. The water in which the alkali was dissolved was drawn to the surface by evaporation, resulting in a concentration of the alkali in the surface layers. When the objects were fired, they hardened as a result of the alkali fusing with the quartz and the copper. The *Qom* technique, referred to earlier, involved cementation, first by making the moulded faience object, and then submerging it in a glaze mixture and firing it. The third technique of faience production speaks for itself: it involves the *direct application* of a glaze slurry to the moulded object followed by firing. However, Vandiver (1998) has pointed out that it can be difficult to prove with certainty which production technique was used to make faience based on microstructural characteristics. Henderson has noted metallic inclusions, such as tin and antimony in Bronze Age Swiss and early Iron Age Balearic faience (Henderson 1988b, 438, figs. 1a and 1b; Henderson 1999). Their occurrence is a means of characterising the copper minerals used to produce a turquoise colour in the vitreous phases of faience. A 'transitional' material between glass and faience, 'glassy faience', exists (Lucas and Harris 1989, 164–5; Shortland and Tite 1998; Nicholson 2000; Santopadre and Verità 2000; Shortland 2000) in which there is a high proportion of glass in the quartz-rich object. Some glassy faience predates glass.

Moulded faience objects were made in a single process, which, unlike moulded glass, did not require subsequent working. Faience was evidently manufactured at Kerma in the Sudan c. 2000 B.C. Amongst the evidence were faience moulds, an indication of primary faience production. On the basis of the distribution of faience, its production debris and the historical evidence for its production, Shortland, Nicholson and Jackson (2001, 154–6) noted that it is likely that temples controlled some faience workers (in Egypt) but that there was also an independent private sector that produced faience found/used in a wide range of contexts, including those of a low social status. Despite the fact that the first faience was made much earlier than glass and that opaque moulded glass is quite similar in appearance to faience, faience continued to be made well after glass was introduced. Tite *et al.* (2007) emphasised the conservatism in the production of plant ash faience glazes with distinctively low calcium and high potassium oxide contents from the Egyptian Middle Kingdom until at least the Eighteenth Dynasty, presumably a reflection of a repeated series of technological processes using similar raw materials from similar geological sources. Relatively recent discoveries (Nicholson 2002, 2003) confirm that faience vessels were still being made as late as the Ptolemaic–early Roman periods at Kom Helul, Memphis, in Egypt. The comprehensive evidence consisted of saggars, three-pointed stands and furnaces, some of which were preserved to at least 3 metres in depth. At Kom Helul it is thought that a stack of saggars containing the faience vessels was placed inside the furnaces. The continuing 'late' manufacture of faience must indicate that the tradition of making it was firmly established in Egypt and that it filled a ritual niche that the production and use of, for example, faience ushabti figurines, that could not be replaced by glass (see Fig. 1.2).

Another material, which appeared about the same time as the first glass vessels (c. sixteenth and fifteenth centuries B.C.), were glazed ceramics. The glaze is often not particularly durable, so it is difficult to analyse it scientifically. Scientific analysis has been carried out on pottery glazes from Tell 'Atshana (Peltenburg 1969), Ras-Shamra-Ugarit, Meskene, Kition (Matoïan and Bouquillon 1999), Kish (Hedges and Moorey 1975), Failaka (Pollard and Højlund 1983), Tell al-Rimah (Pollard and Moorey 1982), Tell Brak (Henderson 1999b) and (of a somewhat later date) Kish, Nineveh and Nippur (Hedges 1976, 1982). For the glazes that are not too weathered, it is clear that, in general, they are of the expected plant ash composition with elevated potassium and magnesium oxide contents (Tite, Shortland and Paynter 2002). However, the glazes from Failaka, analysed using atomic absorption spectroscopy, have anomalously high magnesia and low potassium levels. Pollard and Højlund (1983, 199) have suggested that this was probably due to the use of a high magnesia raw material. Unexpectedly, an example from Tell Brak indicates that natron may well have been used as the alkali source (Henderson 1997, table 6b, sample Br16). Matoïan and Bouquillon (1999, 74 and 2003) are of the opinion that there was a specialised production of glazed pottery at Ugarit. This is partly based on its rarity in the southern Levant and partly because it was unknown in Egypt. Similar vessels have been found on sites on the southern and south-eastern coasts of Cyprus. Matoïan and Bouquillon (2003, 344) note that the mineralogical characteristics of the pottery bodies are likely to be those of the environment of Ras Shamra-Ugarit.

Tite et al. (2002) have pointed out that glass and metal working would have involved hot viscous fluids and therefore that metal workers may have played a part in the emergence of glass working. A further important process that the metal and glass technologies have in common is annealing, and this may have constituted another input from metal technology. A suggested trigger for technological change/advancement, which may have led to the sharing of technological knowledge, was put forward by Tite *et al.* (2002, 588): 'it can be argued that the discovery of techniques necessary for hot-working glass was the result of political upheavals that occurred in Egypt and the Near East leading to changing controls over artisan organisation.' This may well be part of the explanation, although given that the manufacture of the earliest glasses in Bronze Age Mesopotamia and Egypt was very much an elite pursuit, whereas metal and faience production were not necessarily, and that glass production was described in great detail in cuneiform texts, unlike metal and faience production, the explanation is probably more complex than this. In conclusion, between c. 4500 and 2500 B.C., any glass produced formed part of something else and was not a separable entity in its own right. This is important, because until there was a realisation that glass could be manufactured and used to make separate objects, no progress would be made in the manufacture of glass objects. Once it was realised that raw (unworked) glass could be manufactured, it would have acted as a technological trigger to further developments; it may simply have been the 'first' sight of a *translucent* block of *coloured* glass that triggered further developments. Whether it would initially have been perceived as a man-made coloured form of rock (crystal), or something entirely new, because it was formed in an entirely different way and had a number of different properties, it is difficult, if not impossible, to assess. The corollary of this is that if a glass melt was allowed to cool too slowly past the transition temperature (T_g), it would have produced what might have been regarded as a kind of coloured stone—coloured crystalline silica. Indeed it is inevitable that the silica-rich crystals resulting from devitrification would have been produced during the early stages of (unsuccessful) glass production. Subsequently devitrified glass may have provided a visual link between semi-precious 'stones' and true glass; it is no coincidence that early glass was referred to as a 'stone' (see Chapter 5).

Thus, there are several aspects to the production of the first glass that can be claimed as forming part of a paradigm shift. First, to deliberately produce a block of brilliant translucent glass that reflected and refracted light was clearly a break from what had gone before technologically. The addition of trace levels of colourants to the glass, such as cobalt and copper, was a highly controlled new technological process. Moreover, the creation of an opaque glass that involved the addition of crystalline materials to the glass melt which were retained at high temperatures, or the development of crystals out of solution by heat treatment (and its link to annealing), were also new processes (such as the copper droplets in Fig. 1.4). They were especially important because, together with, or instead of, the addition of trace levels of transition metals that dissolved in the glass, the crystals coloured the glass. The social and ritual significance of (glass) colour in ancient societies cannot be underestimated. Glass coloration can be regarded as a primary driving force behind the high levels of innovation in making the first glass (Duckworth 2011). Ancient opaque glasses are what today would be labelled as *glass ceramics*: the first opaque glasses were therefore the earliest glass ceramics. However, over and above all of these innovative processes, which by themselves can be regarded as major variations and innovations from existing technologies, was the fusion of glass from its primary raw materials. When compared with metal and pottery manufacture, there is little resemblance between glass raw materials (plant ash and silica) and glass. In every respect the manufacture of the first glass can therefore be regarded as a paradigm shift. In the next section, the difficulties of interpreting evidence for primary and secondary glass production are discussed.

1.5 Evidence of Production Sites

The discovery and excavation of glass furnace sites might be expected to provide 'fixed points' in the investigation of glass production of all ages. However, the interpretation of archaeological excavations is rarely straightforward. The evidence for glass making and working can be variable and inevitably incomplete. This is hardly unexpected, given that this is true of any archaeological excavation. A first crucial question to be addressed is whether primary or secondary glass production has occurred. Evidence for primary glass production can consist of evidence of fritting, including overheated frit and fritting ovens (see Chapter 11); the structure and scale of glass furnaces can also provide an indication (Henderson 2000, 39–40). If a fritting oven is discovered, then this is clear evidence for primary production. Even though frit could potentially also be imported, this does not detract from the evidence of primary production. The recent identification of the evidence for fritting in thirteenth century B.C. Qantir (Rehren and Pusch 2005) and fourteenth century B.C. Tell el-Amarna (Smirniou and Rehren 2011) are rare examples. Tank furnaces that can be up to 6 metres long (see Chapter 7) can potentially be used for reheating glass or for fusing primary raw materials. Scientific analysis, especially involving the use of isotopic analysis in an environmental context can establish whether glass was fused using local raw materials and whether mixing has occurred. There is also the possibility that a tank furnace was used for both primary and secondary glass production.

The evidence for glass working, including blowing, moulding, shaping, manipulating and decorating, leaves a range of evidence, including glass furnaces incorporating annealing chambers, separate annealing ovens, relic deposits of fuel, moulds, crucibles, drops, dribbles and pulls of glass. Pulls of glass were extracted so as to test the working properties of the glass as it was heated up and to observe its colour. In an ideal world, tools for working glass, such as blowing irons and pontil rods, would be found. Given that the floors of glass workshops are likely to be clean, it is unlikely that when deserted, a range of useful evidence for past activities would remain. If glass is found, unless fused to the furnace or nearby surfaces, it is most likely that it would be redeposited and may even derive from other glass workshops or nearby furnaces. There may be associated layers of material that can be shown to result from the glass workshop that provide evidence for the colour and chemical composition of the glass being produced and (just possibly) the kinds of glass vessels made in the workshop. Nevertheless, there is always the possibility that even the archaeological layers connected to glass production contain imported scrap glass (for recycling), including vessel fragments. So it would be easy to make an assumption that glass vessel fragments found in or around glass-working workshops or furnaces were made there, but this assumption needs to be scrutinised very carefully.

Scientific analysis set in an environmental/ecological context can provide evidence that the glass was fused using local raw materials. However, glass objects may well have been made from imported raw glass (see Chapter 11). Experimental archaeology combined with scientific analysis can help to provide evidence for the kinds of by-products that can result from glass making, the extent to which they are likely to survive unaltered and how easily they can be connected to the glass being made at the site (Paynter 2008).

1.6 Conclusions

The manufacture of glass represents a significant development in terms of material transformations beyond the manufacture of a vitreous component in faience. It involved the full solution of colourants in the glass, producing a translucent highly coloured, smooth, refractive material quite unlike anything else that had been made before. Its first appearance, and subsequent uses, can be regarded as a paradigm shift. Furthermore, heat treatment of glass to promote the growth of calcium antimonate crystals and opacity in coloured glass produced a material that was very close in appearance to semi-precious stones. In modern terms, it would be classed as a glass-ceramic, and the mid-second millennium B.C. saw the first examples in the world being made. The heat treatment of glass to produce opacity was a new departure in technology. It may possibly have been 'borrowed' from a similar procedure involving heat treatment, metal annealing, although the latter did not produce the same obvious visual results. The idea of heat treating the glass, by reheating it to produce opaque glasses, probably formed part of an evolution in the understanding of how molten glass behaved at various temperatures, including the cooling rate of the melt at the appropriate rate past the transition temperature (T_g) so as to produce glass rather than silica crystals. In the initial stages of making the first glass, it is inevitable that glassmakers cooled the glass melt too slowly and that coloured crystalline silicates instead of glass were produced. The reason why this occurred was that the rate at which the melt was cooled led to a solid (crystals) rather than a (super-cooled) amorphous liquid-glass forming. In scientific terms, the properties of glass and crystalline silicates are quite different, as measured by volume per unit mass, the coefficient of expansion and the specific heat capacity (Guinier 1984, 168–9, fig. 8.7). Although the creation of glass was still possible by remelting the silicate crystals produced, their creation must have contributed to the description of glass as a stone, the crystals having greater similarity to minerals, with internal refractivity, than to amorphous glass. The transformation of these silicates into glass must have provided a great sense of control in the transformation of natural 'rocks' into glass (synthetic stones). A third new process involving heat treatment was the annealing of fully formed glass objects at relatively low temperatures, leading to durable and longer-lasting glass vessels and other objects.

The synthesis of opaque glass to imitate semiprecious stones was accompanied by a range of other groundbreaking innovations. These included the production of the first core-formed vessels, the first mosaic glass vessels using sections of opaque glass assembled in a mould and the first mass production of core-formed vessels (by the fourteenth century in Egypt); see Chapter 5. The fifteenth and fourteenth centuries B.C. therefore saw several contemporary seminal innovations in glass technology. From at least c. 1370 B.C., larger furnaces and associated structures must have been built in Egypt, not only to accommodate a sufficient volume of glass to make larger numbers of objects but also to have afforded sufficient control of the heat source to be able to fully melt raw glass and to be able to provide a consistently high temperature of c. 1100°C and above so that plant ash glass would remain fluid enough to gather it on a core. To wind a decorative glass filament or cable around the vessel and to marver the decoration into the vessel surface (by rolling it across a flat metal surface) would have involved reheating it several times. Finally, an annealing oven or chamber allowing plant ash glass vessels to cool slowly through the critical annealing temperature of 529°C and an annealing range of between about 500 to 575°C for a soda-lime glass (Brill 1988, 279–81) would have been necessary from an early stage. A change in glass viscosity (measured in units of poise) occurs between the temperature at which the glass is melted (about 2–3) and the annealing point (13). All of these high temperature operations would have required a clear understanding of the calorific values produced by combusting a range of fuel types.

A range of characteristic smells would have been produced by burning different fuel types at different stages of the production process and also during the process of glass melting. For example, during the fritting process, if sufficient sulphates were present in the raw materials, the evolution of sulphur dioxide gas would have produced a characteristic smell of rotten eggs for that stage of the process! The colour of the glass melt and of the flame produced would similarly have provided ways of assessing the stage that the chemical

Glass as a Material

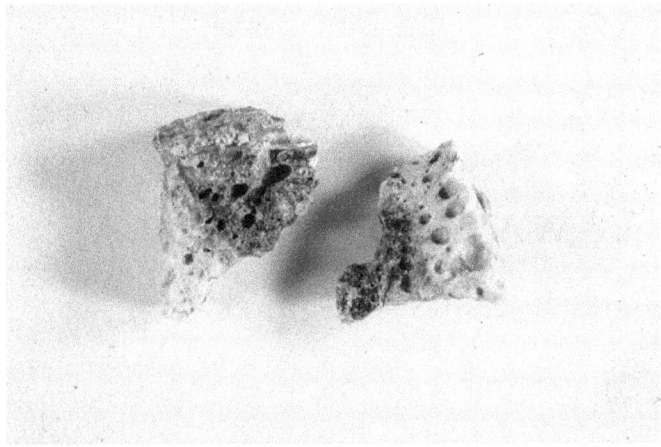

1.5. Two small pieces of overheated frit found at the ninth-century glassmaking site of al-Raqqa, Syria (photo: J. Henderson). The left hand piece is 2.5cm wide.

reactions had reached. For example, the cuneiform texts mention the production of *Dusu-coloured* glass (Oppenheim *et al.* 1970, 47–48): once a mixture of three raw materials begins to glow green, it is taken out of the kiln and ground finely (this is presumably frit, a sintered combination of glass raw materials; see Fig. 1.5). It is then placed in the kiln again and heated until it glowed yellow at which point glass is allowed to form. Observed colour changes in the glass batch are also described in later (medieval) texts (e.g. *Theophilus presbiter*).

The ritual associated with the use of specific raw materials at particular times of year would have provided a driving force for each link in the *chaîne opératoire* in glass production. Therefore, a combination of natural raw material properties and man-induced forces led to the production of the first true synthetic material.

TWO

WAYS TO FLUX SILICA

ASHES AND MINERALS

2.1 GLASS RAW MATERIALS

The selection of raw materials for the manufacture of glass was the first in a series of interrelated production processes. The type of raw material selected, and potentially its geographical origin, had a critical effect on the fusion and working properties of the glass made from it. Chemical analyses of raw materials, historical descriptions of glass production and, to a lesser extent, evidence from archaeological excavations provide us with the evidence for the raw materials used to make ancient glasses. What follows in this chapter and the next focuses mainly on the types, occurrence and chemical compositions of glass raw materials. In this chapter, emphasis is placed on research into high soda halophytic plants because Mesopotamian and Islamic glass (discussed in Chapters 5, 6, 9 and 10) were made from them. Other ashes used in glassmaking (wood and fern) are discussed in section 2.4 and the glass compositions made using them in Chapter 4. Historical descriptions of glass production are given in these and subsequent chapters (5–10) that focus on three case studies. A less detailed consideration of glass colorants, much of which is published elsewhere, is given in Chapter 3. This chapter and the next provide background information for other chapters in the book, especially for the three case studies presented as three pairs of chapters (Chapters 5 and 6, 7 and 8 and 9 and 10). A wide range of glass compositional types is described in Chapter 4; these were made from a range of raw material types. The main primary raw materials used to make ancient glasses were alkalis, silica, calcium and, more recently, lead.

2.2 Halophytic Plant Ashes

2.2.1 Occurrence and Range of Types

The source of alkali that seems to have been used to make the earliest glass, until approximately 800 B.C., was the ashes of halophytic plants of the Chenopodiaceae family. It was also used intermittently to make Roman and Sasanian glass (presumably at 'inland' locations in the Middle East) and was again the dominant alkali used in glass production from c. A.D. 800 until the seventeenth century in the Middle East. The plants that were the source of the alkali grew, and still grow, in semi-desert environments, often on the edges of deserts, in saline maritime environments and in inland salinas. They tolerate high levels of alkalis in the soil and, by transpiration, concentrate alkaline salts, such as sodium carbonate, in their tissues. One of many ways that these plants tolerate high alkali levels is by accumulating alkali salts in their leaves, which they later shed. For example, plants of the genera *Salsola* and *Sueda* with succulent leaves do this; excessive levels of the salts become toxic for the plants, and thus leaves of *Salsola vermiculata*, for example, become loaded with alkalis, turn black and are shed. *Halocnemum* and *Salicornia* release large quantities of alkalis by discarding leaves and parts of their fleshy cortex (Batanounay 2001, 155, 157, table 6). There is a wide range of halophytic plant genera, and scientific analysis can establish which are most likely to have been suitable for making glass (as discussed later in the section 2.2.3). In addition to *Salicornia* and *Salsola*, plants belonging to the genera *Anabasis*, *Arthrocnemum*, *Halopeplis* and *Hammada* (*Haloxylon*) may have been used.

The crucial characteristic of halophytic plants that makes them suitable for glassmaking is the presence of sufficient quantities of sodium carbonate when ashed. It is this compound that interacts with silica and reduces its melting temperature from between 1710°C and 1730°C to a more manageable temperature of c. 1100° to 1200°C for soda-lime glasses. Other salts that may be present in plant ashes and will also interact readily with silica are bicarbonates, sulphides, sulphites and hydroxides. Sodium sulphate may also be present. However, it is slow to react with silica below 1200°C, and thus for efficient fusion, it needs to be reduced to a sulphite or sulphide by carbonaceous material or other reducing agents (Turner 1956b, 291). Sodium chloride (common salt) and potassium chloride are also present in plant ashes, but these interact minimally with silica, react very slowly and are excluded from the glass melt as a scum (together with sodium sulphate) or are volatilised without decomposition.

Amouric and Foy (1991) have noted that sodium-rich plants used for glassmaking could have derived from a range of locations, including Syria, Sicily, France (Narbonne, Aigues-Mortes), Spain (Alicante, Malaga) and North Africa (Carthage). As part of her excellent study of French medieval glass production,

Foy (1989) notes that halophytic plants grow beside the lakes in the Camargue region of southern France and in Spain, especially in the Ebro delta. *Salsola prostala* and *Salsola soda* grow on the Rhone delta (the Camargue) and on the delta of the river Durance. *Salsola tragus* and *Salsola kali* are found in maritime environments. Although it is difficult to guess which species was used, a plant known (colloquially) as 'Rochetta' was apparently imported in the thirteenth century from Syria. It is quoted as containing up to 45% 'sodium' (Foy 1989, n. 22), although this figure would also include non-glass-forming compounds and it is difficult to generalise about the chemical compositions of halophytic plant ashes (as discussed later in 2.2.3.4). By the fourteenth century, Foy considers that soda for glassmaking was produced from plant ashes in the Languedoc, with *Salicornia* being used in the regions of Montpellier and Nîmes (Foy 1989, 34, n. 26). By the end of the Middle Ages, it is likely that plants, and especially *Salicornia*, growing adjacent to lakes in the Languedoc were used for glass production. 'Rochetta' was certainly imported through Marseilles in the thirteenth century (Foy 1989, 33), about the same time that the Venetians started to make their own glass from imported plant ashes. In the sixteenth century, Antonio Neri, who was under the patronage of Don Antonio de Medici, considered imported 'Rochetta' from Syria to be the best source of soda, producing the finest white ash, and noted that it was better than imported *barilla* from Spain (Neri 1612, 1–2). Other authors also mention that 'Syrian' ashes are of the highest quality (Ashtor and Cevidalli 1983). Already in the tenth century, Arabic authors mention that *ushnān* (another colloquial term for the vernacular *keli*) was exported from Aleppo in (inland) Syria. They may be referring to *Hammada scoparia* (Ashtor 1992, 482, 487–8).

2.2.2 Technological Considerations

Social factors affecting the use of plants during the manufacture of ancient glass are not easily definable or quantifiable as part of a *chaîne opératoire*, but they are at least as important as technological ones. Technological choices would have been influenced by either the use of established traditions or the process of manufacture. Such a tradition was described in c. 1300 by Abū'l Qāsim, who was a member of a potting family of Kashan in Iran (Allan 1973), where soda-rich plant ashes were used for the manufacture of pottery glazes and glassy frits in pottery bodies:

> The fifth (substance) is *shakhar*, which they call *qali*, which is made by burning pure, fully-grown *ushnan* plants.... The best *shakhar* is that which has a red-coloured centre when broken, with a strong smell.

This ancient description underlines the fact that both colour and smell were important characteristics when it came to selecting the appropriate plant as an alkali source. In other parts of the Middle East, different characteristics were perhaps considered important when determining which plants to ash. In a video filmed in 1969 directed by Robert Brill, titled *The Glassmakers of Herat* (Afghanistan), the glassmaker selected congealed lumps of ash from the market by taste. The sweetest was selected, perhaps because it contained the lowest proportion of sodium chloride, a substance that could be described as having a bitter taste.

Another way of investigating the use of plant ashes in ancient glass manufacture is the scientific analysis of ashed plants. There are several reasons for doing this. The first is to assess which genus or species was used for making ancient glass. The second is to investigate to what extent the chemical compositions of ashes were transferred directly into the glass. A third reason is to investigate whether the soil geochemistry in which the plants grew, and therefore the chemical composition of the plant, can provide provenance information about the glasses made in specific locations (the second and third reasons are discussed in more detail in Chapter 11). The implicit reason for trying to understand the chemical characteristics of plant ashes is to estimate which genera or even species were used. Furthermore, there is possible scope for contributing to the provenance of glasses by determining the isotopic compositions of (ashed) plants deriving from a range of geologically contrasting contexts across the landscape and comparing the results with isotopic results in ancient glasses (see Chapter 11).

The relationship between the chemical composition of plant ashes and that of glasses made from them is complex. An early attempt by Brill (1970) to fuse glass from raw materials provided grounds for optimism. However, a variety of factors affects the relationship between plant ash and glass composition. First, as already noted, not all plant ash components necessarily interact readily with silica. In addition, the chemical composition of glasses can be determined by the following:

- Ash compositions differ according to the specific plant species used (including, potentially, their mixtures).
- Variations in ashing procedures may produce different ash compositions because plant ash compositions change with increasing temperatures (discussed later in the chapter).
- Variations in the drift and bedrock geologies, and ultimately the soil geochemistries in which the plants grow, affect the chemical and mineralogical composition of the plants and their ashes.
- Purification of the ashes by dissolution, distillation and recrystalisation can potentially reduce levels of insoluble components such as calcium and magnesium carbonates, making the glasses 'purer' and less durable.
- The melting conditions during glass production can also affect its chemical composition.

Soda glass with compositional characteristics such as elevated magnesium and potassium oxide levels was manufactured from plant ashes (Sayre and Smith 1961, 1967; Sayre 1964; Henderson 1988a; Freestone 1991). Until recently, a restricted amount of research had been carried out to address the points just listed. In addition to those already mentioned, chemical analyses of plant ashes have been published by Rye (1976), Besborodov (1975), Verità (1985) and Verità and Tontinato (1990).

2.2.3 Chemical Analyses of Plant Ashes

2.2.3.1 The Choice of Scientific Techniques for the Analysis of Plant Ashes

Techniques such as atomic absorption spectroscopy and inductively coupled plasma emission spectrometry (ICP-AES) provide a quantitative bulk analysis for major, minor and trace inorganic elements (see Appendix). However, using these techniques dissolves the sample. A result is that crucial information about the crystalline identify of salts in the plant ashes is lost. This information is important because not all (alkali-rich) components react with silica, as discussed earlier, and those that do not may boil off, can be excluded from the batch as a scum or excluded as a result of the fritting process. Although not easy to interpret, the results from X-ray diffraction spectroscopy (XRD) analysis can provide a means of determining the ratios of different salts in (undissolved) plant ashes that could potentially interact with silica. This technique can be used to identify alkali-rich compounds, such as sodium carbonate and sodium chloride; the presence of these cannot be detected using ICP-AES, and so both techniques are necessary.

The determination of radiogenic isotopes using thermal ion mass spectrometry (TIMS) can also help link the location where the plant was growing to the location where the glass was made. Lime-rich soil is drawn up into the plant, and the geological age of the lime in the soil determines its strontium isotopic signature. When the glass is fused, this will be retained in an unaltered state in the glass (Henderson *et al.* 2005; see Chapter 11 and Appendix). We used the following techniques in our study: ICP-AES, ICP-MS, XRD and TIMS (see section 2.2.3.4).

2.2.3.2 Turner's Research

As is the case with much of W. E. S. Turner's research, his work on plant ash compositions (1956b) and their relation to ancient glass production was an important starting point. The Spears mission to Syria in 1943 gathered fifty samples of *keli/ushnān*, and Turner (1956b, Table VI) published their average composition (see Table 2.1); it is possible that plants of different species were mixed. It was stated that such plants yielded 25% of their weight as ash.

Table 2.1. The average composition of 50 samples of Syrian *keli* in parts per 1,000 (after Turner 1956b)

Soluble in water (600)		Insoluble in water (400)	
Na_2CO_3	450	$CaCO_3$	340
NaOH	25	'$CaPO_4$'	40
KCl	45	$MgCO_3$	10
K_2SO_4	30	C	10
Silicates	50		
Phosphates	50		

The data in Table 2.1 show first that plant ashes contain a range of components other than sodium carbonate (Turner 1956b, 291). Second, an important distinction has been made between soluble and insoluble components. This is important because plant ashes can be purified by the reduction or removal of insoluble salts. This was certainly something that the Venetians did (see Chapter 9), and there is a Tuscan recipe dating to c. 1450 in which it is described with clear terminological links to Venetian procedures (Zecchin 1990, 217–19), although it is uncertain whether earlier glassmakers purified ashes. The process would reduce the levels of the insoluble salts of calcium and magnesium oxides. Because calcium oxide plays an important role in ancient glass – that is, making it durable (a network stabiliser) – this could potentially have had a major impact on the glass produced. Indeed, crizzling of seventeenth-century Venetian glasses can be largely attributed to the purification of ashes and to excess alkali levels (Newton and Davison 1989, 141; Freestone 2001, 620). In contrast, during the manufacture of Chinese glazes, the aim instead was to *reduce* the levels of alkaline elements (potassium and sodium) in the ashes (Chen Xianqiu et al. 1986, 237, table 3; Wood 2009) to produce a calcareous wood ash, a process that was still used in twentieth-century Japan (Leach 1940, 160). For example, in one sample washing, the ashes reduced the potassium oxide (K_2O) levels sharply from 10.9% to 2.1% and thereby increased the level of calcium oxide from 33.6% to 40.5%, but, for some reason, did not increase the magnesium level.

If the composition of the *keli* given in Table 2.1 is thought to be an alkali source for glass production, it is surprising that there is only 1% of magnesium carbonate because the levels of magnesium oxide in ancient glass, even if diluted still further by the addition of silicon, would be expected to vary between 3% and 7% (Brill 1970, table 1; Brill 1999b; Henderson et al. 2004). Some of the samples reported by Turner may have contained far higher magnesia levels, but by averaging the results this would have been masked.

The results in Table 2.1 suggest that 93.5% of components will react with silica. Even if K_2SO_4 and $CaPO_4$ do so only in a limited way, this still leaves

c. 90% of the components to react. In ancient glass, we must not forget that there may have been other potential sources of calcium (e.g. in seashell fragments), magnesium (e.g. dolomitic limestone) and potassium (e.g. potassium-rich feldspars). However, based on this average *keli* composition, sufficient sodium and calcium would be present in the plant ashes to provide the levels of soda (c. 12–14%) and calcium oxide (c. 6–9%) found in plant ash glasses. Nevertheless, 34% of $CaCO_3$ is an excessive level for glass, even when reduced by the proportion of ash used.

2.2.3.3 Brill's Work

The second main set of plant ash analyses was published by Brill (1970, table 2). It consists of results for the ashes of halophytic plants, seaweed, and tamarisk from a range of contexts and for a range of species and genera. Discussion of these results is provided here to illustrate the progress in this research area.

The first set of sixteen analyses was published by Brill in 1970, and another thirty-five was published in 1999 (Brill 1999b, 482–6). Brill's 1999 data provide similar information to that of his 1970 data set.

The interpretation of the data published in 1970 was set within the context of the translated cuneiform texts describing the manufacture of Bronze Age vitreous materials (Chapter 5). Brill's ashed plant samples consist of *Salicornia* sp. from Aci Gölü, Turkey; a desert bush from near Cairo, Egypt; seaweed from Acre Bay, Israel; 'soda plant' (*ŏsnān*) said to have derived from Shareza, thought to be of the genus *Salsola*; a lump from Qom, Turkey, described as potash by Hans Wulff; a lump described as calcined potash by Hans Wulff; tamarisk (possibly *Tamarix meyeri* or *Tamarix tetrandra*) from near the Belus River, Israel; a bush from sand dunes near Caesarea, Israel; possibly *Haloxylon salicornium* or *Rhanterium epapposum* from near Jezaziyat, Iraq; bracken leaves from Knightons, Alford, Surrey, England; *chinān* (plant) purchased in the *souk* in Baghdad; a chunk of *tezāb* from a soap shop in Kandahar, Afghanistan; a lump of *eshghar* from near Herat, Afghanistan said to have been made from *chub-i-balut*; a sample of *chinān* plant from Qasr al-Hayr al-Sharqi, Syria, said to have been used for the production of *qilī* produced by local workers. Ashtor (1992, 482) considers that *qilī*, *keli* or *kali* (*ushnān*) is *Hammada scoparia*. It is clear from these samples that Brill was well aware of the link between the use of high soda plant ash for the production of both soap and glass; the first hard soap was produced in the Islamic world using soda-rich plants (Ashtor 1992, 481).

Brill's analyses of these alkali-rich materials reveal that all contain relatively high alkali levels. However, in some, the potassium oxide levels are significantly higher than the soda levels. These are the samples from Shareza, Kandahar, the desert bush from near Cairo, the bush from Caesarea, bracken from Surrey and seaweed from Acre Bay (and the 'soap'). Moreover the seaweed from Acre Bay has a mixed-alkali composition with 14.1% soda and 18.1% potassium oxide; so neither the potassium nor the sodium levels dominate; bracken from the Weald contains 26% potassium oxide.

Samples of *eshgar*, *qilī*, the 'soda plant', two specimens of *chinān*, a sample of so-called potash and another of so-called calcined potash all contain elevated soda levels and relatively low potassium oxide levels. This indicates that these materials may have been suitable as a flux in glass production.

However, the levels of MgO, CaO and Al_2O_3 are more variable. The MgO values range from very low to between c. 4% and 12%. There is a high CaO level in one *qilī* sample (17.7%); others contain between 3.5% and 9.54% CaO. The alumina levels for samples in this group lie between 0.42% and 1.82%. The two soda-rich plants (?*Salicornia* sp. and *H. salicornium*) are compositionally distinctive from all of the other samples because, for some reason, they contain much higher levels of alumina. The 'potassium-rich' samples contain between 18.1% and 37.2% potassium oxide.

Brill (1970, 125, n. 4) provided a suggested 'composite' composition in the form of compounds, which satisfied both the valence requirements and mass balance. However, he was careful to point out that many other composite compositions were possible. Attributing the Cl/CO_2 to particular compounds is a problem. For example, if an excess of sodium chloride (NaCl) is present, it will reduce the overall soda levels available for the production of glass because, as noted earlier, it is minimally reactive with silica: the absolute proportion of sodium carbonate is especially significant here.

The following plants would have been suitable for glassmaking based on the compositions of plant ashes Brill (1999b) published: 'shinai' from Iraq (sample no. 4401), *Haloxylon recurvum* from Pakistan (sample nos. 4405, 4421 and 4422) and Salicornia from Doñana, Spain (sample no. 4415). However, in these instances, the calcium oxide (CaO) levels would be too low to produce a durable glass unless an additional source was used. Samples of *ghar* from Pakistan (sample nos. 4426, 4428, 4429, 4432 and 4433) have soda:potassium oxide ratios varying from c. 2:1 to 6:1 with calcium oxide levels at between 0.45% and 7.57%. (Some of the results are for materials bought at the *souk* were intended for glazing.) Nevertheless, these are notable variations that would have had a significant impact on the melting behaviour of glass or on the glaze produced. It is not clear whether *Ghar* is a generic term for the ashes of a range of plant species or only for *H. recurvum* and whether this is the same as the *sajji* plant (Brill 1999a, 214–15).

2.2.3.4 *Other and More Recent Research*

A small number of chemical analyses of plant ashes have been published by Verità and Tontanino (1990). These provide evidence that there were other possible geographical sources of halophytes in the Mediterranean that could have been used in ancient periods and can be added to the sources we already know about from historical sources in Alexandria and Spain, as well as to the French (Mediterranean) sources published by Amouric and Foy (1991) discussed earlier.

The published analyses of these alkali-rich materials by Turner, Brill and Verità and Tontanino therefore set the scene for further work to be carried out on specific halophytic plant species in more restricted areas. This work will now be discussed.

Tite *et al.* (2006) focused mainly on the analyses of one species of plant, *Salsola kali*, with sixteen specimens from Greece (Attica), Crete, and the United Kingdom (Pembrokeshire and Mull). One analysis of *Salsola* sp., two analyses of *Suaeda* sp. and one analysis of *Anabasis articulata*, all harvested in Egypt, were also reported by these authors. Analysis of these plant ashes was performed with energy-dispersive X-ray fluorescence. Following Turner (1956b), in the absence of (quantitative) XRD data, a number of assumptions were made about the combination of moles of chlorine and sulphur trioxide with the alkalis (sodium and potassium) present and hence the relative proportions of interactive carbonates present (*ibid.*, 1288–9). They readily admit that, in the absence of quantitative XRD data, there are limitations to this approach. The outstanding problem, also encountered by Barkoudah and Henderson (2006), is that when there is a relatively high chlorine content, it is difficult to demonstrate what proportion is combined with sodium and what proportion with potassium, and this is clearly important for glass technology. Tite *et al.* (2006, 1288) made certain assumptions as to what extent chlorine and sulphur combine with potassium and sodium. Naturally, if a plant ash contains a small amount of sodium and a high level of potassium, it would not have been used for the manufacture of soda glass – and there is a possibility that most, if not all, of a low level of sodium is present in the form of a minimally reactive sodium chloride. If there is a higher level of sodium, then potentially only a proportion could form a chloride. A possible way of showing what proportions of alkali-rich salts are present in ash samples is by using XRD spectroscopy. Barkoudah and Henderson (2006, 313–15) used this technique, but even here they encountered problems, as described later in Section 2.3.3. Tite *et al.* (*ibid.*) present a set of results for *S. kali* showing that the potassium levels are normally higher than sodium levels for plants growing in both the Mediterranean area and other parts of Europe. This, they interpret, for their coastal plant samples, as being due to the relative contributions of (ion-containing) seawater, rainwater and hinterland runoff. On this evidence, this characteristic appeared to them to be a genus-specific, or possibly a species-specific, characteristic.

Barkoudah and Henderson (2006) took a different approach. Because relatively few compositional data for halophytic plants had been published, their

2.1. *Anabasis syriaca*. Reproduced by kind permission of Professor Y. Barkoudah.

study, which focused on Syrian plants, involved harvesting a wide range of halophytic plant species from semidesert environments and a salina. It had been suggested that halophytic plants of the genera *Salsola*, *Salicornia* and *Hammada* (*Haloxylon*) were used for the manufacture of plant ash glasses (Amouric and Foy 1991; Ashtor 1992, 492). However, there had not been a focused investigation to test scientifically whether other alkali-rich plant genera might also have been used to make glass, so it was investigated here. Several other research questions were addressed:

- Do the ashes of particular plant species or genera have diagnostic chemical compositions?
- Do ash samples of the same plant species growing in the same place display a constrained 'local' compositional signature?
- Do the chemical compositions of ashes of (the same species of) plants growing in contrasting geological environments (e.g. on basalt, on limestone and in a salina) reflect the geological variations?
- How do ashes of plants growing on a transect across the same geological zone vary in composition?

At the time, in the absence of *proven* Mesopotamian Bronze Age (plant ash) glassmaking centres, plants were harvested from locations for which there was historical or archaeological evidence for Islamic plant ash glass production (al-Raqqa, Aleppo and Damascus; see Chapter 9).

Twenty-five plant samples were collected. These were samples of *Anabasis syriaca* (Fig. 2.1), *Arthrocnemum strobilaceum*, *Atriplex leucoclada*, *Chenopodium album*, *C. murale* (Fig. 2.2), *Girgensonia* sp. (Fig. 2.3), *Halopeplis* sp. (Fig. 2.4), *Haloxylon articulatum* (Fig. 2.5), *Noaea micronata*, *Salsola jordanicola* (Fig. 2.6), *S.* sp., *Sueda* sp. and *S. vermiculata*. ICPAES and ICPOES were used to analyse

2.2. *Chenopodium murale*. Photo: J. Henderson.

them in Nottingham University. By analysing a range of *Salsola* species – *S. vermiculata*, *S. jordanicola* and *S. sueda* – the possibility was investigated that species other than *S. kali* (Fig. 2.7 and 2.7a) might have been used. A second batch of Syrian plants was collected, ashed and analysed using Inductively Coupled Plasma Atomic Emission Spectroscopy at the University of Melbourne. These results are (currently) unpublished and are partially referred to in the discussion that follows. Samples of *S. kali*, and more samples of *S. jordanicola*, *H. articulatum* and *A. syriaca*, were collected to complement the first (published) batch. X-ray diffraction was performed on a number of specimens of plants and ashes from the second batch. Because the compositions of ashes change with increasing temperature, analyses were performed on both dried plants and their ashes. The plants and plant ashes chosen for XRD analyses were those that contained sodium oxide levels between 11.7% and 42%. XRD was performed on the following species: *Salsola vermiculata*, *Arthrocnemum strobilaceum*, *Halopeplis* sp., *S.* sp., *Sueda* sp., *S. jordanicola* and *Anabasis syriaca*.

Locations for plant harvesting were partly selected according to geological contrasts observed on a 1:20,000 German geological map. Each location from which plants were collected was also recorded using a global positioning system. Locations are shown in Figure 2.8.

2.3. *Girgensonia* sp. Reproduced with kind permission of Professor Y. Barkoudah.

The plants were collected from a range of semi-desert locations and also from the edge of an evaporation lake (or salina) southeast of Aleppo at Jaboul. The locations were Aleppo, located on the border between Paleogene and Neogene rocks, and Jaboul, which sits on a basalt–Paleogene border. Heracla and al-Raqqa near the Euphrates are part of a region of tertiary limestone. (The Euphrates valley cuts through Tertiary sediments and volcanics and has extensive deposits of Quaternary alluvium). The transect along which plants were sampled extended along the Balikh valley. It is characterized by Neogene (including gypsiferous) limestone. To the south, the area around Damascus is dominated

2.4. *Halopeplis* sp. Reproduced by kind permission of Professor Y. Barkoudah.

2.5. *Haloxylon articulatum*. Reproduced by kind permission of Professor Y. Barkoudah.

by cretaceous limestone and quaternary basalt rocks, especially to the south, whereas the area to the west of the Anti-Lebanon is Jurassic limestone.

2.3 Major Elemental Compositions of Syrian Halophytic Plants and the Implications for Ancient Glass Production

The results and interpretation of the first study of a range of Syrian plants have already been published (Barkoudah and Henderson 2006), and so they are summarised here. However, the interpretation of plant ash compositions and the relationship between ash-composition glass made from them are dealt with in considerably more detail here.

2.6. *Salsola jordanicola*. Reproduced with kind permission of Professor Y. Barkoudah.

Table 2.2. Chemical analyses of the ashes of three plant species from Syria (after Barkoudah and Henderson 2006, table 1)

	Salsola vermiculata	Halopeplis sp.	Anabasis syriaca
Na_2O	19.4	23.2	31.6
MgO	2.26	7.05	5.12
Al_2O_3	0.12	0.66	0.22
P_2O_5	1.64	0.84	0.57
K_2O	25.5	8.2	6.6
CaO	4.17	8.91	5.28
TiO_2	0.01	0.04	0.01
MnO	0.005	0.008	0.007
Fe_2O_3	0.07	0.37	0.12

The published results (Barkoudah and Henderson 2006, table 2) indicate that the plants sampled are either soda-rich (*Anabasis, Arthrocnemum, Halopeplis* and some specimens of *Salsola*; see Table 2.2), potassium-rich (e.g. *A. leucoclada, Chenopodium murale, C. album, Girgensonia* sp., *Noaea micronata*) or are rich in both alkalis (e.g. *Halopeplis* sp. and other specimens of *Salsola*; see Table 2.2). However, it is significant that plants that have mixed alkalis may contain potassium partly or predominantly in its unreactive form (such as potassium chloride), so that a soda- or a mixed-alkali glass might result. Rehren (2008) has noted that a high level of chlorine in a glass melt can lead to a reduced level of potassium. Ashes of *Atriplex leucoclada, Chenopodium murale, C. album, Girgensonia* sp. and *Noaea micronata* contain high potassium oxide levels and

(a)   (b)

2.7a. *Salsola kali*. 2.7b. *Salsola kali* (detail). Reproduced with kind permission of Prof. Y. Barkoudah.

insufficient soda to have been used for making ancient soda glasses. The ashes with high soda and high potassium levels could also be distinguished chemically by their associated trace and minor elements, such as phosphorus and manganese. As far as we know, plant and tree ashes containing high potassium concentrations were not used for making glasses in the ancient Middle East. Although glasses containing levels of c. 4% potassium oxide and above have been found in Afghanistan (see Chapter 4), these are not sufficiently high to suggest that tree ashes were used to make the glass. Another complex factor is that the same species of plant can contain substantially different levels of alkali. For example, samples of *Salsola vermiculata* contained soda levels of 19.4%, 21.2% and 28.7%, and samples of *S.* sp. (*sueda?*) contained 11.7%, 16.9% and 19.5%. The plant species that were found to contain the highest soda levels were *Arthrocnemum strobilaceum* (42%), *Salsola jordanicola* (28.6%), *Halopeplis* sp. (23.2%) and *Anabasis syriaca* (31.6%). Examples of plants with high levels of potassium oxide are 40.2% in *Atriplex leucoclada* and 44.6%, 36.1%, 31% and 35.1% in *C. murale*. Examples of the relative levels of alkalis in mixed-alkali plants were 32.7% K_2O and 18.3% Na_2O in *Halopeplis* sp. and 29.4% K_2O and 16.9% Na_2O; 29.2% K_2O and 15.1% Na_2O and 33.7% K_2O and 11.7% Na_2O in *Salsola* sp. (?). No mixed alkali glasses have (yet) been found in the ancient Middle East, although there is increasing evidence that they were in use during the late Bronze Age and early Iron Ages further west in the Mediterranean basin. Mixed-alkali glasses are characterised by low concentrations of magnesia and calcium, and so if plant ashes had been used to make them, either the ashes were purified or the plant ashes contained low levels of Mg and Ca levels (see Chapter 6).

2.3.1 A Relationship between Halophytic Plant Ash Compositions and Plant Species?

It would be interesting to be able to identify the plant species or genera that were used in glassmaking by chemically analysing selected plant ashes. For this to be possible the chemical composition of the ash and its modified presence in the glass would need to be diagnostic for the plant species or genus used. A level of predictability in ash composition was important for glassmakers given that relative levels of each element could affect the behaviour of the glass being melted. The length of time the glass could be worked, the working/softening temperatures and the melting temperature of the batch could all be affected by variations in ash composition. Sodium and potassium are fluxes, whereas high levels of magnesium and calcium in the ash/glass batch would increase the glass melting temperature and produce a 'shorter' glass (see Chapter 1). Using the data published by Barkoudah and Henderson (2006, table 2), it can be stated that the compositional variation of major and minor components (sodium, magnesium,

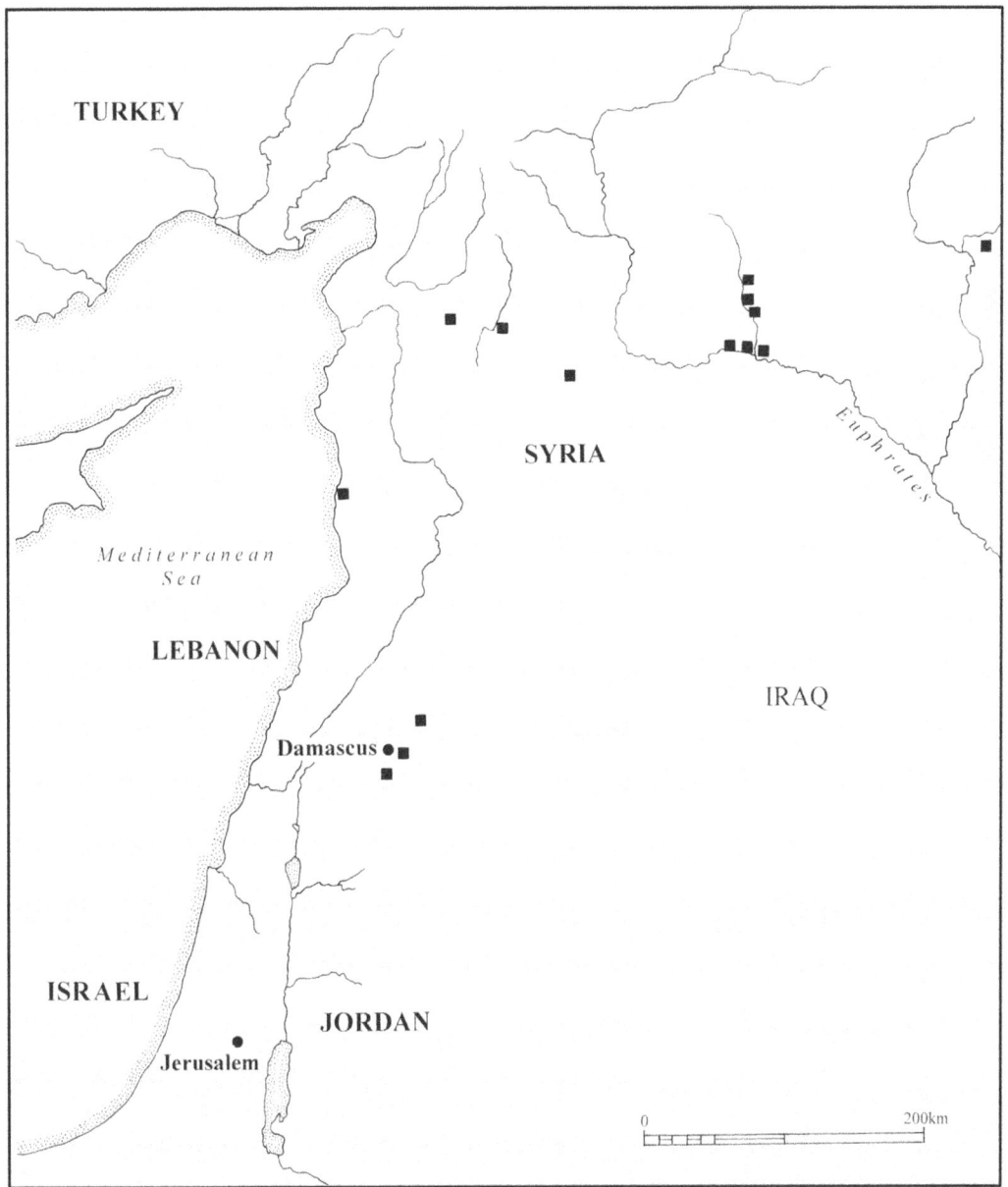

2.8. Map of the sampling locations in Syria and Lebanon from which halophytic plants were collected.

calcium and potassium) for each plant genus was 10% to 30% relative, and therefore relatively consistent. For example, this can be seen for the relative levels of magnesium and calcium oxides in most plants of the genus *Salsola* in Figure 2.9. Levels of trace components in plant ashes were found to be far more variable. However, if sufficient samples of ashed plants were to be analysed from contrasting geological locations, in the long run it might be possible with

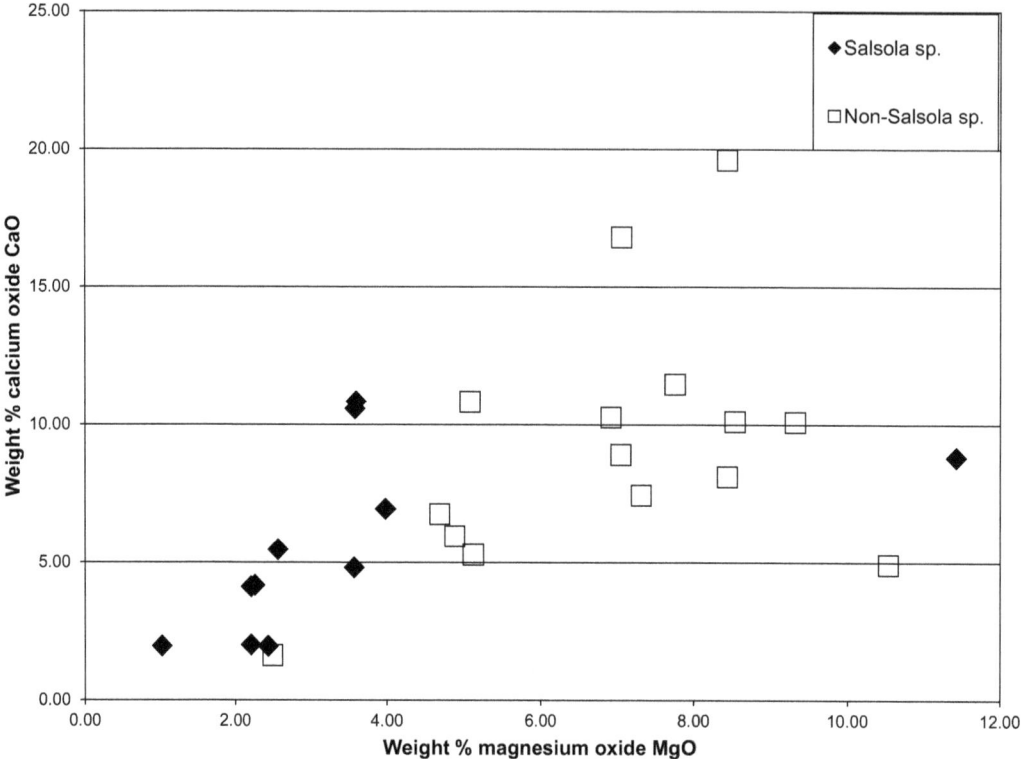

2.9. A bi-plot of weight % magnesia versus calcium oxide in ashes of *Salsola* and non-*Salsola* plants (data from Barkoudah and Henderson 2006).

more confidence to suggest which species were used to make ancient glass (see Chapter 11).

By focusing only on the relative levels of calcium and magnesium oxides in soda-rich ashes (Fig. 2.10), we can make a number of points. First, apart from a single sample of *Salsola* sp. sample, the rest of the *Salsola* results are united because they fit to the positive correlation. Moreover, the two results for *Haloxylon* fall away from the correlation. All four samples of *S*. sp. *sueda* have a relatively constrained range of magnesium oxide contents, whereas other species within the genus are more variable. The results for *S. vermiculata* are especially highly correlated. Bearing in mind that all of these ash samples are soda-rich and could therefore be used as a glass flux, an important consideration is their calcium oxide contents. From Figure 2.10, it can be seen that specimens of *A. Syriaca*, *S. sp. Sueda* and *S. kali* and *Haloxylon articulatum* are sufficiently calcium-rich to make glass using two ingredients: plant ash and silica. However, samples of *S. vermiculata*, *S. jordanicola* and some specimens of *A. syriaca* and *S.* sp. *sueda* contain calcium levels that would necessitate the use of an additional calcium-rich raw material. Indeed, this may be the reason for the apparent use of a different calcium source in the manufacture of al-Raqqa type 4, as discussed later. Although this is clearly a useful way of characterising ashed plants, the

WAYS TO FLUX SILICA

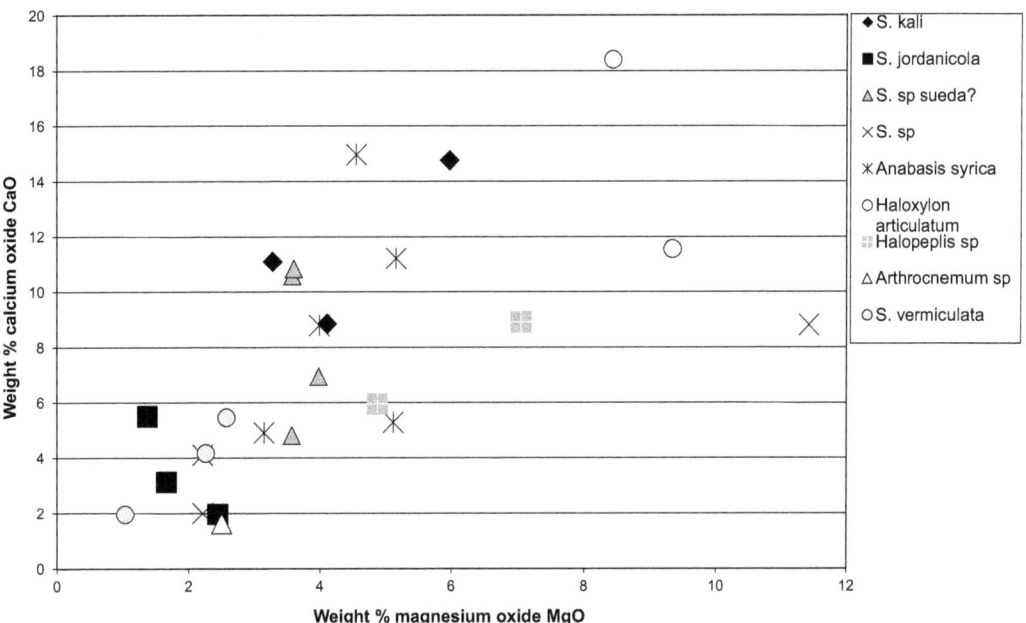

2.10. A bi-plot of weight % magnesia versus calcium oxide in ashes of various species of *Salsola, Anabasis, Haloxylon, Halopeplis* and *Arthrocnemum* (unpublished data combined with data from Barkoudah and Henderson 2006).

positive correlation between magnesium and calcium oxides found in ashes (Fig. 10.10) is not especially common in ancient plant ash glass compositions. If plants with this correlation were used for glassmaking, in many cases the calcium: magnesium ratios must have been disrupted during the process of glassmaking.

A plot of magnesium and potassium oxides levels in all sodium-rich plant ashes analysed provides evidence for the likely plant ash species used to make glasses (Fig. 2.11). Given that many ancient plant ash glasses exhibit a clear positive correlation between potassium and magnesium oxides, it is significant that almost all ash compositions of the genus *Salsola* fall on this positive correlation line (see Chapter 11). The species involved are *S. kali, S. jordanicola, S. vermiculata* and *S. sp. sueda*. One *S. kali* sample from Dumeir falls away from the correlation. However, a sample of *S. jordanicola* from Dumeir does fit the correlation. With such small sample numbers, it is difficult to explain this anomaly. Of the two samples of *Halopeplis sp.*, only one fits the correlation line. The single sample of *Arthrocnemum sp.* falls close to the correlation line. Although both sodium-rich, the ashes of *Anabasis syriaca* and *Haloxylon articulatum* do not fit the *Salsola* correlation, which suggests that they may not have been used to make ancient plant ash glasses. However, there is another important consideration. The relative levels of magnesia and potassium oxides in the ashes may have been disrupted in the process of making glass. It can be

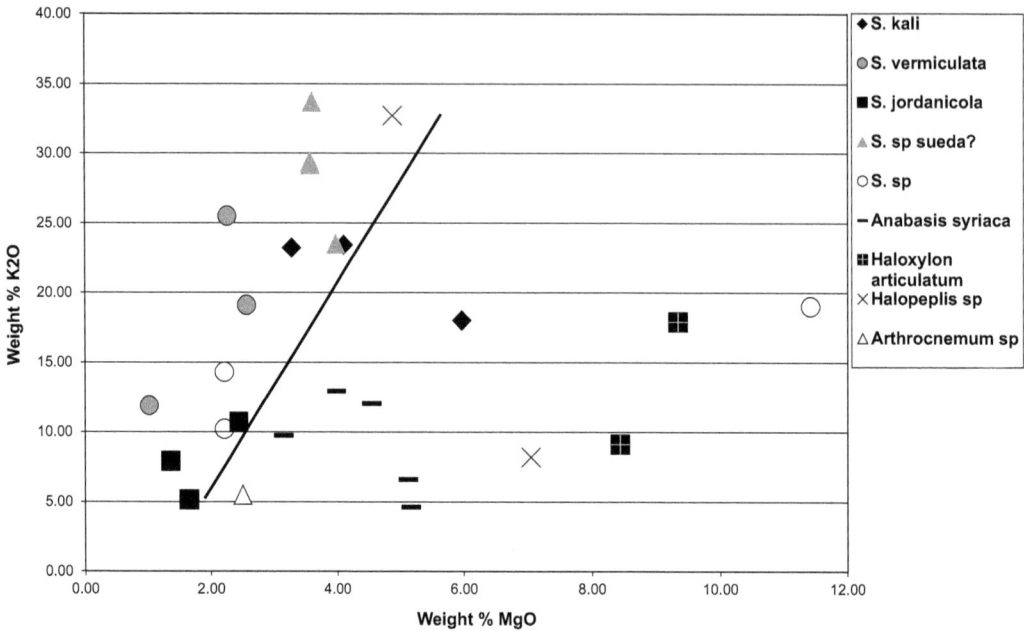

2.11. A bi-plot of weight % magnesia versus potassium oxide in sodium-rich plant ashes of the genera *Salsola, Anabasis, Haloxylon, Halopeplis* and *Arthrocnemum* (unpublished data combined with data from Barkoudah and Henderson 2006). The '*Salsola* correlation' line is given.

shown experimentally, for example, that high chlorine levels in a glass melt can reduce the potassium levels (Tanimoto 2007). Nevertheless, depending on the significance of this factor, the positive correlation could still be retained, and in the absence of other sources of potassium and magnesium in plant ash glasses it clearly is. This is discussed in more detail in Chapter 11.

Another consideration is the high potassium levels that have been found in plant ashes (Barkoudah and Henderson 2006; Tite *et al.* 2006), especially in *S. kali* and *S.* sp. *sueda*. It may be that the presence of elevated chlorine in maritime environments led to the reduction of the potassium levels in glass melts, leading to the levels that are found in plant ash glasses (Rehren 2000). However, inland halophytic plants were also used, and some contain relatively low potassium oxide levels.

The geological context in which plants grew clearly contributed to the chemical composition of their ashes. This can be illustrated with reference to ash samples of *Salsola*, a soda-rich ash and *Chenopodium*, a potassium-rich ash. Nearly double the level of calcium oxide has been found in *Salsola* ashes in plants growing on gypsiferous limestone of the Balikh valley, northern Syria when compared with Salsola ashes from nongypsiferous areas of northern Syria. Moreover, a single sample of *Chenopodium* growing in the basaltic area around Damascus had double the level of calcium oxide than found in samples growing

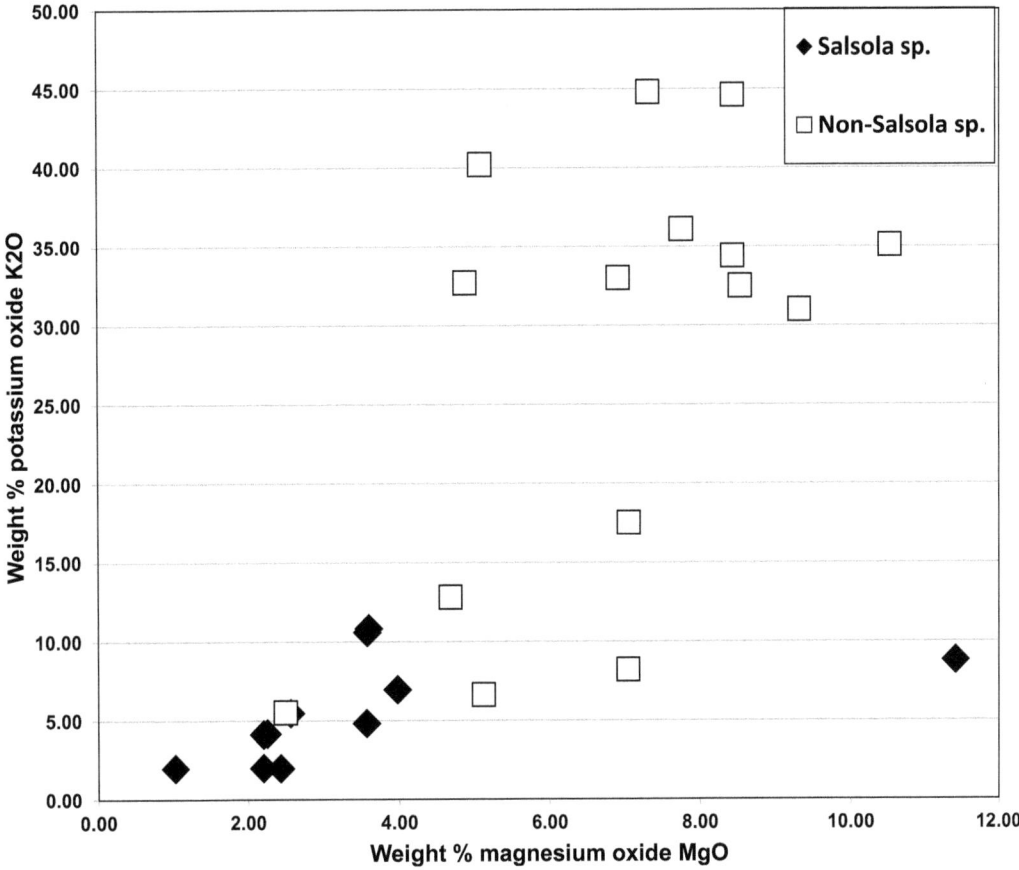

2.12. A bi-plot of weight % magnesia versus calcium oxide in ashes of plants of the genus *Salsola* compared with ashes of non-*Salsola* plants (data from Barkoudah and Henderson 2006).

in the rest of Syria. Samples of *Anabasis syrica* from the basaltic area of Damascus and the (Jurassic) Anti-Lebanon were compared with samples from the limestone area north of Damascus. These plants contained similar levels of soda of c. 30% and similar major and minor components (apart from potassium); relative levels of alkalis and tolerance to them are controlled by the plant physiologies. These are small sample numbers, but it is clear that variations in surficial geology can contribute significantly to the calcium levels in *Salsola* ashes, although not apparently those of *Anabasis*. The effect that geological variations have on the relative levels of alkalis is probably negligible: they can be explained by the overriding affect of the plant physiologies. *Salsola* plants growing on the gypsiferous limestone of the Balikh valley and on the Jaboul evaporite deposit were found to contain comparable ratios of sodium to potassium. The absolute levels of sodium in Balikh samples were equal to or greater than in samples from the Jaboul. Therefore, although more intensive research needs to be carried out to investigate these findings using a much larger data set, it is apparent that the calcium levels found in ashes do reflect gross differences in the geology. This is discussed in more detail on Chapter 11.

A comparison between the chemical compositions of two ashed *S. vermiculata* plants growing in the ditch of the Aleppo citadel shows a generally low level of variation for major and minor components, with some increasing levels of variation for trace components. For major components, the relative levels between the two plant ash samples are 8.5% for Na_2O, 24% for CaO and 25% for K_2O; for a minor element, the level is 12% for MgO; and for trace elements, 44% for P_2O_5, 20% for Cu 20%, 49% for Sr and 17% for Zn. If we compare the variations for the 'same' location in Aleppo with an ash composition for the same species of plant from the Jaboul evaporitic deposit, some 60 kilometres away, apart from the relative level of P_2O_5 of 28%, the levels increase to 32% for Na_2O, 53% for CaO, 53% for K_2O, 54% for MgO, 57% for Cu, 484% for Sr and 245% for Zn. We can therefore have some confidence that local variation is considerably less than for interregional variation.

The analyses of samples of *S. kali* ashes published by Tite *et al.* (2006, table 2) provide some comparative data from other geographical areas. The data fall in a rather broad scatter, irrespective of origin, with few falling on the '*Salsola*' correlation line for magnesium and potassium oxides in Figure 2.11. The technique of analysis used by Tite *et al.* (*ibid.*, 1288) involved the averaging of three results. They analysed 2-mm-diameter areas within dried pellets of ash using energy-dispersive X-ray fluorescence (XRF) analyses. To make a more valid comparison between the Syrian results and analyses of plant ashes from other Mediterranean areas, the same analytical technique would need to be used and preferably one that provides a bulk analysis by dissolution such as ICPAES rather than a less representative XRF surface analysis and one that provides trace element results.

2.3.2 Glass Coloration and Halophytic Plant Ashes

As suggested by Geilmann and Brückbauer (1964), transition metals in plant ashes (e.g. Fe, Mn and Cu) could potentially produce coloured glass. The occurrence of minor and trace levels of compounds in plants is determined by a range of complex factors. It can be species-specific, as found in plants known as hyperaccumulators (Bhatia *et al.* 2003). However, the levels found in plant ashes are low. Low levels of manganese oxide found in faintly tinted and colourless Bronze Age colourless glasses from Tell Brak and Nuzi of c. 0.02% to 0.05% (Brill 1999a, 40–42; Brill 1999b, 39–42) are likely to have originated in plant ashes and levels above this added deliberately. Levels of Mn and P are positively correlated in plant ashes (Barkoudah and Henderson 2006, 311, fig. 3), so if plant ashes provided the main source of manganese in lightly tinted Bronze Age glasses then the same correlation should be observable. It is in some cases,

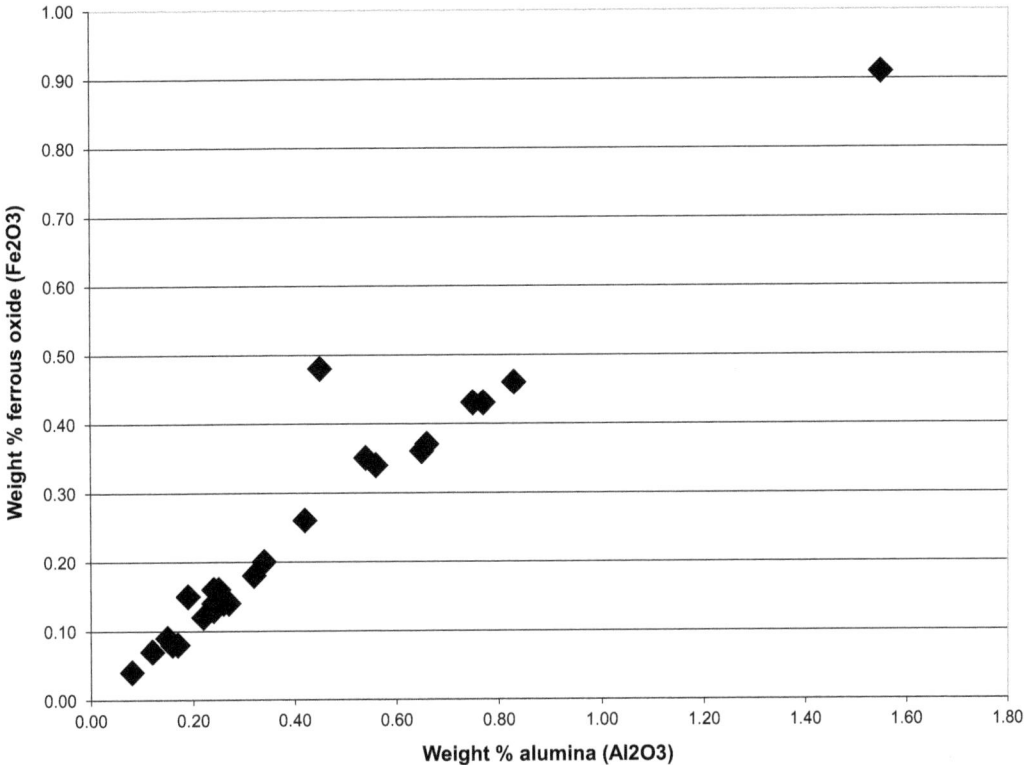

2.13. A bi-plot of weight % ferric oxide versus aluminium oxide in ashes of *Salsola* and non-*Salsola* plants (data from Barkoudah and Henderson 2006).

but not in all, so the glass melting behaviour and other possible manganese impurity sources could disrupt the correlation. Almost all colours of Islamic plant ash glasses contain relatively high levels of manganese oxide of between c. 0.2 and 2.0% MnO (Brill 2001, 29; Henderson *et al.* 2004, 461, fig. 9), so it must have been added deliberately during primary glass production.

The strong positive correlation between iron and aluminium in all plant samples analysed of $R^2 = 0.9424$ provides evidence for this common geological association in the soils in which plants grow (Fig. 2.13). Both components are generally at a lowish level in plant ashes; a second source of both is the silica source. A positive correlation between alumina and iron oxides is commonly observed in ancient glass and until now has been attributed to iron and alumina impurities in the silica source alone with contrasts attributed to the use of different silica sources (e.g. Henderson 2006; Mirti *et al.* 2008). With this new evidence from plant ash analyses, it is clear that a proportion of iron and aluminium was introduced in the glass melt in plant ashes. To decide whether a pure source of silica (such as crushed quartz) was used, the levels of other impurities known to be associated with silica (e.g. Nd) should be taken into account. The same positive correlation between iron oxide and alumina may be present in both the silica and in plant ashes: if this is the case the affect on the glass composition would be summative. Nevertheless, to be able to define the primary source of both elements would be significant for ancient glass technology. If a deviation from the plant ash correlation is observed, this could suggest that a second source of iron and aluminium was contributed by silica.

Another source of transition metals in glasses is ancient and modern pollution. Barkoudah and Henderson (2006, 313) showed that heavy and transition metals (copper, zinc, tin and lead) are accumulated in halophytes. It was shown that the lead and zinc levels present in plants growing in the Islamic industrial complex in al-Raqqa, Syria, were correlated. Lead-rich glazes (and possibly glasses) were made and metals probably smelted there. These results could therefore be the result of ancient pollution. The lowest levels of heavy and transition metals were found in the least polluted rural areas of Syria and the highest in ancient urban areas. These metallic elements would have been passed on to glasses made from ashed plants that grew in polluted areas. The levels of heavy metal impurities that have been observed in natron glasses have been explained as a result of glass recycling (Henderson 1993, 247–59; Wedepohl *et al.* 1997). We now have an additional/alternative source of the trace levels of heavy metals found in plant ash glasses; glass recycling is not the only explanation for their presence.

2.3.3 Alkali Levels in Halophytic Plants

One technique used to investigate the presence and relative levels of potassium and sodium salts in plant ashes is XRD (see Appendix). One reason for using this technique is to determine the relative levels of reactive sodium and potassium. At temperatures above 400°C sodium oxalate ($C_2Na_2O_4$) in dried plants breaks down to sodium carbonate, with the evolution of carbon monoxide; at about 600° to 650°C the sodium carbonate would break down to sodium oxide and carbon dioxide.

It was found that all of the samples with high soda (11.7%–33.8%), as determined by ICPAES, contained sodium oxalate. XRD analysis of the high soda plant ashes (see Fig. 2.14) showed

1. that potassium chloride was an important component of *Salsola vermiculata*, *S.* sp. (*sueda*?), and *S. jordanicola* and
2. even though sodium oxalate had been found in dried plants, the first impression from the results was that there was an *apparent* absence of simple sodium-rich compounds suitable for glass production.

The best interpretation of an *apparent* absence of simple sodium-rich compounds is that ionic substitutions occurred during ashing, particularly between carbonate phases. In addition, some compounds were not well crystallized.

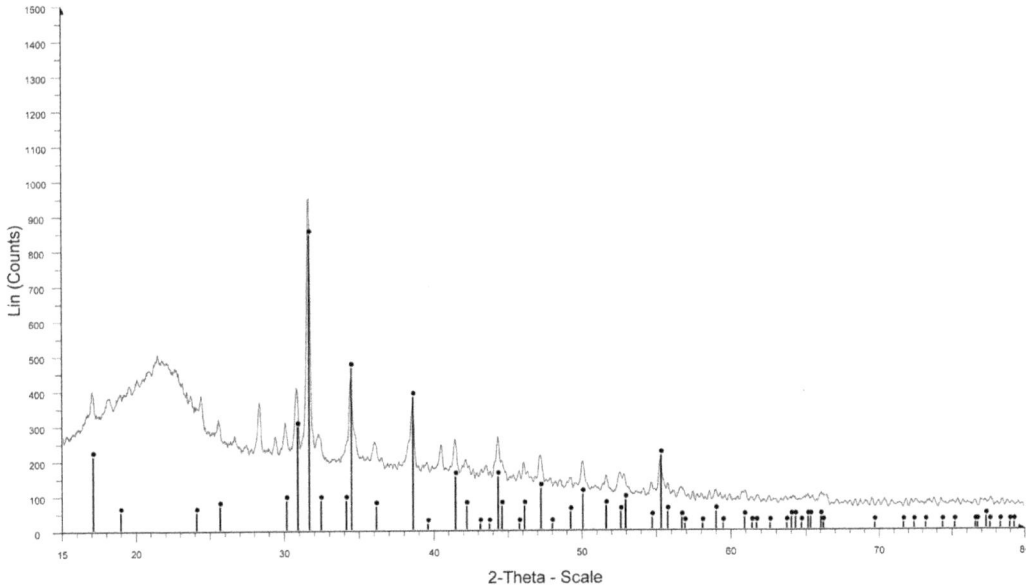

2.14. An X-ray diffraction pattern derived from the analysis of ashes of *Salsola jordanicola*.

Examples of the substitutions are $Na_2Ca(CO_3)_2$ (nyerereite) and $Na_2Ca(CO_3)_2$ (natrofairchildite). These compounds would be suitable as sources of both sodium and calcium oxides in glass production. However, the sodium and potassium chloride detected would be relatively unreactive with silica (Bateson and Turner 1939), so they would have formed part of an unreactive scum that forms during glass melts – or they would be volatilised. The solubility limit of NaCl (sodium chloride) in a soda-lime glass is 1.4% (*ibid.*). Brill (1970, 124) noted a high percentage of potassium (17.2%) in *tezāb* ash from Kandahar, Afghanistan, which could potentially explain the occurrence of plant ash glasses with relatively high potassium levels in Afghanistan (see Chapter 4) but that only 75% of the analytical concentrations ended up in the final glass. His interpretations of this included the absorption of water or a high percentage of carbon in the ash. However, it can now be demonstrated that at least part of the 25% balance is likely to be unreactive salts, such as potassium chloride.

It is therefore clear that it is the percentage of *reactive* potassium that is important, not the absolute levels detected. Results for three specimens of *S. vermicula* from two locations in Syria were found to contain between 11.9% and 25.5% 'potassium oxide' (determined by ICPAES), and KCl was identified in the samples by using XRD. Given the complex salts that are formed during ashing and the difficulty of identifying them, the possibility of producing quantitative XRD results, giving the percentage proportion of each compound detected in plant ashes, is slim. The presence of unreactive compounds such as chlorides could disrupt the ratio between potassium and sodium that found their way into the glasses.

2.3.4 Calcium Levels in Halophytic Plants

Apart from the presence of reactive sodium salts, the other key component in plant ashes that would have had a critical affect on the chemical composition of the glass produced, as well as its (related) working properties and durability, was calcium. Its level would have been determined by which plant species or colony of plants was harvested. There is also increasing evidence that plant ash glasses made on the Levantine coast involved the use of beach sand with a (possibly additional) calcium source introduced with seashell fragments in sand (see Chapter 10). There is certainly evidence that both Islamic and Bronze Age plant ash glasses contained relatively low levels of calcium oxide levels (see Figs. 6.1 and 10.10 later in this volume; Henderson 1997; Henderson *et al.* 2004; Kato *et al.* 2010).

However, the chemical analyses of some ashes of *A. strobilaceum*, *S. vermiculata* and *S. jordanicola* revealed that insufficient calcium was present for glassmaking. These plants were soda-rich; had they been used for glassmaking, a separate calcium-rich raw material would have been required. Indeed there is some compositional evidence for the use of a separate calcium-rich raw material in Islamic plant ash glasses from Syria. Figure 10.5 is a bi-plot of phosphorus pentoxide versus calcium oxides in plant ash and natron glasses from al-Raqqa. It shows the expected distinction between purer natron glass, in which aragonite in seashells provides the calcium source (with low phosphorus), and plant ash glasses with higher levels. The type 4 plant ash glasses contain lower calcium oxide levels, and some samples elevated phosphorus, compared with the type 1 plant ash glasses. There is a weak positive correlation between the oxides – also found in type 2. This suggests that bone ash was used as one of the calcium sources (Henderson *et al.* 2004, 459). An additional one may be suggested on the basis of a positive correlation between alumina and calcium oxide especially in the 9th century glasses (Fig. 2.15), perhaps a mineral such as a feldspar (Tal *et al.* 2004, 64, fig. 10).

Had the plant ash been purified, the levels of both magnesium and calcium oxides would have been reduced in the glass. This resulted in low values in colourless *cristallo* plant ash glass made in sixteenth- and seventeenth-century Italy (Verità 1985, table 4; Henderson 2006, 301, table 4, illus. 9), although there is no evidence for the practice in most earlier glasses. Low calcium levels are a characteristic of Islamic glass from Iran (especially colourless glass; see Chapter 10), but it also contains high magnesia levels.

Therefore, irrespective of what potential changes in glass compositions may have occurred during the glassmaking process (Rehren 2008), to be able to

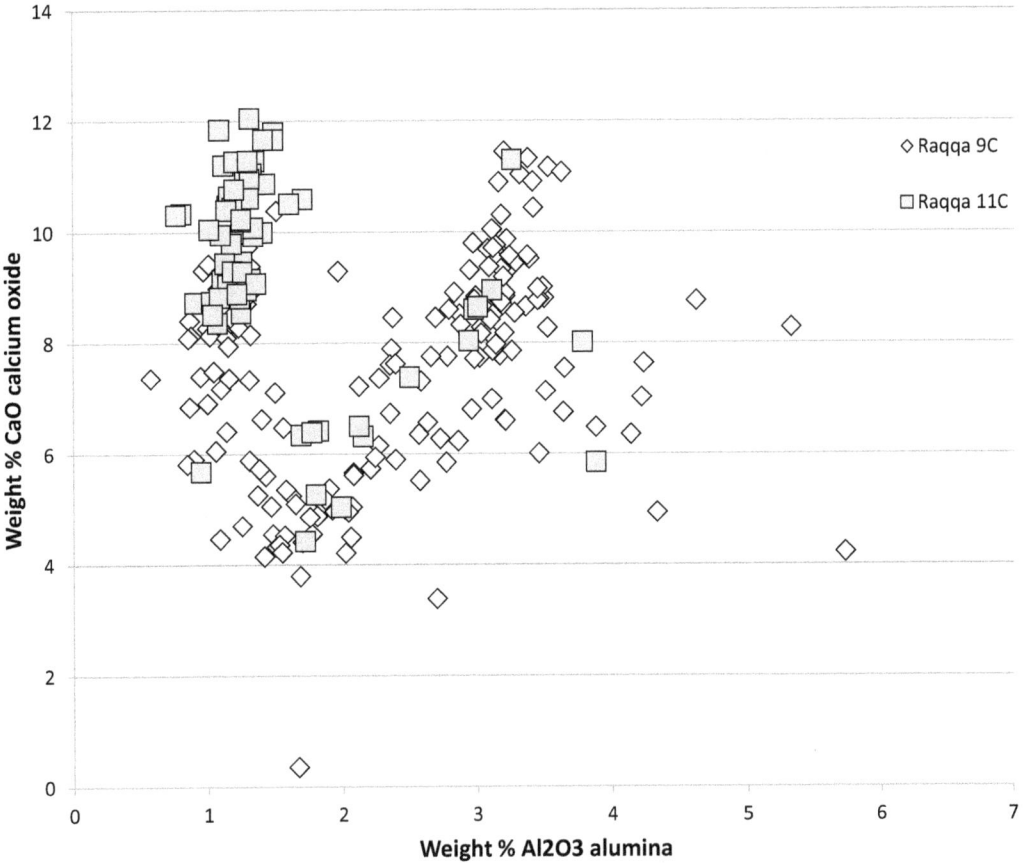

2.15. A bi-plot of weight % calcium oxide versus aluminium oxide in ninth- and eleventh-century plant ash glass from al-Raqqa, Syria. Data from Henderson *et al.* 2004.

predict the levels of calcium in the plants used in glassmaking would have been important to both glassmakers and glassworkers. This could have been achieved by returning to the same colony of plants at the same time of year, because a predictable calcium level would be produced by ashing plants from the same geological location.

It was found that plants growing in a northern Syrian area of gypsiferous limestone (Ca_2SO_4) along the Balikh valley running north of al-Raqqa to the Turkish border contained sufficiently high calcium levels to make glass from them. Some of the rest of the *Salsola* specimens analysed did not; had they been used in glass production, a separate source would have been necessary. Stern and Gerber (2004, 137) have claimed that there is insufficient calcium in plant ashes to provide enough for ancient soda-lime glass. These plant ash results show that in many cases this is not true. A consistent positive correlation between magnesium and potassium oxides is observed in *Salsola* samples irrespective of their origins (see Fig. 2.12). This correlation (and others) in plant ashes is

Table 2.3. Major elemental compositions of the ashes of three species of forest plants from the United Kingdom (after Jackson et al. 2005, table 1)

Element oxide	Beech *Fagus syvatica*	Oak *Quercus* sp.	Bracken *Pteridium aquilinium*
Na_2O	0.4	0.69	0.26
MgO	6.87	4.39	5.4
Al_2O_3	0.76	1.93	0.35
SiO_2	14.62	6.78	21.07
P_2O_5	12.87	2.88	9.27
K_2O	18.8	14.47	42.52
CaO	31.06	65.41	10.49
TiO_2	0.06	0.08	0.02
MnO	6.35	0.37	0.68
Fe_2O_3	0.65	2.41	0.35

produced by a combination of the plant physiology, bedrock geology and the surficial geological environment; the consistent ratio between magnesium and potassium oxides is present even though the plants were collected from widely separated locations with contrasting geologies. Had river or coastal sand been used as a silica source, an additional calcium source could be freshwater or marine shell fragments.

2.4 Forest Plant ash Compositions

We know that medieval glassmakers north of the Alps used a variety of alkaline ashes, including those of oak (*Quercus* sp.), beech (*Fagus syvatica*), pine (*Pinus* sp.) and bracken (*Pteridium aquilinium*) to make glass with a ratio of 2:1 by weight of ash:sand. Examples of the chemical compositions of the ashes of beech, oak and fern ashes are given in Table 2.3. Stern and Gerber (2009, 117) discuss the potentially devastating effect on woodland of the amount of wood consumed by glass production. Using Wedepohl's (2003) assumption that a 'production unit' produced 4 to 10 tons of glass per year, between 2 ha and 50 ha of wood would be consumed per year, depending on whose calculation is used. This would depend on the species, the forest density and the ecological habitat.

Just as for halophytic plants used for the manufacture of Bronze Age, Sasanian and Islamic glasses in the Middle East discussed earlier, glasses fused from beech, oak and bracken ash used to make 'forest glass' in central and western Europe were affected by the inhomogeneity of the ashes and the presence of volatile and insoluble compounds. As also noted for halophytic plants, various factors can affect the compositions of the ashes made from them: the time of year that they are harvested, the tree or plant part used and the geology of the soil in which the plant or tree is growing. For example, Miewes and Beese (1988) note that the CaO:K$_2$O ratio is 13 times higher in the bark than in the trunk of the tree. Hartmann (1994, 117) lists an approximate ash composition for beech (trunks and boughs) as 23% K$_2$O, 48% CaO, 12% MgO, 11% P$_2$O$_5$, 3–6% MnO and 2,500 ppm Ba and oak as 30% K$_2$O, 44% CaO, 5.5% MgO, 11.6% P$_2$O$_5$, 3.6% MnO and 11,000 ppm Ba.

Moreover, Jackson and Smedley (2004, 40) have noted that the size of the melt is important and therefore that the total amount of alkali available is as well, because it will determine how much alkali will react with free silica. However, if we hope to suggest, on the basis of ash composition, where the wood is from, the considerably wider geological variations across the forested areas of western Europe compared with the limestone steppe areas of the Middle East would make it impossible using chemical analysis alone. There is greater scope using isotope analysis (Meek *et al*. 2012).

Jackson *et al*. (2005, 791) have noted that elemental concentrations in seasonally renewable plants like bracken can change from year to year. In addition to the factors listed earlier, they also found that the compositional variations were dependent on variations in climate and variations in ground cover. Jackson *et al*. noted that bracken harvested in two consecutive years from the same location had significantly different concentrations of total alkali and iron. This would have a significant impact on the working properties of the glass and on the glass colour. Factors such as ground cover and variations in climate in the Middle East may not be as significant as they are in temperate environments.

Ashes of oak, beech and bracken can contain elevated levels of potassium oxide, magnesia, phosphorus pentoxide, sulphur trioxide, manganese oxide and calcium oxide, producing glasses enriched in these oxides; they may also introduce traces of boron, zinc, rubidium, strontium, copper, nickel and barium (Turner 1956b; Sanderson and Hunter 1981; Smedley et al. 1998; Jackson and Smedley 2004; Jackson et al. 2005, tables 1 and 3; Stern and Gerber 2009, table 2; Wedepohl and Simon 2010). There is no question that ashes of each species had a highly variable chemical composition, and as already noted, different plant or tree parts have different chemical compositions (Turner 1956b, Bezborodov 1975; Henderson 1988a; Smedley et al. 1998). Moreover, Sanderson and Hunter (1981) and Hartmann (1994) showed that the chemical compositions of beech and oak ash taken from trees growing in neighbouring locations could not be distinguished. From this it follows that it is difficult to be able to state which species of tree or other forest plant has been ashed and used to make glass. Nevertheless, bracken ash tends to be characterised by its silica and potassium oxide levels and its calcium-to-potassium-oxide ratio (Stern and Gerber 2009, 113, fig. 2). Fern tends to contain more potassium and less lime than beech ash (Wedepohl 2003). Some of these characteristics can be seen in Table 2.4, the chemical compositions of three experimental glasses made from beech, oak and bracken ashes. Three (medieval/postmedieval) authors who describe the use of

Table 2.4. The chemical composition (weight % oxide) of glass made from beech, oak and bracken ashes, analysed using wavelength-dispersive X-ray fluorescence (after Jackson and Smedley 2004)

Oxides	Beech	Oak	Bracken
SiO_2	48.97	46.41	62.03
Na_2O	0.77	0.42	0.83
K_2O	11.91	5.63	21.33
CaO	18.03	35.53	6.38
MgO	4.29	2.35	3.58
Al_2O_3	1.35	6.25	0.59
Fe_2O_3	0.66	1.08	0.26
P_2O_5	9.23	1.75	4.01
TiO_2	0.02	0.07	0.01
MnO	3.7	0.2	0.36
SO_2	<0.01	<0.01	<0.01

glass raw materials to make northern 'forest' glass, Georgius Agricola, Theophilus Presbiter and Heraclius, all recommend the use of beech ash (Smedley and Jackson 2002, 23), but presumably this is only when glassmakers could obtain it.

The work by Misra *et al.* (1993) makes it clear that we are dealing with a complex situation, and anyone claiming the contrary is oversimplifying it. For example, they have demonstrated that increasing the temperatures at which wood ashes are heated in furnaces can lead to a reduction in the levels of, for example, calcium, boron, manganese, sulphur and especially potassium. Volatilisation of potassium began at between 800° and 900°C; copper and boron started to volatilise at c. 1000°C. However, this is also dependant on the species of wood involved (Misra *et al.* 1993, table 6). In addition, fritting wood ash (with silica), reduces the amount of alkali loss. The control of the furnace temperature and whether fritting was practiced are therefore factors that would both have had an impact on the chemical compositions of the glass produced and potentially on its melting behaviour. The leaching of terrestrial plants is a further technological consideration (Stern and Gerber 2009, 109–11). This would have removed the all-important insoluble lime component and therefore necessitated a separate addition of lime to the batch. They also suggest that part of the extract, including lime, may have been added later. Hartmann (1994) and Schalm *et al.* (2004) have suggested that lime was added separately. However, the levels of impurities in medieval wood ash glasses suggest that the leaching of ashes was not a common practice, and experimental replication has shown that in any case, sufficient lime could be introduced in wood ashes (Jackson and Smedley 2004; Verità 2005). Godfrey's (1975) consideration of the historical evidence for the use of raw materials in the manufacture of wood ash glass has not revealed that lime was mentioned as a separate raw material. Wood ash was used for the production of a somewhat gelatinous soap in medieval Europe.

2.5 NATRON

Glass produced from the minerals natron and trona (known as 'natron glass') was predominantly manufactured from c. 800 B.C. until about c. A.D. 800 in the West and Middle East (Sayre and Smith 1961, 1967; Henderson 1985; Freestone et al 2000; see chapter 4). The minerals provided a relatively pure source of alkali, compared with plant ash, and the glass made from it would therefore have behaved in a somewhat more predictable way when worked. Moreover, it was denser and generally had higher alkali concentrations than found in plant ash. Both of these factors meant that it was easier to transport and to melt. Natron glass was used to make the full range of glass artefacts and the evidence suggests that it was the first glass type to be used for glass blowing.

In Book XXXVI, 191*ff* of his *Naturalis historia*, Pliny the Elder (23 B.C.–A.D. 79) describes 'the invention' of (natron) glass on the Levantine coast:

> *The beach extends for not more than half a mile, but for many years this area (Sidon) was the sole producer of glass. A ship belonging to traders in soda once called here, so the story goes, and they spread out along the shore to make a meal. There were no stones to support their cooking pots so they placed lumps of soda from their ship under them. When these became hot and fused with the sand on the beach, streams of an unknown translucent liquid flowed and this was the origin of glass.*

With perspicacity, Trowbridge (1930, 96) emphasised that when Pliny recorded the production of natron glass, it was a *tradition*. The passage was later quoted by Isodore of Seville in *Etymologiarum sive originum*, XVI. If a tradition, Pliny's description of its 'discovery' should not be taken too literally. An alternative interpretation could be that a glassy material was formed on the beach that resulted from the fusion of fuel ash and sand, rather than natron and sand. Given that the first natron glass occurs some 700 years earlier, with a continuous use until the time that Pliny was writing, it is most unlikely that if this was indeed a 'discovery', or that it was *reinvented* in the area of Sidon. Perhaps he was describing its discovery some 700 years before his birth. It is worth noting that one of the earliest uses of natron was during the Fourth Egyptian Dynasty (2575–2465 B.C.) for mummification (David 2000, 373). So even though natron was a recognized material, presumably it was tradition that prevented natron, rather than plant ash, from being used as a glass raw material for some 1,500 years.

The sodium-rich minerals natron and trona occur at Wadi el Natrun on the edge of the Egyptian Western Desert, some 100 kilometres northwest of Cairo. They occur as evaporites in a 50-kilometres-long Wadi el Natrun depression

(Turner 1956a, table IV). The minerals that occur there are what are known as evaporates and form seasonally as water evaporates from the surface of alkali-rich soils. Shortland (2004) has described the separate lakes that currently make up Wadi el Natrun and has discussed the complex combination of sodium-rich minerals that occur there. He noted that the mineral natron ($Na_2CO_3.10H_2O$) was quite rare and that most of the sodium that would have been useable in glass production was in the form of a relatively pure sodium source, trona, a sodium sesquicarbonate ($Na_2CO_3.NaHCO_3.2H_2O$), the occurrence of which was noted by Brill (pers. comm. 1984; Brill 1988, 268, table 9-9) and Henderson (1985, 273; 2000, 26). Other minerals identified in samples removed from Wadi el Natrun by Shortland (*ibid.*, Table 2) were halite (NaCl) and burkeite ($Na_6CO_3.2SO_4$).

Trona is formed seasonally and can be gathered up relatively easily. Harris (1961) noted that ancient Egyptian texts refer to Egyptian natron deposits; Strabo (*Geography* XVII, I, 22, 23) and Pliny (*Natural History* XXXI, 106–11) mention these deposits. Pliny (*Nat. Hist.* XXXI, 191) and Isiodore *Orig.* (*Etymologiarum sive originum* XVI, 16.6) refer to *glabeas nitri* and *glebas nitri*, the traders in natron. However, perhaps the actual mineralogical identity of what they were referring to was trona rather than natron. Therefore, it would seem that 'natrun' was/is a blanket term that refers to evaporite deposits that are rich in sodium carbonate. Moreover, it is likely that the proportions of the two minerals, natron and trona, at Wadi el Natrun were different when Pliny and Strabo were writing. Although most researchers have focused on Egypt as the main source of sodium-rich minerals, other sources occur. Again Pliny (*Nat. Hist.* XXX1, 106–11) refers to a deposit in Asia Minor.

Ignatiadou *et al.* (2005) noted that an evaporitic lake near the ancient city of Chalestra in Macedonia, Greece, produced *nitrum chalestricum*. This is the natron source that Pliny mentions as an important one in his *Naturalis historia* (XXXI, 107–8), in which he describes its medical uses. He also describes its formation:

> *Soda forms in it about the rising of the Dog Star for nine days, ceases for nine days, comes to the top again and then ceases.*

This passage emphasises that soda formation depends on the levels of precipitation. The start of the hottest part of the summer occurs when the Dog Star, Sirius, appears, and therefore it was at this time when trona would be have precipitated from the alkali-rich lakes. The seasonal harvesting of trona would apply to other trona deposits too and would be determined by heat intensity and evaporation rates. The actual location of ancient Chalestra is unknown. However, the mineralogical identity of evaporated brines from Lake Pikrolimni, 20 kilometres northwest of Thessaloniki, showed that, just as at Wadi el Natrun, burkeite ($Na_2CO_3.2Na_2CO_4$), trona ($Na_2CO_3.NaHCO_3.2H_2O$) and halite (NaCl) were present. Both burkeite and trona would be suitable for glassmaking. Dotsika *et al.* (2009, 142) noted that the 'progressive concentration of brines in alkaline lakes leads to a preferential precipitation of sodium carbonate, followed by sulphates and chlorides'. By using a combination of XRD, evaporation simulations and a thermodynamic model, Dotiska *et al.* (2009) have confirmed that trona, burkeite and halite are deposited from the brines. However, their discovery at Lake Pikrolimni does not prove that they were used as an alkali source in ancient glassmaking. The research nevertheless reminds us that sources other than Wadi el Natrun in Egypt may have been used as a source of trona for ancient glassmaking, and Pliny's mention of such a source is no coincidence.

2.6 Reh and Oos

Reh is an alkali-rich mineral that occurs quite commonly in India. It is a combination of sodium carbonate, sodium sulphate and sodium chloride and contains low levels of calcium and magnesium (Wadia 1975). Brill (1999b, 481) published chemical analyses of a sample of *reh* found near Shikohabad. XRD showed the presence of $MgAl_2(SO_4)_4.2H_2O$ (Brill 1999b, 481). Singh (2005, 140, table 5.2) notes that some Indian 'salt-effected' soils contain sodium-potassium-calcium feldspars. Dussubieux and Gratuze (2003, 139) have provided the following chemical composition of reh: c. 10% Al_2O_3; <1% MgO; c. 3% CaO; >1.5% K_2O. Glasses made from this source of alkali are therefore characterised by a high content of alumina. Brill (*ibid.*) managed to produce glass by heating reh and sand together: the result contained relatively high alumina (9.91%) but also had the other compositional characteristics of soda-alumina glasses (see Chapter 4). Apart from the high alumina level in reh, there are other compositional differences from natron. Whereas natron is associated with low levels of both magnesia and potassium oxide, at below c. 1%, reh also introduced up to c. 1% magnesia, and in some cases there are apparently significantly higher levels of potassium oxide, of c. 3% to 6%, than is found in natron (Robertshaw et al 2006, 98, fig. 2; Dussubieux *et al.* 2008, 2). Nevertheless Brill's (1999a, 212; 1999b, 481) analysis of extracted *reh* which was then used to make glass only gave a level of 2.28% potassium oxide. This is still higher than would be introduced by natron; an additional potassium source would be necessary to form some high alumina glasses.

In the early nineteenth century in India, glassmakers apparently collected and crystallized crude alkali-rich mineral deposits throughout the year (Jaggi 1977, 193). Ethnographic work by Kock and Sode (1995; 2001) revealed that reh was still being used in the 1990s. Clearly, as for other mineral and plant ash alkali sources, a full analytical survey of various reh deposits needs to be carried out to determine the vertical and lateral variations in composition within a single deposit but also to determine variations in chemical and mineralogical compositions of reh deposits across the Indian landscape. In his ethnographic research, Brill (2001) also discovered that a similar evaporite source of alkali was used in India, called *oos*, but to the author's knowledge, this has not been defined chemically.

2.7 Nephaline

Another potential mineralogical source of alkali that could have been used in (Indian) glass production is nephaline syanite, a sodium, potassium, aluminium silicate (Bhardwaj 1979, 68; Henderson 1985, 274). Moreover, in recent work on a frit-like material found at Roman Sagalassos, Turkey, its use as a flux was suggested tentatively ('but without soda'!) in combination with quartz and feldspar, but primarily as an aluminium source (Degryse and Poblome 2002, 350).

2.8 Conclusions

Both plant ashes and minerals were used to flux silica in the manufacture of glass in antiquity. Halophytic plant ashes were used to make the earliest glasses in Mesopotamia, and they were then used in Egypt to make glass. Natron was introduced around the tenth century B.C. for the manufacture of glass, and there was a period of transition as it replaced plant ashes at a time shortly after the collapse of Bronze Age civilisations in the Mediterranean and at about the same time that iron was introduced. In this chapter the complex characteristics of plants and plant ashes found in semi-desert environments in the Middle East and temperate woodland environments in Europe and soda-rich minerals from Egypt and India have been discussed. There is still much research to carry out in the investigation of reh and oos, both into their distributions and also into their chemical and mineralogical compositions.

In antiquity, it is likely that plants for glassmaking were mainly harvested in the late summer because they have their greatest biomass at this time. Those who selected the plants may have been part-time farmers or, in the Middle East, Bedouins. Halophytic plants are still collected today in Syria in the late summer–early autumn for the preparation of soap. Sufficient levels of sodium compounds that would flux silica, and perhaps sufficient calcium levels, would probably have determined which plants were selected and from which (geological) location. If plants with insufficient calcium were harvested, calcium would have been added separately as a third primary raw material. This does not seem to be the case for woodland plants in which sufficient calcium levels are present (Jackson *et al.* 2005). However, chemical analysis of halophytic semi-desert plants shows that there was not always sufficient calcium in the ashes to make a durable glass. Strong communications between glassmakers and plant harvesters seems likely. Even so, glassmakers would still need to test the ash by fusing small quantities with silica to see whether it had appropriate working properties.

The complex relationships between the species of the plant used, its ash composition and the composition of the glass made from it are difficult to disentangle (see Chapter 11). Although experimental synthesis of glass can contribute to our understanding of the mechanisms of glassmaking, there is always the possibility that more than one methodology can lead to the same result. It is also difficult to identify alkali-rich phases found in plant ashes using XRD (Jackson et al. 2005; Barkoudah and Henderson 2006). Chemical analyses of plant ashes can sometimes provide evidence for which plant genera were used, and particularly which plant ashes were *not* used, to make glass (see Chapter 11). Nevertheless the experimental synthesis of glasses is one way of trying to understand the complex relationships between the chemical compositions of plant ashes and silica and the glasses made from them.

THREE

SILICA, LIME AND GLASS COLOURANTS

3.1 Silica

Silica is a glass-former, a compound that forms the basic building block of the majority of ancient glasses. It is present in ancient soda-lime glasses at levels of c. 65% to 70%. The elements that form silica (SiO_2), silicon and oxygen, are two of the most abundant elements in the earth's crust. They combine as free silica or combine to form silicate minerals, the principal constituent of rocks. Silica is a polymorph that has three principal crystalline forms (with the same chemical composition): quartz, tridymite and cristobalite. Quartz is far more abundant than the others, forming 12% of the earth's crust. Tridymite and cristobalite are high-temperature forms of silica that occur rarely in certain volcanic rocks and can be produced by heating quartz to high temperatures. They can therefore be found in, for example, pottery and other high-temperature products. Quartz exists as its α form at room temperature but is converted to its β form at 573°C. Above 1470°C, quartz slowly forms cristobalite, which, if heated at temperatures between 870°C and 1470°C, slowly converts to tridymite.

The crystalline forms of silica are mainly composed of SiO_4 tetrahedra. In these, a silicon atom is surrounded by four oxygen atoms. Each oxygen atom is shared with another silicon atom, creating a three-dimensional framework. Each of these three crystalline forms has different densities. The melting point of silica is between 1710°C and 1730°C; if cooled rapidly, it produces vitreous silica, but clearly its melting temperature was not achieved during ancient glassmaking: its melting temperature was reduced with a flux (see Chapter 2).

Silica also forms part of noncrystalline and microcrystalline siliceous materials, such as chalcedony and opal, with varieties such as flint and chert. The sources of silica for glassmaking could be quartz pebbles, which would be crushed, riverine and coastal sand deposits or inland geological deposits. High-purity quartz can either occur as vein quartz, which is an intrusion of silica in igneous and metamorphic rocks, or fill cavities as a result of the precipitation of silica-rich (hydrothermal) groundwater. Vein quartz can be colourless, milky white or pink; rock crystal is the colourless quartz found in cavities. Milky quartz and rock crystal have different microstructures: water-filled vacuoles characterise the former and microcrystalline aggregates the latter. The weathering of these quartz-rich deposits leads to deposits of quartz pebbles and gravel along the courses of rivers and to quartz-rich sands.

A further source of silica for making glass is chert, such as suggested for the manufacture of fourteenth- to seventeenth-century Venetian glasses (Zecchin 1987). Danièle Foy's (1989, 30–1) consideration of silica sources used in medieval glass furnaces in France includes evidence for the use of 'grès erratiques', very pure quartz 'galets', calcaires crétacés and river sand. In this period in France, the location of a glasshouse was not necessarily dependant on a local silica source.

The purity of sand is greatly influenced by the type of rock from which it is derived and the distance over which the material has been transported before deposition. Sediments containing sand that derive from the weathering of crystalline rocks tend to have a higher feldspar and heavy mineral content than those derived from arenaceous sedimentary strata (rocks composed of sand). If the latter are reworked, minerals such as feldspar can be removed, producing purer sand. The depositional environment is also important in determining the composition of sands. Deposition in current-swept waters produces characteristically well-sorted, clay-free sands. Aeolian sands generally have rounded grains and are well sorted, whereas fluvio-glacial and glacial sands generally have angular grains, are poorly sorted and are of a more variable composition.

The environments in which sands are formed will determine the grain sizes, which, in turn, will affect the melting behaviour of the glasses made from them. The glass batch is affected because the packing will vary according to the raw material particle sizes involved, producing variations in batch compositions, potentially leading to inconsistencies between batches and unpredictable glass melting behaviour (Cable 1958; Boffe and Letocart 1962; Smedley and Jackson 2002). By standardising the procedures for preparing glass raw materials, glassmakers could at least minimise this potential source of variation.

Heavy minerals that occur in sands are inert in the ambient aqueous conditions of geological cycles. These include iron oxide (e.g. magnetite; see Figs. 3.1 and 3.2) and iron titanites (e.g. ilmenite) and silicates such as zircon (see Fig. 3.3). Rare earth elements (REEs) can be found in zircons and specific phases such as monazites (Patyk-Karar et al. 1999). Cerium- and yttrium-type elements display different chemical behaviour in terrestrial rocks (Allègre and Michard 1974), so this has potential to provide chemical distinctions between glasses containing them. Hoskin and Ireland (2000) are sceptical that REEs in zircons

3.1. A backscattered electron micrograph of a high iron and alumina mineral inclusion (pale grey) labelled c in beach sand from Beirut, Lebanon (×50; photo: E. Faber).

can be used as a (geological) provenance indicator and by extension this indicates that this approach may have a limited archaeological application. Moreover, given sufficient geological contrast between silica sources, there is a place for their characterisation using neutron activation or LA-ICP-MS (see Appendix). The concentrations of europium (Eu) and cerium (Ce) relative to other REEs are

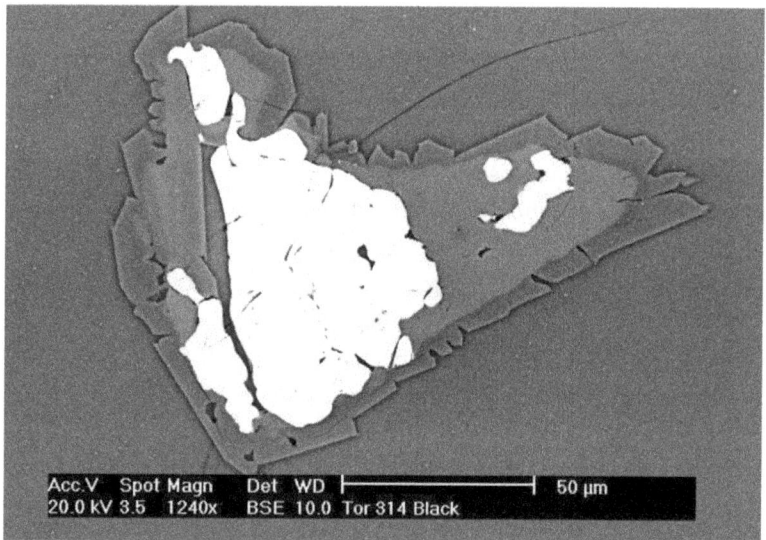

3.2. A backscattered electron micrograph of an iron oxide inclusion introduced with sand in an eleventh-century intensely dark green glass mosaic tessera from the west wall of the basilica of St. Maria Asunta on the island of Torcello in the Venetian lagoon (×1240; photo: E. Faber).

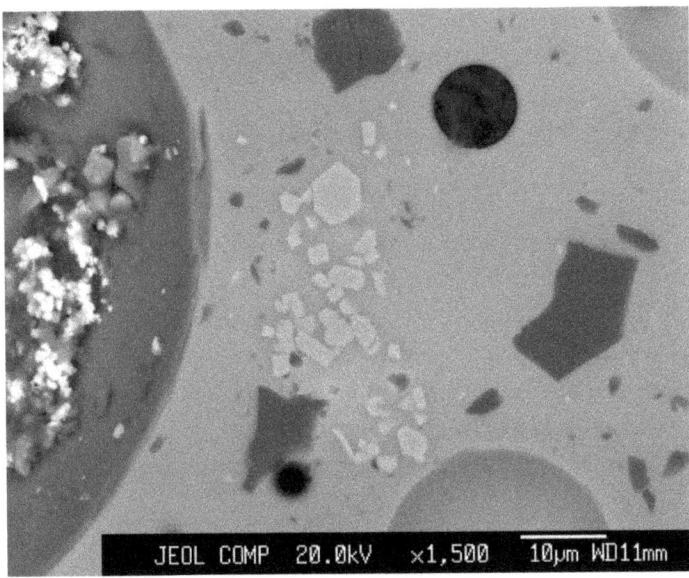

3.3. Five-sided zircon crystals (pale grey) in the glassy matrix of ninth-century faience from Mantai, Sri Lanka (×1500; photo: E. Faber).

thought to reflect the particular parameters of the chemical system in which minerals containing REEs were formed (Loveland 1989) and are known as the *europium and cerium anomalies*.

These anomalies are produced by fractionation during various geological processes. Europium variation results from fractionation during igneous processes (with minimal cerium variation). These produce basaltic and gabbroic rocks with REE signatures enriched in heavier elements; rhyolitic and granitic rocks will be enriched in lighter elements. Cerium fractionation (with limited europium variation) occurs during a range of processes leading to cerium variations. These processes are cementation (the formation of sandstone), lithification (formation of deep-sea clays) and precipitation (such as the formation of carbonates) all occurring in marine environments. These REE anomalies will be reflected in the compositions of different sand sources used in glassmaking. For example, the chemical analysis of tenth- to thirteenth-century Egyptian and South African glass beads indicate that REE determinations could provide a means of distinguishing between silica sources and help to source glasses (Saitowitz *et al.* 1996). The relative levels of zirconium and titanium, thought to be associated (only) with the sand used, have been used as a means of characterising sources of sixteenth- and seventeenth-century European glasses (Šmit *et al.* 2004). Other impurities that occur in sand include shell fragments (see Fig. 3.4), which introduce the calcium-baring biogenic form of the mineral aragonite ($CaCO_3$) to a glass melt (Lucas 1948). Feldspars often contain high alumina levels and otherwise can have variable compositions. They can be potassium- or sodium-rich. Sodium-rich feldspars are also rich in calcium and barium. Their chemical compositions can form a compositionally continuous series between end members. Plagioclase feldspars, for example, have sodium-rich albite as one end member and calcium-rich anorthite as the other end member, with four identifiable intermediaries. Another potential impurity in silica is clay. Aluminium levels in ancient glass have, in the past, mainly been considered as having derived from feldspars in the silica used. This is certainly true, but alumina can also be introduced in plant ashes when making plant ash glass (see Chapter 2), and there is a possible contribution at the edges of melts from the walls of crucibles in which glasses were melted (Merchant *et al.* 1997). The low level of alumina found in many late Bronze Age glasses does indicate that a pure source of silica was used. Trace levels of lead can be introduced in the glass melt by glass recycling (Henderson 1993). However, the presence of lead-rich minerals in sands can also introduce low lead levels.

Olivine ($(Mg, Fe)_2 SiO_4$), chromite ($FeCr_2O_4$), epidote ($Ca_2(Al, Fe)_3(SiO4)_3OH$, ilmenite ($FeTiO_3$), sphene or titanite ($Ca.TiSiO_5$), mica, calcite ($CaCO_3$), zircons ($ZrSiO_4$) and other mineralogical impurities can occur in sands. These minerals can therefore produce iron, chromium, titanium, alumina and calcium oxide impurities in glasses. Both Goerk (1977, 48) and Highley (1977, 9) have noted that titanite and epidote can introduce small quantities of calcium in the glass. Inevitably Turner (1956b, 281–2, table III) was one of the first to determine the mineralogical composition of sands in the context of ancient glass production. He studied samples from Tell el-Amarna, from the Theban shore of the river Nile opposite Luxor and from the river Belus by Haifa in Israel. All of the samples that he studied contained variable levels of silica, calcite, feldspars, pyroxenes and ilmenite. The Theban sand contained mud and mica. The sand from Tell el-Amarna contained between 50% and 55% quartz, 30% and 33% calcite, 5% feldspars, 5% pyroxenes and 1% ilmenite. The mineralogical and chemical analyses of nine sand samples from beaches by the Belus River and at Akko and Caesarea, Israel, published by Brill (1999a, 208–9; 1999b, 478–9) provide clear evidence for the variable, but significant, levels of shell, feldspars and bone fragments mixed in with quartz sand. These analyses provide the expected explanation for the high calcium oxide levels found in the natron glasses made on the Levantine coast.

Freestone *et al.* (2000) showed that, in general, late antique natron glasses fused on the Levantine coast contained low levels of zirconium (c. 60 ppm) and high levels of strontium (c. 400 ppm), the strontium deriving from shell fragments in the sand. Inland sands would be expected to contain lower strontium (c. 150 ppm), associated with limestone, and higher zirconium (c. 160 ppm) as Freestone *et al.* (*ibid.*, 73) have suggested is the case for inland Egyptian sands. Nevertheless, the discovery of Carthage glass with high zirconium and strontium shows that the simple distinction may not hold.

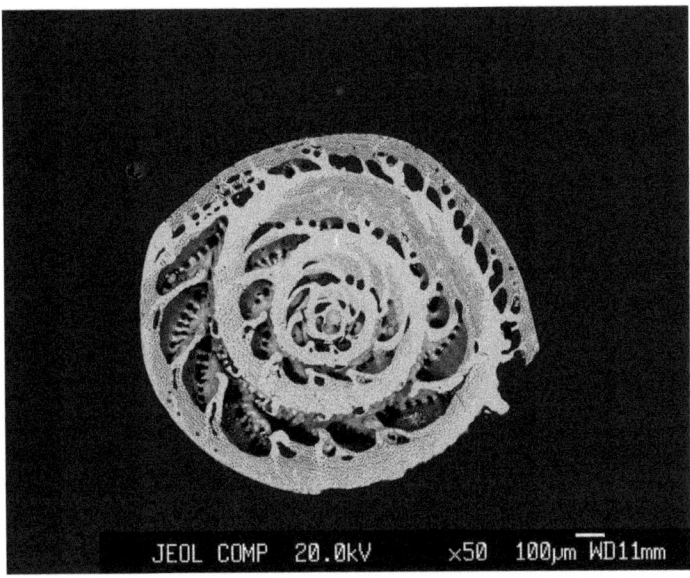

3.4. A backscattered electron micrograph of a section of a shell found in a sand sample from the beach at Tyre, Lebanon (×50; photo: E. Faber).

The mineralogical analysis of sands from the mouth of the river Volturno (Silvestri *et al.* 2006, 422, table 1) demonstrated that trace levels of Ba and Rh were introduced as part of the alkaline feldspars present and that Sr was introduced with the carbonates and to a lesser extent calcium plagioclase. Vanadium, chromium and zirconium were found to be associated with accessory phases and lithic fragments. These sands do, nevertheless, contain very high levels of alumina of between c. 6% and 12% (Brill 1999a, 209; 1999b, 475) for natron glass.

So-called naturally-coloured glass is a pale green colour, caused by iron-rich impurities in the glass raw materials. Although the incorporation of other colourants and the use of decolourants could eclipse the colouring effects of iron, finding a relatively pure source of silica was, at different periods of glass manufacture, an important consideration. Sands can be beneficiated by simple washing or sieving. Silica can also be ground so as to promote as full a fusion as possible in the glass batch to optimise the surface area to volume ratio. However, sufficiently high temperatures will also bring about a comprehensive fusion. Pliny the Elder notes that sand from the river Volturnus (discussed later) was prepared by grinding it 'in a mortar or a mill' (*Naturalis historia*, XXXVI, 194). Silvestri *et al.* (2006, 426) state that if the sand was ground in a mortar and pestle, an excessive decrease in the carbonate (shell fragments) fraction would result, so the batch would need to be 'adjusted'. Moreover, Pliny mentions the addition of lustrous stones and shells to the melt (XXXVI, 192). Whatever the procedure, Pliny's reference to sand preparation allows one to consider what effect this would have on the resultant glass. It is difficult to know precisely what effect it would have on the components introduced with silica in the glass melt and could reduce the lime content, but not if the dust was retained. However, the consistent chemical composition of much Roman natron glass shows that a controlled and repeated procedure must have been adhered to during its manufacture. If this involved a separate addition of shell fragments to the batch, then specific proportions must have been repeatedly used.

Pliny the Elder mentions a source of glassmaking sand at the mouth of the river Belus (Nahr Naaman) in Book XXXVI of his *Naturalis historia*. This river flows into the Mediterranean between Haifa and Acre in today's Israel (Turner 1956b, Brill 1988, 265–7). Josephus (*Bell. Jud.* II, 189–191) appears to have described the same location as a source of sand for glassmaking (Trowbridge 1930, 98):

> *a remarkable place no larger than one hundred cubits. This is round and hollow, and yields such sand as glass is made of.*

Tacitus (*Hist.*, V, 7, 2) is another author who makes reference to the use of Belus for making glass:

> *The river Belus also flows into the Jewish sea. About its mouth is a kind of sand which is collected, mixed with nitre, and fused into glass. This shore is of limited extent, but furnishes an inexhaustible supply to the exporter.*

The Roman geographer Strabo (c. 63 B.C.–A.D. 3, or 17/19 at the latest) refers to a suitable source of sand for making glass further to the north near Sidon in today's Lebanon (Trowbridge, *ibid.*). Some years after this, Pliny (N.H. 36, 193) refers to Sidon as *artifex vitri*, 'the maker of glass'. Moreover Strabo (*Geographica* XVI, 2.25) states that there were three main areas of glass production in his day: Syro-Palestine, Egypt (particularly the region of Alexandria) and Italy. This being the case, one would therefore hope that the sand used in each area would be compositionally and/or isotopically distinct, which would carry through into the glasses produced.

As noted earlier, an Italian source of clean white sand was also referred to by Pliny as being located on the Campanian littoral between Cuma and Literno near the mouth of the river Volturno.

> *But now [first century A.D.] in Italy, a white sand occurs in the river Volturnus: it is found along six miles of the seashore between Cumae and Liternum. The softest of the sand is ground in a mortar, or mill. It is then mixed with three-quarters of a part soda, either by weight or volume. After fusion, it is taken in its molten state to other furnaces. There it becomes a lump called 'sand soda': this is remelted and ends up as pure glass – a lump, that is, of clear glass.*

The coastal strip is fed by the drainage basins of the Pleistocene Roccamonfina volcano and the Meso-Cenozoic calcareous-dolomitic formations of the Appenine mountains (Gandolfi and Paganelli 1984). Vesuvian and the Phlegraean Fields are close.

Studies of Volturno sand suggest that it is unsuitable for glassmaking because it contains high levels of alumina (Turner 1956b; Brill 1999b, 475; and see above). However, given its extent, a comprehensive survey to determine its compositional variation would help to test Pliny's claim that sand from the Volturno petrographic province was used in antiquity.

Pliny's reference to 'sand soda' presumably refers to the production of frit, where the manufacture of a vitreous state is arrested, producing a granular material. Later, Isidore of Seville, born c. 560 in Spain, in his *Etmologiarum sive originum* (XVI) also refers to a soft white sand being ground in a mill and then combined with '3 parts of natron by weight or measure' to produce sand-soda (*ammonitrum*), which 'when heated again, it became pure, clear glass'. Intriguingly, Pliny also refers to the use of sand being treated in the same way as that from the Belus in two other (glassmaking) zones: Gaul and the Iberian Peninsula (N. H. XXXVI, 3 and 190–4; V.75).

Currently there is no archaeological evidence for primary glass production in Gaul or Spain, but this is not proof that it did not occur there. Therefore, the historical references to Roman production in the Syro-Palestinian area, Alexandria, Italy, the Iberian Peninsula and Gaul Roman glassmaking zones still need to be investigated using scientific investigations: one isotopic study has indicated that there was at least one western production zone (Degryse and Schneider 2008).

The scientific analysis of Austrian Iron Age glass armlets demonstrated clearly that two compositional groups of natron glasses existed, one dating to c. 260–220 B.C. (La Tène C1) with high zirconium levels of between c. 150 and 500 ppm (with between 200 and 600 ppm strontium) and the other to the second–first century B.C. (La Tène C1b–D1) with zirconium levels of between c. 50 and 150 ppm and strontium levels of between 600 and 1600 ppm, which is unusually high (Karwowski 2004). These results clearly show that different sand sources were exploited in these two periods and possibly that the raw glasses used to make the glass armlets were made in different locations, although neither fit the typical results for the Levantine coast (see Chapter 8). The positive correlation of zirconium with strontium in the earlier group is the opposite of what might be expected geologically. It could potentially indicate that some of the glass was manufactured in Europe and not necessarily that all was imported from the Mediterranean, although this would involve the importation of natron. An isotopic approach could help to shed light on this question (Degryse *et al.* 2009; Gebhardt 2010, 12).

Casellato *et al.* (2003) investigated sand sources that could potentially have been used for the production of glass from the fourteenth-century site of Germagnana, Tuscany (Mendera 1991) and the renaissance (sixteenth-century) glass production site of Gambassi, Piazza del Castello in Veldelsa, Florence (Mendera 1991; Bianchin et al. 2005). The type of glass being made and (apparently) worked at Germagnana was a wood ash glass; that at Gambassi was described as a 'sodic-calcic' (soda-lime) glass (*ibid.*, Table 1). However, its relatively high potassium, magnesium and phosphorus levels suggest that the glass was made from wood ashes. It is also characterised by relatively high aluminium and iron levels. It is, of course, interesting to analyse mineralogically and chemically such potential sand sources because their location might be a reason for locating a glass-production site there. In the seventeenth century, we have historical references to glassmakers exploiting different sand sources depending on the quality of glass that they were making (Henderson 2006, 307). Undoubtedly glassmakers would have been well aware of variable sand qualities and the resulting glasses from an early phase of glass production.

In view of this, it is worth describing some of the scientific research into potential sand sources. Bianchin *et al.* (2005) used X-ray fluorescence spectroscopy (XRF) and Inductively Coupled Plasma spectrometry to analyse 'archaeological sands' from an underground vaulted room, from a kiln and from room D (*ibid.*, 42–3) and sand samples from two sand quarries 'in the neighbourhood' of Gambassi. The 'archaeological' quartz sands contained high aluminium, calcium (calcite), and iron with low levels of magnesium and potassium. The aluminium was present as sodium feldspar (albite), potassium feldspar (orthoclase) and sodium potassium feldspar (anorthoclase). In contrast, the quarry sands tested were characterised by high aluminium and low levels of other components (apart from silica). When compared with the chemical compositions of the Gambassi raw glass, it was found that the archaeological sands contained calcium levels that were too high, although subsequent purification could have reduced this level. Therefore, there was not a clear link between the glass and the sands tested.

The *age* and depositional history of silica will be reflected in the levels and nature of impurities such as titanium, neodymium and zirconium. Its geological age (which is a reflection of its source) will be reflected in ^{143}Nd/^{144}Nd ratios introduced in the glass from the silica. Such ratios are unaffected by the manufacture of glass at high temperatures and can help to provenance glass (see Chapter 11).

3.2 Calcium and Aluminium

As noted in Chapters 1 and 2, calcium is an essential network stabiliser in glass that provides glass durability. The source of calcium in most natron glass is thought to be shell fragments. In his *Naturalis historia*, Pliny (Book XXXVI, 92) mentions that shells were added to the glass melt separately, but such a separation from sand would not be necessary if there was a consistent proportion in sand deposits. The source of calcium in most plant ash glasses is generally considered to be the plants used. Another possible source of calcium is bone (see Chapter 2). Bone fragments have been found in ninth-century glass frit from al-Raqqa, Syria (Henderson 1995) and calcium phosphate in Byzantine opaque red slabs of glass from Petra in Jordan (Marii 2007). Another source of calcium is dolomitic limestone: this would produce a positive correlation between calcium and magnesium oxides in glass. Research into relic stones (unmelted or partially melted batch ingredients) and fully melted glass at the fourteenth-century Monte Lecco glass factory on the northern slopes of the Ligurian Apennines, northern Italy, has indicated that dolomitic limestone was used as a calcium source (Basso et al. 2008); presumably the mixed-alkali plant ashes would also have been a source of calcium. Glass production in sixteenth- and seventeenth-century Europe also seems to have involved the use of dolomitic limestone ($CaMg(CO_3)$) as a source of calcium (Henkes and Henderson 1998, 97–8).

Glasses with alumina levels of up to 12% alumina have been found in India, Africa and the Far East (Brill 1987; Lankton and Dussubieux 2006; Robertshaw et al. 2006; Dussubieux et al. 2008). These levels have mainly been attributed to the use of granitic sands and possibly reh, a source of alkali (see Chapters 2 and 4). Most natron glasses contain alumina levels that can be attributed to feldspar minerals in the sands used, as noted earlier. Feldspars found in sand produce levels of alumina in Roman natron glasses of between c. 1.7% and c. 3.5% (Jackson et al. 2003; Foy et al. 2004), differing levels apparently helping to distinguish between sand sources/production zones (Freestone et al. 2000; Foy et al. ibid.). Sands were also used for the production of some plant ash glasses (Henderson et al. 2004); lower alumina levels suggest the use of a purer silica source (Henderson 2009) and the plant ashes themselves contributed alumina to the glasses, as discussed earlier.

3.3 Colourants and Opacifiers

3.3.1 What Causes Glass Colour?

Colouration is a key characteristic of glass and may have been the primary reason for making it in the first place (see Chapter 1). Glass colour is due to a range of complex factors. For example, in the right environment, transition metal ions, such as cobalt and copper, can produce deep blue and turquoise blue colours in translucent and opaque glasses. The physical reason why we observe colour in glass is because the glass absorbs part of the visible wavelengths of light as a result of its interaction with colourant oxides in the glass; the reflected balance of light wavelengths form the colour that we see (Doremus 1994, 306). Transition metal ions such as Cr^{2+}, Mn^{2+}, Mn^{3+} Fe^{2+}, Fe^{3+}, Co^{2+}, Cu^+, Cu^{2+} can provide deep colours in translucent ancient glasses. The depth of that colour depends on their linear absorption coefficients (the relative strength of absorption; Bamford 1977, 42; Paul 1990, 305 *ff.*), their concentration and their chemical environment. The atomic weight of the alkali present will have an effect on the observed colour: the heavier the alkali (in terms of its atomic number), the darker the colour observed. If, for example, an amount of potassium oxide in a glass is replaced by the same amount of sodium oxide, the colour of the glass will appear lighter (Bamford 1977). However, a number of other factors are involved and are discussed elsewhere in detail (Pollard and Heron 1996, 168; Henderson 2000, 29). These factors are as follows:

1. the preparation of the glass batch;
2. occurrence of transition metal ions such as Cr^{2+}, Mn^{2+}, Mn^{3+} Fe^{2+}, Fe^{3+}, Co^{2+}, Cu^+ and Cu^{2+};
3. the gaseous atmosphere of the kiln at different stages of the melting cycle;
4. the presence of crystalline opacifiers;
5. the chemical environment;
6. the nature of the heating cycle and the maximum temperature achieved.

Thoroughly mixing the glass batch in which the colourant raw material is present (preferably in a finely ground state) will reduce the amount of colour streaking. Relatively high melting temperatures not only produce a full reaction between the primary raw materials but are also important to ensure total dissolution of the colourant in the melt.

The gaseous atmosphere of the furnace, which may change at different points in the heating cycle, has a critical influence on the final colour of the glass that is produced. An *oxidising* atmosphere is one in which oxygen predominates, whereas a *reducing* atmosphere is oxygen-deficient (i.e. carbon dioxide and carbon monoxide predominate) and is often smoky. For medieval potassium (forest) glass, the presence of manganese and iron oxide impurities in the melt had the 'unintentional' result of producing a wide range of glass colours: pale blue, green, purple, colourless, brown and pale yellow (Newton 1978; Sellner *et al.* 1979; Brill 1988; Royce-Roll 1994). The process was described by Theophilus Presbiter in his *De diversis artibus* (Dodwell 1961; Hawthorne and Smith 1963). The furnace described by Theophilus has been compared with contemporary and later ones by Horat (1991). In Theophilus's Book II, chapters 7 and 8, he emphasises the importance of time to the development of various glass colours. The reason he did this is that the atmosphere in the glass furnace changes with time:

> *If you see [the glass in] the pot changing to a saffron yellow colour, heat it until the third hour and you will get a light saffron yellow. Work up as much as you want of it in the same way as above. And if you wish, let it heat until the sixth hour and you will get a reddish saffron yellow. Make from it what you choose.*

But [alternatively] if you see [the glass in] any pot happening to turn a tawny colour, like flesh, use this glass for a flesh-colour, and taking out as much as you wish, heat the remainder for two hours, namely from the first to the third hour and you will get a light purple. Heat it again from the third to the sixth hour and it will be reddish purple and exquisite.

The description of changing glass colours over time is a reflection of how the gaseous atmosphere of the furnace changed from a slightly reducing one to a slightly oxidising one (Dodwell, 1961). Royce-Roll (1994), who created a small-scale reconstruction of Theophilus's medieval forest glass furnace and melted glasses in it, was able to produce the full range of glass colours using beech wood ashes and sand as raw materials. The ashes of beech trees are especially rich in manganese, producing a purple colour in glasses under oxidising conditions and a weak yellow colour under reducing conditions. A flesh or pink colour can result if there is a mixture of reduced and oxidised manganese ions present. Yellow and colourless glasses contained higher iron levels than found in the purple and flesh-coloured glasses. It is also worth noting that with a slightly reducing atmosphere, when there is a mixture of ferrous and ferric ions, a green colour will result (Schreurs and Brill 1984). This is the familiar so-called natural bottle-green colour. Theophilus also mentions the deliberate addition of copper to produce turquoise blue and under reducing conditions an opaque red colour. Sellner *et al.* (1979) demonstrated scientifically that this range of glass colours could be produced. The presence of both manganese and iron oxides in 'naturally coloured' late Roman (Brill 1988, 272), Islamic (Henderson *et al.* 2004, 460–1, fig. 9) and postmedieval, including *cristallo*, glasses (Henderson 2006, 308, table 4, type III, illus. 11) has led investigators to emphasise how important the gaseous atmosphere of the glass furnace was in determining the final colour of the glass. Ultimately these different glass colours are produced because the colourant ions occur in different states of coordination.

The theory that provides an explanation for these colouring effects is called the 'ligand field theory.' When a transition metal is in an ionic state, the ions surrounding the colourant, *the ligand*, depend on both the field strength and negative charge (provided by the oxygen). In transition metals, one of the energy shells (in this case, the 3d sub-shell) is only partly filled with electrons. It is this that produces some of the colouring characteristics in glass. When co-ordinated with other ions, such as Si^{4+}, the energy levels of the d electrons in transition metals are split (distorted) by the electric field produced by the co-ordinating ions. This splitting is sensitive to the arrangement of surrounding ions (*the chemical environment*); the result determines the glass colour. When higher energy level orbits in the colouring ions are unoccupied, the electrons in lower energy level orbits absorb different wavelengths of light quanta to move up to the higher energy level quanta. It is this last energy transition that causes the glass or glaze to appear to be a particular colour.

The heating cycle of a melt determines the temperatures at which the raw materials in the batch partially melt and then melt completely. The heat of the furnace may be held at particular temperatures by using specific fuel types, which, in turn, will affect the furnace atmosphere and the glass colour. The maximum temperature to which a melt is heated to is also critical because high temperatures will allow gases that have evolved during the melting process to be driven off and will produce a less bubbly and streaky glass that is more homogenous. Newton and Davison (1989, 61) claim that ancient glassmakers had difficulty reaching temperatures above 1000°C, but this cannot be the case, because 1100° to 1400°C would have been necessary for an efficient melt for soda-lime glasses (Cable 1998). Indeed, an estimation of the temperatures achieved in the 9th century tank furnaces at al-Raqqa, Syria, based on a study of refractory thermal properties of furnace bricks, indicated that the maximum temperature achieved in the furnace was between c. 1150°C and 1200°C (McLoughlin 2003, 165). Addition of cullet (scrap glass) to the glass batch produces more efficient fusion of raw materials and can reduce the maximum melting temperature needed (as indeed does the addition of lead oxide), but it is clear that temperatures well above 1100°C were achieved by ancient glassmakers – and for that matter, in other ancient high-temperature industries – so this would help to homogenise a glass batch containing colourants.

The kind of fuel used during glass melting determined the calorific value (a measure of the heat produced) and the gaseous atmosphere in the glass furnace. Charcoal has a high calorific value. It is produced when wood is combusted incompletely; in the process, wood is converted into almost pure carbon. Typically the volume of the wood is reduced by c. 30% and its weight by about 25% of the original. Charcoal can burn at about 900°C and, with an air blast, much higher temperatures can be achieved. The actual calorific value achieved is dependent on the density of the wood that charcoal is derived from, hard woods being the densest. Charcoal has a great affinity for oxygen and for absorbing vapours. Even the size of the charcoal fragments was important for a successful firing because of the affect it had on the circulation of gases through the furnace charge. Studies to determine the age and species of wood or other fuels used in glass production are yet to be carried out. A study of the fuel used in iron smelting has even produced evidence of coppicing (Figueral 1992).

3.3.2 The Use of Colourant Materials

Discussion of the use of colourants in ancient glasses was provided earlier, and other publications have already dealt with it in detail (Henderson 1985, 278–86;

Henderson 1989, 33–6; Freestone 1991, 41–4; Henderson 2000, 30–5; Pollard and Heron 1996) so, with some exceptions, it will not be repeated here. However, a detailed description of the occurrence of cobalt-bearing minerals and its use follows. This is provided for two reasons:

1. cobalt blue glass coloration is arguably one of the most striking colours in ancient glasses and glazes and
2. unlike most other glass colourant raw materials, there is a possibility of being able to relate the impurities associated with cobalt in glass to its mineralogical source (Henderson 1985, 278–81).

3.3.2.1 Cobalt Coloration in Glass

Cobalt is the most powerful transition metal colourant. This means that as little as 0.02% CoO (present as Co^{2+}) in soda-lime glass can produce the famous deep blue colour that was so popular in, for example, the late Bronze Age Mediterranean and Iron Age Europe and is present in Chinese blue and white porcelain glaze. Although not reported (yet) in ancient glasses, the Co^{3+} ion can produce a pink colour. Cobalt was used as a colourant in glasses, glazes, faience and enamels. Cobalt blue glass was first manufactured in the Bronze Age in imitation of lapis lazuli, a stone that was considered to have had health-giving properties (see Chapters 5 and 6). Much later, ground lapis lazuli was used as an opacifier in enamel that was painted onto Islamic (Mamluk) glassware, including mosque lamps (as discussed here and in Chapter 10).

Cobalt is found in ancient rock mineralisations (e.g. in mountains) in association with other minerals, such as copper, arsenic, nickel, iron, manganese and zinc. A range of minerals is rich in cobalt:

Erythrite (cobalt bloom), a weathering product of cobalt-containing minerals such as cobaltite ($Co_3(AsO_4)2.8H_2O$), is associated with silver, cobaltite and skutterudite.

Trianite is a copper-bearing cobalt ore ($2Co_2O.CuO.6H_2O$).

Cobaltite contains cobalt in association with sulphur and arsenic and sometimes a trace of zinc (CoAsS).

Skutterudite contains cobalt in association with nickel, iron and arsenic ($(Co,Ni,Fe)As_3$).

Reddish-pink *bieberite* is a cobalt sulphide ($CoSO_4.7H_2O$).

Absolane is a mixture of MnO and heterogenite (CoOOH).

Trace elements associated with cobalt-bearing minerals can be carried through into ancient blue glasses. These include lead, antimony, vanadium and bismuth. Geilmann (1962) was sceptical about the possibility of characterising the sources of cobalt-rich minerals by their associated impurities, but there has been a considerable discussion about this in the literature, and in some instances a strong case for the use of a particular cobalt source in glass can be made. Sayre (1964, 7–8) distinguished between 'western' Roman cobalt blue glasses, containing elevated manganese levels, and those found in Mesopotamia and southwestern Iran which were said to be characterised by arsenic (Young 1956). Associated impurity levels of iron, nickel, copper, tin and lead oxides also supported Sayre's distinction. However, roasting cobalt-rich minerals to purify them (to drive off arsenic and sulphur) can complicate the characterisation.

In Egypt, yellowish, pale purple or pale pink cobaltiferous alums occur in the Dakhla and Kharga (western) Oases; the purple and pink colours would have provided a means of identifying the ores and attracting prospectors. They consist of mixed hydrated iron and aluminium sulphates coquimbite, alumino- and ferrocopiapite and allunogen. These contain cobalt, aluminium, manganese, nickel and zinc found in New Kingdom glass (Kaczmarczyk 1986), a combination also found in New Kingdom cobalt blue faience (Kaczmarczyk and Hedges 1983, 46). One set of samples from the Kharga oasis examined by Shortland *et al.* (2006, 156, table 2) was a pale pink colour. The samples consisted of granular kieserite and/or pentahydrite with crystals of pickeringite. They also contained cobalt, manganese, nickel and zinc. Segnit (1987) noted that trace levels of transition metals partition between ores, pentahydrite containing some manganese and most containing cobalt, nickel and zinc. The bivalent cobalt (Co^{2+}) substitutes for Fe^{2+} in iron-rich alums and for Mg^{2+} in the magnesium-rich alums.

The presence of cobalt in blue Egyptian glasses and other vitreous materials has led to an extensive discussion of its use (e.g. Sayre and Smith 1974; Henderson 2000, 30–1; Rehren 2001; Tite and Shortland 2003). Shortland *et al.* (2001, 154) noted that gifts/taxes were given to the king by the governor of the western oases in New Kingdom Egypt and have, on the basis of this, suggested that the supply of cobalt for faience production from there was through the temples or state bodies. Farnsworth and Ritchie (1938, 160) confirmed the presence of cobalt in New Kingdom Egyptian glasses and suggested that Egyptian alum containing cobalt might have been used as the cobalt source. Garner (1956a and 1956b) and Kaczmarczyck (1986) also suggested this. Lilyquist *et al.* (1993, 42–3, ns. 78 and 94) claimed that the relative levels of impurities in Egyptian cobalt blue glasses are difficult to reconcile with the use of a cobalt-rich Egyptian alum in a plant ash-silica recipe. An intriguing problem is that most (but not all) cobalt blue New Kingdom glasses contain *lower* potassium levels than found in non-blue glasses, and at the same time, they contain comparable magnesia levels. The presence of a high alumina level in cobalt blue Egyptian glass suggests that a cobalt-rich alum has been used and that the aluminium level has diluted (reduced) the potassium level. However, the discovery of similar levels of soda and magnesia in both cobalt blue and non-blue glasses makes this more difficult to reconcile. In 2001, Rehren suggested that the low potassium levels were due to the use of a plant ash with a lower potassium level than that used for the manufacture of contemporary non-blue glasses. However, as a result of experimental work, he has shown that melting conditions can have an effect on potassium levels (Rehren 2008), and the same species of plant can have variable alkali levels (see Chapter 2), so there is not necessarily a clear link between plant ash compositions and the glasses made from them. The use of a specific silica source with high alumina-rich feldspars seems to be an unlikely explanation for the high alumina levels in blue glasses. Shortland *et al.* (2006) carried out a series of experiments in which they mixed synthetic alum with sodium carbonate and precipitated cobaltiferous alums as originally suggested by Noll (1981) and Kaczmarczyk (1986). The precipitates contained different alum components that varied according to the amount of sodium carbonate added. Once enough carbonate had been added to allow 50% reaction to occur, the aluminium, cobalt and manganese levels did not increase. However, the magnesium levels continued to increase up to 100% reaction. Cobalt is correlated with manganese, nickel, alumina and zinc, but not magnesium, in cobalt blue New Kingdom glasses. Magnesium would also be introduced in the plant ash, so the two sources of magnesium explain the lack of correlation with other components introduced with cobalt. The lower potassium levels found in some cobalt blue New Kingdom glasses have also been explained by the suggested use of natron (with low potassium impurity) as a source of sodium carbonate to precipitate the alum combined with plant ash and quartz (Shortland et al. 2005, 164). This would reduce the overall potassium level. However, strontium isotopic analyses of cobalt blue and other colours of Egyptian fourteenth-century B.C. glasses from Tell el-Amarna show that this is unlikely. They indicate that a geologically relatively old calcium source was used, introduced in a plant ash source of alkali (Henderson *et al.* 2010, 14, table 1, fig. 3a), and therefore that a natron glass (made with beach sand and an associated 'young' source of lime) had not been used. Moreover, cobalt blue glass with low potassium levels and other colours of glass with elevated potassium levels have constrained strontium isotope signatures, confirming that the plants used as the alkali source derived from the same or a very similar geological source. Alum may have been roasted before its use, so this would have modified its composition. Lilyquist *et al.* (1993), Rehren (2001) and Jackson and Nicholson (2007) have all noted that not all Egyptian low potassium/high alumina glasses are cobalt blue in colour and not all cobalt blue glasses contain low potassium oxide levels. The use of a different plant species is unlikely to be the only possible interpretation for these compositional variations.

Lucas (1934, 218) noted that Iran had sources of cobalt-rich minerals, and Garner (1956b, 148) stated that arsenic was a compositional characteristic of Persian cobalt sources. Kaczmarczyk (1986, 373) stated that Persia produced a 'purer' cobalt pigment. One area in Iran where cobalt-rich deposits occur is the Anarak district in the central part of the country, at Talmesi and Meskani

(Tarkian *et al.* 1984; Bagheri *et al.* 2007). Two deposits consist of a tertiary nickel-cobalt-silver-bismuth-uranium-arsenide (niccolite, ramelsbergite, safflorite and skutterudite) with bismuth and uranium and some copper-arsenides (domeykite, and koutek). Another tertiary deposit has a copper-iron-sulphide composition (chalcopyrite, bornite, digenite and chalcocite). If the uranium-rich cobalt source was used as a blue colourant in ancient vitreous materials, including glass, one would expect to detect silver, nickel, bismuth, arsenic and uranium along with the cobalt. Thus far, no uranium has been detected in cobalt blue glasses (although it may not have been sought), but arsenic, a volatile element, is sometimes detected.

Another Iranian cobalt deposit occurs at Qamsar 35 kilometres southwest of Kashan. The minor deposits there consist of a 1- to 10-metre-thick magnetite vein interrupting dolomitic limestone that contains some copper ores with secondary copper and cobalt minerals such as erythrite (characterised by cobalt and arsenic, as noted earlier), absolane (characterised by manganese, as noted earlier), nickel sulphides and spinel $(Co, Ni, Fe)_3S_4$. The cobalt source at Qamsar was still being exploited in the 1960s.

Reference to the use of the Qamsar cobalt source was specifically made by Abū'l Qāsim who was a member of an important family of potters at Kashan. He was also a historian in the Mongol court at Tabriz (Allan 1973; Watson 1985, 31–2; Watson 2004, 29). His treatise on the manufacture of stonepaste tiles and pottery was written in AH700/1301AD (Allan 1973; Henderson 2000, 187). The reference is as follows:

> The stone lajvard, which the craftsmen call sulaimani. Its source is the village of Qamsar in the mountains around Kashan, and the people there claim it was discovered by the prophet Sulaiman. It is like white silver shining in a sheath of hard black stone. From it comes the lajvard colour, like that of lajvard-coloured glaze.

The word *lajvard* means lapis lazuli. The use of cobalt to imitate lapis lazuli dates back to the Bronze Age (see Chapters 5 and 6).

A further Iranian cobalt deposit is the Baycheh Bagh mine in Mahneshan of Zanjan, Iran. This is a mixture of smaltite and cobaltite. Another is a cobalt deposit at Eqlid, north and east of Fars, consisting of mixed lenses of iron and manganese, together with copper, lead and zinc. The main deposit here is cobalt.

Kleinmann (1991) published a scientific study on ninth-century blue painted tin glazed wares from Samārrā, Iraq. She is of the opinion that the use of linnaeite (cobalt sulphide) combined with magnetite fits Abu'l Qasim's description best. Kerr and Wood (2004, 664) underline the difficulty of attempting to relate the mineralogically complex Kashan cobalt source to its use as a colourant material in glasses and glazes. If indeed a low manganese cobalt source was used as a colourant in many Mesopotamian glasses, this would suggest that erythrite and linneite (for example) provided the cobalt, at the exclusion of the manganese-rich cobalt ores. Moreover, it is also possible that both manganese- and arsenic-rich cobalt was exploited at the same time. Furthermore, cobalt sources can be (and were) roasted before use, which could have removed most of the volatile arsenic impurity, if not all of it. Despite these considerations, we are still left with the fact that during various periods, cobalt in glasses (and glazes) is characterized by certain associated impurities at the exclusion of others, and this does allow us to suggest which particular cobalt-rich mineral deposits were used.

Cobalt blue glass of a ninth- to tenth-century date from Nishapur, Iran, has a lead oxide to zinc oxide ratio (Brill 1995, 217) that fits exactly that found in many Islamic and Western medieval blue glasses (Gratuze et al. 1995; Henderson 2003). Cobalt colourants used in Abbāsid cobalt glazes are characterised by zinc and iron (Wood et al. 2007), the chronologically earliest being ceramics and glass from ninth-century Samārrā (Zhang Fukang and Cowell 1987). Islamic scratch decorated vessels and blue glasses from the Serçe Limani shipwreck are also characterised by zinc (Whitehouse et al. 2000, 91). The positive correlation between lead and zinc suggests that the same or a very similar geological source of cobalt was used to colour both medieval oriental (Islamic) glass and medieval western glass (Fig. 3.5). The Western glass has a totally different wood ash base composition from plant ash Islamic glass, but the finding nevertheless suggests that that a cobalt-rich colourant with positively correlated zinc and lead impurities was exported from the Islamic world to western Europe (Henderson 1998; Whitehouse et al. 2000). This same zinc-rich cobalt source was even used in French vessel glass (Gratuze et al. 1992) and in fourteenth- and fifteenth-century Limoges enamels (Wypyski and Richter 1997). At Akda'Madeni in Turkey, there is a cobalt-rich mineralisation that is associated with galena and could therefore be the source for the lead-rich cobalt mineral found in many Islamic glasses and glazes. Moreover, the lead isotope ratio in the Islamic glasses tested by Brill has an isotopic match with the source at Akda'Madeni (Henderson 2003, 242). Although this is a single lead isotope result, and we do not have a clear idea of what lead isotope variation there is in the cobalt ore source, it can nevertheless be considered as a possible source for the cobalt characterised by lead and zinc. Gratuze et al. (1995) carried out analyses of cobalt blue glasses of a range of dates, including the glass with cobalt associated with lead. In their work on western European glasses they highlighted discrete chronological changes in the use of cobalt sources characterised by different impurity patterns.

A source of cobalt in the German Black Forest area is said to be characterised by arsenic (Hahn-Weinheimer 1955). In medieval Europe, a cobalt-bearing mineral called zaffre, smalt or Damascus pigment was roasted to remove any sulphur or arsenic before being used by glassmakers and artists. From 1168, the cobalt

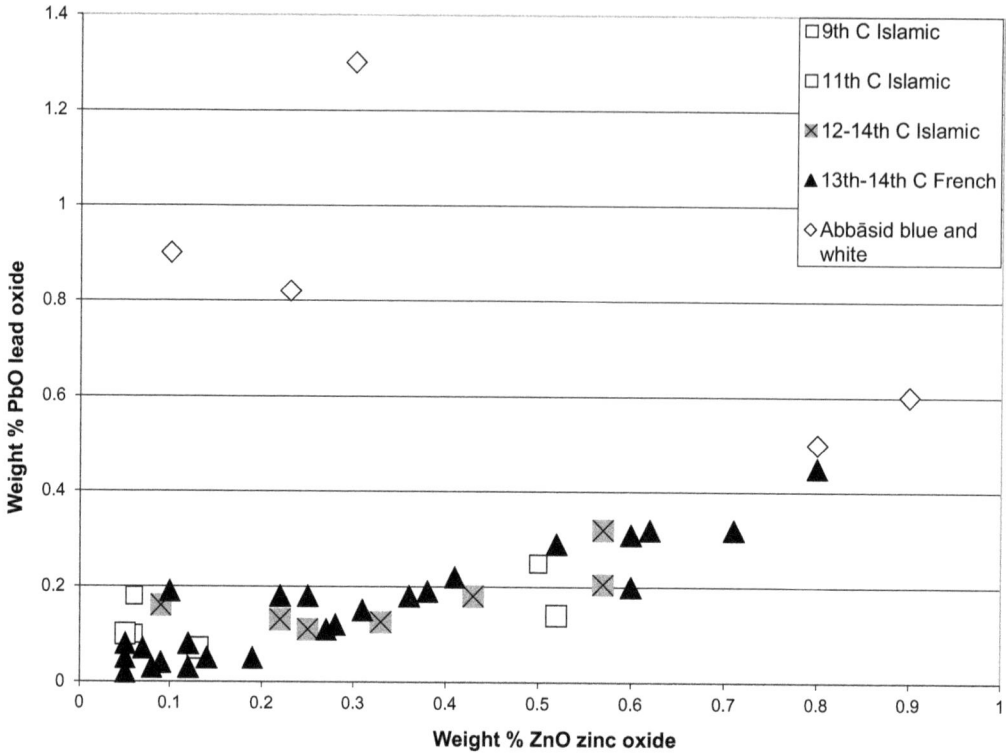

3.5. A bi-plot of weight % lead oxide versus zinc oxide in a variety of glasses from the Middle East and Europe dating to between the ninth and fourteenth centuries and ninth-century Abbāsid blue and white glazes.

source, at the Erzgebirge mining district near Freiburg, started to be exploited. Like the Turkish cobalt source, the cobalt ores in this deposit are associated with zinc blend and galena and might be considered a possible alternative to the Turkish source. At the moment, however, the lead isotope result for the Turkish source apparently shows that, of the two, it is a more likely source.

The second century B.C. saw the late Hellenistic glass industry grow significantly (see Chapter 7), and at this time in Iron Age Europe, the predominant use of an antimony-rich cobalt source changed to that of a manganese-rich source (Gebhardt 1989; Henderson 1991, 128, figs. 6–8; Gebhardt 2010, 6–7, fig. 4). The change in cobalt source also coincided with a change in silica source. So the use of cobalt characterised by manganese from the second century B.C. found in European Iron Age glass is likely to be the same source as that used in 'Roman' glass and enamels (Sayre 1964, 7–8; Henderson 1991a, 287–8, table 1).

Cobalt-bearing minerals are one of few used to colour ancient glasses that can be characterised by associated impurities. In attempting to relate the compositional characteristics of a mineral to its use in glass, a thorough analytical assay of the variations in cobalt-bearing ore bodies is essential. Moreover, it is possible that cobalt-rich areas of the ore bodies used to colour ancient glasses have been worked out. Nevertheless, such a survey is important. This has occurred only in the Iranian deposits. It can provide a basis for claiming when a particular cobalt-rich ore was used. Had such a thorough survey been carried out in the western oasis deposits in Egypt, it would have allowed Shortland and his team to assess whether any of the cobalt ores exploited in antiquity still survive today.

3.3.2.2 Manganese-Coloured Glass

Manganese can either be used as a glass decolouriser or a purple glass colourant. Biringuccio, in his *Pirotechnica* published in 1558, noted that it could be used as a glass decolourant, and there is evidence in glasses dating to as early as the fourteenth century B.C., and through succeeding centuries, that it was used deliberately as a colourant. In the Western Medieval and postmedieval periods, a combination of the furnace atmosphere with the presence of manganese and iron produced a range of glass colours, including purple (Sellner *et al.* 1979; Royce-Roll 1994, Quartieri *et al.* 2005; Henderson 2006). The only glass production site where it has been found in mineral form, destined to be deliberately used as a colourant in glass beads was at la Seube (Hérault), France (Lambert 1972, 96). The mineral deposit there is said to have a composition of 0.49% iron and 52.7% manganese. Manganese-bearing minerals such as pyrolucite (MnO_2) may well have provided an important manganese source in antiquity. The purple colour would be produced by the trivalent Mn^{3+} ion in the glass (Weyl 1937, 118).

3.3.2.3 Copper- and Gold-Coloured Translucent Glasses

Translucent turquoise is one of the commonest colours of Bronze Age glass both in the Middle East and in Europe. Because this same colour is also commonly found in faience, attributed to copper in its oxidised form, there is clearly a link between the use of copper used as a colourant in faience and glass. The translucent turquoise colour of soda-lime glasses is basically due to cupric (CuII) ions (Brill 1970a, 120). Weyl's (1951, 164–5) assertion that this colour is produced only in the presence of lead does not hold up for ancient glass, for which there are many examples with no detectable lead oxide. The development of a blue green colour in glass is aided by a high MgO level (Weyl 1951, 159), and this is a characteristic of Bronze Age glasses made using a plant ash alkali source.

A second translucent glass and enamel colour in which copper is the colourant is 'copper ruby' (Weyl 1951, 428–9; Biek and Bayley 1979). This colour is analogous to 'gold ruby' glasses and is due to submicron (c. 200–400 Å) particles of copper (and gold). The presence of these particles causes light of a particular wavelength (0.53 μm) to be absorbed (Weyl 1951; Doremus 1994, 315). The absorption band exists because of the spherical geometry of the metal particles and their particular optical properties. To form the particles in the first place, glass must be reheated ('struck') at an intermediate temperature (below its melting temperature) to allow nucleation of particles to occur by diffusion of atoms or ions through the glass. Weyl (1951, 399–400, figs. 98 and 99) noted that striking a gold-rich glass at different temperatures had an effect on the crystallisation, producing blue, ruby and purple-amethyst colours at different temperatures, so this would apply to copper-ruby glass too.

The Lycurgus cup is one of the most famous Roman glass vessels. It appears to be a translucent wine red colour in transmitted illumination and a dull pea green colour reflected illumination; this is known as *dichroism*. Its coloration is due partly to the presence of 40 ppm of colloidal gold, and partly to 300 ppm of colloidal silver (Brill 1965; Barber and Freestone 1990; Henderson 2000, 63; Freestone et al. 2007). In reflected light, the particles of gold and silver are coarse enough to reflect light; in transmitted light, the particles scatter light wavelengths at the blue end of the spectrum more effectively than the red end, the result being that transmission occurs and the vessel appears a ruby-red colour.

The mineralogical sources of copper are widespread (Henderson 2000, 210–11) and unlike cobalt it has, so far, not been possible to use associated impurities to suggest the use of a particular type of copper-bearing mineral. However, it might now be possible to do so and to be able to source it (Artioli *et al.* 2008b). The same may be true for gold.

3.3.3 Decolorised Glass

Although colourless glass was produced in the second millennium B.C., it was especially popular in the Hellenistic world from c. sixth century B.C. until c. second century B.C. (Sayre and Smith 1961), during the Roman era, for the production of Sasanian glass and for the manufacture of *cristallo* by the Venetians from about the fourteenth century. Colourless glass could be produced by using silica with relatively low levels of iron, by the use of antimony trioxide (between the sixth and second centuries B.C.) or manganese oxide after the second century B.C. (Henderson 1989, 50–3). However, Gratuze (2009, 13) has now identified a glass bead decolourised with manganese oxide that dates to between 776 and 414 cal. B.C. The Venetians appear to have controlled the furnace conditions in such as way as to produce colourless glass. It can be assumed that the main mineralogical source of antimony was stibnite (SbS_3). The largest numbers of colourless glass samples that have been chemically analysed are for second- and third-century Roman glass (Foy *et al.* 2004; Jackson *et al.* 2004; Silvestri *et al.* 2008). It is clear that tons of raw colourless glass were being shipped by the

Romans, such as amongst the glass discovered in the wreck of *Ouest Embiez 1* (Foy and Jézégou 1998).

3.3.4 Glass Opacification

Whereas translucent (and transparent) glasses allow wavelengths of light to be transmitted, the inclusions in opaque glass have the opposite effect: the wavelengths of light are reflected from the glass by the presence of crystals. Silica crystals that remain partially reacted in the glass after its manufacture and masses of gas bubbles can produce complete or partial opacification. Deliberate opacification is due to the addition of materials called *opacifiers* to the glass or it results from heat-treating the glass to develop opacity ('striking' it). Early scientific work on opaque glasses (Rooksby 1962, Turner and Rooksby 1959, 1961, 1963) using X-ray diffraction spectroscopy defined which crystallites produced opaque colours. They referred to these as 'opalisers' rather than 'opacifiers'. However, here the term *opalescent* will be used only in the context where glasses are a cloudy semi-opaque (opaline) colour because of the presence of crystals that are smaller than those which cause opacity. This effect can be caused by very small (nano-scale) crystals that are only easily detected and identified using techniques such as a transmission electron microscope. Examples have been found in Roman ruby-red glasses (Barber and Freestone 1990) and in later Venetian 'chalcedony' glasses.

A series of other publications have followed the seminal work by Turner and Rooksby on opacification (Moretti and Hreglich 1984; Henderson 1985; Rehren 1997; Mass et al. 1998; Henderson 2000; Tite *et al.* 2008; Rutten *et al.* 2009). Common opacifiers used in early ancient glasses are calcium antimonate ($Ca_2Sb_2O_7$ and $Ca_2Sb_2O_6$) to produce white opaque glass (see Figs. 3.6 and 3.7), lead antimonate ($Pb_2Sb_2O_7$) to produce opaque yellow glass; from the second century B.C., tin was used in the form of tin oxide (SnO_2, cassiterite) to produce opaque white glass and lead-tin oxide (Pb_2Sn_2O7) to produce opaque yellow glass (see Fig. 3.8). Cuprous oxide (Cu_2O) and iron produced an opaque red-coloured glass. The use of lead isotopes has been used to investigate where the lead for the manufacture of Egyptian opaque yellow glasses were derived from (Shortland 2006). Relatively recently, it has been noted that silica was used as an opacifier (Verità *et al.* 2002).

Calcium phosphate crystals have been found in Byzantine opaque glasses from Petra, Jordan (Marii 2007); finely dispersed in turquoise tesserae from Armorium, Anatolia, dating to c. 900 (Wypysi 2006, 187, fig. 4); in opaque white Islamic enamel (Freestone and Stapleton 1998, 125, table 3), in later glasses in the West (Turner and Rooksby 1961, 2; Turner and Rooksby 1963, 306, Werner and Bimson 1963, 303; Besborodov 1975, 73) and it was used in Italy as an opacifier

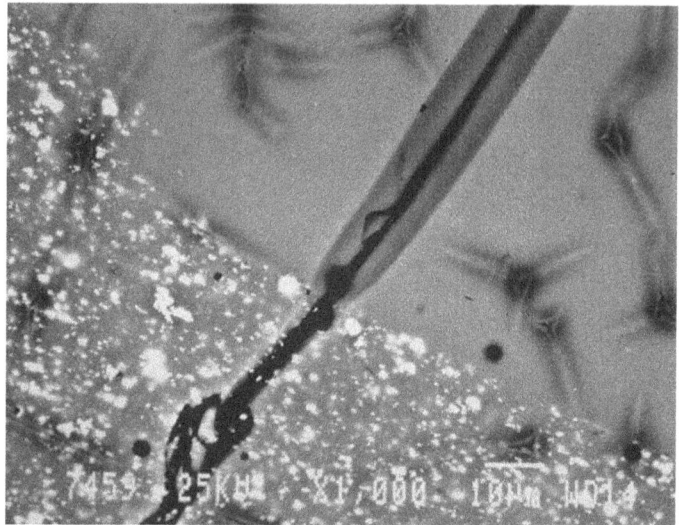

3.6. A backscattered electron micrograph of a sample of glass consisting of opaque white calcium antimonate ($Ca_2Sb_2O_7$) crystals suspended in a translucent glass attached to a translucent cobalt blue glass full of bubbles. From a fourth-century B.C. compound eye bead found at Bury Ball, Devon, England (×55; photo: E. Faber).

from c. fourteenth century (Verità 2000, 68). Towards the end of the seventeenth century in the West, lead arsenate ($3Pb_3(AsO_4).PbO$) was used to produce opaque white glass (Verità 2000, 64 Wypyski 2002; Moretti and Hreglich 2005, table 3); arsenic occurred increasingly as an impurity in glasses from this time onwards.

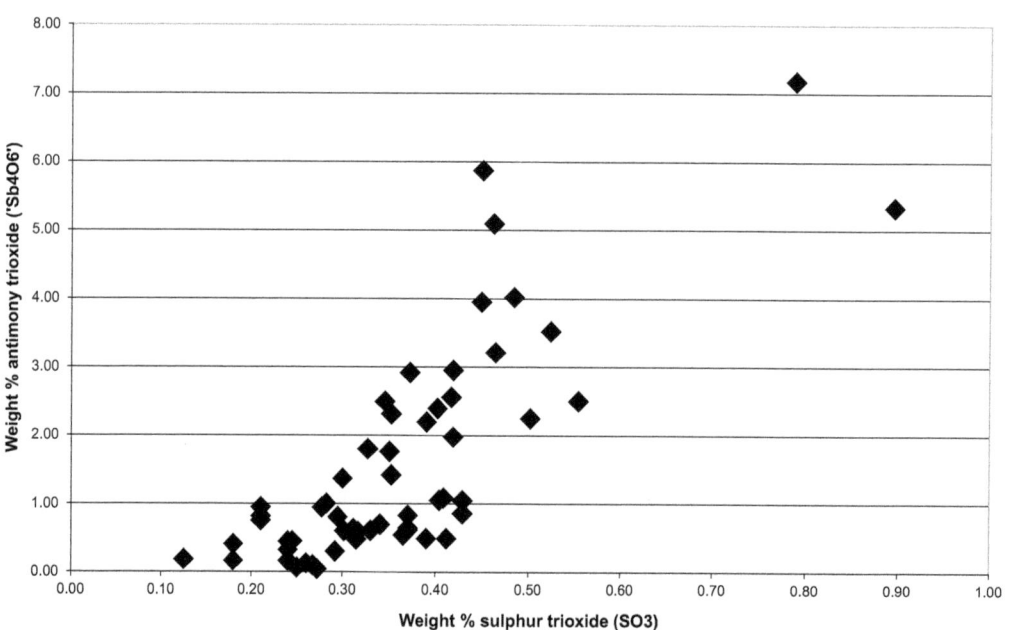

3.7. A bi-plot of calcium oxide against sulphur trioxide in second- and third-century Roman glass tesserae. The positive correlation indicates that the mineral stibnite (Sb_2S_3) was the antimony source used.

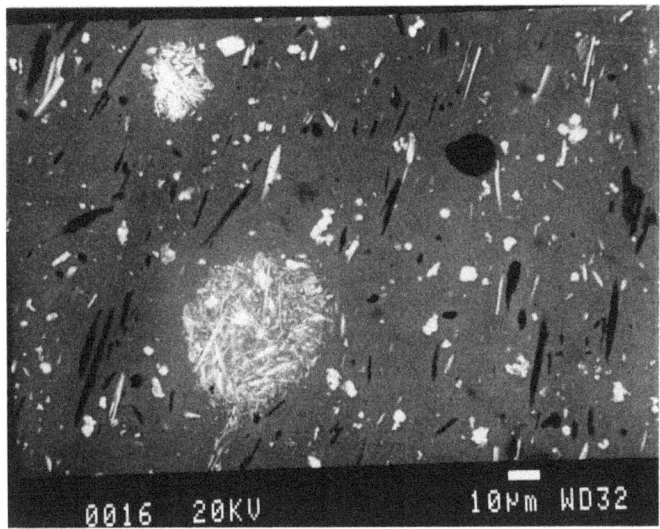

3.8. A backscattered electron micrograph of second-century B.C. chunk of opaque yellow raw glass from Staré Hradisco, Moravia, showing agglomerations and individual lead-tin oxide ($PbSnO_3$) crystals (appearing white) accompanied by elongated darker grey (soda-lime) crystals that have formed due to heat-treating the glass (×1000).

Sodium phosphate was also used as an opacifier. From the sixteenth century, calcium fluoride was used in China as the main white opacifier (Henderson et al. 1989), and it was introduced in Italy in the nineteenth century (Verità 2000; Fig. 3.9). Calcium antimonate was used to opacify nineteenth-century glasses, and other contemporary crystallites have also been detected, such as calcium stannate ($CaSnSiO_5$; Andreescu-Treadgold and Henderson 2006, 134).

Opaque red glasses and enamels have seen quite a high volume of research devoted to them. The following is a summary of what has been discovered.

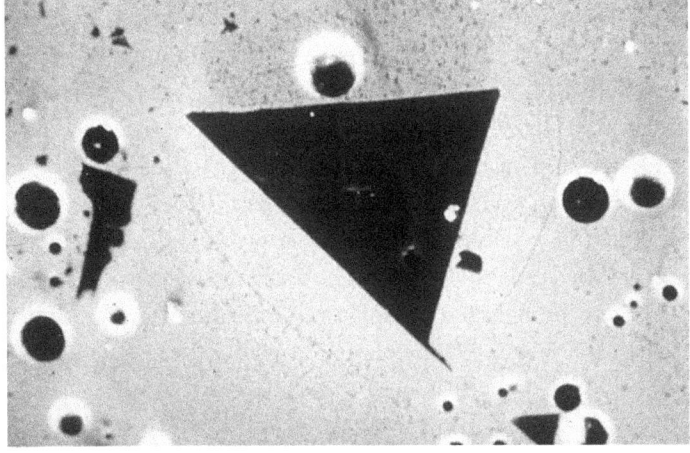

3.9. A backscattered electron micrograph of calcium fluoride (CaF_2) crystals in a sixteenth-century A.D. Chinese cloisonné enamel (×1500).

A bright sealing-wax red or a dull brownish-red colour can, under the right circumstances, be produced by the presence of copper I (cuprous) oxide (Cu_2O) and metallic copper (Hughes 1972; Freestone 1987; Guido et al. 1984; Brun and Pernot 1992; Andreescu and Henderson 2006; Fig. 3.10). The red colour is produced when the glass has an appropriate chemical composition and there is a tight control over the furnace atmosphere in which the copper-rich glass is melted. The characteristic branching dendritic crystal form of Cu_2O (cuprous oxide) leads to a bright sealing-wax red colour; a dull opaque brownish-red colour is due to the presence of micron-sized Cu_2O crystals. Lead improves the solubility of copper in soda-lime glass, and iron and antimony can both act as reducing agents (Guido et al. 1984), leading to successful 'striking' of the crystals out of the glass matrix. As noted earlier, striking the glass involves reheating it at temperatures at which crystals form.

It is notable that pre-Roman Iron Age sealing-wax red glasses contain more lead oxide than the earlier and later dull brownish-red ones (Hughes 1972; Henderson 1989, 48; Henderson 1991b, 603–5, table II). Through the chemical analysis of pre-Roman opaque red enamels that were used to decorate horse riding equipment, it has been possible to show that different proportions of lead and copper oxide correlate to different tribal regions in the United Kingdom (Henderson 1989, fig. 2.6) and the proportions of lead and copper in sealing-wax enamels of various dates from other parts of Europe also form separate groupings according to their origin (Brun and Pernot 1992, fig. 4). These results appear to provide evidence for the existence of different regional technologies although further rigorous scientific analyses need to be carried out to test these results.

There are other compositional characteristics of opaque red glasses and enamels that help us to reconstruct the production technology involved. One unexpected characteristic of red enamels and glasses is the presence of elevated magnesia and potassium levels in dull brownish-red Roman enamel (Henderson 1991a, 289, table 2, fig. 1). Although there are a small number of other examples of Roman plant ash glasses, it appears that, for some reason, plant ashes only (or possibly a mixture of plant ashes and natron) were used for making dull opaque red glasses and enamels. Amongst all of the enamels analysed so far, this characteristic has only been found in opaque brownish-red enamels and tesserae. It is likely that these tesserae derived from ransacked Byzantine or Roman 'monuments' and were being reused. The elevated magnesium and potassium oxide levels have been found in some reddish-brown glass beads found at the early medieval site of Wijnaldum, the Netherlands (Henderson 1999b), although in this case, as with the opaque white glasses from the site, elevated magnesia levels were associated with only slightly elevated potassium oxide levels. This may be due, in part, to the mixture of natron and plant ash glass.

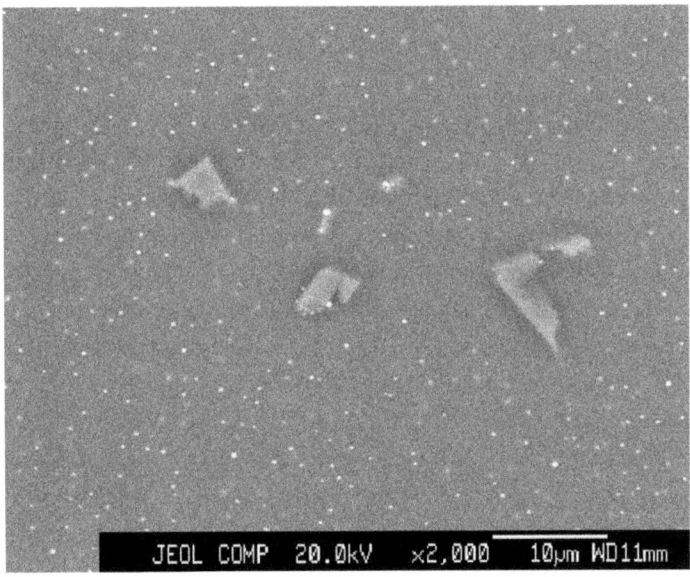

3.10. A backscattered electron micrograph of metallic Cu drops in ninth-century opaque red glass from Mantai, Sri Lanka (×2000; photo: E. Faber).

Some opaque white Roman enamels contain elevated potassium oxide, but not elevated magnesia (Henderson 1991b, 602–3, table 1).

As a complete break from the traditional natron technology of making opaque red glass, the use of plant ashes must have played an important role in the development of the colour. By incorporating such ashes in the batch, initially the carbon content of the glass would be increased, thereby increasing the proportion of reducing gases such as carbon monoxide, in the crucible in which the glass was being worked and aiding the precipitation of the red crystals. A further means of producing a dull red opaque enamel is by the use of crushed iron oxide (hematite) mixed with glass (Freestone 1991, 51–2, fig. 3.12; Verità 1995, 91–2; Freestone and Stapleton 1998, 123, table 3). Verità noted that the composition of the glass mixed with the hematite was the same as in the vessel it was used to decorate, indicating that the enamellers worked with the artisans producing the vessels, in this case beautifully decorated thirteenth- and fourteenth-century beakers of the 'Aldrevandin' type.

3.4 Conclusions

Glassmakers must have become aware of the ways in which the environment from which they extracted raw materials ultimately determined the working properties and colour of the glass that they made. It is easy to forget today, in an industrialised environment, that all of the processes in the manufacture of glass were probably connected seamlessly and that it is more likely, for example, that the groups who fused the raw glass understood the reasons why plants with a specific appearance were harvested from a specific zone or location at a particular time of year.

The first time that *deliberately* made glass was produced using a successful combination of raw materials, it would probably have been semi-transparent, brightly coloured, smooth and (eventually) cool. However, the use of opacifiers was something very new, as was the achievement of high enough temperatures (in the context of glass technology) to melt sufficiently large quantities of silica to make glass and vessels. All the techniques of glassworking were completely new, including the retention of the glass at high enough temperatures and its heat treatment at lower temperatures. They should not be underestimated as major steps forward in the understanding of how glass behaved at high temperatures in the late Bronze Age, including core forming, trailing and subsequent manipulation of trails, and marbling. The later introduction of 'new' colourants and opacifiers built on the technological foundations laid down by early glassmakers and glass workers.

FOUR

GLASS CHEMICAL COMPOSITIONS

In seminal articles, Turner (1956a and 1956b) and Sayre and Smith (1961) demonstrated that a variety of glass chemical compositions were used in the past and, to a lesser extent, that the compositions were relatable to the geographical origins of the glass. We are now aware that quite a wide range of factors can modify the chemical and mineralogical compositions of glass raw materials (see Chapter 11). However, the principal compositional types that Sayre and Smith defined remain unchallenged. These are high magnesia glass (c. 2500–800 B.C. and c. 800– A.D. 1700), referred to henceforth as 'plant ash glasses'; low magnesia glass (c. 800 B.C.–800 A.D.), referred to henceforth as natron glasses; and high lead glass (c. A.D. 1000–1400). Sayre and Smith also noted that 'high' antimony glasses occurred in contexts dating to between c. 600 and 200 B.C. As more coherent glass compositional groups from well-dated archaeological contexts are published, it has become possible to attempt to answer research questions that can contribute to mainstream archaeology, such as questions relating to provenance, distribution and trade.

Major, minor and trace components may be introduced in the glass batch by using primary raw materials, colourants, opacifiers, decolourants, clarifiers, scrap (vessel) glass or deliberately mixed chunks of raw glass (see Chapters 2 and 3). As discussed in greater detail in Chapter 11, glass chemical compositions are the summed result of a potentially wide range of decisions and actions on the part of a number of groups associated with glass manufacture. All such decisions would have been taken within the ritual, social, economic and political contexts of the society concerned. Chapters 5 through 10 provide detailed case studies of the production contexts of Bronze Age, late Hellenistic/early Roman and Islamic glass, including how scientific research has helped to place the respective glass technologies and chemical compositions into a broader context. Here, however, the range of glass chemical compositions that have been defined, including those reported by Sayre and Smith, are discussed. At the beginnings and ends of periods when particular glass chemical compositions dominated there were periods of technological transition. These transitions, involving new raw materials, must have involved 'trials by fire' and experimentation during which glassmakers and glassworkers became familiar with the unfamiliar properties of new glass types.

In summary, here are some of the reasons for carrying out scientific (including chemical) analysis of glass:

- To investigate the links between scientific analysis and religious, social, economic and political contexts in which glass was made and used, involving a consideration of the archaeological contexts in which glass was found; compositional groupings have been produced both by 'natural' factors and those imposed by man (Henderson 1989, 30).
- To provide a provenance for glass and to investigate the evidence of glass recycling.
- To investigate trade and exchange in raw materials, raw glasses and glass artefacts linked to social, ritual and economic value of artefacts in 'foreign' contexts partly by a comparison with the use of other materials. This would create connections between primary and secondary production centres, centres of trade and areas where glass was used.
- To investigate and, where possible, reconstruct the different linked technological processes in the *chaîne opératoire* (originally applied to the study of flint technology: Bar-Yosef *et al.* 1992). These include the identification of the type and source of the raw materials used ranging from alkalis, silica, lead, colourants, opacifiers and clarifiers to the sources of the clays used to construct the glass furnace bricks and to make the crucibles, the fuels used and the artefacts manufactured. The decisions made, whether by individuals or groups of people, may well be informed by links with, for example, farmers, metal smelters or potters. This investigation should consider workshop organisation and industrial specialisation in both primary glass production (glass with particular working properties, coloration) and secondary glass production (coloration and specific vessel, bead or other artefact forms and decoration).
- To attempt to link technological and scientific investigations of glass to ethnographic studies of its production. This helps to investigate the *chaîne opératoire*, for example when considering rituals associated with production and use, the contexts of tradition, the potentially different social groups involved, the decisions made, the reasons for specialisation, industrial innovation and the links with other industries. This aspect of research has been investigated in far more detail and more extensively in 'traditional' potting (Day 2004, Day *et al.* 2010) and metal smelting (Phenyo *et al.* 2009) than in the context of glass production.

- To investigate the relationships between industrial and archaeological evidence for glass production and historical sources (Martinón-Torres and Rehren 2008).

4.1 The Middle East and Europe c. 2500 B.C.–A.D. 1700

4.1.1 Plant Ash Glasses c. 2500–c. 800 B.C.; A.D. 800–1700

As noted in Chapters 1, 2 and 5, the earliest true glass was a plant ash glass made from two basic raw materials, ashes of salt-tolerant (halophytic) plants and silica (see Table 4.1). The initial stages of plant ash glass production would probably have been linked to the early manufacture of soap and detergent because both require the ashing of alkali-rich plants, a practice that is still in use today (Barkoudah and Henderson 2006). This link probably existed in the Islamic period, a second major period of plant ash glass production. Indeed al-Magdasi, a tenth-century Arabic geographer, refers to al-Raqqa as being a famous centre for soap production but fails to mention that large-scale plant ash glass production occurred there in the ninth century (see Chapter 9). The reference demonstrates that plant ashes were still being supplied to al-Raqqa in the tenth century, even if not for glass production.

The geological conditions in which the plants grew made a major contribution to their chemical and mineralogical compositions. An assessment of whether different geologies as reflected in the chemical compositions of plant ash glasses and their provenance in given in Chapter 11. Modification of the chemical composition of plant ash glasses by the crucible in which they were melted may have occurred in Egypt (Rehren and Pusch 2005; Walton *et al.* 2009, 1500–1), although isotopic and trace element results indicate that this had a limited affect on glass compositions (Henderson *et al.* 2010). Sayre and Smith (1961) and Sayre (1967) noted that plant ash glasses were characterised by high contents of magnesium and potassium oxides and coined the acronym HMG (high magnesia glass). A number of (HMG) late Bronze Age glasses from Tell el Amarna, Egypt and Tell Brak, Syria have been plotted in Figure 4.1. Nikita and Henderson (2006) confirmed that such glasses were in use in late Bronze Age Greece, a period when the first glass vessels were being made in Mesopotamia and Egypt. Shortland *et al.* (2007) using trace elements were able to observe differences between Egyptian and Mesopotamian Bronze Age plant ash glasses. Henderson *et al.* (2010) showed that isotopic data provide evidence for more than one production zone for plant ash glasses in Mesopotamia whereas trace element results do not. Gan Fuxi (2009b, 64, table 2.4), has published chemical analyses of plant ash glasses from Kiziltur, Xinjiang dating to c. 1100–800 B.C. that

Table 4.1. Examples of plant ash glass compositions (weight % oxide)

Country	Egypt	Greece	Syria	United Kingdom
Site	Amarna	Elateia	Tell Brak	Fishbourne
Date	14 C B.C.	13–12 C	14 C B.C.	1 C A.D.
Description	Raw turquoise	Dark blue bead	Cobalt blue fragment	Emerald green fragment
Na_2O	16.18	17.52	16.14	16.5
MgO	3.95	2.25	5.47	2.5
Al_2O_3	0.42	1.94	0.76	1.8
SiO_2	62.18	69.33	63.34	63.2
P_2O_5	0.15	0.08	0.27	1.1
SO_3	0.42	0.17	0.23	0.4
Cl	0.8	1.19	0.45	1
K_2O	2.16	0.55	4.42	1.9
CaO	9.72	5.35	4.98	7.2
TiO_2	0.02	0.08	0.03	0.1
Cr_2O_3	ND	ND	ND	ND
MnO	ND	0.11	0.06	0.6
Fe_2O_3	0.21	0.49	0.79	1.3
CoO	ND	0.04	0.24	ND
NiO	ND	0.06	0.06	ND
CuO	1.96	0.05	0.32	2.5
ZnO	ND	ND	0.07	ND
As_2O_3	0.07	ND	0.07	ND
Sb_2O_3	ND	ND	ND	ND
SnO_2	ND	ND	1.67	ND
BaO	ND	ND	ND	0.1
PbO	ND	ND	0.05	0.1
Source	unpublished	Nikita and Henderson 2006	Henderson et al. 2010	unpublished

Site	Seleucia: Tell Umar	al-Raqqa	al-Raqqa	Lincoln	Shahr-i-Banu	
Date	4–6 C? A.D.	9 C	9 C	17 C	7–13 C A.D.	c. 2–6 C A.D.
Description	Purple ?chalice	Green flask	Green furnace glass	Colourless goblet	Aqua vessel	Average composition
Na_2O	13.1	13.1	10.85	11.2	15.3	15.2
MgO	4.7	6.3	3.74	3.4	4.14	3.8
Al_2O_3	1.48	1	1.32	0.7	2.67	2.2
SiO_2	68.6	67.6	69.3	71.9	62.45	67
P_2O_5	580	0.1	0.31	0.2	ND	
SO_3	0	0.3	0.15	0.3	ND	
Cl	0	0.7	0.65	0.7	ND	
K_2O	2.58	3	2.26	1.8	5.03	2.9
CaO	6.3	6.9	9.38	8.9	9.55	6.8
TiO_2	0.081	0.02	0.09	0.1	0.1	
$Cr2O3$	34.4	ND	ND	ND	0.005	
MnO	1.96	0.53	1.07	1	0.03	
Fe_2O_3	0.61	0.26	0.63	0.5	0.59	1
CoO	9	0.02	ND	ND	ND	
NiO	25	ND	ND	ND	ND	
CuO	36.4	0.03	0.08	0.1	0.03	0.1
ZnO	ND	ND	ND	0.1	0.04	
As_2O_3	ND	ND	ND	ND	ND	
Sb_2O_3	ND	ND	ND	ND	ND	
SnO_2	5.2	ND	ND	ND	ND	
BaO	238	0.05	0.05	0.1	0.01	
PbO	ND	ND	0.08	ND	0.001	
	Mirti et al. 2008	Henderson et al. 2004	Henderson et al. 2004	Henderson 2006	Brill 1999b	Lankton and Dussubieux 2006

ND = Not Detected; P, Cr, Cu, Sn and Ba are $\mu g\,g^{-1}$.

would probably have been transported along the Silk Road from Mesopotamia. The production context and detailed discussion of the provenance of plant ash glasses is discussed in Chapters 5 and 6 respectively.

Although plant ash glass was largely replaced by natron glass by the eighth century B.C., there are nevertheless examples from the island of Rhodes dating to the seventh or sixth century B.C. (Beltsios *et al.*, 2012 and others from central European sites of seventh-century B.C. date; (Braun 1983).

There are few examples of plant ash glasses from the Middle East and the West that date to between the third century B.C. and the first century A.D. Plant ash glass was used (and presumably produced) by the Romans for the manufacture of first-century A.D. emerald green glass (and small numbers of cobalt blue glasses), such as that found at Fishbourne, a Roman place in southern England (Henderson 1996). First-century Roman examples of plant ash glasses have also been found at the Magdalensburg, Austria, where they were used for the manufacture of 'black' beakers (Cosyns *et al.* 2006). Between the second and sixth centuries A.D., plant ash glass occurs in the area of Iran and Iraq, the Sasanian Empire, where it must have been fused from raw materials (Brill 1999b; Mirti *et al.* 2008). Mirti *et al.* (2008, 440) identified two types of plant ash glass from Seleucia and Veh Ardašīr, central Iraq, based on a combination of trace elements (zirconium and hafnium) and calcium:aluminium ratios. On the basis of this they have suggested that two different silica sources were used to make the glasses.

The plant ash glasses from Fishbourne and the Magdalensburg are too early to have been made by the Sasanians. A single translucent purple armlet fragment from late first century B.C.–early first century A.D. contexts of the Iron Age site of Hengistbury Head, Hampshire, contains levels of magnesium and potassium oxides (1.4% and 1.3%) that suggest that a proportion of plant ash glass has been added to natron glass (Henderson 1987, table 16, fiche 6: A10–11, analysis 13). Arletti *et al.* (2008) have also published results for mainly first-century Swiss glasses with elevated magnesia and potassium oxide levels. Their interpretation of these levels is that impure sands were used, but the use of an alkali plant ash to make them seems a more likely interpretation. Indeed, the relatively low magnesia and potassium contents in these and other (e.g. Fishbourne, England) first-century Roman glasses when compared with Bronze Age and Islamic examples could suggest that plant ash glass, or, less likely, the ashes themselves, were mixed with natron glass. Quite why this occurred is still to be investigated. Additional second- to fourth-century examples of black *unguentaria* plant ash glass have been found at Elkab in Egypt (Cosyns *et al.* 2006). Moreover, Mirti *et al.* (1993) have reported some examples of predominantly fourth-century plant ash glass from the Roman site of Augusta Pretoria (modern Aosta) in northern Italy. Throughout the Roman and Byzantine periods, a rather dull, opaque, red-coloured glass was used for the production of enamels and

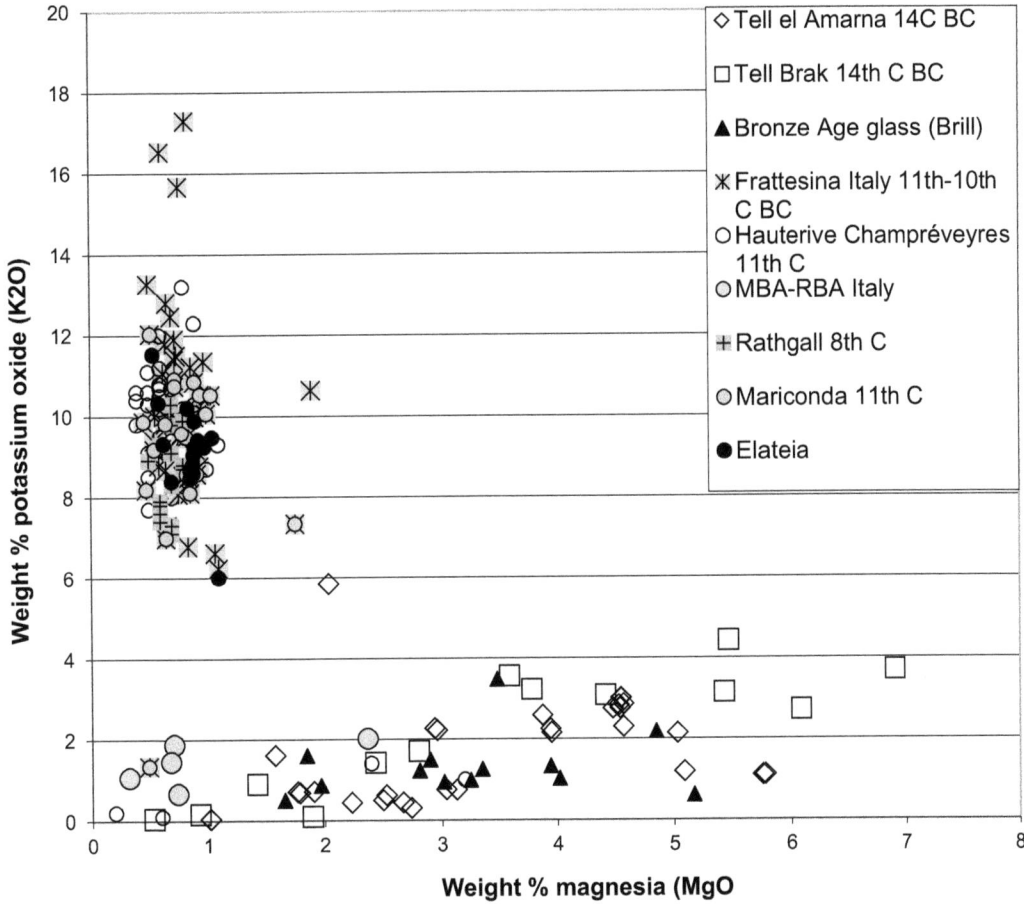

4.1. A bi-plot of weight % magnesia versus potassium oxide in Bronze Age glasses from the Middle East and Europe (MBA = Middle Bronze Age; RBA = Recent Bronze Age; these Elateia glasses date to the eleventh–tenth centuries BC). For data sources see Chapter 6.

glass mosaic tesserae (see Chapter 3), with elevated potassium and magnesium oxide levels (Freestone *et al.* 1990; Brun *et al.* 1991; Henderson 1991a and 1991b; Shugar 2000; Andreescu-Treadgold and Henderson 2006; Santagostino Barbone *et al.* 2008) indicating that plant ashes formed part of the original batch. The inclusion of plant ash and charcoal in the glass batch would have helped to produce the essential reducing conditions necessary to produce the reduced form of copper (Cu I) crystals for the red colour.

In a different cultural context, Brill (1991) has published compositional results for a variety of glass samples of natron and plant ash compositions from Meroitic Sudan dating to between A.D. 250 and 300, including results for cut colourless vessels and gaming pieces. Whether these plant ash glasses were made in the Sudan or imported from the Sasanian empire is still to be investigated. A more detailed consideration of plant ash glasses is given in Chapters 6 and 10.

Table 4.2. Examples of mixed-alkali and high potassium European glass compositions (weight % oxide)

Country	Italy	Italy
Site	Frattesina	Mariconda
Date	11–10 century B.C.	11–10 century B.C.
Description	Translucent blue bead	Opaque white bead
Na_2O	0.96	5.8
MgO	0.61	0.65
Al_2O_3	2.84	1.85
SiO_2	67.96	85.14
P_2O_5	0.21	0.17
SO_3	0.03	0.03
Cl	0.01	0.06
K_2O	16.53	6.99
CaO	2.05	1.66
TiO_2	0.06	0.06
Cr_2O_3	ND	0.03
MnO	0.05	0.03
Fe_2O_3	0.43	0.56
CoO	0.02	ND
NiO	ND	ND
CuO	5.39	1.04
ZnO	ND	ND
As_2O_3	ND	ND
Sb_2O_3	ND	ND
SnO_2	ND	ND
BaO	0	0.05
PbO	ND	0.04
Source	Towle *et al*. 2001	Towle *et al*. 2001

(ND = Not Detected).

4.1.2 Mixed-Alkali Glass and Its Compositional Variations (c. 1100–c.750 B.C.)

The proto-glass (glassy faience) found in Italy and France dating to the Middle Bronze Age, and probably earlier, developed into glass with the glassy component having similar compositional characteristics. The glass is characterized by containing mixed alkalis (sodium and potassium oxides), low calcium oxide (c. 4%), low magnesia and can contain mineralogical impurities such as quartz and feldspars (Henderson 1988a; Santopadre and Verità 2000; Angelini *et al*. 2004). It has been labelled LMHK – low magnesium, high potassium – glass because these characteristics clearly distinguish it from both natron (low magnesium, low potassium) and plant ash (high magnesium, elevated potassium) glasses (Henderson 1988a) – see Table 4.2. The key production site of Frattesina in the Veneto has been surveyed and excavated and has produced the largest amount of raw and artefact LMHK glass to date (see Chapter 5). The total alkali is almost always about the same at c. 14% to 16% oxide. Although the majority of the glasses are of a mixed-alkali composition, a plot of soda versus potassium reveals the compositional range. At one end are the earliest high potassium oxide glasses in the world, with contents of up to 18% (see Figs. 4.1 and 6.17). The first example published was a plain annular glass bead from a tenth-century B.C. burial on the Greek island of Thasos (Henderson 1992). Since then, other examples have been found (Towle *et al.* 2001, nos. 235, 236 and 294). In China the high potassium glasses also contain low levels of magnesium and calcium oxide (Li Qinghui *et al.* 2009, table 21.6, and see sections 4.12 and 4.13 later in this chapter) and make them indistinguishable from the early European examples. Current research indicates that the Chinese examples are later than the European ones, but this may change. So it is difficult to know whether high potassium glass was made in both areas independently or whether it was exported along the Silk Road from west to east – or east to west.

In the West the distribution of these mixed-alkali glasses stretches from Italy northward through Switzerland into Germany (Hartmann *et al.* 1997) and westward into France (Gratuze *et al.* 1998), England and Ireland (Henderson 1988a and 1988b). Until relatively recently, the occurrence of LMHK glass on Thasos was thought to have been an isolated occurrence in Greece, but more examples in early Iron Age contexts have been found farther south in Greece at the cemetery of Elateia (Nikita and Henderson 2006). The latest dated occurrence of LMHK glass is from eighth-century B.C. contexts at the site of Rathgall in Ireland (Raftery 1987; Henderson 1988) where gold was applied to a bead made from LMHK glass, so the bead clearly had a high social and ritual value. The use and perhaps the production of mixed-alkali glass therefore continued into the period when the next major glass technology, natron glass, was introduced. For a more detailed consideration of mixed-alkali glasses, see Chapter 6.

4.1.3 The Transition from Plant Ash Glass to Natron Glass, c. A.D. 800

Around the tenth century B.C., some glassmakers started to use a mineral (natron) instead of plant ash an alkali source. It is, however, no coincidence that mixed-alkali glass was introduced at a time when the manufacture of plant ash glass was dying out. At this time of transition in the eleventh to tenth centuries B.C., beads made from both mixed-alkali and plant ash glasses occur in Italy and to the north across the Alps (Henderson 1988a and 1988b; Gratuze *et al.* 1998). Natron is a purer source of alkali that occurs in evaporite deposits (see Chapter 2). It is considerably denser than plant ashes and therefore occupied a smaller volume in the fritting oven, in the 'founding' glass furnace (whether fritted or not) and in glass crucibles (if they were used for fusing the glass). Furthermore, unlike plant ash, natron did not need to be gathered from large areas and prepared by ashing. Its relative rarity did nevertheless mean that it was necessary to move it from, for example, Wadi el Natrun to primary production centres, such as on the Levantine coast as described by Pliny (see Chapters 2 and 3). If the Phoenicians made natron glass the raw materials would probably have been fused on a relatively small scale in crucibles. Import of natron to the Levantine coast would have occurred; the selection of an evaporite deposit high in trona rather than sodium chloride would have been critical (see Chapter 2); as for plant ashes, it was necessary to find a material that fluxed silica effectively. Undoubtedly, to begin with, there would have been an experimental phase during which the search for usable natron deposits occurred.

The question is, why was natron not used instead of plant ash to make glass earlier? Natron had been used for embalming bodies in Egypt since c. 2000 B.C. (David 2000). One explanation could be that the earliest glass was made in Mesopotamia, where there are fewer suitable evaporite deposits. An evaporite deposit at Jaboul southwest of Aleppo in northern Syria may have offered appropriate mineral deposits. It is no coincidence that the probable primary manufacture of mixed-alkali (LMHK) glass in large quantities in northern Italy from the eleventh century B.C. occurred just as the manufacture of plant ash glass was starting to be less common and natron glass was on the point of being introduced. This was also a time when the Mycenaean palace economy collapsed. Therefore, it appears that LMHK filled a lacuna left by the disruption of the supply of glass (beads) from the Aegean and the Middle East to Europe (Henderson 1988b). Indeed, the collapse of the great Bronze Age palatial economies of Mycenaean Greece, Egypt, Mesopotamia and Hittite Turkey, along with a wide range of other socio-economic changes (including the breakdown of trade and exchange networks) triggered the introduction of natron glass for the first time.

4.1.4 Natron Glasses c. 800 B.C.–c. A.D. 800

The transition to natron glass must have involved experimentation with the new alkali source, perhaps producing glasses of unusual chemical compositions. Some of the earliest natron glass objects are in the form of tenth-century B.C. Egyptian core-formed vessels (Schlick-Nolte and Werthmann 2003). These vessels were found in the burial of Nesikhons, which dates to 975/974, providing a *terminus ante quem* for the vessels. They are mostly monochrome dark brown, translucent green, opaque yellow or translucent blue coloured glasses. Their natron composition is characterised by low calcium oxide levels of between 1.3% and 4.8% (a characteristic of later Egyptian natron glasses). As noted by Sayre and Smith (1961), by 800 B.C. natron glass was in use across large areas of the Mediterranean and into western Europe. Even though this observation is based on relatively small numbers of chemical analyses, it holds true today. Natron glass is characterized by low levels of magnesium and potassium oxides (see Table 4.3). Figure 4.1 is a bi-plot of these oxides and serves to underline the

Table 4.3. Examples of natron glass compositions (weight % oxide)

Country	Iraq	Israel	Israel	United Kingdom
Site	Nimrud	Eli'ezer	Dor	London
Date	9 C B.C.	8 C A.D.	5–6 C A.D.	1 C A.D.
Description	Blue plaque	Green furnace glass	Green furnace glass	Average compositions
Characteristics	Average composition	Levantine II	Levantine I	HIMT
Na_2O	16.94	12.14	16.61	18.54
MgO	1.7	0.57	0.5	0.92
Al_2O_3	0.55	3.29	3.26	2.47
SiO_2	69.85	75.98	67.3	66.38
P_2O_5	0.1	<0.1	0.14	0
SO_3	0.54	0.15	0.3	ND
Cl	ND	0.7	0.7	ND
K_2O	0.95	0.4	1.05	0.57
CaO	6.15	5.96	10.08	5.99
TiO_2	<0.1	<0.1	0.12	0.33
Cr_2O_3	ND	ND	ND	ND
MnO	0.1	0.1	<0.1	1.9
Fe_2O_3	0.28	0.45	0.3	1.15
CoO	<0.1	ND	ND	ND
NiO	ND	ND	ND	ND
CuO	1.74	ND	ND	ND
ZnO	<0.1	ND	ND	ND
As_2O_3	ND	ND	ND	ND
Sb_2O_3	0.92	ND	ND	ND
SnO_2	<0.4	ND	ND	ND
BaO	ND	ND	ND	ND
PbO	ND	ND	ND	ND
Source	Reade et al. 2005	Freestone et al. 2000	Freestone et al. 2000	Freestone et al. 2005

Country	Lebanon	Syria	Italy	China
Site	Beirut	al-Raqqa	Torcello	Mai'e Pur
Date	1 C B.C.–1 C A.D.	9 C	11 C	616–1127
Description	Green furnace glass	Green tubular vessel foot	Green tessera	White fragment
Characteristics	similar to Egypt II	variation on Levantine I	Levantine II	similar to Egypt II
Na_2O	18.4	13.1	17.82	17.98
MgO	0.546	0.52	1.08	0.46
Al_2O_3	2.35	3.32	2.74	1.66
SiO_2	71.57	69.9	65.43	70.15
P_2O_5	0.072	0.07	0.17	0.09
SO_3	0.457	0.15	0.24	ND
Cl	0.683	0.78	0.66	ND
K_2O	0.706	0.51	0.62	0.53
CaO	7.67	11.03	6.45	7.7
TiO_2	0.043	0.06	0.42	0.12
Cr_2O_3	ND	ND		ND
MnO	ND	0.03	2.17	0.63
Fe_2O_3	0.02	0.43	2.61	0.59
CoO	0.3	ND	ND	ND
NiO	ND	ND	ND	
CuO	ND	ND	0.06	ND
ZnO	ND	ND	ND	ND
As_2O_3	ND	ND	ND	ND
Sb_2O_3	ND	ND	0.08	ND
SnO_2	ND	ND	ND	ND
BaO	0.02	0.05	0.06	0.04
PbO	ND	ND	ND	0.01
Source	Unpublished	Henderson et al. 2004	Andreescu and Henderson 2006	Li Ganghui et al. 2009

ND = Not Detected.

simple compositional differences between natron glasses with low levels of both, plant ash glasses with elevated levels of both and glasses made from a mixture of natron and plant ash glasses.

There is evidence that natron glass was used during the Middle Eastern Iron Age at, for example, Nimrud and Hasanlu, the latter further characterised by low boron levels (Reade *et al.* 2005; Brill 1999b, IID, nos. 5420 and 5431). It was also used by the Etruscans and Villanovans (Towle 2002; Towle and Henderson 2008) in Italy, by the Phoenicians, in the early, middle and late Iron Age Europe (Braun 1983; Henderson 1983; Henderson 1989), in the early Warring States period (c. 500 B.C.) in China (Gan Fuxi *et al.* 2009), by the Hellenistic Greeks, the Romans, the Byzantines, the Muslims in the Umayyad caliphate (up to c. 750AD), and by the Anglo-Saxons and the Vikings. Natron glass was therefore distributed widely. There is some evidence for an ongoing but decreasing use of plant ash glass in the early Iron Age, as noted earlier. Natron glass has therefore been found as far as China in the east, Ireland in the west, Africa in the south (Brill 1999b) and Scandinavia in the north.

A recently published article (Purowski *et al.* 2012) has reported what are probably natron glasses but they contain low magnesium oxide levels and 'medium' levels of potassium oxide (i.e. higher than found in most natron glasses) and have been labelled 'LMMK'. (The elevated potassium levels may be due to the exploitation of sand that contains potassium feldspar.) These glasses, which are also characterised by low calcium oxide levels (Purowski *et al.* 2012), have been found in Poland and date to Hallstatt C and the start of Hallstatt D (seventh–sixth centuries B.C.). They are in the form of beads and are often a translucent pale blue colour. Pusowski *et al.* also note that the low magnesia and calcium oxide levels and the occurrence of silica crystals in these glasses provide a link with mixed-alkali glasses (discussed and in Chapter 6) that had largely gone out of use in the tenth to ninth centuries B.C.

By the late Hellenistic–early Roman period, we know that natron and sand were melted together in huge tank furnaces of up to 6 metres in length in Beirut (Kowatli *et al.* 2008). However, for the period between the eighth and the first centuries B.C., no archaeological evidence of primary production sites has yet been found. Because its primary production then would have been on a smaller scale than that demanded for the first-century B.C. production of cast and blown vessels, it is likely to have occurred in crucibles. It is therefore possible that some natron glass before the first century B.C. had a composition that was modified by interaction with the crucible lining (if present) or the crucible wall. This could increase the levels of (e.g.) alumina and calcium oxides.

As noted in Chapter 1, soda-lime-silica glass has a minimum liquidus temperature (the absolute melting temperature of glass) of 725°C for a soda-lime-silica glass with a composition of 21.9% soda, 5% calcium oxide and 73.1% silica (Morey 1964, fig. 20, table 13, 33). The chemical composition of ancient natron glass is, within limits, fairly constant (see Table 4.3) but tends to contain lower soda, higher calcium oxide and lower silica than the latter composition. Nevertheless, there is some compositional variation according to the date of production and the production zone. With the discovery of primary glassmaking centres with the remains of glass furnaces in which natron glass was fused from raw materials in Egypt and the Levant (Gorin-Rosen 1995; Nenna *et al.* 2000; Nenna *et al.* 2005; Kowatli *et al.* 2008), it has become possible to relate the chemical compositions of raw natron furnace glass to the place where the glass was melted. Thus, Nenna *et al.* (1997 and 2000) were the first to define compositional types of Egyptian natron late Roman glass that were typical of the region, one with a very low calcium oxide level. This was followed by Freestone *et al.* (2000), who identified compositional types of natron glass found in Levantine furnaces dating to between the sixth and eighth centuries (see, e.g. the distinction between first-century A.D. and eighth-century A.D. furnace glasses in Fig. 8.4).

These compositional results for natron glass demonstrate how important the discovery of primary glassmaking sites is to the study of glass production and distribution in the ancient world. Without such sites, the baseline data for glass known to have been fused from primary raw materials in specific locations would be lacking. It is assumed that such glass contained a minimal amount of (recycled) glass made at different locations. The results provide a compositional provenance for the glass made there. Glass provenance is discussed in more detail in Chapter 11: isotopic results provide a *geological provenance* for both furnace and nonfurnace glasses.

One type of natron glass is characterized by unusually high levels of iron, manganese and titanium, labelled HIMT (Mirti *et al.* 1993; Freestone 1994; Freestone *et al.* 2005; see Table 4.3). This natron glass type is thought to have been manufactured in Egypt using sands characterized by high levels of these impurities (Freestone *et al.* 2005; Leslie *et al.* 2006). Foster and Jackson (2009) have shown that two types of HIMT glasses can be identified amongst 'naturally coloured' late Roman vessel glass from Britain and that it was probably the most common compositional type by the fourth century in Europe. They discuss the results for an impressive total of 344 samples. Type 1, introduced at least as early as A.D. 330, is characterized by containing (on average) double the levels of iron, manganese and titanium than found in type 2, higher levels of magnesia, barium, chromium and zirconium, and slightly higher levels of alumina. Foy *et al.* (2003) have demonstrated that there are naturally occurring levels of copper, lead and antimony in (unrecycled) raw HIMT, and so if this is assumed to be characteristic of all HIMT glasses, and higher levels are detected than the 'natural' level, then they can be used as a guide to the extent that it has been recycled. Price (2000) has noted that fourth-century glass technology underwent important changes with a shift to vessel formation and decoration involving a quicker and simpler single process, and so this can also now be linked to the introduction of HIMT (Foster and Jackson 2009, 195).

Glass vessels dating to the fourth and fifth centuries from Mayen, western Germany, have chemical compositions that are similar to HIMT glasses but in several ways are also distinct, suggesting to Hartmann and Grünewald (2010) that this could constitute evidence for local production. This may be the case and it may also have involved mixing and remelting of imported glass.

The scientific analysis of raw chunks of glass found at Byzantine Sagalassos, Turkey, showed that they had been produced by mixing HIMT glass with a more standard type of natron glass (Degryse *et al.* 2006 and see Chapter 11). Glasses of a similar composition have been found in eighth-century contexts at the Red Sea port of Raya, but these contain significantly lower manganese oxide levels (Kato *et al.* 2009, 1705) at 0.1% MnO.

Natron glasses with high iron levels have been found in early Iron Age European contexts north of the Alps and in Italy. The seventh-century B.C. cemetery of Chotin produced a number of these glasses where it is clear that iron – or perhaps iron-containing slag – was added to colour the glass an intense deep brown colour (Brill 1999b, 59, nos. 3462, 3463 and 3467; Towle 2002). Two glasses from Chotin were also characterized by high levels of potassium oxide (7.76% and 7.9%) and phosphorus pentoxide (1.88% and 1.64%). The high phosphorus pentoxide would suggest that a plant ash rich in potassium (although with low soda levels of c. 2%) was the source of the alkali and elevated magnesia of 2.01% and 3.4% confirms this. The seventh century (Hallstatt C, c. 750–650 B.C.) is a time when natron and plant ash glasses are found on Greek and central European sites (Braun *ibid.*, tables. 12, 13 and 14); other glass analyses from Chotin published by Brill are natron glasses. By Hallstatt D times (c. 650–550 B.C.), no plant ash glasses are apparently in use in central Europe (Braun 1983, table 17 and 199). One of the natron glasses from Chotin (Brill 1999b, 59, no. 3467) even contains a high iron level (15%) comparable to those found in the high potassium glasses from the site. Like the other two high iron glasses it contains an elevated manganese oxide level (0.53%), suggesting that the manganese was introduced with the iron. However, if an iron slag is the source of the iron, the alumina levels might be expected to be higher.

A study of natron glass manufactured in the Levant over the first millennium A.D. shows that, with some exceptions, the alumina levels rose from as low as 2% in the first century A.D. to c. 3.5% to 4.5% by the seventh century A.D. and that the CaO levels rose from c. 6.5% to c. 9.5% over this period. Moreover, during this period soda levels fell from c. 17%–20% to c. 15% in some glasses (Henderson 2002b). The latter is illustrated in Figure 4.2. Variations in the levels of alumina and calcium oxide reflect the changing exploitation over time of sand sources in the Levant. With time, increasingly less pure sand was used with

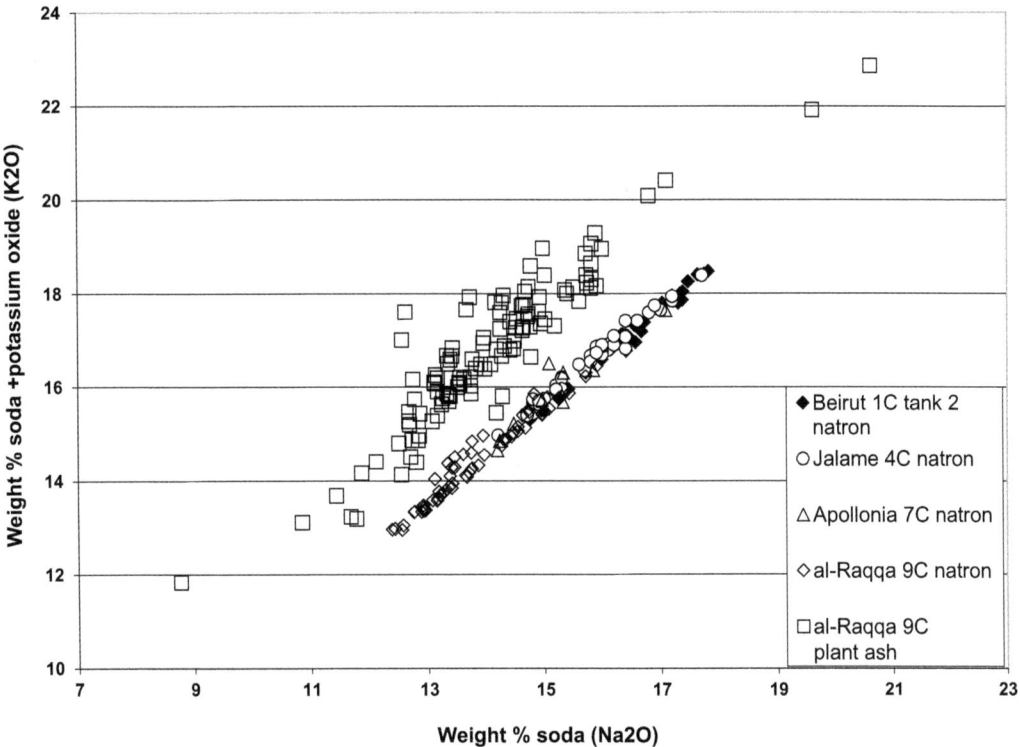

4.2. A bi-plot of weight % soda versus soda+potassium oxide in Middle Eastern furnace glasses dating between the first century B.C. or A.D. (Beirut) and the ninth century A.D. (al-Raqqa).

generally higher alumina levels. Had the sand been prepared before melting (as Pliny the Elder described; see Chapter 3), with the removal by centrifuge, grinding or filtering of alumina-rich feldspars and of shell fragments, it may not be possible to prove that this has happened scientifically. A thorough analytical survey of existing sand deposits in the Levant might help to locate sands used in antiquity with these differing mineralogical characteristics. However, the sand deposits may have changed to the extent that the exercise would not help. The determination of neodymium isotopes of a range of sand samples along the Levantine coast is more likely to provide provenance information for glass made in the area (see Chapter 11).

4.1.5 The Demise of Natron Glass Technology: The 'Reintroduction' of Plant Ash Glass Technology from c. A.D. 800 in the Middle East

From c. A.D. 800, the Muslims (re-)introduced plant ash glass technology on a massive scale as part of a phase of experimentation and innovation in a range of technologies (se Chapter 9 and 10), and natron glass technology began to be eclipsed (see Table 4.1). However, compositional evidence suggests that the technological transition from natron glass occurred more quickly than north of the Alps. The change from natron to plant ash glass in the ninth century A.D. must have caused a total change in a whole range of glass manufacturing procedures, just as it had in the eighth to seventh century B.C., when there was a change from plant ash to natron glass. For the ninth century A.D., Foy and Nenna (2001, 26) have suggested that climatic factors, particularly an increase in the 'pluviosite' (heavy rainfall), reduced the amount of natron in the evaporite deposits of North Africa, presumably because less evaporation occurred, less natron was formed, and the amount of natron available for glassmaking was reduced. This had a direct affect on the supply of natron to the Syro-Palestinian area, a major zone of natron glass production. The impact that this had on glass production at the time cannot be underestimated. As the supply of natron became constrained, the melting and working temperatures of natron glass would have increased. It would have created an enormous extra demand for the most expensive raw material, fuel (Henderson 2003).

The soda levels in glasses made in the Levantine coastal area started to fall off as early as the sixth century. In Figure 4.2, results for two kinds of glass are presented: compositional results for ninth-century plant ash glass from al-Raqqa, Syria (Henderson et al. 2004) and natron glass from a first-century B.C.–first-century A.D. furnace from Beirut (Kowatli et al. 2008; unpublished data), fourth-century Jalame, Israel (Brill 1988), seventh-century Apollonia, Israel (Freestone et al. 2000) and ninth-century al-Raqqa, Syria (Henderson et al. 2004). It is clear that the natron and plant ash glasses are separable using this plot because plant ash glasses contain higher potassium oxide levels. That natron is a purer source of alkali than plant ash is illustrated clearly by the more constrained compositional variation in the natron glasses plotted. Furthermore, as we would expect, the levels of potassium and sodium oxides increase together (they are positively correlated). One other clear result is that most of the ninth-century al-Raqqa natron glass falls at the low end of the natron correlation; some of the results plotted also extend into the glasses containing higher soda levels. Three samples from seventh-century Apollonia and a single one from Jalame fall below a total alkali value of c. 15.3% and soda value of c. 14.8%. All of the first-century Beirut glasses fall above this line.

Because natron glass technology was being replaced by plant ash glass technology in the ninth century, the occurrence of glass compositions indicating that a technological transition was occurring, with low total alkali contents, is precisely what we would expect. Moreover, although the glasses from al-Raqqa were found in ninth-century contexts, the natron glasses are all vessel fragments and may have been in circulation for a while (perhaps up to 30–50 years) before

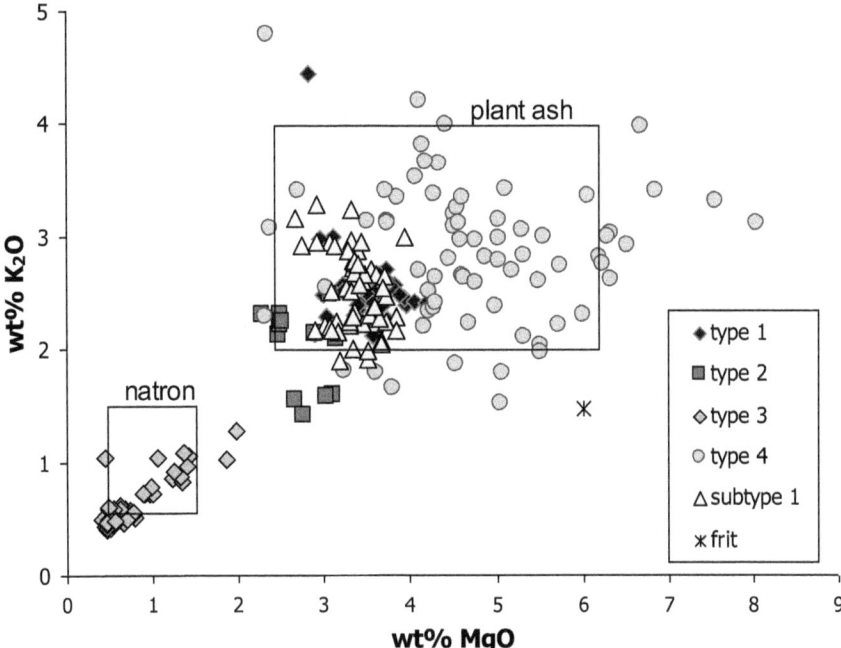

4.3. A bi-plot of weight % magnesia versus potassium oxide in ninth-century glass from al-Raqqa, Syria.

being deposited in al-Raqqa. There is no evidence that natron glass was fused at inland locations in the Middle East. A bi-plot of weight % magnesium versus potassium oxides in ninth-century al-Raqqa glasses is given in Figure 4.3. It can be seen that natron glasses were made from purer raw materials, leading to a more constrained variation in these two impurities. Moreover, there are al-Raqqa glass compositions that fall between plant ash and natron glass (type 2), suggesting that a mixture of different glass types occurred at this time of technological transition (Henderson *et al.* 2004) or that plants used to make them were growing in a contrasting geochemical environment. Strontium isotope results have proven that the natron glasses were made with sands from the Levantine coast (Henderson *et al.* 2005b), and so it is clear that the raw glass used to blow the vessels found was derived from there.

Because soda was the principal flux used in the glass, the attempts to use a lower proportion of natron would have caused the melting temperature of glass to increase. This would have produced two results: first, it would have increased the demand for fuel, the most expensive raw material; second, glassblowers, in particular, would have been forced to retime their operations. Since timing is something that is developed and learned by glassblowers until it becomes second nature, this must have made glassworking difficult. The glass would have melted at higher temperatures and become what is known as 'shorter' (i.e. the glass would be workable for a shorter time). There would therefore have been a period of readjustment to the wholesale adoption of the plant ash alkali source. Scientific analyses of unworked raw and other glass have provided the evidence of experimentation with raw materials around the end of the eighth and in the early ninth centuries (Henderson 2003; Henderson *et al.* 2004). It must have been a difficult time for glassblowers, during which they experimented and modified the timing of their procedures. The supply chains feeding the glassmaking furnaces would have changed dramatically. Plant ash would have originated from semi-desert, coastal and other environments instead of the evaporite deposits that derived from places like Egypt (see Chapter 2). Natron is a far more concentrated form of alkali; gathering an equivalent volume of halophytic plants from semi-desert environments would have involved many more man-hours. As with natron glass, the primary manufacture of plant ash glass occurred in tank furnaces – a classical invention – (Henderson 2000, 42; Goren-Rosen 2000) rather than in crucibles in beehive-shaped furnaces.

Quite a wide compositional range of ninth-century plant ash glasses has been identified; by the eleventh to fourteenth centuries, the compositional variation of Islamic plant ash glass from Syria and some other parts of the Middle East had become somewhat more restricted, making the glassworking properties more predictable. The range of ninth- to fourteenth-century Islamic plant ash compositions is discussed in detail in Chapter 10.

An indication that natron glass was still in circulation in the Mediterranean by the eleventh century is provided by the chemical compositions of glass mosaic tesserae from the basilica of Torcello in the Venetian lagoon (Andreescu-Treadgold and Henderson 2008). The results include both natron and plant ash glass tesserae. Moreover, the results reveal that plant ash and natron glasses were mixed. The glassmakers or glassworkers were extending what was left of the natron glass with plant ash glass. This may simply have occurred because the scrap natron glass was available. It is, however, significant that specific colours of glass tesserae (e.g. opaque red) were made of the mixed natron-plant ash composition but apparently not made from plant ash or natron glass. This shows that specialisation in making coloured glasses existed and perhaps that different colours were made in different places (see Fig. 4.4). The chemical analysis of two sets of stained glass windows, one from Ducio's late thirteenth-century window in the Duomo of Siena, Italy, and the other from the fourteenth-century presbytery of the Santa Maria de Pedralbes church on the edge of Barcelona, revealed that no natron glasses had been used: all was of a plant ash glass composition, the Sienese glass generally containing higher calcium oxide levels (between 8.55% and 10.55%) than that from Barcelona (Gimeno *et al.* 2008).

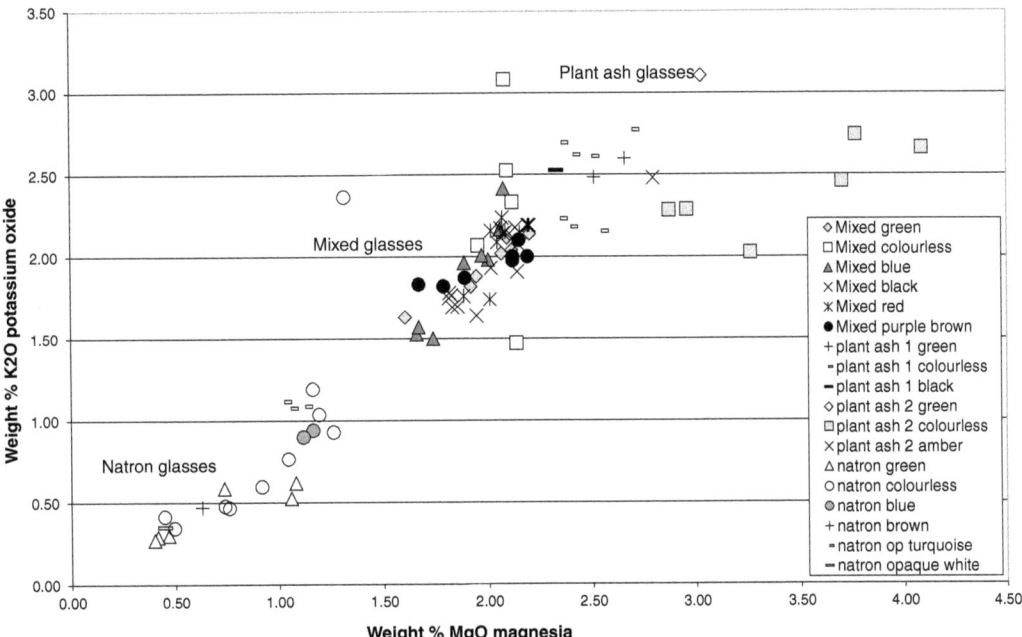

4.4. A bi-plot of weight % magnesia versus potassium oxide in eleventh- and twelfth-century glass mosaic tesserae from the west wall of the basilica St. Maria Asunta on the island of Torcello in northern Italy showing the relationship between colour and composition. Data from Andreescu-Treadgold and Henderson 2006.

Plant ash glass was initially imported from the Middle East by the Venetians, remelted and made into vessels and perhaps tesserae. From the thirteenth century onwards, plant ashes were imported from the Middle East to make the glass in the Veneto (see Chapter 9). No glass furnaces for the manufacture of plant ash glass have yet been discovered in the Veneto dating to before the thirteenth century (Andreescu-Treadgold and Henderson 2006). The plant ash glasses manufactured by the Venetians using imported plant ashes were used to make the famous *cristallo* and *vitrum blancum* glasses (Verità 1985, Verità and Tontinato 1990). The production of *cristallo,* a water-white transparent glass, involved the 'purification' of ashes by reducing the levels of certain insoluble components, calcium carbonate, magnesium carbonate and calcium phosphate. The ashes used for the production of *vitrum blancum* glass were evidently not purified in this way. From the sixteenth century onward, Venetian glassmakers migrated to the Low Countries and other parts of northern Europe where they made plant ash glasses. Such glass was used for the manufacture of decorated vessels in the Venetian fashion (*Façon de Venise*). However, several compositional subtypes of *Façon de Venise* glass have been identified (Henderson 2006) and not all are of a plant ash composition (discussed later). Trace element analysis of Venetian and *Façon de Venise* glasses show that it is possible to distinguish between them (De Raedt *et al.* 2001; Šmit *et al.* 2004). This is largely due to the compositional distinctions amongst the silica sources used.

4.1.6 The Demise of Natron Glass Technology: From Natron Glass to Wood Ash Glass North of the Alps

At about the same time as natron was being replaced by plant ashes as a source of alkali in Middle Eastern glass c. A.D. 800, albeit with a small proportion of natron glass still in circulation, a similar thing happened in the West. The Carolingians may have been partly responsible for introducing this change from natron glass, but in any case, the change appears to have occurred more slowly than in the Orient, over a period of c. 200 years, but starting at about the same time. From the eleventh century, a massive amount of glass was needed for glazing churches and cathedrals. A search for an alternative alkali source occurred north of the Alps, and the result was the exploitation of a local source: wood ash. So far, evidence for a compositional transition has been found especially in the Czech Republic (Staré Město), Denmark (Hedeby), the United Kingdom and Germany (Bezborodov 1975, table VIII, no. 398; Dekówna 1978, tables 2 and 4; Dekówna 1980, tables 41, 44, 46, 48, 50 and 52; Henderson 1991c, 1993a; Wedepohl *et al.* 1997). Objects made from glass with a transitional composition include window and vessel glass. Moreover, the glass beads found in a tenth-century Viking princess's burial on the Isle of Man at Peel Castle had wood ash, natron glass and mixed wood ash-natron compositions (Henderson 2002a, table 1). Compositional markers for the use of wood ash are elevated levels of magnesia, phosphorus pentoxide, potassium oxide and barium oxide; the use of beech ash is reflected in increased levels of manganese oxide (Turner 1956b; Besborodov 1975; Sanderson and Hunter 1981; Henderson 2006; Stern and Gerber 2009); an impure source of sand sometimes introduced high levels of iron and alumina in medieval glass. Natron glass with low potassium levels was therefore being extended with the addition of wood ash glass (see Fig. 4.5). Although wood ash could have been added to natron glass, a more controllable process would be the mixture of wood ash and natron glasses. The resulting glass would have more predictable working properties.

The effect of a change from natron to wood ash must have made a great impact on glassmaking technology. Just as in the Middle East when plant ash almost totally replaced natron, the glassmakers and glassworkers must have performed a series of experiments to determine the working properties of the transitional (mixed) natron-wood ash glasses and also the new wood ash glasses.

A change from natron to wood ash would probably have been even more challenging than the change from natron to plant ash. Wood ash has highly variable compositions producing glasses with variable working properties (Besborodov

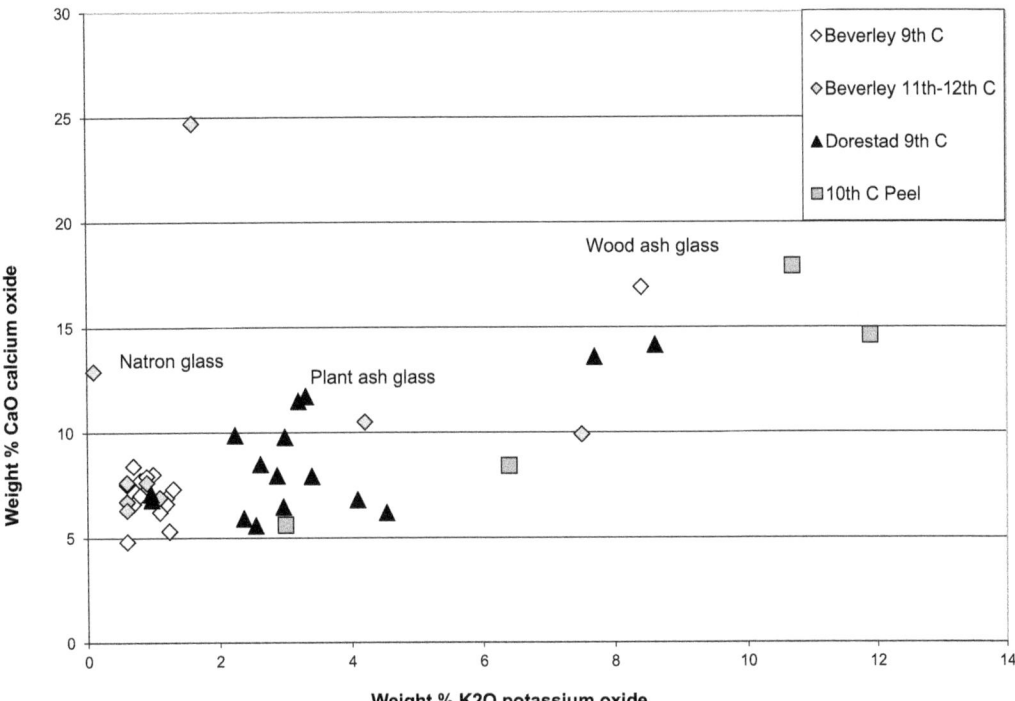

4.5. A bi-plot of weight % potassium oxide versus calcium oxide in early medieval (ninth–tenth century) and high medieval (eleventh–twelfth century) glass from Beverly and Peel, United Kingdom, and Dorestad, The Netherlands. Data from Sablerolles and Henderson 2012 and Henderson 1993a.

1975; Sanderson and Hunter 1981; Henderson 1988a; Jackson and Smedley 2004). As with plant ashes, a number of factors determine wood ash compositions: the tree species, the geology in which the tree is growing and the time of the year that the wood is gathered (Henderson 1988a). Moreover different parts of the tree can have different compositions (see Chapter 2). Initially it must have been a nightmare dealing with an unpredictable glass, especially for glassworkers. The transitional compositions are best interpreted as a period when glassmakers carried out experiments to test the properties of the new alkali source used to make glass. Unfortunately, there is minimal evidence for eighth- to ninth-century glass furnaces. That at Glastonbury is an exception (Bailey 2000). However, it is not clear what kind of glass was being made or worked there.

The change would have been especially marked when it came to blowing the glass, including the production of muff window glass, when the ways in which the glass behaved was critical for success. The total and relative alkali levels, as well as their balance with more refractory compounds such as silica, calcium oxide and alumina, would have determined the temperatures at which glasses became soft and the length of time that they could be worked.

4.1.7 Wood Ash Glasses and the Use of Other Alkali-Rich Plants (c. A.D. 800–1700)

As already noted, the chemical compositions of wood ashes vary widely. Nevertheless, wood ash glasses can be characterised by elevated levels of potassium oxide, magnesia, phosphorus pentoxide, sulphur trioxide, manganese oxide and calcium oxide (see Table 4.4); ashes may also introduce traces of zinc, rubidium, strontium, copper, nickel and barium (Brill 1999b, section XI; Wedepohl 2000; Fleury *et al.* 2002; Brill and Pongracz 2004, table 9; Henderson 2006, table I). Attempts to distinguish between the chemical compositions of ashes of different tree species using atomic absorption spectroscopy (Sanderson and Hunter 1981) showed that there were major overlaps between them (see Chapter 2). Despite the compositional variations of the ashes used, scientific analysis of wood ash glass from different medieval furnace sites in England shows that they can be distinguished from each other (Meek *et al.*, 2009). The compositional distinctions appear to be regional, with glass derived from groups of furnaces, such as in the Weald in southern England, being distinguishable from furnace glass in Staffordshire (Little Birches) and northern England (Hutton and Rosedale), (Meek *et al.*, 2009, fig. 2; Meek *et al.* 2012). Therefore, although the chemical compositions of raw furnace glasses result from a combination of the raw material types, the geological/geographical sources of the raw materials, the techniques used to prepare them, the melting regime in the furnace and, potentially, the proximity of the glass to the crucible wall (Merchant *et al.* 1997; Paynter 2008), it is still possible to identify differences between glasses melted in furnaces *within the same region* (e.g. the Weald). The most likely interpretation of this is simply that the homogenising affect of the melting process produces a *melt composition* and each melt is a reflection of the particular factors (listed earlier) affecting the furnace conditions at the time. This is despite the fact that the chemical compositions of some ashed plants from the same location can even vary from year to year (Jackson *et al.* 2005). Such constrained compositional groupings are presumably what Brill and Pongracz (2004, 131) have identified as being reflected in a 'glazing campaign' with respect to the production and insertion of stained glass windows of Saint-Jean-des-Vignes in Soissans, France. Barrera and Velde (1989) noted chemical groups of French medieval glass vessels that were both site and regionally specific, but in the end, an isotopic approach may be one of few ways of provenancing such glasses with any degree of confidence (Meek *et al.* 2012). It would be interesting to be able to compare the chemical compositions of glasses that result from several different melts from the same medieval wood ash furnace: there is no absolute guarantee that two different melt compositions from the same furnace would be the same. However, the isotopic results show that there are regional distinctions that are likely to

Table 4.4. Examples of European medieval/postmedieval compositions
(weight % oxide)

Country	United Kingdom	United Kingdom	United Kingdom
Site	Lincoln	Lincoln	Lincoln
Date	16 C	17 C	17–18 C
Description	Green lamp	Goblet	Bottle
Characteristics	Wood ash	Low alkali-high calcium	Mixed alkali
Na_2O	1.2	5.5	8.6
MgO	8.1	5.1	1.7
Al_2O_3	1.5	3.0	1.0
SiO_2	49.0	63.5	69.7
P_2O_5	5.7	1.7	0.5
SO_3	0.1	0.2	0.2
Cl	0.4	0.5	0.6
K_2O	13.8	3.9	11.5
CaO	17.7	15.6	5.4
TiO_2	0.1	0.1	0.1
Cr_2O_3	ND	ND	ND
MnO	1.1	0.1	0.8
Fe_2O_3	0.5	1.2	0.5
CoO	ND	ND	ND
NiO	ND	ND	ND
CuO	ND	ND	ND
ZnO	ND	ND	ND
As_2O_3	0.1	ND	ND
Sb_2O_3	ND	ND	ND
SnO_2	ND	ND	0.1
BaO	0.2	0.1	0.1
PbO	0.1	0.5	0.2
Source	Henderson 2006	Henderson 2006	Henderson 2006

ND = Not Detected.

reflect contrasting soil geochemistries in which the trees grew that were used for making the glass.

A second explanation for the distinctions in chemical compositions between some furnace glasses is that alkali sources other than wood ash were used. Beech and oak ash were used, but it is also known that bracken (Besborodov 1975; Jackson et al. 2005) and seaweed (Sanderson and Hunter 1981; Dungworth et al. 2009) were used. Moreover, there is always the possibility that ashes of different tree species were mixed, making it even more difficult to attempt to determine which plant species may have been used. By using experimental archaeology, Jackson and colleagues have produced what they have defined as ideal ash compositions for bracken, oak and beech, leading to glasses baring those compositional characteristics (Jackson et al. 2005). Furthermore, Besborodov (1975, Table 5, 183) has pointed out that there are many other possible sources of alkali-rich plants that could be used for glass manufacture. However, given the degree of compositional variability of ashes, even of the same species (Sanderson and Hunter 1981; Henderson 1985, 1988, 77) it will always be difficult to establish, with confidence, what species of terrestrial or maritime plant have been used to make glass from them. In any case, variations in the geological environment in which plants grow clearly have a dominant effect on the ash composition. The likelihood of being able to identify the species or genus of plant used to make glass based on glass chemical compositions is considered in more detail for Middle Eastern semi-desert plants in Chapters 2 and 11.

Wedepohl (2008) has subdivided the period that 'wood ash' glass was produced into three:

1. A.D. 780–1000: this is a period when the glass had a relatively low potassium level; sodium was added.
2. A.D. 1030–1300: in the northwest, this period saw an enormous production of wood ash glass to feed demand for church and cathedral window glass.
3. A.D. 1300–1500: Wood ash glass continued to be produced and increasingly used to manufacture glass vessels.

Medieval wood ash ('forest') glass technology continued until the seventeenth century north of the Alps in central and western Europe (Velde and Barrera 1986; Henderson 2006). The technology was also in use south of the Alps, but probably to a more limited extent, such as at the site at Germagnana, northern Italy (Mendera 1991). South of the Alps beehive-shaped furnaces were used to fuse glass raw materials. North of the Alps tunnel-shaped forest glass furnaces with a central firing trench and sieges along the side to support the crucibles were used all over Europe (Crossley and Åberg 1972; Lappe and Möbes 1984; Horat 1991). In Charleston's (1978) review of ancient glass furnace types, he distinguished between 'northern' and 'southern' types. There do not appear to be any examples of forest glass furnaces south of the Alps, but by the sixteenth century, when Italian glassmakers migrated to northern countries (Charleston 1984; Henderson 2006; Willmott 2002, 10), they took their furnace technology with them and built beehive-shaped 'southern' furnaces in northern Europe.

By the sixteenth century, in addition to the high potassium glasses, a high lime-low alkali glass had been introduced (Hartmann 1994; Mortimer 1995; Schalm et al. 2004). This glass type was commonly used for the manufacture of glass bottles that were in increasing demand; by the late seventeenth century, apothecary bottles (type VII in Figures 4.6–4.9). Both wood ash glass and the low alkali-high lime wood ash glasses were evidently made using sand of a rather poor quality with high levels of iron, aluminium and titanium oxides (Henderson 2006, table 4; and Figs. 4.8 and 4.9). It is clear that low alkali-high lime glasses

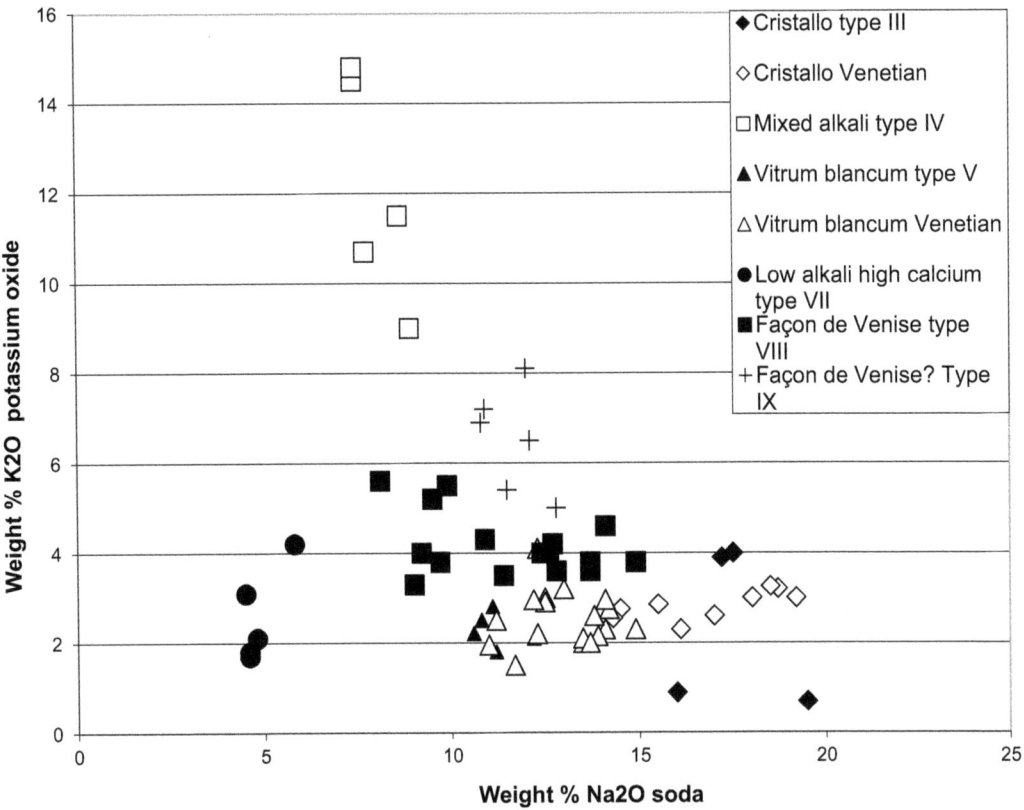

4.6. A bi-plot of weight % soda versus potassium oxide in postmedieval (sixteenth–seventeenth century) glass from Lincoln, United Kingdom. *Cristallo* and *vitrum blancum* are glass types made in Venice. The balance are glasses found in Lincoln. The other chemical types are those defined in Henderson 2006.

were manufactured in a range of late medieval (fifteenth- and sixteenth-century) glassmaking centres, such as Gerterode, Limlinggerode, Hüttplatz/Gererode and Wingerode in the Eichsfeld region of central Germany (Hartmann 1994). A third glass type, with mixed alkalies, has been interpreted as having been made from seaweed ashes (Dungworth and Loaring 2009).

By the seventeenth century, there is an historical reference to the use of different sand sources for making glass: in his *Art of Glass* (1662), Christopher Merrett distinguished between fine white sands from Maidstone in Kent (that would contain low iron) and sand from Woolwich in London. Woolwich sands would probably have been used to make 'coarse' green bottle glass referred to in the *Books of Rates and Customs* for the port of Exeter for the years 1610 to 1629 (Allan 1984, 263). By the seventeenth century, in addition to the *cristallo* and *vitrum blancum*, plant ash glasses manufactured in Italy and a range of other compositional types were in use across Europe. A glass type with high calcium oxide levels was used for the production of goblets and römers of typical

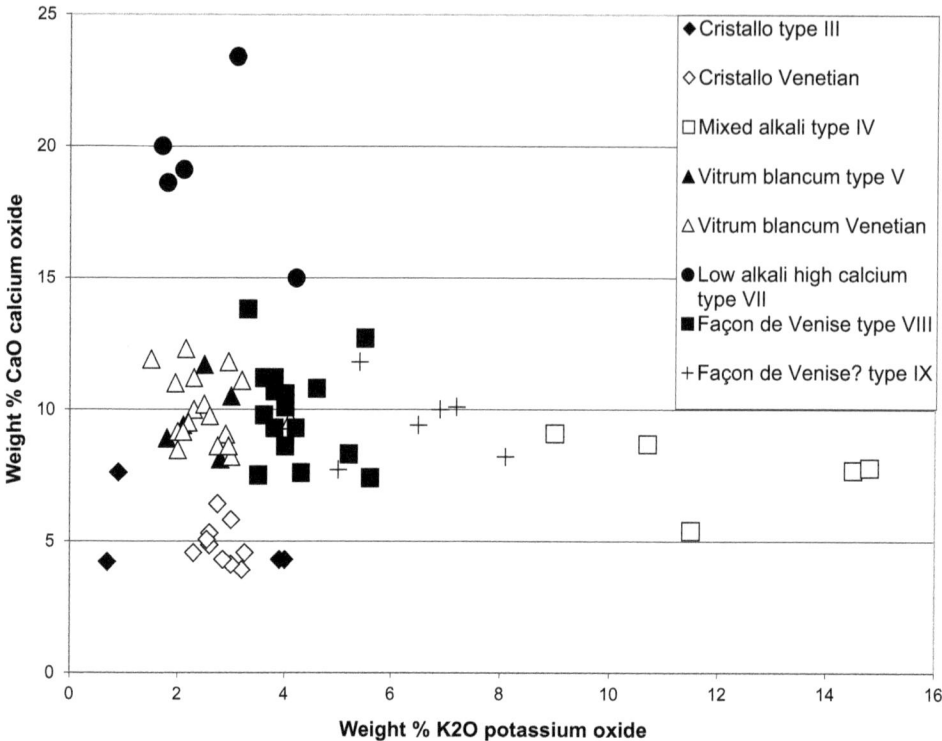

4.7. A bi-plot of weight % calcium oxide versus potassium oxide in postmedieval (sixteenth–seventeenth century) glass from Lincoln, United Kingdom. *Cristallo* and *vitrum blancum* are glass types made in Venice. The balance are glasses found in Lincoln. The other chemical types are those defined in Henderson 2006.

European types (See Fig. 4.7 and Henderson 2006, type VII tables 1, 3 and 4). This is similar to the later high calcium glass (Henderson 2006, type VII) used for the manufacture of apothecary bottles into the eighteenth century. A further mixed alkali glass type, with much higher total alkali levels (Henderson 2006, type IV), was used for the manufacture of goblet forms with bucket shaped bowls of a type ordered from Venice by John Greene of London. Although a clear colourless glass was produced which allowed the colour of the liquid to be observed, its chemical composition was not a typical Venetian one (See Fig. 4.6). It is therefore possible that the raw glass used to make the vessels ordered by John Green was not of Venetian origin, that the vessels were made using a 'second rate' glass type and that Greene was duped by his suppliers.

4.1.8 Lead Oxide-Silica Glass in the West

Ninth-century Muslim glassmakers may be credited with the independent manufacture of lead oxide-silica glasses in the West (Sayre and Smith 1961). Al-Biruni, an Islamic scholar, wrote a description of the production of lead-rich

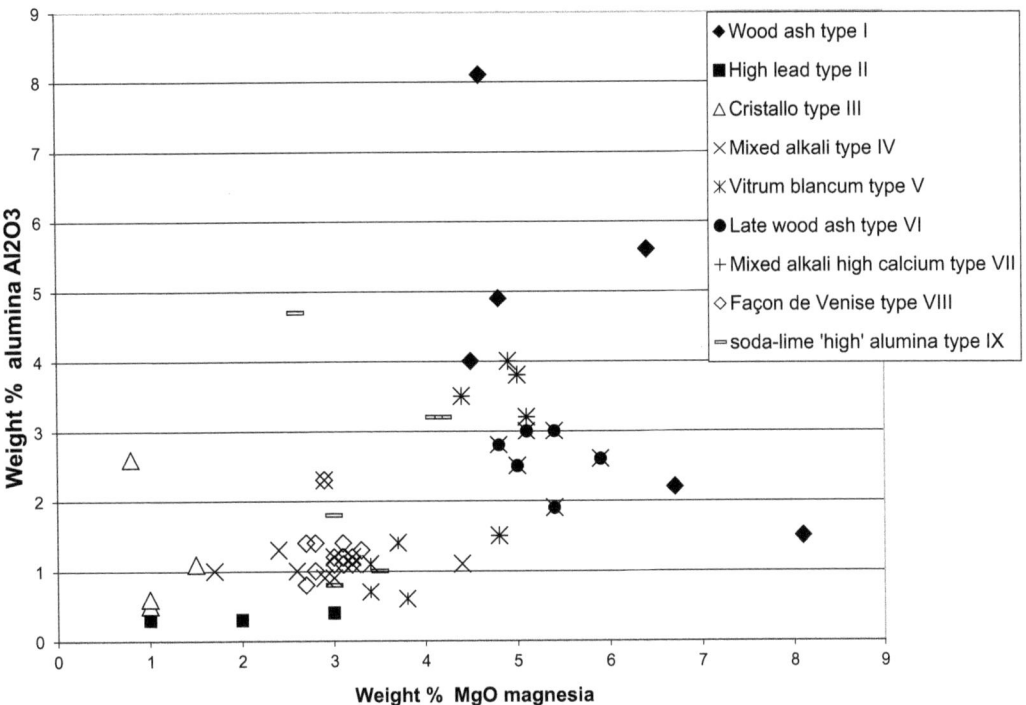

4.8. A bi-plot of weight % alumina versus magnesia in postmedieval (sixteenth–seventeenth century) glass from Lincoln, United Kingdom. *Cristallo* and *vitrum blancum* are glass types made in Venice. The balance are glasses found in Lincoln. The other chemical types are those defined in Henderson 2006.

glazes that dates to 990. However, the earliest examples of lead-silica glasses appear in Southeast Asia (see Section 4.4.7) and are likely to have been introduced in the West by technology transfer along the Silk Road (In-Sook Lee 1993, n. 167) (see Table 4.5).

It is no coincidence that the first large-scale production of lead-rich glazes, along with the large-scale (re-)introduction of plant ash glass, occurred in the ninth-century Islamic world (see Chapter 10). Furthermore, there are Carolingian lead-calcium-silica glass linen smoothers that have been found on several sites in central western France (Gratuze *et al.* 2003, 102). Gratuze *et al.* have made a convincing case that glassy slags produced from lead-silver cupellation (Henderson 2000, 238) from the site of Melle in France were used to make the linen smoothers (Gratuze *et al.* 2002; Gratuze *et al.* 2003).

The relatively low melting point of lead silica glasses would have made it ideal for the manufacture of Islamic glass vessels, but there is currently no evidence for this amongst Middle Eastern compositions. It was, however, used by them for the manufacture of windows and beads. In the West, the glass was used for the manufacture of highly coloured glasses, including emerald green, yellow and red vessels in northern Europe (Wedepohl *et al.* 1995) containing lead oxide

Table 4.5. Examples of western lead oxide–silica glasses (weight % oxide)

Country	Syria	Germany
Site	al-Raqqa	Lübeck
Date	11 C	1–14 C?
Description	Emerald green window	Yellow fragment
Na_2O	0.39	0.05
MgO	0.15	ND
Al_2O_3	1.67	0.6
SiO_2	28.5	30.2
P_2O_5	0.06	0.05
SO_3	0.31	ND
Cl	0.12	0.06
K_2O	0.52	0.13
CaO	0.35	0.06
TiO_2	0.15	0.05
Cr_2O_3	ND	ND
MnO	0.1	0.07
Fe_2O_3	0.34	0.32
CoO	ND	ND
NiO	0.02	ND
CuO	0.82	ND
ZnO	0.2	ND
As_2O_3	ND	ND
Sb_2O_3	ND	ND
SnO_2	0.23	ND
BaO	0.11	ND
PbO	66.07	67.8
Source	Henderson et al. 2004	Wedepohl et al 1995

ND = Not Detected.

levels of between 54.4% and 84%. The glass is softer than soda-lime (plant ash and natron) glass and could be cut and carved more easily. Such glass was used for the manufacture of glass beads, arm rings and finger rings in the west and has been found in a number of Viking age sites in, for example, Russia, Poland and England (Besborodov 1957; Besborodov 1975; Dekówna 1979; Henderson 1986; Bailey 1987; Bailey 2000, 139). The recent discovery of a thirteenth-century glass furnace at Sigtuna in Sweden where lead-silica glasses were worked and possibly made (Henderson and Faber 2008) shows that its use extended into Scandinavia.

In 1670, Ravenscroft obtained a patent for high lead ('flint' glass) manufacture in England (Charleston 1984, 109). Flint glass and lead-rich crystal glass were in use in England and central Europe in both the seventeenth and eighteenth centuries (Brain 2002; Kunicki-Goldfinger et al. 2005). Examples of both lead oxide-silica and lead oxide-silica-potassium oxide glasses were found in a compositional survey of seventeenth and early eighteenth century Lincoln glass (Henderson 2006, tables 1 and 2).

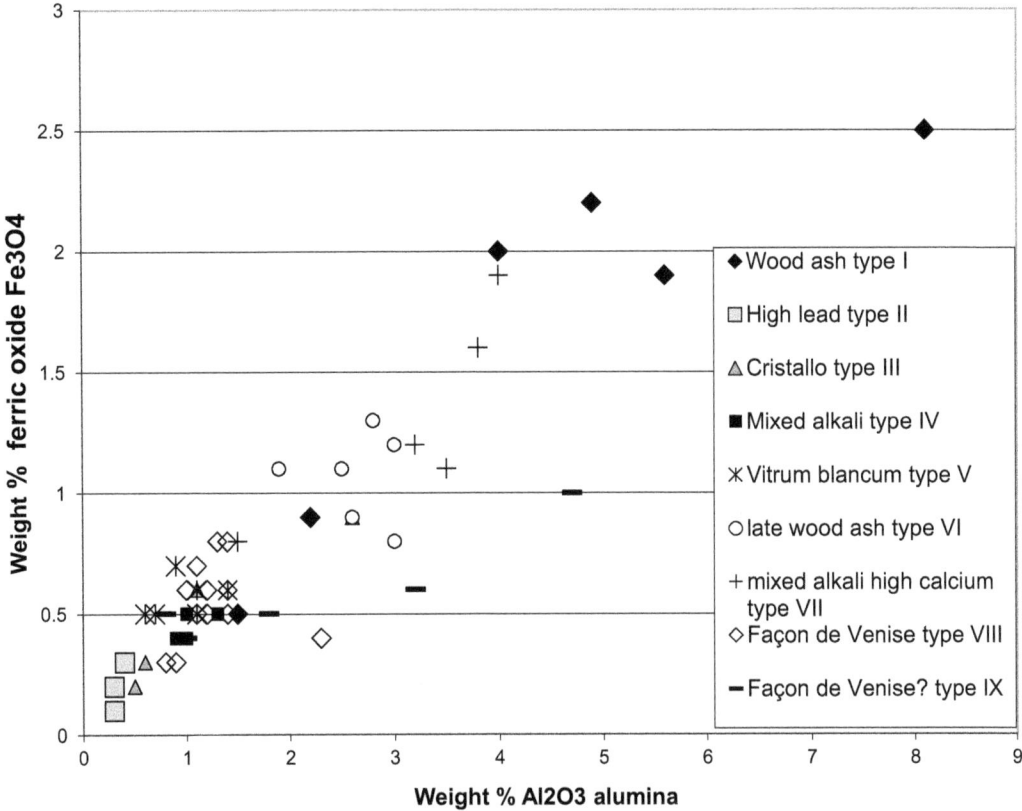

4.9. A bi-plot of weight % ferric oxide versus alumina in postmedieval (sixteenth–seventeenth century) glass from Lincoln, United Kingdom. *Cristallo* and *vitrum blancum* are glass types made in Venice. The balance are glasses found in Lincoln. The other chemical types are those defined in Henderson 2006.

4.2 India, Pakistan and Sri Lanka

4.2.1 The Compositional Types

Compared with the increasingly large compositional databases for Western glasses, there are relatively few published chemical analyses of Indian glasses (Bhardwaj 1987; Singh 1989; Brill 1999b, section XIII). Unfortunately, a number of the glasses for which chemical compositions have been published are poorly dated or have been given a broad date range. This makes it difficult to compare meaningfully such data with better-dated glass from other parts of the world.

The compositional data for Arikamedu glass is an example of this. It dates to between at least the first century B.C. and as late as the tenth century A.D. It was an important trading site and has produced comprehensive evidence for glassworking in the form of raw glass and malformed glass beads (Stern

1987). Stern (*ibid.*, 30) notes that the anonymous author of the Periplous Maris Erythraei, probably written about A.D. 70, refers to large amounts of unworked glass being shipped to the Kerala coast and that it could therefore be the glass found at Arikamedu. Brill (1999b, VIII, E, 337–8) has provided a date range for the samples he published of first century A.D. or later (Brill 1999a, 141). Clearly carefully controlled excavations of the site would have produced reliable dates for the use of each compositional type. Despite this, 22 chemical analyses of glass from Arikamedu are worthy of mention, if only because they exhibit a wide compositional range (compositional analyses of many of these types are given in Tables 4.7 and 4.8). These consist of

1. plant ash glasses
2. potassium- silica glasses with low calcium oxide
3. potassium- silica-lead oxide glasses
4. soda-alumina glasses
5. 'natron' glasses with very low calcium oxide levels

Brill (1987) has noted that Indian plant ash glasses are characterised by high alumina levels (4% and above) and low calcium oxide levels (below 5%). When compared with a large number of Middle Eastern plant ash glasses found at al-Raqqa of ninth- and eleventh-century dates (Henderson *et al.* 2004), this characterisation holds up. Such glasses have been found at Bihar, Kausambi, Rupar and Sar Dheri in Pakistan and in northern India dating to between c. third century B.C. and third century A.D. (Brill, *ibid.*). Brill's (1972, 1989) consideration of Islamic glasses from Afghanistan revealed that they were characterised by levels of potassium oxide exceeding 4% and were probably made there. For example, five of the twelve compositional analyses of vessel fragments from Shahr-i-Banu, north of Tashkurghan, dated between the seventh and thirteenth centuries, contain more than 5% potassium oxide (Brill 1999a, 145; 1999b, 342–3). An earlier publication for glasses of central Asian origin (Besborodov and Abdurazakov 1964) dating to between the eighth and thirteenth centuries that derive from sites near Tashkent, Samarkand, Bukhara, Fergana and Dzhambul include plant ash glasses with potassium levels above 4%. These glasses and others found on sites such as Lou Lan and Pendjikent, also on the Silk Road (Brill 2009a, tables 3.3.1, 3.3.2 and 3.3.5), would seem to form a central Asian glassmaking tradition. It is difficult to suggest the precise reason why they contain higher potassium oxide levels. Amongst the possible reasons are the geology where the plants grew, the plant species used or conceivably the way in which the plants were ashed. All could severally or individually explain it.

Lankton and Dussubieux (2006) have published summary data for a wide range of dated glasses from Asia and provided a comprehensive discussion of these results in a broad cultural context. They report (*ibid.*, 127) that 90% of the data that they considered falls into four main compositional types:

1. soda-alumina glasses ('mineral-soda glass with high alumina and lower level of lime');
2. 'soda-lime glass with a mineral alkali source and lime levels similar to or higher than alumina';
3. plant ash glass ('plant ash glass with a plant ash alkali source'); and
4. potassium (oxide)-silica glasses with low calcium oxide ('potash glass with varying levels of CaO and Al_2O_3').

Lankton and Dussubieux (2006) also mention plant ash glasses with high CaO and Al_2O_3 and mixed-alkali high alumina red and opaque orange glass. In their Table 2, they subdivide their category of 'potash glass with varying levels of CaO and Al_2O_3 into (1) that with moderate CaO and Al_2O_3, (2) that with low Al_2O_3; (3) that with low CaO, and (4) that with higher Na_2O. In this table, they also list "Arikamedu" green, red and black soda glass. It is therefore clear that a range of raw material combinations was used to melt these glasses.

Excavations at Hastinapur, from a phase apparently dating to between 600 and 300 B.C., produced glass with a high potassium composition. Even though the high silica level in this sample suggests that it is a weathered glass, it is unlikely that the potassium level would be significantly lower (Brill 1999b, XIII, 335, sample no. 433). Another sample from the site, also apparently of this date, contains high aluminium and iron (*ibid.*, sample no. 2903). If the dating for it is correct, then an even earlier example of a high alumina (8.8%) glass derives from Rupar in the Punjab dating to between 1000 and 700 B.C. (Brill 1999a, 142, sample no. 2900; Brill 1999b, 333). The discovery of high potassium glasses from Southeast Asia that definitely date to 400 B.C. (discussed later) gives some credence to its early occurrence in India. However, it is difficult to know whether the glass type was all made in India, in Southeast Asia, or in both areas. Another potentially early example was found at Kausambi (Brill 1987). Additional examples of high potassium glasses from Arikamedu are said to date to first–second century A.D. (Lal 1987, 51, table 2).

4.2.2 Soda-Alumina Glass: Technology and Trade (Beads and Bangles)

Evidence for the production of raw soda-alumina glass was found by Sode and Kock on a site dating to the late twentieth century (Kock and Sode 1995; Sode and Kock 2001). Fieldwork was conducted in small glass workshops in villages northeast of Agra in the province of Uttar Pradesh, northern India (Sode and

Kock 2001). The villages lay in an alluvial plain south of the river Ganges. The main location studied was around the town of Firozabad where 400 factories mainly produced glass bangles and beads, which had apparently been made in the area since 1450 (*ibid.*, 157). The area was ideal for glass production because evaporite deposits rich in alkali carbonates – reh (see Chapter 2) – occurred there, and local sand contained lime. So the principal raw materials for glassmaking were at hand. Moreover, when Sode and Kock visited the area in 1994, they discovered the remains of three tank furnaces for the manufacture of raw glass (Kock and Sode 1995, 157). Apparently, in addition to bead and bangle production, bottles and other vessel forms had been blown in the area.

The use of a relatively impure sand source containing high alumina has been put forward as the most likely explanation of the occurrence of between 5% and 15% alumina in ancient soda-alumina glasses (Brill 2001a). Sands from India and Sri Lanka quite commonly contain high alumina levels. In other ways, the sand is similar in composition to granite; it contains low levels of calcium and potassium and high levels of iron, titanium and rare earth elements (Dussubieux 2001; Dussubieux and Gratuze 2003). An example is the sand from the Jamuna River at Agra and Allahabad (Bhardwaj 1987b; Brill 1987, 7). Nevertheless, aluminosilicate impurities in the alkali source could also produce high alumina levels, possibly in addition to or instead of those in the sand (Robertshaw *et al.* 2006, 102, 104). That the alkali reh could be the source of high alumina was demonstrated by Brill (1999b, 481, sample no. 4443), who melted a sample of reh to produce a glass with high alumina (see Chapter 2). If correctly dated, the earliest example of a soda-alumina glass in the form of greyish beads was found at the burial site of Dzharkutan, Uzbekistan, dating to between the sixteenth and tenth centuries B.C. (Abdurazakov 2009, table 8.1). Abdurazakov (2009, 218) is of the opinion that high alumina glasses originated in Bactria.

Dussubieux *et al.* (2008) analysed 347 samples of trade-wind beads and other glass samples from sites in Sri Lanka, west and south India and Africa using LA-inductively coupled plasma emission mass spectrometry (see Appendix). They found that c. 46% were of the soda-alumina composition. The low levels of both potassium (c. 1.2%–2.9%) and magnesium oxide (c. 0.5%–1.7%) levels in the soda-alumina glasses, with soda levels mainly falling between 15% and 20%, suggested to them that a mineral alkali source, such as reh, was used to make them (Dussubieux *et al.* 2008, 2).

The soda-alumina glasses that they found could be subdivided into two subgroups according to relative levels of uranium, barium, strontium and zirconium. The subgroup with low uranium (c. 7–c. 80 ppm) and high barium (c. 400–c. 3000 ppm) dates to between the fourth century B.C. and the fifth century A.D. and is found in southern India and Sri Lanka, with few exceptions.

There is evidence for trade between these areas as early as the third century B.C. However, examples of the subtype have also been found in Southeast Asia (Dussubieux and Gratuze 2003 and see later discussion in this chapter). Evidence for glassworking involving the occurrence of thousands of glass beads, glass fragments including lumps of raw glass (weighing a total of about 54 kg) and crucibles have been found at Kopia in Uttar Predesh, India (Roy and Varshney 1953) now more securely dated to post 200 A.D. (Kanungo and Brill 2009, 16). Although this suggests that glassworking occurred on a large scale at this site, without the discovery of primary glassmaking furnaces and frit the evidence cannot be claimed as being for primary glassmaking. Glassworking evidence has also been found at Giribawa and Mantai on Sri Lanka. Mantai dates to c. ninth through twelfth centuries (pers. comm. Prof. J. Carswell): large amounts of raw glass of a wide range of colours, including opaque orange, opaque red, translucent turquoise, translucent purple and translucent green have been found.

The second subgroup of soda-alumina glass contains high uranium (15–300 ppm) and low barium (100–700 ppm). It was in circulation between Madagascar, East Africa, including Kenyan sites and the west coast of India (including the trading site of Chaul) between the ninth and nineteenth centuries A.D. Because the glass beads analysed have such diagnostic compositions, also reflected in relative levels of strontium and zirconium, it is possible to demonstrate cultural contacts.

Additional soda-alumina glass (with 10.4%–12.5% alumina) has been excavated from Vankali on Sri Lanka (Brill 1999a, 143; volume 2, 335), said to date to c. A.D. 1200–1250. Evidence for the use of soda-alumina glasses is provided by the chemical analysis (using electron probe microanalysis) of ninth-century raw glass from Mantai on Sri Lanka (Henderson, in press). The determination of Nd and Sr isotope ratios in a single sample of raw soda-alumina glass from the site (unpublished) shows that the sand was from a young volcanic source and therefore supports the analytical evidence for the use of granitic sands that occur on Sri Lanka and India and adds to the evidence for its manufacture in this zone. Two 'black' armlets from Samudera-Pasai, twin cities on Sumatra, are of the high alumina-natron composition (McKinnon and Brill 1987, 2, 8, table Iiii). Samudera was an Islamic trading site dating to at least c. thirteenth century, when Marco Polo visited it, and the sixteenth century, when it was destroyed.

India is one possible source for the recent discovery of 'black' armlets of Mamluk and Ottoman dates discovered at two sites in central Jordan: Tell abu Sarbut and Khirbat Faris. Moreover, their compositions suggest that they were made from a mixture of high alumina natron glass and plant ash glass (Boulogne and Henderson 2009); see Figure 10.12.

4.3 Africa and Madagascar

4.3.1 The Compositional Types

Some of the earliest scientific analyses of African glass (beads) using neutron activation analysis and X-ray fluorescence formed part of Claire Davison's Ph.D. thesis (1972). More recent scientific analyses published by Brill (1999b, 410–11), Lankton and Dussubieux (2006), Dussubieux *et al.* (2008) and Robertshaw *et al.* (2003, 2006) have allowed Davison's results to be interpreted in a more comprehensive way.

Natron-alumina glass has been found on Madagascar (Robertshaw *et al.* 2006) and confirmed to be of the high uranium and low barium subtype in glasses found on East African sites (Dussubieux *et al.* 2008). The glass type has also been found on South African sites (Robertshaw *et al.* 2003; Robertshaw *et al.* 2010). For the ninth century and later examples, it is notable that the glasses contain low manganese oxide levels of between c. 200 and 600 ppm (Robertshaw *et al.* 2006, table 1) with generally slightly higher levels in the examples published by Dussubieux *et al.* (2008, table 3) of between 400 and 800 ppm. This is a technological contrast with some Middle Eastern Islamic plant ash glasses, which tend to contain a minimum level of 0.1% manganese oxide (see Chapter 10).

The chemical analysis of 29 glass beads from Mahilaka and two from Sandrakatsky on the island Madagascar dating to between the ninth and fifteenth centuries A.D. revealed that plant ash and natron-alumina glasses were used to make them (Robertshaw *et al.* 2006). The plant ash glasses contained higher levels of calcium and magnesia, and some contained higher levels of strontium (*ibid.*, figs. 2–4) than the natron-alumina glasses. All three of these components are generally thought to have been introduced in the plant ashes (see Chapter 2). The claim (*ibid.*, 98) that the Madagascar plant ash glass beads are distinct from 'Middle Eastern and Egyptian' examples in terms of their manganese and alumina levels (*ibid.*, 99–100) is not absolutely true. Although some contain MnO levels as low as c. 200 to 600 ppm, a number also contain comparable levels to those from the Middle East. As noted elsewhere (Andreescu and Henderson 2006, 127) levels of manganese oxide of 0.4% and above appear to represent deliberately added levels. Nevertheless, manganese impurities below this level characterise Middle Eastern glasses, and these are present in the Madagascar plant ash glasses, too. Another suggested difference between the two geographical sources of plant ash glasses is the alumina levels. Although it is true that most plant ash glasses analysed from al-Raqqa, for example, contained lower alumina levels, eight contained levels of 4% and above, with one containing 5.8%. Moreover, one of these 'high alumina' glasses was a raw furnace glass, and so clearly glass with high alumina compositions was made there (see Chapter 10; Henderson *et al.* 2004). Of the nine lead-rich plant ash glasses from

Madagascar, only two contain alumina levels of above 5.8%. These alumina levels are not necessarily positively correlated with the published levels of neodymium (Robertshaw *et al.* 2006, table 1), which, for the higher alumina glasses, is somewhat unexpected because both would be expected to derive from the sand. However, if we consider the relative levels of calcium and aluminium oxides, with only a few exceptions, a distinction between Middle Eastern on the one hand and Madagascan/Indian on the other does hold up as originally suggested by Brill (1987; see Fig. 4.10).

Another approach taken to the Madagascan data is a comparison of rare earth anomalies (see Chapter 3; Robershaw *et al.* 2006, 104–5). They discovered clear differences in the Ce and Eu anomalies between the soda-alumina and plant ash glasses from the site. Their results indicated that the main variation was along the igneous rock trend with three groups of results falling around the reference sample for granite, basalt and intermediate igneous rocks and diorite and andesite, respectively. It is suggested that the granite values may result from several regions in South Asia and Sri Lanka, the intermediate igneous rock values from Indonesia and the basaltic values from west-central India. However, a more rigorous interpretation will result from testing far more samples. A good start would be to apply it to well-dated raw glasses from primary glassmaking factories. Then at least the production centre is known and the use of raw materials from a constricted range of possible sources (including local) with associated REE values would be involved. The beads tested by Robertshaw *et al.* (2003) could have derived from a range of sources, as the authors noted. Another, compositional type that is apparently only found in West African glass is high lime-high alumina (HLHA) glass.

4.3.2 High Lime–High Alumina Glass

This type of glass contains between c. 11% and 31% calcium oxide, between c. 11% and 18% alumina, between c. 3% and 8% potassium oxide and between 0.5% and 7% soda (see Table 4.6). Its earliest occurrence is at sites in Ile-Ife (southern Nigeria) where other technological developments occurred and where there was an oral tradition for glassworking (not glassmaking). Here crucible fragments, blocks of dichroic glass and glass beads were found. Davison's (1972) PhD thesis, included HLHA glass compositions, including results for glass attached to crucible fragments. The earliest date for the HLHA glass in the ninth century (Lankton and Dussubieux 2006, 126), and its use continues as late as the thirteenth century. Currently, this unusual glass type is mainly known from West Africa. Its first appearance in the ninth century is probably no coincidence: just as in the West and in the Middle East, where glass technologies underwent a transition from natron to plant ash types (Henderson 2002), it is apparent that such a transition affected Africa too.

Table 4.6. The composition of high lime–high alumina glass (weight % oxide)

Country	Africa
Site	Igbo Ukwa
Date	c. ninth century
Description	Olive bead
Characteristics	High calcium and aluminium
Na_2O	3.33
MgO	0.12
Al_2O_3	12.66
SiO_2	59.35
P_2O_5	0.31
SO_3	0.05
Cl	ND
K_2O	3.3
CaO	16.48
TiO_2	0.07
Cr_2O_3	ND
MnO	0.07
Fe_2O_3	3.09
CoO	ND
NiO	ND
CuO	ND
ZnO	0.009
As_2O_3	ND
Sb_2O_3	ND
SnO_2	ND
BaO	ND
PbO	ND
Source	Brill 1999b, XVIC no. 5551

ND = Not Detected.

4.4 China and Southeast Asia

4.4.1 The Compositional Types

Early Chinese glasses are compositionally distinctive because the principal alkali used was potassium rather than soda, and also because barium oxide is found in them at significant levels (see Table 4.7). The use of barium is not found anywhere else in the world. Moreover, there was an early production of high lead glass in China. Examples of Chinese barium-lead, lead-silica, high potassium and soda glass with high alumina are given in Table 4.7.

Although China was technologically highly advanced, and in many ways far more so than the West from an early period, indigenous glass technology appears to have started somewhat later than in the West (Hsueh-Man 2002 and 2002a; Gan 2009a). This can probably be explained by the fact that jade

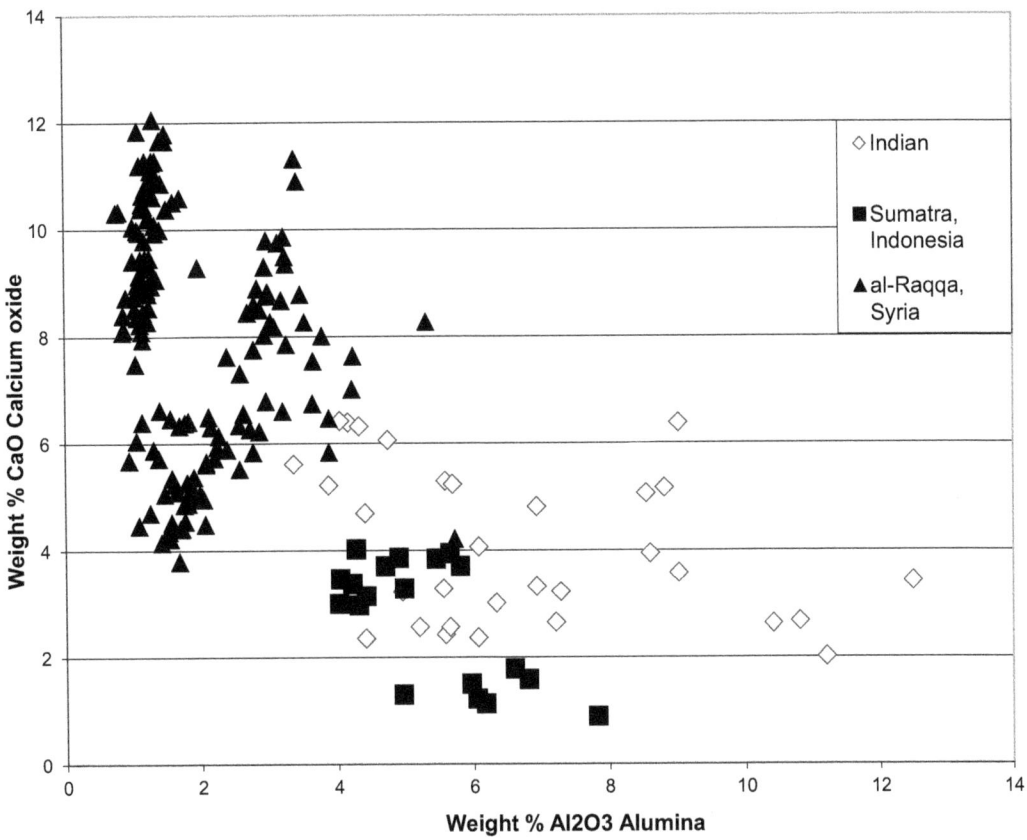

4.10. A bi-plot of aluminium versus calcium oxides in glasses from India, Sumatra and northern Syria (al-Raqqa). Data from Brill 1987, McKinnon and Brill 1987, Henderson et al. 2004. The positively correlated group of al-Raqqa glasses containing between c. 2% and 4% alumina are of the experimental type.

was used as a translucent coloured material for the manufacture of jewellery instead of glass, and vessels were made out of wood or metal. In the medieval period glass was often associated with wealthy Buddhist monasteries, was given a uniquely prestigious status and was treated as a rarity (Hsueh-Man 2002a, 71, 73). Nevertheless, as Brill (1991a, 34) has pointed out, the turbidity of early barium-rich Chinese glasses had a similar appearance to jade, and it might therefore be suggested that the glass was an attempt to copy jade or at least increase the amount of translucent materials that had a jadelike appearance in circulation. The earliest glass from China seems to date to c. 1000 B.C. One of the earliest wide-ranging reviews of Chinese glass compositions was edited by Brill and Martin (1991) with more detailed reviews by Gan Fuxi (Gan 2005; Gan Fuxi et al. 2006, Gan Fuxi 2009a, 2009b) and Li Quinghui et al. (2009b). The Silk Road played an important part in the transmission of materials, leading

Table 4.7. Examples of glass compositional types found in China (weight % oxide)

Country	China	China	China	China
Site	Ka-La-Ke'er	Pingba	?	Chaoyang
Date	420–907 A.D.	6 C A.D.	7–9 C A.D.	7–9 A.D.
Description	Opaque green fragment	Bead ?green	Emerald green fish	Green bead
Characteristics	Mixed alkali-High calcium	Potassium-silica-lead	Potassium-calcium	Lead-potassium
Na_2O	9.94	1.78	1.24	0.29
MgO	2.82	1.1	4.33	0.1
Al_2O_3	4.25	2.41	1.25	ND
SiO_2	60.73	47.91	62	26.32
P_2O_5	ND	ND	ND	ND
SO_3	ND	ND	ND	ND
Cl	ND	0.66	ND	ND
K_2O	8.76	7.36	13.6	10.09
CaO	9.06	ND	15.5	0.13
TiO_2	0.65	ND	0.1	ND
Cr_2O_3	ND	ND	ND	ND
MnO	0.03	ND	0.014	ND
Fe_2O_3	1.29	0.1	0.95	0.16
CoO	ND	ND	ND	ND
NiO	ND	ND	ND	ND
CuO	1.78	0.63	0.55	0.13
ZnO	ND	ND	0.3	ND
As_2O_3	ND	ND	ND	ND
Sb_2O_3	ND	ND	ND	ND
SnO_2	ND	ND	0.08	ND
BaO	0.52	ND	0.01	ND
PbO	0.07	36.68	0.05	50.31
Source	Li Quinghui et al. 2009	Gan Fuxi 2009a	Brill et al. 1991a, no. 3343	Gan Fuxi 2009a

ND = Not Detected.

to the discovery of both Roman and Sasanian glass vessels in China (Braghin 2002).

A recent consideration of the earliest Chinese glass and its compositional relationship with Western glasses was published by Gan Fuxi et al. (2006). Brill (1991, 32–41; table 4) identified a range of glass compositional types: lead-barium glasses (see section 4.15); potassium-silica (see section 4.16); potassium-lime glasses (see section 4.17); mixed-alkali glasses (see section 4.5.6), lead-silica glasses (see section 4.5.7) and soda-lime glasses. Gan Fuxi (2009a, 8) has divided the development of Chinese glass compositions into five stages:

1. c. 800–c. 400 B.C. Potassium-lime-silica glasses
2. c. 400–c. 200 B.C. Barium-lead-silica and potassium-silica glasses
3. c. A.D. 200–c. 700 Lead-silica glasses
4. c. A.D. 600–c. 1200 Potassium-lead-silica glasses
5. A.D. c. 1200–c. 1900 Potassium-calcium-silica glasses

Table 4.8. Examples of glass compositional types from Southeast Asia (weight % oxide)

Country	China	Inner Mongolia	Korea	China
Site	A'Ke-Si-Pi-Li castle	Zhalainuer tomb	Zhongyang dong	Hazar-tam or Saqal-tam
Date	206 B.C.–A.D. 220?	2–1 century B.C.	100 A.D.	2?–6 century A.D.
Description	Eye bead: green	Green bead	Blue bead	Aqua vessel
Characteristics	Barium-lead	Lead-silica	Potassium glass	Soda high alumina
Na_2O	3.36	7.72	0.89	16.31
MgO	ND	ND	0.42	2.79
Al_2O_3	6.94	3.22	3.48	10.08
SiO_2	57.04	49.91	73.47	57.43
P_2O_5	0.98	ND	ND	0.55
SO_3	ND	ND	ND	ND
Cl	ND	ND	ND	ND
K_2O	0.51	0.49	14.9	4.8
CaO	3.41	2.91	1.42	5.78
TiO_2	ND	ND	ND	0.15
Cr_2O_3	ND	ND	ND	ND
MnO	ND	ND	ND	0.06
Fe_2O_3	0.84	0.37	2.38	1.94
CoO	ND	ND	ND	ND
NiO	ND	ND	ND	ND
CuO	1.08	0.26	0.62	0.01
ZnO	ND	ND	ND	ND
As_2O_3	ND	ND	ND	ND
Sb_2O_3	ND	ND	ND	ND
SnO_2	ND	ND	ND	ND
BaO	5.42	0.08	ND	0.08
PbO	23.77	35.04	0.01	0.02
Source	Li Quinghui et al. 2009	G Fuxi 2009a	G Fuxi 2009a	Brill 1999b, XIVB no 6122

ND = Not Detected.

Chinese *soda-lime glasses* of both natron and plant ash compositional types date to the eastern Zhou dynasty (1050–221 B.C.; Gan Fuxi 2005, table 14.6). Brill showed that soda-lime glasses were in use between c. 500 and 800 and c. 1600 and 1900 A.D. (Brill 1991a, table 4). He also published compositional results for Chinese potassium-silica glasses (Brill 1999b, 360*ff.*; see 13), and late examples of lead-potassium glasses (Brill 1999b, 356, nos. 1570, 1563, 1560 and 4102). Lead-soda glasses were not included in Brill's comprehensive survey (see section 4.5.5).

Soda-alumina glass was used in Southeast Asia from the mid-first millennium B.C. and into the late first millennium A.D. in a range of countries such as Sumatra from the sixth century A.D. until at least the fourteenth century (McKinnon and Brill 1987) but also in a range of other places as far east as Japan (Brill 1973; Brill 1995; Brill *et al.* 1995, Dussubieux and Gratuze 2003; Lankton and Dussubieux 2006); See Table 4.8. In Korea, it was used between the second century B.C. and the sixth century A.D. In recent times high alumina glass was being made in Uta Pradesh (Sode and Kock 2001). Granitic sands were available in Southeast Asia and efflorescent salts were available in South Vietnam. Malleret (1962) suggested that this glass was manufactured using these raw materials at Oc Eo, although there is no proof of primary glass production there (Dussubieux 2001, 207). Nevertheless, in combination these raw materials could be used to make high alumina glass. Lankton and Dussubieux (2006, 134) noted that high alumina blue, opaque red, opaque orange and 'black' (presumably deep translucent brown, purple or green) glass compositions could be subdivided according to their colour.

Examples of plant ash (soda-lime) glasses assumed to have been imported from western Asia along the early Silk Road dating to the spring and autumn periods of the western Zhou dynasty (as early as 1000 B.C.) have been found in the Kiziltur cemetery, Xinjang in western China (Gan Fuxi 2009b, 64 and 67, Table 2.4). Other examples have been found in Southeast Asia dating to the second century A.D. and in eleventh to sixteenth century contexts in Sumatra (McKinnon and Brill 1987). Because halophytic plants are widespread, the possibility that production of plant ash glass occurred in these areas cannot be excluded. Some of the plant ash glasses from Kota Cina (Sumatra) have typical Middle Eastern compositions, others are also characterised by low calcium oxide levels and a third subtype have an unusual balance of magnesia and potassium with less than 1% magnesia but c. 2.2% to 2.3% potassium oxide (McKinnon and Brill 1987, table 1ii). The latter also have low calcium oxide levels, suggesting that they have been made from purified plant ashes (*ibid.*, 6).

Until recently, no natron glasses found in South and Southeast Asia could be dated to before the first century B.C. Dussubieux and Gratuze (2003, 68) have pointed out that Asian natron glass contains ten to twenty times more uranium oxide than that found in the Mediterranean so the primary production of natron glass in Southeast Asia is a possibility. Examples of natron glass found in China almost certainly made in the Middle East – probably the Levant – have been published by Gan Fuxi *et al.* (2009).

There is limited research into the occurrence of opacifiers in Chinese vitreous materials. However, opacifying materials were used that are unique to Southeast Asia. By the Tang dynasty (618–907), calcium fluoride (fluorite, CaF_2) was used to produce an *opalising* effect in imitation of jade and this material continued to be used much later as an opacifier in sixteenth- and seventeenth-century cloisonné enamels on metalwork (Henderson *et al.* 1989). Fluorite, criolite and sodium fluorosilicate were used as a means of introducing fluorine as an opaliser in nineteenth- and twentieth-century glasses (Turner and Rooksby 1959, 27, table 1; Moretti and Hreglich 2005). Turbidity (not the same as opacity) has been observed in soda-lead-barium glass (Brill 1991b, 34, fig. 19) and is probably due to (localised) devitrification.

4.4.2 Lead–Barium Oxide Glasses

This glass type was a purely Chinese innovation. It dates to the Warring States, Han and Eastern Han periods in eastern and central China (Beck and Seligman 1934, 982; Seligman *et al.* 1936; Gan Fuxi *et al.* 1978, 99; Zhang Fukang 1983; Meiguang *et al.* 1987; Jiazhi and Xianqiu 1987; Gan Fuxi 2009a, table 1.6) and is thought to have been made in central China (Gan Fuxi 2009a, 25). A number of results, published by Gan Fuxi (2005, table 14.7), have lead oxide contents of between c. 18% and 44% and barium oxide contents of between 6.6% and 21.5%. Alumina levels are variable and can be as high as 9.5% and are generally correlated with elevated iron oxide levels, presumably due to the use of an impure sand source. Magnesium and calcium oxide levels are low; the contents of potassium and sodium oxides are also generally negligible. Nevertheless, three published glass compositions contain between 5.29% and 9.3% soda, so these can be classified as soda-lead-barium glasses. The turbidity that is visible in these glasses can be ascribed to the presence of barium disilicate crystals, apparently the result of devitrification (Brill 1991b, 34, fig. 19). The mineral barite, barium sulphate, does not melt well; a more likely raw material to have been used is barium carbonate; its mineral form is witherite. Barium-rich glasses also occur in other parts of South-east Asia such as in Japan and Korea (In-Sook Lee 1993, 164–166).

4.4.3 Potassium-Silica Glasses

The chemical compositions of some of the earliest Asian and Southeast Asian potassium glasses are essentially potassium-silica, characterised by negligible soda (<1%), low levels of magnesium and calcium oxides (mainly 2% or less), variable levels of alumina (0.8%–6.5%) and relatively high iron of up to 3% (Shi Meiguang *et al.* 1986; Brill 1999b, 360–2; Gan Fuxi, 2005, table 14.8; Gan Fuxi 2009a, table 1.4; Li Qinghui *et al.* 2009b, 402, 407, table 21.5). High levels of alumina and iron (also found in high barium glasses) suggest that an impure sand was used to make them. The earliest high potassium glasses from Southeast Asia date to Warring States sites (475–221 B.C.; Gan Fuxi 2009a, tables 1.4 and 1.6; Shi Meiguang *et al.* 1986; Gan Fuxi 2009b, table 2.6; Li Qinghui 2009b, 402, 407, 409). Others have been found in Western Han (206 B.C.–A.D. 25) and Eastern Han (A.D. 25–220) contexts. Examples of the latter are some from Fengmen-ling M26, Guangxi associated with mixed-alkali glass beads (Li Qinghui 2009b; Li Qinghui 2009b, table 21.5).

Zang Weiyong (2005, tables 10.1 and 10.2) has also published chemical analyses of potassium glasses with low impurities from the Boshan glass factory, and there are others, perhaps dating to the Yuan dynasty (Yi Jialiang and Tu Shujin 1991, 99). There are later Chinese examples that date to eighth-twelfth centuries A.D. (Brill *et al.* 1991, table 2, no. 5805). Farther west examples have been found at the site of Ban Don Ta Phet in Thailand dating to c. 400 B.C. (Glover 1990; Basa *et al.* 1991, table 3; Glover and Henderson 1995); other Thai examples date to c. A.D. 400 (Basa *et al.* 1991). Examples dating variously to between the first century A.D. and fourth to fifth century have been found in Korea (In-Sook Lee 1993, 168, 171). Therefore, the earliest examples dating to c. 400 B.C. have been found in both China and Thailand.

Without primary archaeological evidence for their manufacture, it is difficult to be certain where the first potassium glasses were made, but a strong contender is the southern Chinese areas of Guangdong and Guangxi (Gan Fuxi 2009a, 25). It is nevertheless significant that potassium-silica (and mixed-alkali) glasses with the same compositional (major and minor oxide) characteristics have been found in Mediterranean sites dating to c. 1000 B.C. (see Chapter 6), and so this could have been a potential source for the Chinese examples (see Fig. 4.11). Any links between the West, central Asia and China require further investigation (see tables 4.2 and 4.7)

So it is currently unclear whether potassium-silica glasses were made in both Southeast Asia and India (In-Sook Lee 2004; In-Sook Lee 2009, 186) or only in one of these areas and traded to the other. This is partly because some of the dates for Indian glasses are uncertain, so it is difficult to determine where the earliest examples come from. In addition to China and Thailand, examples have been found throughout Southeast Asia, including in Japan, Korea and Vietnam (In-Sook Lee 2009, 185). Many of the reported analyses of potassium-silica glasses are of beads that could be easily traded.

4.4.4 Potassium-Lime Glasses

Gan Fuxi (2005, table 14.12) has published early examples of high potassium-high lime glasses, as distinct from high potassium-low lime glasses. The potassium oxide levels range from 12.19% to 23.11% and the calcium oxide levels between 6.31% and 11.36%; the alumina levels vary between 0.06% and 7.52%. Two samples listed constitute a subtype – potassium-calcium glasses with lead oxide levels of 14.35% and 38.57%. Low levels of magnesium in these glasses suggest the use of a pure (mineral) form of potassium, such as saltpetre. Moreover, the chemical composition of sixteenth- and seventeenth-century Chinese cloisonné enamels show that many are of a potassium-lime composition but also with high lead oxide contents (Henderson *et al.* 1989, table 1). Potassium-lime glasses with high levels of lead oxide were therefore produced in the sixteenth century or later.

Lankton and Dussubieux (2006) have subdivided the compositions of high potassium glasses found in South and Southeast Asia according to variable

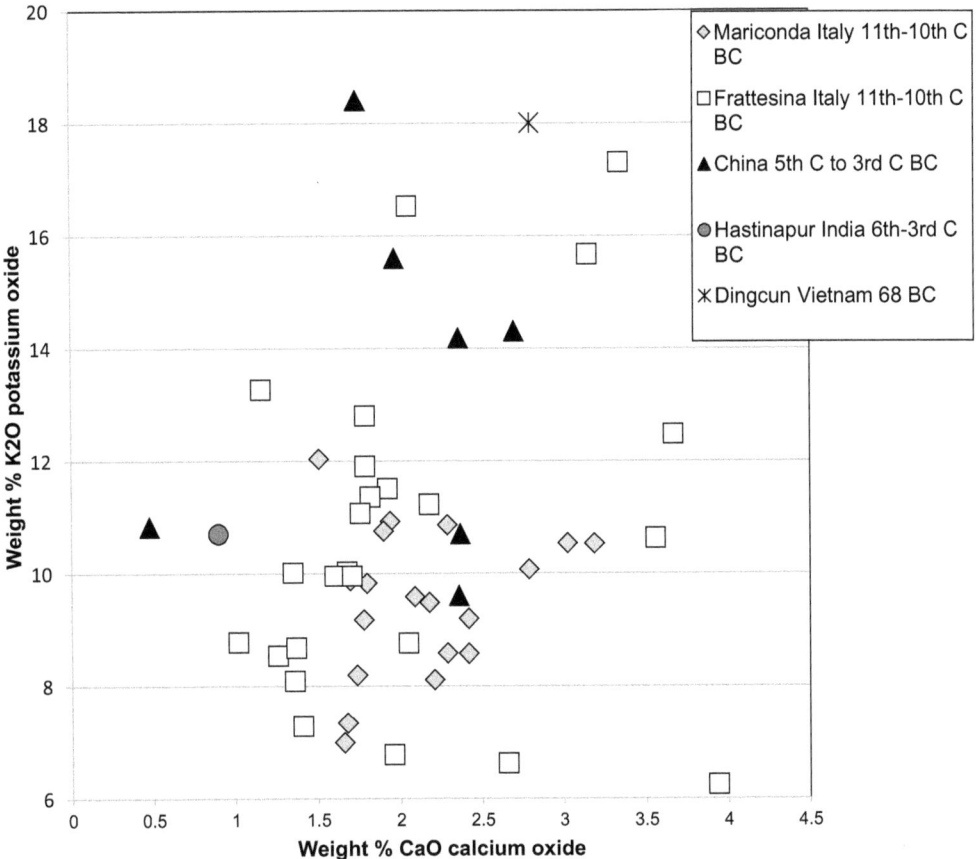

4.11. A bi-plot of calcium oxide versus potassium oxide in high potassium and mixed-alkali glasses from the Mediterranean and China. Data from Gan Fuxi 2009a and Towle et al. 2001.

calcium oxide and alumina levels. The subgroups are characterised by the following:

1. low alumina levels (c. 1%) that date to c. fourth–third century B.C. found at Ban Don Ta Phet central Thailand (Glover and Henderson 1995, 151); other examples date to the time of the Sa Huyna culture in Vietnam with a broad date range of between c. 1000 B.C. and and c. A.D. 100 and
2. glasses containing very low calcium oxide levels of c. 1%. This subgroup is found on Dong Son sites in northeast Vietnam and in Korea, Japan and China. It may date to between the third century B.C. and the fourth century A.D.

4.4.5 Lead-Potassium and Lead-Soda Glasses

The three principal components of these glasses are lead oxide, potassium oxide or soda and silica (Gan Fuxi 2005, table 14.10). They contain low or negligible levels of calcium oxide. A single example of a turbid yellow glass with a lead oxide-potassium oxide composition was found at Kota Cina in Sumatra (McKinnon and Brill 1987, table liv), although no date is given for it.

4.4.6 Mixed-Alkali Glasses in Southeast Asia

Mixed-alkali glasses are taken to be those that contain more than 7% of both oxides. Brill published two examples of this compositional type: one, a bead, dates to fourth–third century B.C. (Brill 1991b, 53 table 1, no. 54) and the other, a hairpin head, dating to between the seventh and ninth centuries (Brill 1991b, 54, table 1, no. 1585). Both of these examples of mixed-alkali glass compositions are distinct from mixed-alkali glasses mainly found in eleventh–tenth century B.C. contexts in Europe. These glasses contain significantly higher levels of both magnesium and calcium oxides and are therefore likely to have been made from a different alkali raw material. Mixed-alkali glasses have also been found in Thailand with these compositional characteristics, but they are partly distinctive because they contain higher levels of soda (11%–14%) than potassium oxide (c.4%) and all are opaque red *mutisalah* bead types, presumed by Glover and Henderson (1995, 152) to be of an Indian manufacture (see Fig. 4.11). Much later, sixteenth- and seventeenth-century Chinese cloisonné enamels in a range of brilliant colours are of a mixed-alkali-lead oxide-calcium oxide composition (Henderson *et al.* 1989).

4.4.7 Lead Oxide-Silica Glasses

This is a distinct compositional type with most glasses containing lead oxide levels above 35% and several examples containing 50% to 70%. The principal other component is silica (Gan Fuxi 2005, table 14.10; Gan Fuxi 2009a, table 1.8). This lead-silica glass is therefore basically of the same composition of that found in the West from c. A.D. 800. The earliest Chinese examples date to after the Han period, perhaps as early as A.D. 600, the Chinese Tang dynasty (618–907; Gan Fuxi 2009a, table 1.8). Lead-silica glasses have been found in contexts dating to the Kofun and Nara periods in Japan (A.D. 250–538 and 710–794, respectively). Two Korean samples date to between the first century B.C. and first century A.D. and a third, well-dated sample dates to the seventh century A.D. or earlier (In-Sook Lee 1993, 166–7, 171). Therefore, there is no doubt that the earliest lead-silica glasses come from Southeast Asia and only appeared in the west around A.D. 800, presumably as a result of technology transfer along the Silk Road. Where it was first made in Southeast Asia is still to be confirmed.

FIVE

EARLY GLASS IN THE MIDDLE EAST AND EUROPE

INNOVATION, ARCHAEOLOGY AND THE CONTEXTS FOR PRODUCTION AND USE

Chapters 5 and 6 are the first of three pairs of chapters. Each pair deals with specific case studies. The first chapter in each pair deals with the archaeological, social, economic, political, ritual and historical background; the second deals with the scientific research on the glass from the period and area. This chapter focuses especially on two contrasting examples of early glass, in Mesopotamia and Italy. These have been selected because they provide interesting and important evidence for the earliest glass in both areas. The archaeological, social, economic, political, ritual and historical background is provided as an aid to understanding why the glass appeared when it did. Both Chapters 5 and 6 should be read in conjunction with Chapter 1, especially in relation to technological innovation, and with Chapters 2 and 3 on raw materials.

5.1 The Social and Political Contexts of Early Glass Production in the Middle East

The locations of the principal sites mentioned in the text are given in Fig. 5.1.

The earliest glass appeared sometime in c. 2500 B.C. Political centralisation occurred in the late third millennium B.C. under two city-dynasties: Akkad in northern Babylonia in the twenty-fourth and twenty-third centuries and Ur in the south in the twenty-first century. Along with political centralisation, administrative and ideological centralisation also occurred. In the later Sumerian King List the dynasty of Akkad was regarded as a family of city-rulers in Sumer and Akkad. Under Sargon the city-states that characterised Babylonia in the period that preceded it ended temporarily. In the north, the Akkadian kings reached Tuttul (al-Raqqa), the cult centre of Dagan, a focus for northern and central Syria; Mari and Ebla were destroyed. At Nagar (Tell Brak) the Akkadians erected a monumental building stamped with the name of Naram-Sin. 'Court sponsorship led to technological and stylistic excellence' (Van De Mieroop 2004, 65), including the refinement of naturalism in stone and copper sculpture, the latter involving a very early use of the lost-wax technique. The empire of Akkad in northern Mesopotamia ended c. 2190 B.C. In this area, people speaking the Hurrian language founded a state called Urkesh and Nawar. Perhaps we should not be surprised, therefore, that the first true glass appeared during a time of centralisation in Mesopotamia, especially because glass was an elite material forming part of a technological level of excellence.

The Third Dynasty of Ur began c. 2100. At this time, Sumer and Akkad flourished economically with a high level of urbanisation. The state's assets included manufacturing workshops; labour was divided between those who did it all year round and those who did it for half the year. There was greater internal coherence in the Ur III state than in the Akkadian state. The kings became deified, their children became high priests and priestesses in major cults and the cults were established throughout the state. This provided a focus of centralisation. However, local hierarchies survived quite independently. The Ur III state did not reach northern Syria but had diplomatic relations with the area. There was a reduction in the number and size of settlements in northern Syria at the time and there are references to people from Tuttul, Ebla, Urshu – and from Byblos. The Amorites, a Syrian force of semi-nomads, were hostile invaders who contributed to the end of the Ur III state; another group was the Elamites from the east.

The early second millennium saw the growth of a number of competing dynasties, some more successful than others, with accomplished rulers who extended their political control over wide areas. Northern Mesopotamia was ruled by Shamshi-Adad who seized the throne of Assur in 1808 B.C., then Babylonia was seized by Hammurabi, ruling there between c. 1792 and 1750 B.C. In northern Mesopotamia when Mari was captured, Shamsi-Adad ruled an area from Assur on the Tigris to Tuttul on the Balikh. He restored the Ishtar temple said to have been built 500 years before. Shamsi-Adad died in 1776, and the area of northern Syria became a number of small independent states.

Historical sources for the early centuries of the Hittites come solely from the colonies of Assyrian merchants in the region, which consisted of a network of small kingdoms. A central Anatolian state was apparently created by Hattusili I in the early to mid-seventeenth century B.C., making the city of Hattusa his capital. Hattusili I invaded the kingdom of Yamkhad, which, in the mid-seventeenth century, controlled northwest Syria. He sacked Alalakh, although Aleppo was not sacked. The next king, Mursili, sacked Babylon and Aleppo, which created power vacuums: the states disintegrated after the founders' deaths, although 'the changes that these men initiated laid the foundations for the territorial states in these areas in later centuries' (Van der Mieroop 2004, 99).

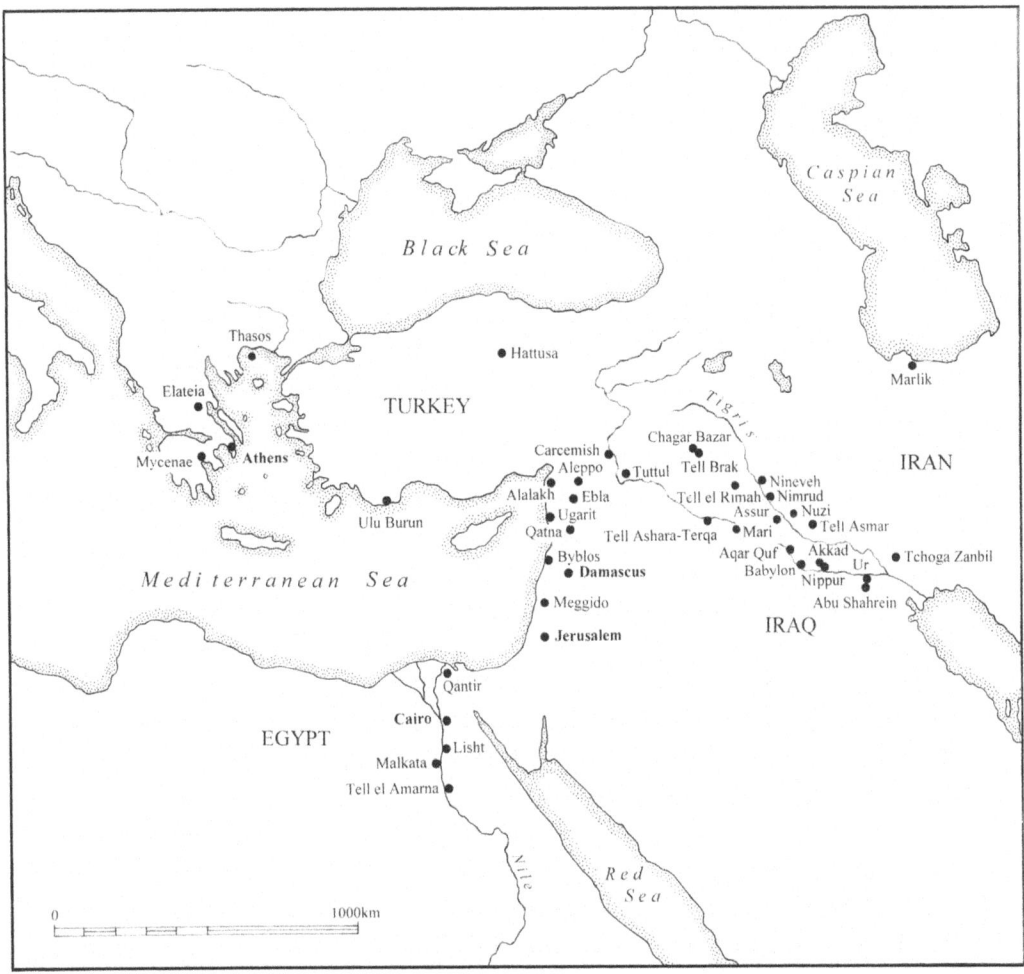

5.1. Locations of principal sites mentioned in the text.

In 1595 B.C., King Mursili of the Hittites sacked Babylon (Porada *et al.* 1992). In Mesopotamia and Anatolia, there then followed c. 200 years of political fragmentation amongst the Hittites from c. 1590 until c. 1400. Van der Mieroop refers to this as the 'Dark Age'. Lack of centralised power led to a discontinuation of administrative and scribal practices.

People with Hurrian names had been present in northern Syria and Mesopotamia since the mid-third millennium. According to written documents, Hurrian minor states existed in north and east Assyria and upper Mesopotamia c. 2200 B.C. (Wilhelm 1995, 1244). Hurrian names of rulers and a substantial percent of the population from the Zagros Mountains to the Mediterranean had Hurrian names. There were a number of small states in northern Syria before the mid-second millennium. In the early second millennium, the area was politically divided between individual states, including Qatna, Yamkhad, Tuttul and Apum. All suffered decline in the sixteenth century. However, when the Mitanni state arose in the fifteenth century the region was unified. The Mitanni invaded the Old Hittite kingdom several times. Their move southwards may have pushed Syrian-Palestinian peoples into Egypt where they formed the so-called Hyksos dynasties of the early sixteenth century. By the early fifteenth century they owned a territorial state called Mitanni, and they were dominant amongst the Hittites and in Kizzuwatna (southwest Anatolia); several rulers of Syrian-Palestinian states had Hurrian names. Moreover, in the early fifteenth century, the state of Mitanni appeared in northern Mesopotamia under the Hurrians. Excavations of 'Hurrian' sites have provided evidence for a social hierarchy as indicated by large-scale elite buildings (Akkermans and Schwartz 2003, 346). The Hurrians may have introduced the horse and chariot (Moorey 1986, 1989, 75–7) and possibly horsemanship throughout the Near East. Cyprus and the Aegean became regular trading partners, and trade contacts with Egypt intensified: it is possible that innovations in boat construction were linked to this. At its height the Mitanni ruled over most of north and central Syria from their capital, Wasshukkani, possibly modern Tell al-Fakhariya in northern Syria (Lemche 1995, 1203).

Between 1500 and 1200 B.C., the Near East formed part of an international system stretching from western Iran to the Aegean and from Anatolia to Nubia. Large states were equals and rivals; the Syro-Palestinian area was made up of smaller states with allegiances with greater powers. The great states were Kassite Babylonia, Hittite Anatolia, Egypt and the Mitanni, followed by Assyria (in northern Mesopotamia and Syria). Assyria was originally a province of the Mitanni state, which gained independence and then replaced its original master as the regional power. On the eastern fringe was powerful Elam, to the west Mycenae. In their midst was Syria and Palestine, with small city states such as Jerusalem, Byblos, Ugarit, Damascus and Aleppo. The area was sandwiched between Egypt and Mitanni, Egypt and Hatti with Assyria in the background. 'It acted as a buffer between these states and as a place where they could interact competitively, both directly and by proxy' (Van de Mieroop 2004, 126). Moreover, the great states prevented this area from unifying. Rulers in Syro-Palestine were regarded as 'servants'. This is underlined by references in the Amarna letters (c. 1365–1335 B.C.), a collection of 350 letters written by King Akhenaten to his 'vassals' in Syro-Palestine and those who he regarded as equally great, the kings of Babylonia, Assyria, the Mitanni, Hatti, Alashiya (probably Cyprus) and Arzawa (in south-west Anatolia). Such tablets have been found elsewhere, such as at Ugarit on the Syrian coast, which, until c. 1330, was subject primarily to Egypt and after this date to Hatti (Van Soldt 1995, 1259). Ugarit was clearly an important trading location, occupying a crossroads between Mesopotamia, central Anatolia and Egypt. Tablets found there reveal links with other Syrian harbours, Cyprus and Crete as well as inland sites such as Carchemish and Emar. Many glass ingots have been found there (Matoïan 2000). The so-called Great Kings had diplomatic links at this time that included the exchange of goods, messengers and women.

The goods exchanged between kings formed part of the much larger trade network (Akkermans and Schwartz 2003, 352). The reciprocal exchange of prestige goods led to greater mutual respect, prestige and brotherhood (Van der Mieroop 2004, 132). It is worth considering how a prestige good, such as glass, can be identified. Stein and Blackman (1993, 50) use the following ways of defining a prestige good:

(1) the presence of the appropriate craft specialist in the texts as a dependent of a "great institution,"

(2) textual evidence for state procurement of the necessary raw materials,

(3) a high price/value equivalent for the raw materials or finished products, and

(4) records indicating that these particular items were presented as gifts to high-status individuals of known social rank.

Because the production of glass was carried out within the ambit of the palace, the necessary raw materials would have been procured by the state. Glass was used for part of a political allegiance, and it can therefore be considered a politically charged commodity (Brumfield and Earle 1987, 5).

Egypt dominated the production and use of gold, which was in great demand: it was often exchanged for horses, copper and other goods such as textiles and fine oils. Exceptional evidence of seaborne trade in such materials (including the largest number of late Bronze Age glass ingots) is provided by the Ulu Burun shipwreck, which was discovered off the coast of Turkey and dates to c. 1300 B.C. (Bass 1986, 281–282, figs. 15 and 16; Pulak 1988, 14; Nicholson *et al.* 1997; Pulak 1997, 242; Pulak 2001; Pulak 2005, 34*ff.*). Its cargo included ten tons of copper ox-hide ingots which, according to the results of lead isotope determinations, mainly derived from Cyprus, 25 kg of two-handled copper ingots, 120 smaller plano-convex 'bun' ingots of copper and the remains of tin ingots, including four-handled types. About 170 glass ingots in the shape of cupcakes, approximately 15 cm in diameter and 6.5 cm thick, were found; most are of a cobalt blue colour, but there are also translucent turquoise green, amber and purple examples. The 'local' produce, presumably picked up along the way at various ports, included ebony logs, cedar logs from the Lebanon, ivory tusks and hippopotamus teeth from Egypt and murex shells from various Mediterranean coastal locations, which would have been used to make purple dye. Other manufactured goods included beads of gold and four faience drinking cups in the shape of a ram's head rhyton. A massive number of other beads were also found: c. 75,000 made of faience and c. 9,500 made of glass (Ingram 2005).

A jeweller's hoard included scraps of gold, silver and electrum. Moreover, a scarab bearing the name of Nefertiti and cylinder seals from Babylonia, Assyria and Syria were found. It is thought that the vessel was travelling from east to west along the Mediterranean (Bachhuber 2006).

The texts provide ample evidence for a clear social division in the late Bronze Age, between those associated with the physically separated palatial area and those who fell outside this direct area of control. The social hierarchy within the palace consisted of those involved in working the estates at the bottom, and those with greater skills higher up: specialist craftsmen (including glassmakers and glassworkers), administrators, those associated with cults (such as priests), scribes and near the top the military, including prestigious charioteers. The Akkadian word *ummânu* refers to a range of high-status people including sages, scholars and other specialists such as craftsmen. The sages advised the king and were equivalent to the viziers of the Arabs. In the context of Mari c. 1780 B.C. (Rouault 1977, XVIII) Mukannišum was the advisor to King Zimrilim. Bottéro (1992, 248) describes the *ummânu* in the following way:

> Their role was, however, considerable, in such an 'industrialized' society, devoted to the production and transformation of usable goods according to traditional procedures that were efficient and highly developed.

The Sumerian superlative for such a person was *apkallu*. They were super-technicians and geniuses who were attached to their patron, Ea. They were considered to be responsible for the invention of technologies and great technical achievements. At Ugarit (several hundred years later) the role of 'vizier' was played by the sākinu (in Ugaritic), or prefect. At Ugarit the social structure apparently included 'guilds' that had their own overseer in the service of the palace, and specialist craftsmen (Van Soldt 1995, 1260). Those who worked outside the palace hierarchy still provided agricultural produce. Over time the degree of servitude increased, resulting in some escaping and operating outside the city-states as nomads. Having provided discussion of the social, political and ritual contexts in Mesopotamia, we can now consider its occurrence on a selection of Mesopotamian sites.

5.2 The Occurrence of Early Glasses in Mesopotamia

The occurrence of early glass before c. 1500 B.C. in the Middle East is limited to a small number of sites, and in some cases it is difficult to be sure that the glass derives from well-dated archaeological contexts. On the basis of his extensive knowledge of lexicography, Oppenheim (1973) was of the opinion that glass was invented in northern Syria. Intriguingly, Barag (1985, 35) notes that words meaning 'glass', which regularly appear in Middle-Babylonian and Neo-Assyrian glassmaking texts of the fourteenth–twelfth and seventh centuries B.C.

(see Chapter 3), can be found in earlier contexts. An inventory dating to the Third Dynasty of Ur (twenty-third century B.C.) mentions a 'glass' (Sumerian an-zaḫ) bowl. Early second millennium word lists also include those meaning glass that are used in the glassmaking texts. Barag (*ibid.*) suggests that mention of a 'glass' bowl at such an early date is probably a reference to a glazed vessel, but this date is even rather early for glazed vessels, which made their first appearance in the late fifteenth–fourteenth centuries B.C., with the earliest extensive evidence deriving from stratum II at Nuzi dating to the first half of the fourteenth century (Peltenburg 1987; Moorey 1994, 159; Tite *et al.* 2002, 590). Although it is a slim chance, there is, nevertheless, a possibility that very early glass vessels will be discovered. Indeed, Moorey (1989, 2001) is of the opinion that the crucial period for many technological innovations that occurred at the time was in the first rather than the second quarter of the second millennium B.C. (i.e. between 2000 and 1750 B.C.). The introduction of core-formed vessels was a 'revolutionary change' (Barag 1985, 36) leading to the manufacture of glass vessels from about the fifteenth century B.C.

Beck (1934) and Lucas and Harris (1989/1962) have listed 17 objects that could date to before 1500 B.C., although for some the dating is not secure. With the examples of questionable date removed, Moorey's (1985) consideration is a good place to start a survey of early glass. Using archaeological evidence, before c. 1500 B.C. in the Middle East, glass was only used to make small decorative objects, such as beads. Many beads were made out of faience (Peltenburg 1987) and what is described as 'frit', although use of the word 'frit' here is not (necessarily) intended to denote (*sensu stricto*) the semi-fused raw materials of glass production. Indeed, other vitreous materials were used to make beads, such as faience and vitreous faience (see Chapter 1).

The following description of the occurrence of Bronze Age glass in Mesopotamia is afforded space here because it includes the earliest glass in the world and because it is in this area that the most significant developments in glass technology occurred. Discussion of the sites, although not intended to be exhaustive, is included to provide an idea of the geographical extent, variety of vessel and object forms and range of dates involved.

1. The earliest bead is from Tell Judeideh in the Amuq plain in Syria. It was attributed to 'phase G', which dates to the earlier third millennium B.C. (Braidwood 1960, 341, fig. 258). It is a pale yellow-green short oblate bead. It is noted that the glass from which it was wound contained a lot of cord and seeds and was pockmarked.
2. A glass pin head from Nuzi in Iraq was found in grave 5A of pit L4 attributed to stratum IV (Starr 1939, 515) associated with "Old Akkadian" texts dating to c. 2350–2150 B.C. Apart from being described as large, there is no other information about it.

3. A pale blue-green (presumably turquoise) chipped glass rod was found in the main level of the northern palace beneath walls of a ruined building at Tell Asmar (ancient Eshnunna) in Iraq, in the fill of room E 16:16. This corresponded with the period of desertion of the Akkadian palace and which according to Moorey (1985, 196) can be dated specifically to the Guti period or more probably the dynasty of Akkad. It dates to the twenty-third century B.C. (Barag 1970, 133). Beck (1934, no. 7, figs. 2 and 3) noted that the rod was probably moulded, only had a few small bubbles and was 'surprisingly free of striae or inclusions of quartz or dirt'. For Beck to have observed these features, the glass must have been translucent and relatively free of weathering.
4. A bubbly translucent blue chunk of raw glass, which was apparently broken off a larger lump, was excavated from Abu Shahrein (Eridu; Barag 1970, 133; Barag 1985, no. 179, 35, 111, pl. 20, pl. A). It was referred to by the excavator (Hall 1930, 213–14) as 'opaque blue vitreous paste which I recognised a true glass'. It dates to the dynasty of Akkad or the early Ur III period and therefore not later than the twenty-first century B.C. Beck (1934, no. 8, figs. 4 and 5) suggests that it was 'a manufacturer's piece of material'. Garner's (1956a, 147–9) chemical analysis established that it is a soda-lime glass probably made using plant ash and was coloured by cobalt oxide (see Chapter 6).

These few well-dated glass objects are some of the earliest in the world, so it is surprising that the twenty-third century rod from Tell Asmar was made out of high-quality glass. This indicates that, by the twenty-third century, early glassmakers had mastered the technique of fusing glass and holding the melt at sufficiently high temperatures to drive off the gas bubbles that form during the fusion of glass raw materials – presumably plant ash and silica. The lump of glass from Eridu is of a lower quality but, nevertheless, to add very low levels of cobalt oxide to the melt required great skill. It is therefore evident that the period when glass was invented must have occurred somewhat earlier – the rod from Tell Asmar, in particular, showing that the technology of glassmaking was 'fully formed' in the area where it was made. We can therefore expect more glass finds from earlier in the third millennium to come to light, to add to the bead from Tell Judeideh.

Glass vessels make their first appearance on north Mesopotamian sites in the fifteenth or possibly in the second half of the sixteenth century. They have been found on sites such as Tell Brak, northern Syria (Oates *et al.* 1997), Chagar Bazar in the Khabur Valley, Ashur and Nineveh in Assyria, Tell al-Fakhar near Kirkuk and Tell Rimah in the Sinjar area (see fig. 5.2). At Chagar Bazar, polychrome vessels have been found dating to the later fifteenth or fourteenth century B.C.; mosaic vessels that are possible exports from Marlik in Iran dating to Iron I c. 1350–1000 B.C. were also found. Sites in the Mitannian area of northern

5.2. A fourteenth-century opaque turquoise bottle from Tell Brak, Syria, decorated with combed opaque white and yellow glass strands (reproduced with kind permission of Dr. J. Oates).

Mesopotamia have produced many early examples of glass. These have been found on sites such as Alalakh, Nuzi, Assur, Tell el-Rimah and Tell Brak. Mosaic vessels are a type that is characteristic to this period and area. They have been found at Rimah, 'Aqar Qūf, both in Mesopotamia, and at Marlik in Iran. These vessels have no parallels in Egypt.

Tell Brak is one of only six inland sites where Bronze Age ingots of glass have been found (fig. 5.3). The other five sites are Alalakh (Tell Atchana), Kar-Tukulti-Ninurta, Nuzi (although this may be either raw glass or a broken ingot; Barag 1970; Barag 1985, 107–13; Peltenburg 1987, 17, table 2; Moorey 1994, 202–3), Ugarit (Matoïan 2000; Matoïan and Bouquillon 2003) and Tell Ashara-Terqa (Rouault 1993, 11; Rouault 1998, 316). Excavations at Tchoga Zanbil have produced a large number of glass cylinders that are thought to have been used for architectural decoration. At Tell Ashara-Terqa, an ingot was excavated, together with a range of other objects including glass beads (Matoïan and Bouquillon 2007). Although incomplete the ingot weighs 5.1 kg and is the largest yet found; its size shows that glass was being made on a somewhat larger scale than the evidence has suggested thus far. It has a thin grey surface layer, below which a yellow-coloured layer surrounds a pale blue-coloured core containing millimetre-sized white glass nodules. It has two well-preserved parallel faces, the longest being 27 cm. It is thought that its overall shape was a (cut) semi-circle and that it was cast. Other late Bronze Age glass ingots are square in cross-section and considerably smaller or 'cup-cake' shaped. The Tell Ashara-Terqa ingot is therefore unique.

Much early glass was found in an area that was under the direct influence of the Hurrian kingdom of the Mitanni. However, excavations at Ugarit, a site outside the direct influence of the Kingdom, have produced the largest amount of Bronze Age glass of any Middle Eastern site, as discussed below.

The first mosaic glass vessels were made in the fifteenth century B.C. An especially important fourteenth-century group was found during excavations of the Kassite royal palace at Aqar Quf (Dur Kurigalzu) west of Baghdad (Von Saldern 1970, 207, 213–15, nos. 3–6). These include polychrome plaques fused in the mould from sections of various cane colours – red inlaid with white, turquoise blue, red and possibly yellow. The plaques have decorations that include stars, eye-like discs, lozenges and hawks with inlaid eyes and engraved feathers. The forms of vessels made in this way are goblets, bowls and beakers. Some examples are decorated with a zig-zag pattern of contrasting colours. There is an important collection of mosaic glass from Assur dating to the thirteenth century. Here a wide range of colours motifs and techniques has led to the suggestion that they were made locally (Von Saldern *ibid.*, 209). The discovery of shops that specialised in selling glazed ceramics at Assur tends to support this suggestion.

At Tell al-Rimah, there was an active frit and faience industry (Oates 1968, 132–3). Here the glass is from Middle Assyrian levels with a terminus date of c. 1250 B.C. The objects found consist of core-formed vessels, including straight-sided button-based goblets, pendants, nude female plaques, demon masks and polychrome beads. However, no primary evidence for the glass industry was found at the site.

The following is a more detailed description of a selection of sites in which the glass has been found. It is given here because it is in this area that the first glass vessels were found and where glass was first made. The descriptions also highlight the kinds of archaeological contexts that early glass has been found in and some of the questions that are (inevitably) associated with its discovery.

5.2.1 Alalakh, Tell Atchana (Plain of Antioch), Turkey

This site is outside Mesopotamia and yet has produced a number of Mesopotamian vessel fragments. Woolley (1955, 369) regarded it as a Hurrian city with fluctuating links to the Hittites to the north and Amorites to the east. The earliest level in which glass vessel fragments appear is Level VI, therefore apparently dating it to at least the late sixteenth century B.C. and possibly earlier (Barag 1970, n. 93). However, there are doubts about the stratigraphy of this site and its interpretation, and such an early date must be called into question: the earliest accepted date is now the later sixteenth century (McClellan 1989). Gates

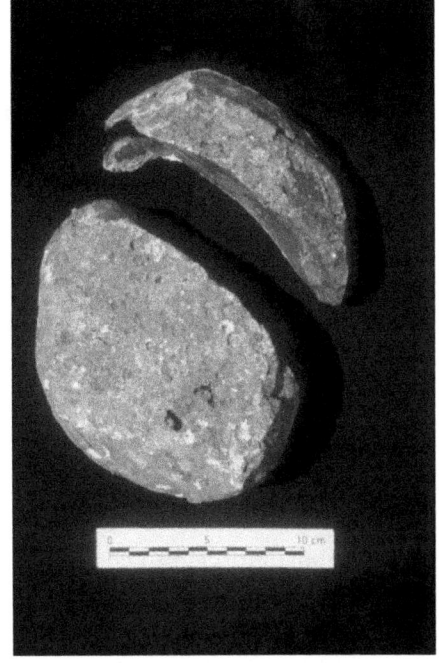

5.3. A fourteenth-century B.C. turquoise blue glass ingot from Tell Brak, northern Syria (reproduced with kind permission of Dr. J. Oates).

(1987, 60–86; Gates 1989, 67–73) uses Egyptian texts, Palestinian destruction layers and ed-Daba dates as a means of dating the site. Glass vessels are also found in the later levels up to Level II, which dates to the thirteenth century B.C.; none were found in Level I. However, some glass found in level II may actually be earlier (Barag 1970 150, n. 94). The range of vessel forms consists of bottles (the most common), a bowl and two straight-sided goblet fragments. Where it is possible to determine the matrix colour of the vessels, it is blue, but many fragments are weathered (Barag 1985, 42–6). One of the roundels has an opaque turquoise body colour (Barag 1985, 44, no. 12). The range of decorative colours consists of opaque yellow and white and translucent dark brown. Decorative techniques included the application of horizontal threads, various kinds of festoons, meanders, feather patterns, eyes and spots. Barag (1970, 151) suggested that some of these vessels were imported from Mesopotamia. A blue rim fragment decorated with a pair of antelope horns in relief is unique. Plaques consisting of moulded nude females were found in Levels VI, V and II.

Three opaque red fifteenth-century 'raw' pieces of glass that were found in level IV in the palace of Niqme-pa, room 30 at the site. The first (Barag 1985, no. 180, 111, pl. 20) has rounded edges and is described by Barag as being a 'fragment of a 'cake ingot'. These were made by pouring viscous glass onto a flat surface and given rounded edges (or poured into a mould). It is an opaque liverish red cuprite glass (Bimson and Freestone 1985, 122) with weathered (green) surfaces. The second is an opaque red lump (Barag 1985, no. 180A, 112, colour pl. C) covered with thick layers of weathering. This was found in the same location as the first. It is a cuprite glass with a similar chemical composition to the first (Bimson and Freestone *ibid.*). The third piece is also a lump with an apparent green colour but is almost certainly the weathered surface of opaque red glass (Barag 1985, no. 180B, 112, pl. 20). Intriguingly, recent excavations at the site directed by Prof. Dr. Aslihan Yener have produced possible evidence for primary glass production.

5.2.2 Ugarit-Ras Shamra, Syria

Matoïan (2000) has published a detailed description of a selection of the vitreous materials that have been found during various excavation campaigns at the site of Ras Shamra-Ugarit on the Syrian coast (Schaeffer 1962). The site, and its principal port, Minet el-Beida, was a crossroads for international exchange. As noted earlier, it was a city-state that acted as a buffer between the great powers of Mitanni, Egypt and Hatti. The sheer quantity of the glassy objects, some 20,000 pieces, is exceptional, with 98% deriving from Ugarit-Ras Shamra. They consist of 18,000 faience objects, 1,000 glass objects, 266 Egyptian blue objects and 27 glazed pots. They date mainly to between c. 1400 B.C. and 1200 B.C. and constitute one of the largest number of glass and faience objects from any late Bronze Age site in the Levant or indeed the Middle East. Any published contribution on such an important range of objects is therefore significant. The glass objects found consist of about 900 beads of various shapes, pendants, inlay elements, gaming pieces, pommels, vessels and fragments of raw materials. Although a wide range of types was found, Matoïan (2000, 28) notes that it is not always easy to be certain of the exact total number of objects involved. An exceptional assemblage of glass beads was discovered in a ceramic jug found in a Late Bronze Age private house in an area labelled 'Centre de la Ville' dating to between the thirteenth and the beginning of the twelfth century B.C. The vessel was filled with one bead of amber, four beads of lapis lazuli and three beads of carnelian, all nonlocal raw materials, and around two hundred beads of vitreous materials, the majority being glass. Some of these were relatively well preserved.

The glass vessels found include a tubular vase (Matoïan 2000, 29, fig. 3), the spherical body of a polychrome vase (Matoïan 2000, 29, fig. 4) and a cylindrical goblet with polychrome zig-zag decoration (Matoïan 2000, 29, fig. 5). Fragments of *kraterikos* vessels from the site, together with parallels from the Levant, Cyprus and Rhodes, are considered to have had an Egyptian origin. Other glass objects found include figural pendants (Matoïan 2000, 29, 32, figs. 6 and 7), a figural animal head (Matoïan 2000, 32 and 33, fig. 9, photo 2), elements of glass inlay, which were found in the royal palace and probably used for decorating furniture (Matoïan 2000, 33, 35 and 36, figs. 14–16, photo 3), the 'head of a sceptre' (Matoïan 2000, 28, 36, fig. 17), a scaraboïde (Matoïan 2000), roundels of glass including 'gaming pieces' (Matoïan 2000, figs. 10–13), monochrome and polychrome moulded and wound beads (Matoïan 2000, 28, 36, 37, 39, figs. 18–28), blocks of raw glass (Matoïan 2000, photo 1) and glass ingots. The colours of glass used were blue, white, yellow, purple, brown and 'black' (presumably a dark translucent brown or green). Many of these colours were used in the beads (Biron *et al.* 2012). The fact that there is such a large number of glass artefacts from the site could suggest that either primary glass production (glassmaking) or secondary glass production (glassworking) occurred there, or conceivably both. However, even though the site was involved in trade and a text was found on the site that states that raw blocks of glass (*meku*) were exported from it (Heltzer 1978, 35, 80), this evidence does not provide a clear-cut case for primary glass production there (Matoïan 2000, 40). Raw glass was even found in the royal palace, which helps to affirm the model that glass was considered to have had a high value and that its distribution and production would have been controlled by elites. The glass ingots that were found could either have been manufactured there or imported. No archaeological evidence for primary glass production has been recognised: scientific analysis and especially isotopic analysis is therefore one of the few ways of investigating the question of local production (discussed in Chapter 6). Two polychrome glass vessels were found at the nearby port of Minet el-Beida, one was a piriform vase, the other a shouldered vase (Matoïan 2000, fig. 1). Significantly Caubet (1995, 2686) has noted that the Levant acted as a 'motor' for production and an independent creative centre, especially for faience manufacture. Faience figures have been found at Alalakh, Ugarit and Ebla (*ibid.*, 2685). The possibility that Ugarit was a creative centre that included primary manufacture of glass (and faience) cannot be ruled out.

5.2.3 Tell Brak, Syria, Northern Mesopotamia

Polychrome (mainly core-moulded) glass vessel fragments have been found at Tell Brak, especially in Level 2, a deposit that resulted from a Middle Assyrian sacking (Oates *et al.* 1997, 81). Level 2 produced more than 73 items of glass (excluding beads). The earliest fragments are from Level 6 and include polychrome beads. Finds from Level 5, which is securely dated to the fifteenth century B.C., include a core-moulded glass bottle. Other vessel forms found were piriform bottles decorated in various ways and footed beakers (Fig. 5.2). Mosaic glass vessels and the remains of a vessel described as being decorated with 'granulation', using tiny glass balls, have also been found. In addition, 13 ingots and ingot fragments were found at Tell Brak, mostly of a pale blue colour (Fig. 5.3). The earliest is of an early fourteenth century, or earlier, date (Oates *et al.* 1997, 28–9, 85–6, fig. 124). There is even possible direct evidence of glassworking (*ibid.*, 86). The ample evidence of glass use and of probable

5.4. Fragments of fourteenth-century B.C. raw glass from Egypt (1:1). Note the conchoidal fractures in the fragment in the top right: this has been struck off a larger piece of raw glass.

glassworking does not, however, in itself constitute evidence of primary glass manufacture from raw materials at Tell Brak.

5.2.4 Nuzi, Near Kirkuk, Iraq, Northern Mesopotamia

Excavations of the remains of the Hurrian city of Nuzi, Iraq, have produced a wide range of glass materials including raw glass (perhaps a broken ingot), fragments representing some forty separate vessels (Starr 1939), thousands of beads and many moulded objects (Starr 1939). The amulets represented Ishtar and Astarte. The glass was found in destruction deposits in (the fill of) rooms, which suggested to Barag (1970, 136) that they would have been used as part of a ritual practice that was carried out in Temple A and in rooms L8, M79. Some 16,000 beads and amulets, including about 11,000 made of glass, were found in stratum II in Temple A, which was dedicated to Ishtar. Glass was also found in 'ordinary' houses. Barag noted that glass was concentrated in central courts (Starr 1939, 92, 108, 128; Barag 1970, 136, n. 26).

The glass was found in Level II, which was originally dated to the second half of the fifteenth century. However, a destruction layer has a terminus date of c. 1350 B.C. or slightly earlier (Stein 1989), so this later date is now accepted (cf. Drower 1973, vol. 3, pt. I, p. 417 and vol. 1, part II, pp. 735–54). The range of vessel forms found at Nuzi is straight-sided goblets, piriform bottles with plain surfaces, piriform bottles with fluted surfaces, a shallow bowl and a possible stand. A wide range of decorative techniques was used on the vessels, with many variations of festoons, meanders, feather patterns, chevrons and applied bichrome rods. A range of moulded star pendants, undecorated pendants (in green glass) and a nude female plaque was found. The main vessel matrix colour appears to have been deep blue (Barag 1970, 136); ranked by (decorative) colour frequency by Starr, blue was followed by white, yellow-orange, brick red and 'black'. Perhaps it is no coincidence, therefore, that three chunks and seven chips of semi-translucent raw dark blue glass were found at Nuzi (Barag 1970, 140–1, no. 19, fig. 16). Barag (*ibid.*, 140) has suggested that the fragments 'All seem to have belonged to the same ingot'. It is unfortunate that the locus of this glass was not marked on the excavation plans. Indeed, apart from this, there is no direct evidence for glassworking (or glassmaking) at Nuzi. However, if the ingot is coloured by cobalt, one crucial question would be whether its chemical characteristics indicate that a Persian rather than an Egyptian or other cobalt source, was used.

5.2.5 Assur, Iraq, Northern Mesopotamia

Although the site was excavated before the First World War (Jordan 1912; Andrae 1935; Haller 1954), it has produced the largest number of vessel fragments of any Mesopotamian site. Today the glass is in the Staatliche Museum in Berlin. However, only information about the vessels from the graves have been published (Haller 1954), with a majority remaining unpublished. The find locations for the glass include a dump and 'frit' objects in the Ishtar temples. There is some doubt about the date of the deposit. The glass from second millennium B.C. tomb groups, especially group nos. 37 and 133, was reasonably complete. It consists of three piriform bottles, a tripodal beaker, a goblet with globular body and a glass pin. Although a range of dates has been suggested for the burials and their contents on the basis of the glass found, Barag (1970, 141–3) has suggested that the vessels date to the fifteenth or fourteenth century B.C. However, Moorey (1994, 197) dates the glass to the thirteenth century or slightly earlier. The colours of the vessel matrices are blue, white and 'brown' with decorative colours consisting of blue, turquoise, yellow, white and red-brown (*ibid.*, 143). Some of the decorative techniques used were similar to those decorating vessels found at Nuzi: festoons, meanders, feather patterns and chevrons. In addition, threads trailed around rims or other parts of the vessel body, eyes, blobs, a *guilloche* design of single threads or twists, petals and the addition of bichrome or trichrome cables were used. Moorey (1994, 197) was of the opinion that there was 'a local industry making glass of all kinds (mosaic and core-formed vessels, inlays and probably plaques) as well as complementary production of artefacts in glassy (sintered quartz) materials.' At Assur, there is a wider range of shapes and a greater variety of decoration amongst the glass vessel fragments and inlay plaques than found at Nuzi; the mosaic glass fragments are especially significant. The discovery of 'cullet', which is presumably raw glass, could indicate local production.

Although the great quantity of finds does not necessarily prove the glass was fused locally, it nevertheless indicates that somewhere glass production occurred on a relatively large scale for the late Bronze Age and that it could have occurred at Nuzi (or Assur).

5.2.6 Tell al-Rimah, Iraq, Northern Mesopotamia

The site was scientifically excavated by David Oates (Oates 1968). The range of vessel forms includes at least two piriform bottles, a beaker or goblet rim, the lower part of a goblet, a complete goblet and fragments of straight-sided mosaic goblets. The decorative patterns used include festoons, meanders, feather patterns, chevrons and eyes. Moulded female plaques and a demon mask made out of blue glass were also found. The glass was found in a range of different contexts, including Level 2 of a temple found on site C, the ruins of a palace on site C and in a store room. The glass found in Level 2 was said to be contemporary with Nuzi stratum II with a suggested date of c. 1300 B.C. and a terminus date of c. 1250 B.C., although this can now be pushed back by some 50 to 100 years. The type of goblet with an elongated body and a solid button base has close parallels amongst Nuzi ware pottery. The vessel type has also been found at Nuzi, Nineveh, Tell Brak, Alalakh and Megiddo.

Another vessel type from the site was formed from sections of polychrome sections of canes (mosaic vessels). The decoration of these goblets or beakers is in the form of wavy bands alternating with blue cane sections. The range of glass colours used was the same as for core-formed vessels: white, red, turquoise and perhaps yellow.

5.2.7 Ur, Iraq, Babylonia

The site was excavated by Leonard Woolley (1965). A single bottle of a 'Kassite period' date was found on site EM, and it almost certainly derives from grave EM 48, Kassite grave 14 (Woolley 1927, 387; Woolley 1965, 87, 105). It therefore dates to c. 1300 B.C. (Barag 1970, 147, n. 74). The bottle is a piriform shape with a pointed base and a long tubular neck. The vessel matrix colour is described as brown decorated with opaque turquoise wavy and chevron threads (BM 120659). A 'ribbed' rod made of nine threads of 'black' glass (BM 128415) was unstratified, found in loose soil covering the EM house, and Woolley (1927, 387) suggests that it is also of the Kassite period. The rod bears a pincer mark at one end that suggested to Barag (1970, 147) that 'the glass was made on the site'. However, the only activity that this suggests is that the glass was melted/worked on or near the site. There is no evidence that it was made (i.e. fused from primary raw materials) there. The 'black' colour of the glass is a deep translucent brown. The colours mentioned on the BM card are 'black and blue'.

5.2.8 Nippur, Iraq, Babylonia

Some moulded blue (lapis lazuli) coloured glass was found in the late nineteenth century bearing the name of the Kassite king Nazimaruttash, which apparently dates to the last quarter of the fourteenth century (Peters 1898, 143, 160, 373–4). Scientific analyses of the blue glass (and a sample of green glass) were performed (*ibid.*, 134). The find spot was described as a 'jeweller's shop'.

5.2.9 Tchoga Zanbil, Iran, Babylonia

The glass found at this site dates to the late thirteenth century B.C. Large numbers of glass inlay rods were found. These were thought to have formed diagonal chevrons of decoration in horizontal panels in the remains of wooden doors. The doors formed part of a ziggurat built by the Elamite king Untashgal. Glass rods were also found in temples and palaces during the excavations of the site (Mecquenem and Michalon 1953, 52–3, fig. 20: 1–2, pl. B:1; Ghirshman 1966, 30, 65, 72, 74, 88, 100, fig. 20, pls. XXV:1, LI: 4, LII: 2–3, LXXVIII: 455, XCVIII: 63a). Each rod was about 27 to 28 cm long. Two portions of inlay rods are in the British Museum (Barag 1985, nos. 19 and 19A, 47, pl. 2). They are described as opaque dark blue and opaque light grey (probably originally opaque blue) with an opaque white spiral decoration marvered into their surfaces. Many other examples are in the Corning Museum of Glass (Goldstein 1979, nos. 5–7). Convex roundels and discoid inlays pieces were also found on the site. Barag (1985, 38) suggested that because so many decorative rods were found on the site, they were made there during its construction perhaps by foreigners or by artisans with Mesopotamian technological knowledge.

Glass objects thought to have been made in Mesopotamia have also been found on Syro-Palestinian sites, such as Alalakh (discussed earlier), Ebla, Hama, Hazor, Megiddo, Beth Shan, Jerusalem, Lachish and Tel Mevorakh (Barag 1970). For example, nude female figurines, clearly of ritual significance, have been found at Alalakh, Hama, Beth Shan, Megiddo and Lachish (Barag 1970, 188) and also in the Hittite capital of Boğazköy in Turkey. Spacer beads, a moulded bead type that has parallel ribs on top and may be perforated longitudinally in two places, are a relatively common Mesopotamian type (Spaer 2001, 58–9). They have been found in contexts dating to as early as the earliest glass vessels (the second half of the sixteenth century B.C.) and as late as the thirteenth century B.C. Barag (1985, 46) claims that 'their exceptionally wide geographical distribution is equalled by no other contemporary class of glass object'. They have been found on a number of Mesopotamian sites, including Nuzi, Ashur, Nimrud (Barag 1985, no. 17, 46, fig. 2, pl. 2, colour pl. A), Nineveh and Nippur. In the Syro-Palestinian area, they have been found at Alalakh (discussed earlier),

Byblos, Hazor, Megiddo, Beth Shan, Gezer, Jerusalem and Lachish, tomb 4004 with a nude female plaque (Tufnell *et al.* 1958, 83, pl. 27:3). They have also been found on Iranian sites, one has been found in Egypt, several in shaft graves at Mycenae and others have been found in a Tholos tomb at Kakovatos and at Boğazköy in Turkey. Having described the occurrence of early glass, we follow with a discussion of innovation in early glass technology in Mesopotamia.

5.3 Innovation and Glass Technology in Mesopotamia

The late sixteenth and fifteenth century B.C. core-formed glass vessels from northern Mesopotamia form a homogeneous group with characteristic trailed-on opaque decoration. The available scientific evidence for these vessels indicates that they were of a soda-lime-silica made from plant and silica (Henderson *et al.* 2010). It is clear that these early glassmakers and glassworkers understood the working properties of the glass, which allowed them to trail the glass around a core or dip a core mounted on a rod into a crucible containing fluid glass. In technical terms, this means that the glassworkers needed a glass that was sufficiently 'long', that is, one that remained soft enough for long enough to be trailed. An experimental period must have occurred during which different proportions of primary raw materials were mixed until a glass with the 'best' properties was obtained. Given the sparse amount of early glass and the small scale of glass production in the late sixteenth century, it is hardly surprising that evidence of early experimentation has not been found. The glass that has survived has a relatively stable chemical composition, although it is notable that fifteenth–fourteenth century B.C. Egyptian glass is generally less weathered than Mesopotamian glass. The archaeological evidence of early glass from the Middle East suggests that the Hurrian Kingdom of Mitanni was responsible for the great leap forward in glass production in two ways. The first was the fusion of sufficiently large volumes of glass, in crucibles inside suitably large furnaces that could reach temperatures of c. 1150–1200°C, to be able to make core-formed glass vessels and to produce glass ingots and canes. The second way was their invention of the core-forming technique of vessel production in the fifteenth or possibly the second half of the sixteenth century B.C.

Barag (1985, 36) notes that 'core-forming reached a high degree of competence immediately it was invented or soon afterwards'. Unusually, this suggests that there was a short experimental phase; it is especially surprising at the 'inception' of a technology. An alternative interpretation is that vessels predating the sixteenth century are yet to be discovered or have not survived. This agrees with Moorey's opinion (2001) that the crucial period of technological innovation was in the first quarter of the second millennium B.C. What he suggests (*ibid.*, p. 4) is as follows: 'What the Mitanni may have facilitated was the scholarly tradition. . . . and the diffusion of a number of influential innovations'.

However, at the moment, what we have are a small number of very early glass artefacts, which include, surprisingly, a piece of 'raw' glass from Eridu. We lack the archaeological evidence for the earliest stages of industrial innovation. The small scale of production would have left a minimal amount of evidence, and even if the evidence is discovered in a workshop context, it would need to consist of a spectrum of by-products so as to prove that primary production occurred. Innovation would certainly not have been instantaneous, and moreover it would have occurred within a particular set of (elite) religious, political, economic and social contexts. Furthermore, we should not underestimate the importance of the complex impact that traditional procedures used in other technologies would have had on the experimental phase. Ideally, when scientifically analysed, each by-product should provide evidence of why it was discarded. To try to answer the question why it was discarded is to get inside the mind of the artisan, which is, of course, impossible. We should try to avoid using the mind-set of modern-day materials science as a means of labelling what was regarded as 'successful' and or 'unsuccessful'. Characteristics that may have been significant include visual relationship to other materials such as colour, depth of colour, brilliance, transparency, opacity, reflectivity, homogeneity, workability, softness and the volume. We know that colour was especially important in the production of early glass because of the links with semiprecious stone colours (see Chapters 1 and 3), but we are in danger of imposing our own interpretations on the evidence if we attempt to attach a degree of significance to other glass characteristics.

As Muhly (1988) has noted, metallurgy took 3,000 years to spread through the Middle East, between 7000 B.C. and 4000 B.C. This rate of 'technology transfer' is exceptionally slow. We may get a hint as to one of the reasons for this from the passage dating to the Old Babylonia period cited by Saggs (1962):

> Let the initiate show the initiate
> the non-initiate shall not see.
> It belongs to the tabooed things of
> The Great Gods.

This reflects the secretiveness and conservatism associated with the ritual tied up with technology. Although this may not necessarily have reduced the speed of technology transfer, it surely underlines the significant role that cult played in making technological processes secretive. This secretiveness (associated with tradition) continued into much later periods. Moreover, it is worth considering the speed at which glassmaking spread. Between the twenty-third and sixteenth centuries B.C., relatively small volumes of glass were used to make beads in northern Mesopotamia. The problem is that evidence for *glassmaking* at this time has so far eluded discovery. Even when glass objects were being produced in large quantities, such as in the late Hellenistic and early Roman periods, it can be difficult to find. In this case it has only just been recognised (see Chapter 7). The craft of *glassworking* to make vessels certainly spread to Egypt by the fifteenth century B.C. However, a study of lexicography led Oppenheim (1973, 263) to attribute the *invention of glass* to the area of the kingdom of Mitanni in the sixteenth century. Lilyquist *et al.* (1993, 43) have suggested on the basis of art historical evidence and the results of chemical analyses that Asiatic craftsmen were present in Egypt from an initial stage during the reigns of Tuthmosis II and III – and that they were involved in glassmaking. Since this publication, scientific evidence now shows that primary glassmaking occurred in fourteenth-century Egypt (see Chapter 6).

Palestinian and Egyptian-style jewellery dating to c. 1525 B.C. from the tomb of Ahhotep has glass inlays. Cobalt- and copper-coloured inlays also used in Egyptian and Near Eastern style jewellery from Wady Qirud (c. 1425), and two Egyptian style button-based vessels from the tomb of Amenhotep II (of Asiatic shapes and sizes) suggested to Lilyquist *et al.* (1973) that Asiatic craftsmen were present in Egypt then. Brill's bivariate plots of oxides in the glasses do not separate samples of button-based vessels from other pre-Malkata glasses; Lilyquist and Brill (1993), suggested that they were therefore made from Egyptian glass. Eyre (1987, 195) was more circumspect and felt that the *borrowing* of technique as a result of the presence of foreign artisans, as opposed to influencing the development of technique, was limited. This is a fine distinction but an important one.

Although Hurrian glassmakers and workers cannot be credited with the invention of glass *per se*, they can probably be credited with the first *controlled* addition of the material which brought about opacification in the glass in it. This was an important step forward in the development of glass technology. To produce opaque turquoise glass, the glassmaker would have added antimony to the melt, which would have dissolved. The glass would have been slowly cooled and then slowly reheated to form the masses of opacifying crystals of (white) calcium antimonate in the glass; this is known as 'striking' the glass. In the turquoise or cobalt blue glasses produced, the white calcium antimonate crystals were set within a (translucent) turquoise or cobalt blue matrix, the net affect being to produce an opaque turquoise or opaque blue glass respectively (see Chapter 3). Such glass could be used to make a vessel or for trailed-on decoration. A second opacification technique involved the direct addition of the ground-up minerals to the glass melt: both lead antimonate producing an opaque yellow colour in the glass and copper, to produce a dull opaque red colour in the glass are examples of this (Bimson and Freestone 1985, 122); see Chapter 3.

A further invention at this time was the manufacture of the first mosaic glass vessels, which involved assembling sections of coloured glass canes in a mould. Another invention was decorating glass vessels with marbling. Lilyquist *et al.* (1993, 10–11) note that a consideration of shape, 'fabric', colour, and the results from Brill's analyses shows the techniques to be 'Middle Eastern' as opposed to Egyptian. Moreover, both of these decorative techniques required the colour contrasts provided by the extended colour range offered by a combination of opaque and translucent glass. Indeed, it can be argued that the invention of opaque glass led to, or inspired, the glassmakers to invent mosaic vessels and marbled decoration. (Mosaic glass vessel production was also a vessel-forming technique.) At the same time, one-piece open moulds were used in glass production to produce spacer beads, nude female plaques and inlay plaques.

Thus, this period in glass production was truly one of invention. In Chapter 1 it is argued that it constituted a paradigm shift. Behind the surge of inventiveness in a range of glassworking techniques, there was also a critical development in pyrotechnology: the construction of a sufficiently large furnace to accommodate a crucible in which a large enough volume of raw materials or frit could be melted would have been essential. In addition, new colourant and opacifying raw materials were used in association with a sophisticated understanding of the heat treatment of glass. The heat treatment of glass to produce some opaque colours, such as red, under reducing furnace conditions and others, such as white, to develop crystals out of the glass, reveals a high degree of sophistication.

The first glass vessels were created by dipping a core mounted on a rod into molten glass – a technique borrowed from bead making. Initial synthesis of brightly coloured glasses to imitate the colours of semiprecious stones, perhaps within a relatively short period of time of glass being invented, led to the use of these same colours in the manufacture of polychrome glass vessels. Opaque white, yellow and turquoise are three of the commonest opaque colours, and they all rely on access to antimony. Moorey (1994, 241) has reviewed the distribution of antimony ores, with sources in three Iranian mines, two near Anarak and another 160 kilometres northwest of Birjand (Harrison 1968, 512). It has also been found in Kurdistan (Partington 1935, 256) and Transcaucasia, where small personal ornaments made from antimony have been found (Selimkhanov 1975). Given that cobalt, an essential colourant in producing 'lapis lazuli from the kiln', also occurs in one of the Anarak mines, this source may have been an important one (and see Chapter 3).

The period of Mesopotamian glass production came to an end in the late thirteenth or early twelfth century B.C. If we use the historical evidence as a guide to glassmaking activity, then it did not start again until the late Neo-Assyrian period in Mesopotamia in the second half of the eighth century B.C. Meanwhile, a very different kind of glass was being used in late Bronze Age Europe.

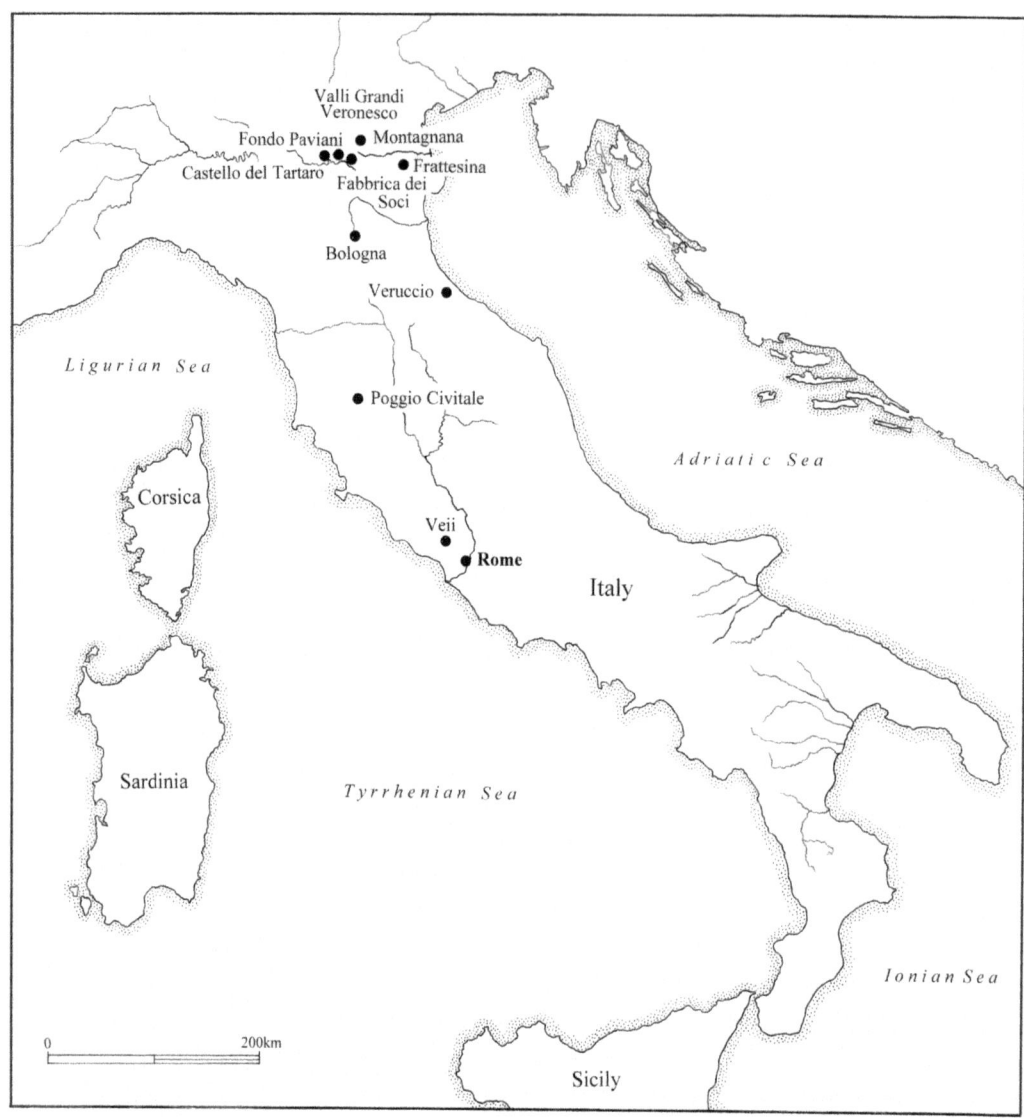

5.5. Map of Italy showing the main sites mentioned in the text.

5.4 THE SOCIAL AND POLITICAL CONTEXTS FOR THE EARLIEST EVIDENCE OF GLASS PRODUCTION IN EUROPE

This section provides the context for the critically important site of Frattesina, in northern Italy, where the largest concentration of glass has been found on any prehistoric European site. Fig. 5.5 provides the locations of the key sites mentioned in the text.

The dates of various sub-periods of later prehistoric Italy and the European equivalents are based on dendrochronological investigations for German and

Table 5.1. The chronology of later prehistoric Italy

Dates B.C. and (Italian) Prehistoric Periods (after Bietti Sestieri 1997 and Peroni 1997, with suggested alternatives as noted in Pearce 2000, n. 1)	European equivalents
c. 1350–1150 recent Bronze Age	BzD c. 14th–13th centuries B.C.
c. 1150–1020 final Bronze Age	HaA1, HaA2, HaB1 13th–11th centuries B.C.
c. 1020 (960/920) Early Iron Age	HaB2, HaB3 10th–early 8th centuries B.C.
c. 780 Advanced Iron Age	HaC c. Early 8th century B.C.–mid-7th century B.C.

Bz = Urnfield; Ha = Hallstatt.

Swiss Lake Village wood samples and high-precision carbon-14 dating (Giardino 1995; Hennig 1995; Bietti-Sestieri 1996). The dates for these periods are shown in Table 5.1.

Bietti Sestieri has pointed out there are still discrepancies between this dating system and some other parts of the Mediterranean, so further research is necessary. During the Bronze and Iron Ages, Italy was made up of a number of regional districts. Each district varied widely in terms of its settlement morphology and the nature of its interaction within and without the district. Bietti Sestieri (1997, 374) stresses the importance of communications over long distances and the location of metal ores as being important factors in the emergence of centres. Northern Italy had relatively easy routes of communication to the north by mountain passes through the Alps, such as the Val d'Adige and the Valle d'Aosta. In northeastern Italy, the Friuli plain allows access to the eastern Alpine zone, the Balkans, Austria and eastern Europe as far as Hungary. In the northwest, the coastal strip allows access to southern France. It is generally agreed that metal ore deposits were important at different times in the following areas: the eastern Alps and Trentino, Etruria, probably western Calabria and Sicily (Marzatico 1997, Bietti Sestieri 1997).

The middle Bronze Age and the early part of the late Bronze Age saw a rather similar range of artefacts in use on the Italian Peninsula, particularly with regard to pottery. This may suggest that a low level of territoriality existed. In the final Bronze Age and early Iron Age, there is far greater evidence for political and territorial division, with the emergence of identifiable regional polities in various parts of Italy. The growth of such polities developed differently in different places and at different times.

Late Bronze Age northern Italy was characterised by intensive occupation in the 'peralpine' lacustrine areas and the northern Po plain. Central northern

Italy was characterised by the so-called *terramare*, villages that either were defended by ditches and banks or were lake dwellings (*palafitte*). Their cultural characteristics were fairly homogeneous, although with some local variations. They had quite different characteristics from the settlements in the Apennine area. The collapse of the palafitte-terramare complex occurred in the Recent Bronze Age. This was especially marked in the southern Po plain. Bernabò Brea *et al.* (1997) have suggested that a number of factors were responsible at a time when the population was growing, including a cooler and wetter climate just before the start of the subatlantic phase and difficulty in obtaining basic resources. The northern Po plain shows less disruption, but a decrease in the population still occurred. Characteristics of the Subapennine culture appeared in the Veneto.

The Friuli-Venezia Giulia regions formed part of the same cultural area as Slovenia and Dalmatia. The Castelliere, embanked densely occupied nucleated villages, are characteristic of this area.

There is evidence that there was a systematic relationship between the mineral resources of Etruria and the terremare groups, the southern Po plain and northern Tuscany (Andreotti and Zanini 1997). Etruria emerged as a territorial and political entity with foreign contacts, internal social hierarchy and settlement hierarchy as early as the late Bronze Age. In the Po plain during the Bronze Age, there is evidence for the emergence of warrior aristocracies, as reflected in burial rituals, especially in the territory of Verona. By the Recent Bronze Age, there was a two-tiered settlement organisation. By contrast, during the middle and Recent Bronze Age in most of central and southern Italy, the archaeological evidence is relatively homogeneous. It has been suggested that the economy in these areas was based largely on transhumance (Barker 1975).

However, in Etruria and in the southeast (in Basilicata and Puglia), complex economic and social developments occurred. These developments are reflected in the occurrence of fortified settlements and in collective family tombs. It appears that individual communities held sway over territories, although they were apparently not permanent and lasted for relatively short periods. Southeastern Italy shows evidence of a Mycenaean presence, and it has been suggested that this was mainly due to the presence of metal ores (Bietti Sestieri 1988). Nevertheless, during the Recent Bronze Age, a site such as Moscoti di Cingoli in Picenum (northeastern Italy) had areas devoted to the production of bone, antler, bronze weapons, tools and personal ornaments and it is therefore similar to sites on the Po (Riva 2007, 89). Amber from the site (and from Santa Paolina di Filottrano), and 'congruous' weighing systems, indicate that trade links existed between Picenum and the Aegean (Riva 2007, 85, 90). During the Recent Bronze Age and early Iron Age, the range of metal artefact types increased dramatically from their use to make prestigious weapons to the manufacture of many other types, such as agricultural implements. Metal hoards from these periods have been found in settlements – metal had become more freely available.

The Final Bronze Age was characterised by the Proto-Villanovan culture for a short period, and it is at this time that Frattesina flourished. Bietti Sestieri (2008) has noted the emergence of two major central places at this time, Frattesina and Bologna. The Proto-Villanovan culture was homogeneous in character and can be recognised over a wide area. Characteristic regional cultures appeared around the interface between the Final Bronze Age and the Iron Age. Bietti Sestieri (1997, 378) has listed the following changes that occurred with their appearance:

1. the need to exploit control and protect metal resources, and routes of distribution, including raw materials and artefacts, which increasingly led to settlements being located on defendable hilltops;
2. the development of settlement hierarchies together with political structuring on local and regional scales and intensification of intraregional communications; and
3. The emergence of linguistic and ethnic identities.

Regional territories and socio-political structures with relatively high densities of population developed at this time, such as in Etruria and Umbria in central Italy and Puglia and Calabia in the south. Metallurgy and the manufacture of copper alloy artefacts clearly played an important role in the emergence of a chiefdom in Etruria. There is growing evidence of local metalworking, the hoarding of copper alloy objects, including ingots, and broken objects. Etruria was the principal focus of metal working as reflected in its resources, but the highest proportion of metal from the Italian Peninsula in the Final Bronze Age and early Iron Age also derived from this area. The period saw innovation in technology, including the production of a high proportion of prestige items (Giardino 1995). It is apparent that much of the metal production was destined for gift exchange or had a prestigious function. An example of the distribution of Etrurian products is that of the pick ingot, considered to be prestigious. It occurs with socketed shovels in hoards found in the Veneto at the settlements of Frattesina, Montagnana and Villamarzana. Pick ingots also occur well outside the Proto-Villanovan Etrurian polity. They performed an important role as a means of metal circulation. They occur in Slovenia, Croatia, Switzerland, southern Germany, north Alpine France and northern France (Bietti Sestieri 1997, 389, fig. 5). However, not all of these pick ingots were made using metal ores found in Etruria: for example, the results of qualitative chemical analyses suggest that some were made using east Alpine metal ores (Borgna 1992; Pellegrini 1992). Intensive research into copper mining and smelting in the southern Alps (Trentino) has produced evidence for more than 100 smelting sites dating to between the thirteenth and eleventh centuries (Marzatico 1997). Pearce and De

Guio (1999) have identified three interlinked stages of metal production there: (1) mining; (2) ore roasting and smelting, ingot production; and 3) melting and casting of tools. The first two of these were apparently carried out by the people of the Luco/Laugen, a culture of the southern Alps; the third by their customers (Pearce and De Guio 1999). This shows how important this area was for metal production at the time.

With respect to Frattesina, Bietti Sestieri (1997, 390) states

> The highest concentration of archaeological materials which indicate a systematic connection with central and northern Etruria appears in the settlement of Frattesina. As we shall see, the site can be identified as an early intermediate centre on the Adriatic and northern route.

In the northern Po plain a number of flourishing Proto-Villanovan settlements appeared with connections to the Tyrrhenian area. Bietti Sestieri (1997) suggests that chiefdom-type structures can be identified in Proto-Villanovan and early Villanovan settlements in this area.

A low population density and low level of Villanovan settlement is evident in the southern Po plain, the Adriatic regions and in Campania during the initial phase of the Iron Age.

The early Iron Age saw the emergence of important proto-urban Villanovan centres. One centre was based around the southern Po plain and along the Adriatic coast and the other in Campania in interior and coastal areas of the southern Tyrrhenian side of the peninsula. The emergence of these centres has been identified with the transition from chiefdoms to early states (Bietti Sestieri 1997, 398). Examples are Bologna and Verucchio in the north; Veii and the territory of Lazio in the south had somewhat different characteristics. Bologna and Verucchio appear to have been successors to Frattesina in that they acquired and processed raw materials and took part in trade with the rest of the Italian Peninsula and transalpine and eastern Europe (Bietti Sestieri 1997, 397). The southern Villanovan centres, on the other hand, appear to have performed 'a strategic role in the relations of Etruria with the indigenous communities of southern Italy, as well as with the Phoenicians and the Greeks in the pre-colonial phase'. Frattesina and the exceptional industrial evidence for industry found there, including the evidence for glass production, are now considered.

5.4.1 Frattesina, Provincia Rovigo, a Key Site for Early Glass in Europe

Bearing in mind that Frattesina flourished at a time when there was a transition from chiefdoms to early states, it is tempting to suggest that such a political change led to intensive industry. The site is located near the (modern) village of Fratta Polesine in the northeastern Po plain (see Fig. 5.5). Occupied between the twelfth and early ninth centuries B.C., it covered an area of more than 20 hectares. It is one of the largest Final Bronze Age–early Iron Age settlements yet discovered in the Veneto. Like other settlements of the period, it was originally located on the south side of a northern branch of the river Po (the Po di Adria) and was close to the Bronze Age Adriatic coastline. Today the site is located some 10 kilometres from the Po and 36 kilometres from the Adriatic coast.

The characteristic that distinguishes this site from other contemporary ones is the range, and particularly the large scale, of specialised craft production. It was involved in acquiring both organic and inorganic raw materials and transforming them into objects so that they could be exported. The evidence for working consists of the production of glass and glazed pottery, antler, bone, elephant ivory (Bietti Sestieri and De Grossi Mazzorin 1995), bronze, iron and amber. Four hoards of metal, including three founders' hoards, were found, which, in addition to broken objects, contained a significant number of pick ingots and socketed shovels (Arenoso Callipo and Bellintani 1994, 26). Separate industrial areas with workshops for glass, antler, bone and ivory-working have been found. Pearce (2000, 119) has suggested that full-time craft specialists were present. The number of metal casting moulds from the site is greater than any other contemporary site (Le Fèvre-Lehöerff 1992). Two cemeteries associated with Frattesina have been excavated, at Fondo Zanotto and Le Narde (De Min 1987; Salzani 1989; Salzani 1990–91). Tomb 227 at Le Narde contained an 'Allerona'-type sword with gold rivets in the hilt, gold decorated buttons and a gold ring. On the basis that in each generation only one man in each lineage was buried with his sword, Bietti Sestieri (1997, 394) has suggested that power had resided in the hands of this man.

Interregional trade is indicated by imported elephant ivory, amber (including beads of 'Tiryns' or knuckle type), ostrich egg shell and Aegean pottery (Jones et al. 2002; Pearce 2007). Thus, the area that Frattesina was connected to by trade was wide. The sherds of Mycenaean IIIC pottery (also found at Fondo Paviani, Montagnana, and Fabbrica dei Soci in the same region) and a significant number of amber beads of the Tiryns type indicate that there was a specific link to the Aegean area. Moreover, evidence for glassworking and amber-working has been found at Fondo Paviani, 30 kilometres upstream from Frattesina (Balista and De Guio 1997). At this time, elephants roamed the eastern Mediterranean, so this would suggest a link also with the eastern Mediterranean. The distribution of 'Frattesina type' glass is discussed later (Venclová et al. 2010). Bietti Sestieri (1997, 396) claims that Frattesina was a 'strategic node' and an intermediate base for trade relations between Etruria, eastern Europe and transalpine Europe. Frattesina may also have performed this function in relation to the Aegean and the East Mediterranean; however, from the thirteenth century Bietti Sestieri (2008) has noted that evidence for trade relations with Mycenaean Greece is replaced by evidence of trade with Cyprus and Phoenicia. Frattesina's nodal rôle developed at a time when the palafitte-terramarre system collapsed. At the start of the twelfth century, there was a collapse in the Valli Grandi Veroneso polity, evidence for which has been found at Fondo Paviani, Castello del Tartaro and Fabbrici dei Soci (Pearce 2000). It is no coincidence that the Mycenaean palatial economy suffered a similar collapse (De Marinis 1975, 47): the end of the terramarre has been put at c. 1200 B.C., which coincides with the collapse of the Mycenaean palace society (De Marinis 1975, 47f.).

5.4.1.1 The Chronology of Frattesina (Twelfth–Ninth Centuries B.C.)

The site dates to the early and advanced phase of the Final Bronze Age (phase 2) and into the Final Bronze Age–early Iron Age transition (phase 3). Here the production of glass beads on a large scale occurred during the Proto-Villanovan phase (eleventh–ninth century; Bellintani 1997). A lot of typical (early Iron Age) Villanovan pottery has been found at the settlement of Villamarzana adjacent to and partly contemporary with Frattesina. Of the four phases of Proto-Villanovan Frattesina, the most important phase is phase 2, dating to the eleventh to tenth centuries B.C. Activities dating to this phase are thought to have developed partly as a result of outside influences, including Aegean metal prospectors.

5.4.1.2 Evidence of the Glass Industry at Frattesina

The site has produced more evidence of the glass industry than any other site in prehistoric Europe up to and including the late Iron Age (up to the first century B.C.). Excavations by Bietti Sestieri at the site uncovered stratified deposits containing glassy materials. Much of the material was discovered as a result of field walking. Nonetheless, the diagnostic chemical compositions of the glassy materials found in both stratified deposits and in field walking is helpful in that it provides a link between the two (see Chapter 6).

The following hues of glass were found at the site: colourless, translucent turquoise, translucent purplish-blue, translucent blue, translucent green, translucent brown, translucent emerald green, opaque red and opaque white. The glass artefacts found include disc ingots of red and turquoise glass; chunks of raw blue, turquoise, pale green and nearly colourless glass and strips of brown and colourless glass. Three crucible fragments with opaque white, red and turquoise glass adhering, a single crucible for working glass, with streaks of red glass adhering, and many malformed beads, primarily of opaque red or translucent turquoise colours, were also found. Moreover, a pottery fragment with turquoise blue glaze decorated with opaque white spots was found. Although the vast majority of the glass from the site is of a mixed-alkali type (see Chapter 6), perversely the glass adhering to one crucible fragments is of a plant ash composition. This is an exceptional range of vitreous materials. No other crucibles with glass adhering have been found on other prehistoric European sites.

Whether all Late Bronze Age mixed-alkali glass found on sites in the rest of Europe was worked or made at Frattesina is open to question, but, given the scale of production there, it is certainly possible that much of it was. Late Bronze Age–Early Iron Age mixed-alkali glass has so far been found in Switzerland, Germany, Bohemia, England, Ireland, France and Greece (Chapter 6). Frattesina was preceded by fourteenth- to twelfth-century Valli Grandi Veronesi centred on large banked and ditched enclosures (terramare) at Fondo Paviani (16 ha), Castello del Tartaro (11 ha) and Fabbrica dei Soci (6 ha). All sites have produced evidence for bronze working and Fondo Paviani evidence for glassworking and amber-working.

De Guio (1991) suggested that Frattesina controlled the flow of Alpine copper. He has suggested that the Altiplano was used for transhumant summer grazing from the Po plain. At the start of the twelfth century, there was a collapse in the Valli Grandi Veroneso polity, and this is evident at Fondo Paviani, Castello del Tartaro and Fabbrici dei Soci, leading to the foundation of Frattesina, a central node for production and distribution of amber and glass – and possibly for copper alloys.

Cultural links existed between Frattesina and northern Europe and with the eastern Mediterranean, as noted earlier. Bietti Sestieri suggests that throughout the Bronze Age, peaceful, rather than exploitative, contact occurred for long periods between northern Italy and distant shores, but it involved only limited numbers of artefacts (Bietti Sestieri 1997, 373) After this peak of glassworking and glassmaking, it is interesting to consider what became of the glass industry in the early Iron Age and Etruscan periods. The chemical compositions of Etruscan glasses fall broadly into the soda-lime (natron) glass tradition and are therefore different from the glass worked at Frattesina on such a large scale (se Chapter 6). It is difficult to find Etruscan sites with comparably high levels of industrial activity. However, one site, Poggio Civitale (Murlo), about 20 kilometres south of Siena, securely dated to the seventh century B.C., has evidence for a range of production technologies (Phillips 1993; Christensen 2000). These include terracotta and ceramic production, antler, bone and ivory working and the production of iron and bronze. Twenty-four glass objects were found, all well dated to before 600 B.C. The majority were beads. Chemical analysis of Etruscan glass beads and vessels has, however, shown that although soda-lime in composition, a range of raw materials of a variety of purities were used to make them, but there is no evidence that mixed-alkali glasses were in use (Towle and Henderson 2008). This could suggest that Etruscan glass was made in several centres, one or more of which could have been located in Etruria. This last suggestion is partially supported by the occurrence of distinctive type of *oinochoai* (Haevernick 1959).

Thus, important changes in the structure of society, reflected in the arrangement of settlements across the landscape, can be linked to the development and fluorescence of glass production in and around Frattesina. It occurred on such a large scale that it is not possible to point to other comparable sites in prehistoric Europe even with the rise of Late Iron Age proto-urban *oppida* north of the Alps.

5.5 Conclusions

The appearance of the first fully formed glass c. 2500 B.C. must have had a powerful impact on its creators. This translucent, highly coloured, lustrous material had been transformed from a red-hot glowing liquid into something that was apparently solid. The production of the first true synthetic material must surely be considered a paradigm shift in the development of ancient technology (as discussed in Chapter 1), and it is not surprising that for the first 1,500 years or so, it was attributed an elite value in whatever way that is measured.

Given the dominant role that ritual played in every aspect of Bronze Age society, the production of the first glass, and the power that the transformation of materials represented, is likely to have been regarded with reverence. Following the initial transformation of materials to produce faience, the first true synthetic material in which the ingredients were fully melted and transformed into glass represented a step change in the relationship of humankind with the environment. Cuneiform texts clearly reveal the complex and intricate rituals that were involved in Bronze Age glass production. The fusion of plant ashes and crushed quartz or sand produced a red, glowing liquid and a radical change in the visual properties of the materials. Is it a coincidence that the name of the plant mentioned in the cuneiform texts, the naga plant (Oppenheim *et al.* 1970, 74), is close to the name of Tell Brak (nagar), in an area where the first glass may have developed – modern northern Syria and Iraq? Perhaps the name nagar was used because there was a memory of the early stages of glass manufacture in the area.

The early phase of Mesopotamian glass vessel production in the fifteenth century B.C. was associated with a number of innovative technological features: (1) the construction of the first furnace that could accommodate a crucible containing a sufficient volume of glass; (2) the deliberate production of opaque glass, either by the addition of suitable opacifying minerals to the glass or by heat-treating it to develop crystals; and (3) the first manufacture of mosaic glass vessels in moulds. All three of these developments were technologically entirely new. However, the deliberate production of opaque glass – known today as a glass-ceramic – was especially significant in that a driving force for making the earliest glass was to imitate (apparently opaque) semiprecious stones, such as lapis lazuli and turquoise. The heat treatment of the fully formed glass to develop crystals and produce opacity is at least as significant a technological development as the production of glass colour because both provide the means of imitating semiprecious stones. Heat treatment occurred at relatively low temperatures, comparable with annealing temperatures of c. 600°C or below, depending on the glass chemical composition. Once the glass had been heat-treated it would have been cooled slowly. Even if the two operations are not quite the same in terms of the temperature regimes employed, the idea of holding the glass at a low temperature to develop opacifiers is similar to holding the glass at a low temperature to anneal it. It is therefore likely that both operations occurred at the same time and that annealing initially occurred 'accidentally'. Whereas the precipitation of crystals from the liquid glass would have been visually obvious, the causal link between the survival of the glass vessel in tact and the annealing process may not have been realised initially. Therefore, when heat-treating the glass to produce opacity ('striking' it), it would have produced both an opaque and a stronger glass. The process of opacification is not mentioned in the cuneiform texts, but by allowing a fully melted homogeneous glass without bubbles to cool slowly (Oppenheim et al. 1970, 35), the process of annealing would have occurred. Moreover, it can be suggested that the process of opacification must have involved reheating glass and may have been responsible for the birth of glass annealing.

The use of cobalt- and copper-rich minerals to produce cobalt blue and turquoise colours respectively was also critical in the synthesis of early glass. Although Moorey (1994, 210) noted the limitations of using the cuneiform texts as a source of information about early glass production, including the fact that there was no mention of the new vessel types that were manufactured in the Bronze Age, the emphasis on producing different glass colours in them was clearly seen as having overriding significance. However, as Von Saldern (1970, 203) has pointed out, there is mention in the texts of the use of 'moulds left in furnaces for seven plus three days', which could possibly relate to the manufacture of moulded, including mosaic, vessels.

It is therefore clear that early glass production involved a range of innovative processes that, by the fifteenth century B.C., led to the manufacture of the first highly sophisticated decorated glass vessels. However, we still lack archaeological evidence for the invention of glass, the early experimental stages of the manufacture of the first glass vessels and the furnaces in which glasses were made from primary raw materials before the thirteenth century B.C., the first two of which apparently occurred in northern Mesopotamia.

SIX

EARLY GLASS IN THE MIDDLE EAST AND EUROPE

SCIENTIFIC ANALYSIS

This chapter builds on the archaeological, technological and other aspects of early glass discussed in Chapter 5. Only a small amount of glass can be securely dated to the period between its invention and c. 1500 B.C. It therefore follows that there are few scientific analyses of such glass. Since 2006 increasing numbers of scientific analyses of late Bronze Age glasses from Egypt and Mycenaean Greece have been published, but there is still a relatively small number of analyses of Mesopotamia glass.

An exceptional piece of early glass, one of the earliest recorded, is from Abu Shahrein (Eridu) in Iraq (see Fig. 5.1). Beck's (1934, 9) discussion of this raw bubbly translucent blue piece includes the suggestion that there was a 'glass factory' near Eridu in the later third millennium B.C. Although clearly the raw glass must have been made somewhere, it could well have travelled over a considerable distance before being deposited at the site – a single piece of raw glass cannot be regarded as evidence of primary glass production at Eridu. Garner's (1956a, pp. 147–8) chemical analysis of the Eridu fragment, by optical emission spectroscopy, is given in Table 6.1.

The presence of elevated magnesia (3.4%) can be attributed to the use of a plant ash source of soda. If accurate, the level of potassium oxide (4.5%) is high for a Middle Eastern plant ash glass (whether of late Bronze Age or Islamic date), and could suggest that the glass was made in an area farther to the north and east, in the area of modern-day Afghanistan. Here many (much later) plant ash glasses with potassium oxide contents in excess of 4% have been found (Brill 1989); see section 4.2.1. Garner's appraisal of the kind of mineral that introduced 0.15% cobalt oxide in the glass (*ibid.*, 147–8) is based on the low level of manganese oxide (0.04%). This low level suggested to

Table 6.1. The chemical composition of blue glass from Eridu, Iraq, c. 2300 B.C. (Weight % oxide)

SiO_2	65.0
Na_2O	17.0
K_2O	4.5
CaO	3.5
MgO	3.4
Al_2O_3	2.5
'Fe_2O_3'	2.4
CuO	0.49
CoO	0.15
BoO	0.15
TiO_2	0.09
MnO	0.04
PbO	0.02
NiO	0.01

him that a cobalt-rich mineral source that was low in manganese would have been added, such as erythrite or cobaltite, both of which are arsenical ores; arsenic was not detected (it is volatile). He suggested that the cobalt mine used was the one at Khemsar near Kashan. However, although Garner refers to the publication of Farnsworth and Ritchie (1938), in which cobalt was detected in Eighteenth Dynasty Egyptian glass, he overlooked their suggestion that cobalt alums from the Dakhla and Kharga Oases in the Egyptian western desert were used in ancient glass production, a suggestion subsequently discussed in much more detail by Kaczmarczyk (1986). This Egyptian cobalt source was found to be characterised by elevated levels of cobalt, aluminium, sulphur, manganese, magnesium, iron, nickel and zinc (*ibid.*, table 34.3). The Erudu raw glass is also characterised by elevated alumina and iron oxide, but, according to the analysis published by Garner, it contains low levels of manganese, nickel and zinc oxides. The elevated alumina and iron oxides could be due to the use of an impure sand source. Nevertheless, the high alumina level is intriguing as a potential indicator of an Egyptian cobalt source. Cobalt blue glass from Tell Brak, northern Syria, dating to the fifteenth–early fourteenth centuries is characterised by low alumina levels (discussed later) which is thought to reflect the use of a relatively pure silica source and a non-Egyptian cobalt source. It is apparent that the time is ripe for a modern chemical analysis to be performed on the Eridu piece, especially to confirm the potassium and arsenic oxide levels and to determine its trace element signature with the possibility of determining the cobalt source used.

A small number of middle Bronze Age glasses have been scientifically examined. The few available results from securely dated glasses indicate that plant ash glasses were being used at the time, such as that found in a nineteenth- or eighteenth-century B.C. context at Tell Dan, Israel (Ilan *et al.* 1993). Other analyses of glass beads, from sixteenth century burials at Deir 'Ain 'Abata, Jordan, also show that plant ash glasses were used to make them (Henderson, 2012).

6.1 Scientific Analysis of Mesopotamian and Syro-Palestinian Glasses (c. 1500–1000 B.C.)

The published quantitative chemical analyses of late Bronze Age glass from Nuzi, Tell Brak, Tell al-Rimah and Tchoga Zanbil, with a summary of results from Ugarit, can hardly be claimed as representative of the industry. Nevertheless, they are valuable. The results show that glass at this time from Mesopotamia, Greece and Egypt was almost exclusively of the plant ash type (see Figs. 4.1 and 6.2).

The microprobe analyses of single examples of each colour of Nuzi glass (Vandiver 1983b, table 1) reveal that the different samples of 'black' (deep translucent brown), opaque blue, translucent blue, opaque red, opaque yellow and opaque white are of an anticipated soda-lime-silica composition with elevated levels of magnesium and potassium oxides, showing that plant ashes provided the source of alkali in their manufacture. Nevertheless, six analyses cannot be claimed as representative of the glass technology there. Vandiver (*ibid.*, 244) suggests that a rather higher ratio of Mg:Ca could indicate that dolomitic lime in the soil was introduced (via the plant ash used) in the glass. For plant species from Syria of the genus *Salsola* alone, we can show a consistently positive ratio between the Mg and Ca (see Chapter 2). However, the plant ashes that showed this correlation and that contained high levels of both were not growing on dolomitic limestone. The reason for this is that the plant physiognomy dictates that a positive correlation between magnesium and calcium is retained, although clearly the drift and bedrock geologies would also determine plant compositions (see Chapters 2 and 11; Barkoudah and Henderson 2006). The Nuzi glasses are of an early- to mid-fourteenth-century date. Brill 1999b (40–1) and Shortland and Eremin (2006) have also published chemical analyses of Nuzi glasses. Lilyquist *et al.* (1993, 41) noted that Nuzi glasses are closest compositionally to pre-Malkata non-cobalt Egyptian glasses and that they contain (generally) low soda levels (see Fig. 6.1). Subsequently it has been shown that, compared with Mycenaean and Egyptian (New Kingdom) glass, Nuzi glass is of a characteristically 'Mesopotamian' type (discussed later). So although Lilyquist *et al.* (1993, 41) were concerned that the high(er) levels of silica they reported in Nuzi glasses were due to the analytical technique that they had used (silica is calculated by difference with atomic absorption spectroscopy), the electron microprobe results for Nuzi glasses published by Vandiver (1983b) and Shortland and Eremin (2006) confirm that their results are acceptable.

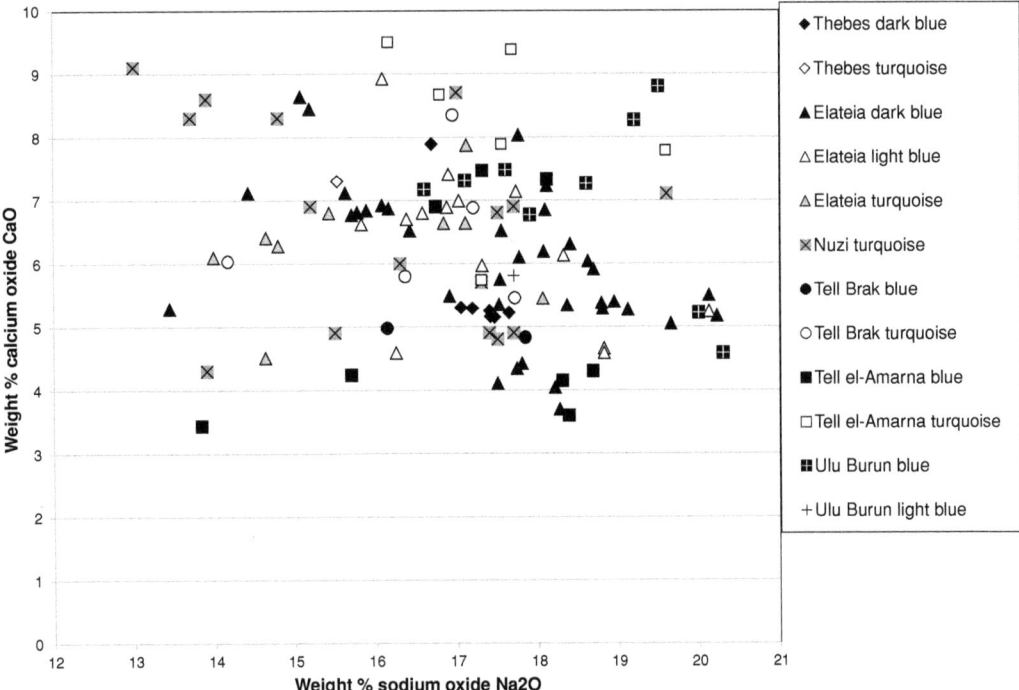

6.1. A bi-plot of weight % soda versus calcium oxide in late Bronze Age cobalt blue and turquoise blue late Bronze Age glass from Greece, Egypt and Mesopotamia. Data from Brill 1999b, Nikita and Henderson 2006, Shortland and Eremin 2006 and Henderson *et al*. 2010.

Figure 6.1 shows that turquoise glasses from Nuzi contain quite a wide range of both soda (13–17.5%) and calcium oxide (4.8–9.1%) levels.

It appears that, by c. 1500 B.C., early soda-lime glass technology was 'fully formed'. Vandiver's (1983b) consideration of Nuzi glass technology led her to suggest that it was already a mature industry by that time and not at an incipient stage (*ibid.*, 239). Vandiver gives a variety of reasons for suggesting this: (1) large numbers of glass artefacts made with great accomplishment and mastery of glassworking were found at Nuzi; (2) there was a limited range of glass compositions (3) the presence of other ceramic materials at Nuzi, including, for example, a glazed tripod and several small glazed vessels from the destruction level of Temple A (Starr 1939, 91, 391–2). Forty-two glazed clay wall nails, three pairs of glazed lions, a glazed boar's head, a glazed ram's head, a wall plaque, an Ishtar statuette and glazed vessels are claimed by Drower (1973) as the earliest use of (green) glassy glazes on clay bodies. However, the glazed pottery from Tell Brak, for example, is likely to be c. 50 years earlier in date. Given that technologies did not develop overnight, and there must have been an experimental phase, Vandiver's first point is incontrovertible. A limited compositional range can be produced as a result of a mature industry, but equally this could also be interpreted as the result of experimentation with a limited range of raw materials at an early stage. In any case the relative levels of soda and calcium oxide are quite wide. There are here the practical considerations as to whether technologies develop quickly, whether 'failed' attempts survive and whether 'imperfect' batches, leading to glass that was difficult to work, were adhered to for a sufficiently long periods for us to be able to detect them in the archaeological record. Moreover, if experimentation produced semi-fused vitreous material, it may have weathered out of existence.

Given the apparent sudden appearance of a 'fully formed' glass technology and its sophisticated range of output, Barag (1985, 37) has suggested that the technology developed rapidly. A period of experimentation lasting a generation (thirty years) might be realistic. However, most archaeological contexts do not allow (such) short periods of time to be dated precisely. Nevertheless, the full reintroduction of plant ash glass (of a similar composition to Bronze Age examples) starting in the early ninth century A.D. has now provided us with the kinds of compositional variation we may be able to expect from a period of experimentation with raw material combinations. Such evidence has been found at al-Raqqa, northern Syria (see Chapter 10). Even though the scale of production, the received technological knowledge and the social contexts in which it took place were entirely different, it is still a valid comparison. Indeed, currently there is little else that can be used as a comparison. To act as a devil's advocate, it would be possible to select a narrow range of glass compositions from amongst those analysed at al-Raqqa, especially for single artefact types, such as armlets or, in this case, the beads analysed from Nuzi – and on the basis of this suggest that the technology was stable and thus there was no evidence for experimentation. Although not particularly early in the developmental sequence, if local experimentation did occur at Nuzi, a much larger database of glass compositions could still provide evidence for it. There may also have been an initial intense phase of experimentation being followed by later periods of experimentation.

A reasonable number of chemical analyses of glasses from Tell Brak Syria, a site located in the Hurrian Kingdom of the Mitanni in northeastern Syria, have been published (Brill and Shirahata 1997; Henderson 1997; Brill 1999b; Shortland 2005; Shortland and Eremin 2006; Shortland *et al.* 2007; Henderson *et al.* 2010). The glasses date to the late fifteenth–early fourteenth centuries B.C., and their chemical compositions are almost exclusively of the plant ash type, with two samples made from a mineral source of alkali (Henderson 1997, 96). The range of artefact types analysed includes the all-important glass ingots of an opaque pale turquoise colour (Fig. 5.3), as well as brown fragments that may be raw (unworked) glass. The balance of artefacts analysed consists of a pendant or plaque, vessel fragments, beads and vessel glaze. All of the raw and ingot glass is of the plant ash glass type. The samples having exceptional compositions, apparently made from a mineral alkali source, are the pottery glaze and opaque white glass used in vessel decoration.

Archaeological and textual material from the Kingdom of Ugarit provide evidence for a flourishing kingdom, at the crossroads of international trade between the eastern Mediterranean and the Near East. About 1,000 late Bronze Age glass objects have been found at Ugarit (see Chapter 5 and Matoïan 2000). Matoïan and Bouquillon (2003) published a small number of semi-quantitative chemical analyses of vitreous objects from Ugarit. They used energy-dispersive X-ray fluorescence system attached to a scanning electron microscope. The elevated levels of alumina detected in the cobalt blue glass ingots and the suggested use of an Egyptian source of cobalt alum is of great interest. Other compositional characteristics detected (elevated manganese, nickel, copper and aluminium) have been found in Egyptian New Kingdom cobalt blue glass. It is possible that raw plant ash glass was manufactured in or near Ugarit to which an imported cobalt-rich material was added. One of the few ways of investigating whether this was the case would be to determine the strontium and neodymium isotope values in the glass: this would provide evidence for the type and relative geological age of calcium oxide introduced in either the plant ashes or shell fragments in sand used and the silica used to make the glass, respectively (discussed later in this chapter and in Henderson *et al.* 2010) and potentially provide evidence for a local signature (i.e. a geological provenance). Blocks of turquoise 'raw material' with a 'sugary' texture from Ugarit are described as possible frit. Matoïan and Bouquillon (*ibid.*) suggest quite plausibly that the range of evidence found could indicate a nearby glass primary production centre. Supporting evidence comes from the use of local clays (based on their mineralogical characteristics) to make *glazed* pottery on the site (Matoïan and Bouquillon 1999; Matoïan 2002–3). If the pottery bodies were made at or near Ugarit, then the glaze could also have been. Ugarit was in a buffer zone between Hittite, Mesopotamian and Egyptian areas. Trade relations would have provided raw materials for primary and secondary glass manufacture, such as ingots of glass and colourant-rich materials. However, without more comprehensive scientific results, or archaeological evidence, it is difficult to be absolutely certain whether primary glass production occurred there. It is claimed that 'non-Egyptian' recipes have also been detected amongst the faience, and, on this basis, it is suggested that local production occurred; it has been suggested that a secondary workshop for Egyptian blue was located there.

Scientific analysis of thirty-five glass beads, including decorated ones, found in the decorated jug noted above has been carried out using proton-induced X-ray emission and electron probe microanalysis (Biron *et al.* 2012). Evidence for the use of two silica sources, one purer than the other, was detected. One compositional group had aluminium levels of between 0.5 and 1.5 weight %, and another had levels between 1.5 and 3.5 weight %. All the purple glasses analysed fall into the latter group. Quite wide variations in the levels of magnesium and potassium oxides in the beads analysed suggest that ashed plants from different geological environments and/or that different preparation techniques were used to make them. The level of variation in the oxide levels in the glass compositions is lowest in the amber-brown and purple glass compared with that found in green and turquoise blue glasses, suggesting that different production centres, different origins and/or glass recycling could be involved. A copper source was used for the coloration of turquoise blue Ugarit glasses just as in Mesopotamian examples, rather than the bronze source usually used for colouring Egyptian turquoise glasses (Shortland et al. 2007). Furthermore, some of the Ugarit glass is very close in chemical composition to glass derived from sites farther east, such as Nuzi and Tell Brak. However, there is no relation between the shape of the beads and the glass composition.

An Egyptian cobaltiferous source could have been used to colour two cobalt blue glasses analysed, one with a Mycenaean shape. The use of such a colourant has already been noted in some Mycenaean glasses (Bass 1986, 282, n. 55) and for a set of blue-grey faience objects from Ugarit supposed to be Egyptian imports (Matoïan and Bouquillon 2003).

The first isotope results suggest that two sources of silica and plants growing on soils containing lime with a relatively wide range of geological ages were used to make Ugarit glasses. It is therefore likely that the glasses were made in more than one location.

Scientific analysis of the large ingot from Tell Ashara-Terqa (Syria) weighing 5.1 kg and probably dating to c. fourteenth–thirteenth centuries B.C. indicates that parts of it is heterogeneous and weathered and that its primary colourant is copper. The contrasting colours that it is made from suggest that glasses of different colours were gathered together at a glassworking site. Indeed it is suggested that the range of durabilities observed also reflect variations in chemical composition (Matoïan and Bouquillon 2007). Moreover Matoïan and Bouquillon have suggested that the existence of such an artefact at the site, with its possible evidence for frit, is suggestive of a primary production centre. Certainly the discovery of this extremely large volume of raw Bronze Age glass is suggestive of such a centre, although no other supporting archaeological evidence appears to have been discovered at the site. An alternative interpretation is that it was imported from a primary production site to be worked into objects at the site.

6.2 Scientific Analyses of Egyptian Glasses

Some of the earliest chemical analyses of ancient glasses were of Egyptian, principally fourteenth century B.C., specimens from Tell el-Amarna. For example, results of analyses using optical-emission spectroscopy were published by

Farnsworth and Ritchie (1938), followed by analytical results combined with a discussion of the technology at Amarna by Turner (1954). In a comprehensive and tightly discussed publication, Lilyquist *et al.* (1993) published new data for Egyptian glass dating to the pharaohs Hatshepsut, Tuthmosis III, Amenhotep II and Tuthmosis IV (*ibid.*, table 2) and compared the results with Brill's analyses of glass from Malkata and Amarna and from the Mesopotamian site of Nuzi. Their balanced discussion tackled the possibility that Mesopotamian and Egyptian glasses could be distinguished by their chemical compositions with the implication that primary production occurred in each zone separately (i.e. that they would provide a provenance for the glass). Subsequently, analysis and interpretation of Egyptian glasses were published by Rehren (1997), Henderson (2000), Shortland (2000), Shortland *et al.* (2000, 2006 and 2007), Mass *et al.* (2002) and Jackson and Nicholson 2007, amongst others.

Rehren and Pusch (2005) were the first to publish evidence of thirteenth-century B.C. Rameside primary glass production at Qantir-Piramesses. More recently still, major, minor and trace data on glasses from Tell el-Amarna, Malkata, Lisht and Nuzi have been published (Shortland and Eremin 2006; Jackson and Nicholson 2007; Shortland *et al.* 2007). All of these glasses were found to be of the plant ash type, with some compositional variations. The most recent evidence for the primary manufacture of glass at Tell el-Amarna, published by Smirnou and Rehren (2011), is in the form of lumps of semi-finished glass. There has been an extensive discussion of the use of cobalt-rich minerals in the coloration of Egyptian glasses. When cobalt is used as a colourant in ancient glass, the cobalt, along with most of the mineralogical impurities associated with it, dissolves fully in the glass. In contrast, the use of other mineral-rich colourants in glass, such as copper, iron and manganese, does not provide the same range of impurities as found in cobalt-rich minerals so there is less potential for being able to locate the mineral sources that were used (Henderson 1985; Gratuze *et al.* 1995; Henderson 2000). A detailed discussion of the scientific analyses of Egyptian blue glasses can be found in Chapter 3 (Section 3.3.2.1). A more detailed consideration of trace and isotopic analyses is presented in the next section.

6.3 Scientific Analyses of Mycenaean Glasses

Compared with the scientific investigations of Egyptian glasses, that of Greek Mycenaean glass has made a slow start. Perhaps this is because the glass objects tend to be small and sometimes poorly preserved. The first quantitative chemical analyses of Mycenaean glass, using neutron activation analysis, were of ornaments, a fragment of the translucent turquoise vessel found in the Tholos B tomb at Kakovatos, glass beads from Ialysos on Rhodes and a specimen from the Smith Collection (Sayre and Smith 1961; Sayre 1964, 9; Sayre 1967, 147). The vessel from Tholos B at Kakovatos has also been analysed by Magou (1992, 79 and n. 38). Another relatively recent study has been the analysis, using atomic absorption spectrometry, of turquoise (copper) blue beads from the Pylona cemetery on Rhodes (Magou 2001, 117, table 1). A further study focused on six pieces of glassy material found in an area considered to be a workshop at the Mycenaean palatial site of Tiryns. These were analysed using atomic absorption spectrometry and wavelength dispersive X-ray spectrometry (Panagiotaki *et al.* 2005). The first consideration of raw materials used to make blue frit, Minoan faience and glass was published by Tite *et al.* (2005).

Many Mycenaean glass objects are either cobalt blue or copper (turquoise) green and, as a result, research has often concentrated on the use of cobalt as a colourant in Mycenaean glass. Sayre and Smith showed that different levels of zinc, nickel and manganese oxides provide a distinction between the similar Mycenaean and Egyptian cobalt blue glass on the one hand and Mesopotamian on the other (Sayre and Smith 1961; Sayre and Smith 1963, 267, table 1; Sayre 1964, 10–11, Table IV; Sayre 1967, 147–9, figs. 2 and 5; Sayre and Smith 1974, 51–4, figs. 3, 5 and 6). Because low levels of cobalt were added to the glass melt carefully and because cobalt does not occur commonly in the Mediterranean, Sayre suggested that cobalt blue glass may have been traded in the form of ingots (Sayre 1967, 146). He did so well before the ground-breaking discovery of the c. 1300 B.C. shipwreck at Ulu Burun off the west coast of Turkey with a cargo that included many cobalt blue glass ingots (Pulak 1988, 14 and n. 67; Pulak 1997). Brill (1986, 282) reported that the ingots of blue glass from the Ulu Burun wreck were of the same general composition as Egyptian core-formed vessels and Mycenaean glass jewellery. Kaczmarczyk and Hedges (1983, 300–1, table XL) discussed the analyses of some cobalt blue glass from Thisbe and in so doing noted that the compositional differences between the thirteenth-century B.C. cobalt blue and copper blue Mycenaean glasses were akin to such differences in Egyptian New Kingdom faience, creating another link between the technologies in the two areas.

Electron-probe microanalysis of six Mycenaean relief plaques at the Corning Museum illustrated compositional similarities and further oxygen-isotope analyses have shown that Mycenaean dark blue glass seems to have been made from a silica source similar to that used in making some Eighteenth Dynasty Egyptian dark blue glass and a Mesopotamian vessel. However, the interpretation of oxygen isotope data is far from simple (see Chapter 11), and currently we are unable to define precisely what variations in oxygen isotope results we could expect from different deposits/ages of silica, the main contribution to the oxygen isotope signature. At present, there are limited results for oxygen isotope determinations for silica collected in the field (Henderson *et al.* 2005a): an environmental approach to isotope research ('ground truthing') is crucial.

It is only relatively recently that sufficient analytical data for Mycenaean glasses from secure archaeological contexts have become available so that it is now possible to compare them meaningfully with the results for Mesopotamia and Egyptian glasses. The majority of these results are for glass found in graves at Elateia, where there is a cemetery that mainly consists of rock-cut chamber tombs. Excavations produced large numbers of glass and faience beads, as well as relief plaques including many glass seals (Dakoronia and Deger-Jalkotzy, 1988, pp. 229–32, figs. 14 and 15; Dakoronia, Deger-Jalkotzy and Sakellariou 1996, pp. v–xxx and indexes I and II; Cavanagh and Mee, 1988, pp. 68 and 95). A smaller number of samples from the palatial complex at Thebes, mainly of translucent dark blue and turquoise jewellery, were analysed. The samples were dominated by translucent dark blue, light blue and turquoise colours. In addition, one translucent purple, three transparent colourless, one opaque white sample and one opaque turquoise sample were analysed (Nikita and Henderson 2006). Chemical analyses were carried out using electron probe microanalysis. The samples were primarily of simple globular or annular beads and the ubiquitous Mycenaean moulded pendants. The pendants were produced in all of the basic geometric shapes when seen in the transverse section and have a range of decorative motifs, including floral, faunal, marine and figural. The dates of the glass samples ranged from the late Helladic III period (about 1425/1390–1390/1370 B.C.), and therefore earlier than Tell el-Amarna glass, to the early Proto-geometric period (about 1000/950 B.C.).

The first clear result from these analyses was the identification of two main compositional types: the anticipated plant ash soda glass, but also some unexpected examples of mixed-alkali glass, which is the largest number from a location as far east as Greece (Fig. 4.1). With the exception of three beads from Thasos, northern Greece, two with a mixed-alkali composition and a third with a potassium-rich composition (Henderson 1992), the only other mixed-alkali examples (from the West) derive from Italy and Europe north of the Alps (see Section 6.6). The mixed-alkali type was labelled LMHK (low magnesia-high potassium) because these characteristics, associated with the type of alkali used, distinguished the glass type clearly from soda-lime natron and plant ash glasses (Henderson 1988a); see section 4.1.2.

6.4 DID PRIMARY GLASS PRODUCTION OCCUR IN DIFFERENT PARTS OF THE BRONZE AGE MEDITERRANEAN?

Lilyquist (1993, 57) has noted that more glass, especially in the form of pendants, multicoloured and shaped inlays (and some vessels) dating to the sixteenth century B.C. has (so far) been documented as having been found in the Mesopotamian rather than in Egypt. This, combined with the occurrence of the earliest glass and the historical evidence for the production of vitreous materials there (Oppenheim et al. 1970), is the principal evidence that glass technology originated there. The earliest dated examples of Egyptian glass were found in the form of Egyptian- and Palestinian-style glass inlaid jewels that formed part of the treasure found in the tomb of Ahhotep c. 1525 B.C.; copper and cobalt blue glass decorated jewels from Wadi Qirud dating to c. 1425 B.C. suggest that Asiatic craftsmen were present in Egypt (Lilyquist and Brill 1993, 23, 43); blue, green and colourless glass beads date to the reign of Hatshepsut (1473–1458 B.C.) with the first glass vessels appearing in the reign of Tuthmosis III (1479–1425 B.C.; Nicholson 1993).

Redford (1990, 46) argues that 'numerous stones-of-casting' mentioned in the Amarna Letters imported from an unknown land after Tuthmosis III's seventh campaign is in fact glass. Given that the word 'stone' was frequently used in place of glass during this period (in the cuneiform texts), and even as late as in the Islamic period, this interpretation is likely to be correct. It is worth emphasising that from Tuthmosis III's lists of booty from his military campaigns and benevolences, it seems clear that the glass is in a raw form. This is important because it suggests that both objects (Lilyquist and Brill 1993, 43) and raw glass were imported at the time. There are also requests in the Amarna Letters for glass tribute to be sent from Syria, so this would date to between the thirtieth year of Amenhotep III to the third year of Tut-ankh-Amon when Amarna was abandoned: the maximum period is twenty-eight years and the minimum seventeen, depending on whether there was a coregency between Amenhotep III and Akhenaten or not (Na'aman 1981, 174; Moran 1992). Afgan lapis lazuli was imported from Syria, either via Assyria or Babylon; large amounts of silver from north Syria, Khatte and Cyprus; malachite from the Phoenician coast and calcite from the Hittites (Redford 1992, 44, 46). Sources for the Amarna period stress the import of glass, which was part of a tax burden of cities (Yurṣa, Ashkelon and Lachish) in the southern coastal plain of the Levant. Four separate rulers from southern Palestine sent raw glass (*eḫlipakku*) on the same occasion to the Pharaoh (*EA* 314, 323, 327, 331). The ruler of Tyre sent *mekku* (*EA* 148:5). The weights are 30, 50 and 100, but it is impossible to estimate what unit or value these had (Na'aman 1981, 175). The tribute and gifts from Palestine included silver, cattle and personnel and, more unusually, raw glass and a chariot; raw glass was exported from the Phoenician coast (Na'aman 1981, 184). All were directed at the Egyptian court.

Although this clearly shows that the imported glass had a high (elite) value, perhaps because its colour and opacity imbued it with a specifically high value due to its ritual associations, it does not necessarily indicate that all raw glass that was subsequently worked at Tell el-Amarna was imported. On the basis of recent archaeological (and scientific) evidence, it can now be stated with some confidence that some was made there as well (as discussed later in the chapter). For some time, there was no clear proof that primary glass production occurred at Amarna (Henderson 2000, 40). There is evidence for large scale glassworking there (Nicholson 1995; Nicholson 2007, 127) and this could suggest that glassmaking also occurred. By the thirteenth century, a frit-like materials was being fused in cylindrical crucibles at Qantir-Piramesses (Rehren and Pusch 2005). There is even evidence of the development of local glassworking techniques (Shortland 2001, 221), whether adopted from Mesopotamian artisans or taught by them, as early as the reign of Tuthmosis III, but archaeological evidence for primary glass-making as early as the fourteenth century B.C. in Egypt was, until recently, lacking. Lead isotope determinations of opaque yellow lead antimonate glasses suggest that lead from the Amarna-Fayum area was used to make lead-rich glasses or to modify low-lead glasses dating to the reigns of Amenhotep II (1427–1401) and Tuthmosis IV (1401–1391) and for glass found at Malkata (Keller 1983; Brill *et al.* 1993, 60–1). It should also be noted that the two Amenhotep II samples and the Tuthmosis IV sample fall slightly outside the main cluster of results and that there does not appear to be a clear statistical distinction between results in group 'E' (English and some European leads) and 'L' (Laurion or 'Laurion-like' leads; Brill *et al.* 1993, figs. 58–60). It is certainly significant that the lead isotope results for a vessel fragment of Egyptian style found in the tomb of Tuthmosis III falls outside the 'early' Egyptian group with the possibility that the lead used was of a Middle Eastern origin, described somewhat ambiguously as "Later" Egyptian glasses and other "Mesopotamian"/Iranian leads (*ibid.*, 63–4, table 3).

A further question is whether the use of an Egyptian lead source in the production of opaque yellow glass there helps to prove that the glass was made in Egypt. First, the interpretation of lead isotope results is not simple: the potential mixing of lead sources and the fractionation of lead can confuse the issue. Moreover, an overriding consideration is the need to determine the variation in lead isotope signatures across individual lead sources that were known to have been exploited in antiquity (Rohl and Needham 1998; Stos-Gale *et al.* 1997; Gale and Stos-Gale 2000; Henderson 2000, 253–4). Some research in characterising lead deposits using isotopes has been carried out, but the problem of mixing remains. Nevertheless, a relatively tight group of eighteen lead isotope results for Eighteenth Dynasty Egyptian glasses suggests that it could represent the true Egyptian source of lead used in Amarna glass. A possible mineral source of lead is galena (PbS) associated with sphalerite (ZnS), which could account for the zinc levels sometimes found in the glasses (Shortland 2002, 524). Bismuth is associated with bindhymite (see Chapter 3), a lead antimonate mineral (Henderson 1985). It has been possible to detect bismuth in lead antimonate crystals in opaque yellow Amarna glass by using Time of Flight-Secondary Ion Mass Spectroscopy, but not in the matrix of the glass (Duckworth *et al.*, 2012). This is the first direct evidence for its use. The apparent presence of 'Egyptian' lead in Mesopotamian glass suggests that some lead was exported to Mesopotamia from Egypt, where it was incorporated in the base glass made there. Overall, lead isotope research of early glasses provides evidence for the use of local lead but not for the primary manufacture of glass.

The chemical compositions of Bronze Age Mesopotamian, Egyptian and Greek glasses are now considered in more detail. The relative levels of sodium and calcium oxides (Fig. 6.1) in these plant ash glasses appear to show a rough distinction between Egyptian glasses on the one hand, with somewhat higher soda levels, and Mycenaean and Mesopotamian glasses on the other. This might suggest that two different production zones were involved. However, there are clearly exceptions, such as (Greek) Elateia samples that fall amongst Egyptian samples and Egyptian samples that fall in the 'Mesopotamian' area. Using these results alone, it could be suggested that the Ulu Burun ingots were made in Egypt. Consideration of positively correlated potassium and magnesium oxide levels, primarily associated with the use of plant ashes used to make these cobalt blue and turquoise glasses (Fig. 6.2), confirms this approximate distinction and reveals several other things:

1. Almost all Tell Brak and Nuzi glasses contain more potassium and magnesium oxides than found in Greek and Egyptian glasses, although this is partly because some Egyptian cobalt blue glasses contain characteristically low potassium oxide levels. This distinction also holds true for the Nuzi cobalt blue glass data published by Shortland and Eremin (2006, table 2), with magnesia ranging between 3.7% and 6.3% and all except one containing between 2.7% and 4.7% potassium oxide.
2. Many Greek glasses contain lower magnesia levels than Egyptian and Mesopotamian glasses.
3. There is a group of five glasses from Elateia, Greece that sits amongst the data points for Tell Brak and Nuzi. This suggests that they were made in Mesopotamia and found their way to Elateia. Significantly, they are not Mycenaean moulded pendants but plain annular beads.
4. A number of Egyptian glasses are characterized by their low potassium oxide levels (many are cobalt blue). However, there are also a number of turquoise 'Egyptian' glasses with higher potassium levels that fall amongst the Mesopotamian data, including almost all of the glass analysed from Malkata with levels of above 1.5% K_2O (not plotted here). This could suggest that the glasses found in Egypt, with higher potassium levels, were made in Mesopotamia and imported as raw glass or ingots to Egypt to be worked into objects.
5. The data for the glass ingots from the famous shipwreck of Ulu Burun indicates that, with the exception of one, they appear to be either of a Mycenaean or an Egyptian provenance, with K_2O levels of c. 1.5% and below.
6. Although the Egyptian and Mycenaean cobalt blue glasses have similarly low potassium levels (below 2%), some Egyptian glasses contain higher magnesia levels.

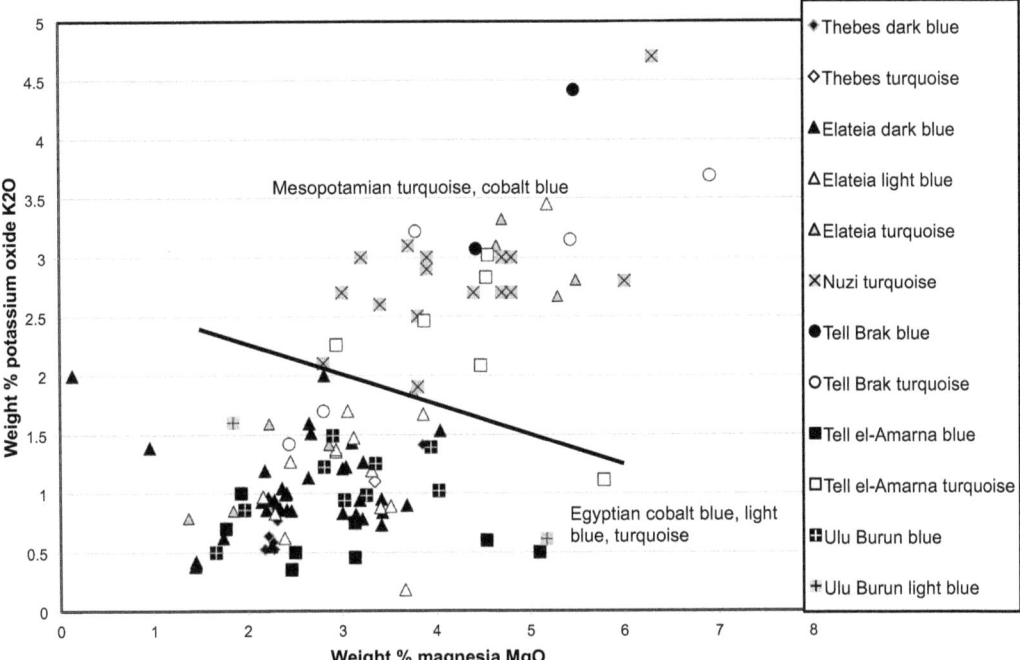

6.2. A bi-plot of weight % magnesia versus potassium oxide in late Bronze Age glass from Greece, Egypt and Mesopotamia. Data from Brill 1999b, Nikita and Henderson 2006, Shortland and Eremin 2006 and Henderson *et al.* 2010.

It is apparent that plant ashes with a variety of chemical compositions, as reflected in the magnesium and potassium oxide levels, were used to make Mycenaean, Egyptian and Mesopotamian base glasses. Although this is not an especially clear pattern, a number of factors may have produced these compositional variations, including the plant genera used, the season in which the plants were harvested, the geochemistry of the soils in which the plants were growing, the way they were ashed, any mixing of plant ashes and the use of colourants. Shortland (2005) has suggested that soil introduced in the glass melt would have led to some of its chemical characteristics. A dominant factor that determined plants ash compositions is the geochemistry of the soil in which they grew – and it is this that would contribute to the major, minor and trace characteristics in glasses made from them (see Chapters 2 and 11). So although the accidental inclusion of soil may have occurred, it is more likely that the use of plant ashes led to the major and minor compositional characteristics associated with the alkali in the glasses. Sayre (1967, 146) followed by Tite *et al.* (2005, 12) have suggested that raw glass was imported from Egypt or the Near East to be made into objects in the Aegean. The results for major and minor element oxides discussed so far indicate that although there are some interesting differences between the chemical compositions of Mycenaean glasses and those of Egyptian and Mesopotamian origins, there are also overlaps, and the case for a separate manufacture of glass from raw materials in Mycenaean Greece still needs to be made.

Chemical analyses of two tubes decorated with opaque white spirals from Tchoga-Zanbil, Mesopotamia (Brill 1999a, 45; Brill 1999b, 46) shows that their relative levels of soda, calcium, potassium and magnesium oxides places them in the 'Mesopotamian' group (see Figs. 6.1 and 6.2). The same is true for three unweathered glass samples from Tell al-Rimah (Brill 1999a, 42; Brill 1995b, 42). If Mycenaean data are added to the plot of alumina and titania published by Shortland and Eremin (2006, fig. 1) the Mycenaean data for non-blue glasses is largely distinctive, mostly containing 0.05% titania, and with higher alumina levels.

Cobalt-coloured glasses were clearly in demand for the manufacture of Mycenaean jewellery. As discussed in Chapter 3, cobalt-rich ores are associated with a range of mineral impurities, such as manganese, copper, nickel, iron, arsenic, sulphur and silver and that these are characteristic of particular cobalt-bearing ore deposits. However, as noted in Chapter 3, it is not necessarily a straightforward matter to relate a cobalt-rich ore deposit of variable composition to the glass that was coloured with it. For example, 'erdkobalt', deriving from the Black Forest in Germany, was roasted in the medieval period to purify it (with the removal of some arsenic) before being used as a colourant (Henderson 1985; Henderson 2000). Even though the levels of some oxidised components may be reduced or lost, we still find that ancient cobalt blue glasses contain impurities that can be derived only from a cobalt ore. Thus, a Persian cobalt deposit, such as the one near Anorak, near Tabriz in Iran, is characterized especially by elevated manganese oxide levels and the cobalt-rich alums of the Dakhleh Oasis in Egypt by elevated alumina, manganese, iron, zinc and nickel (see Chapter 3). Almost all cobalt-rich glasses from Elateia and Thebes, of both mixed-alkali and plant ash compositions, are characterized by high alumina levels, suggesting that an Egyptian cobalt source was used. In Figure 6.3, a group is formed from positively correlated points with samples containing manganese oxide levels of above c. 0.15%. It includes all of the Theban dark blue samples tested. The same general positive correlation is observable in plant ashes, although the manganese oxide levels are far too low to have acted as the only source of manganese (Fig. 6.4). This therefore suggests that either the cobalt-rich colourant was imported from Egypt and added to a base Mycenaean glass, or raw cobalt blue glass was imported from Egypt and made into Mycenaean jewellery in Greece (as discussed earlier).

If primary glass production occurred in late Bronze Age Greece, it is likely to have been carried out in an elite (palatial) context, such as in Thebes. Secondary production of beads and plaques may have occurred elsewhere using exported raw glass.

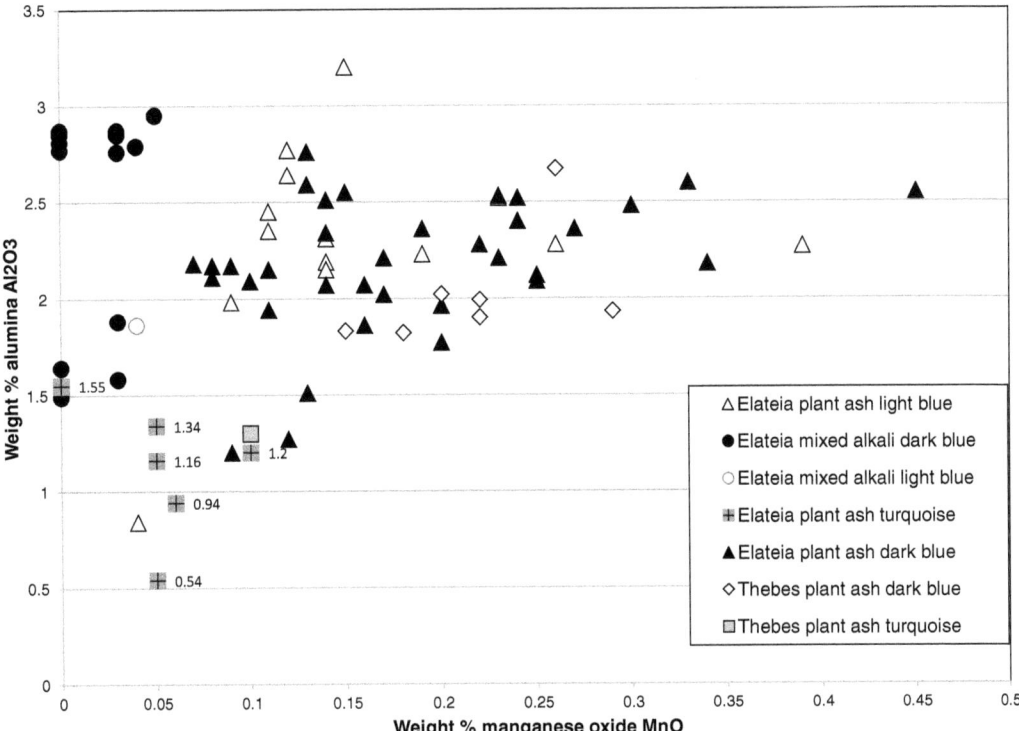

6.3. A bi-plot of weight % alumina versus manganese oxide in early Iron Age glass from Elateia (LMHK) and late Bronze Age glass from Thebes and Elateia, Greece. Data from Nikita and Henderson 2006.

It can also be hypothesized that a major palatial centre such as Thebes would have been both a primary manufacturer and a distributor of raw glass to other cities where beads and plaques were made. The settlement of Elateia, with its associated cemetery in the same region of northern Greece, may well have obtained glass objects from a centre such as Thebes. Suitable halophytic plants such as *Salsola soda*, *Salsola kali*, *Salsola vermiculata*, *Salicornia fruticosa* and *Salicornia europaea* and other species of halophytes are known to grow in the Aegean (Rechinger 1973), and there is every reason to suspect that some, if not all, of these grew in the region in the late Bronze Age. Suitable silica sources were also available to make glass. However, this means only that the raw materials were available for primary glassmaking but does not prove that it definitely occurred. Without the elusive archaeological evidence for primary glassmaking, supported by trace element and isotopic data, the suggestion that primary glass production occurred in Mycenaean centres such as Thebes must currently remain a hypothesis.

Early Proto-geometric (c. 1000/950 B.C.) mixed-alkali (LMHK) cobalt glass from Greece is of two types. Eight of the samples tested contain relatively high alumina levels above 2.6% and manganese oxide levels at below 0.05%.

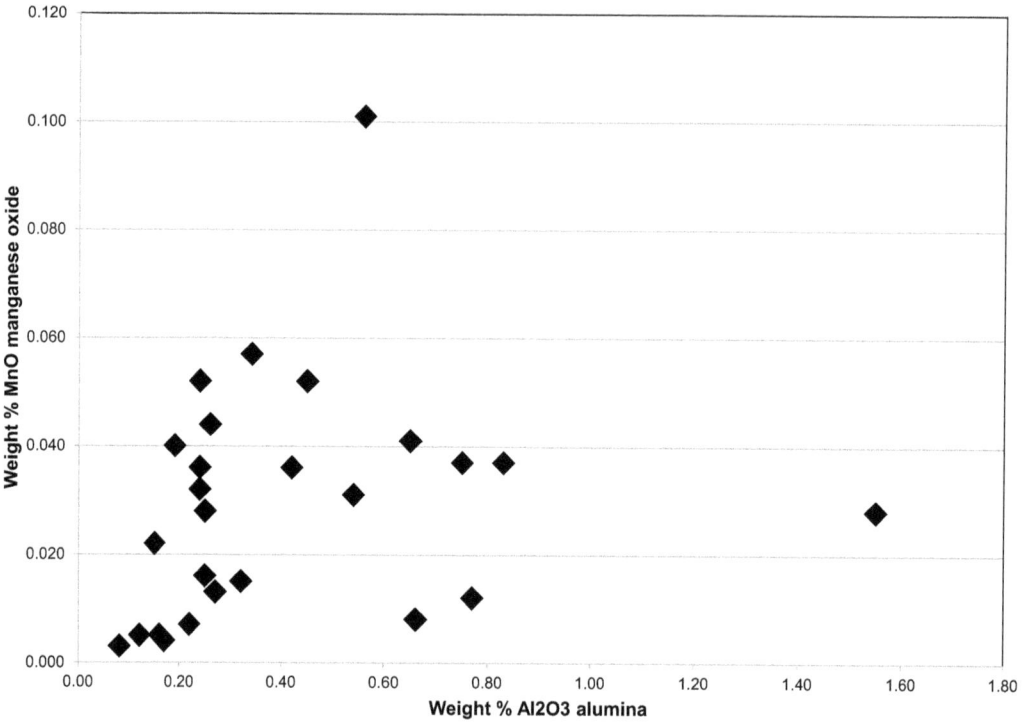

6.4. A bi-plot of weight % alumina versus manganese oxide in Syrian halophytic plant ashes. Data from Henderson and Barkoudah 2006.

Another four samples of cobalt blue LMHK glasses contain alumina levels of between 1.58% and 1.88% (Fig. 6.3; Henderson and Nikita 2006, fig. 19). Similar alumina levels in the plant ash and mixed-alkali glasses can be used as a basis to suggest that the cobalt source was an Egyptian one. However the relative levels of nickel and manganese oxides can be used to show that different cobalt sources were used, with the mixed alkali dark blue glasses from Elateia containing higher levels of nickel oxide (Fig. 6.5). However, the oxides do not discriminate between cobalt sources used in LMHK glasses. Elateia glasses do however contain lower levels of cupric oxide and higher levels of nickel oxide than found in Frattesina glasses (Nikita and Henderson 2006, figs. 25 and 26). Nevertheless, there are exceptions with two Frattesina glasses containing high nickel oxide and three low cupric oxide. Despite these exceptions, which may be indications of trade as shown by (other) evidence (Laffineur and Greco 2005), these differences suggest that the contemporary eleventh–tenth century late Bronze Age–early Iron Age mixed-alkali glasses from Frattesina and Elateia were coloured using two different cobalt-rich colourants. This is a tentative indication that cobalt blue LMHK glasses were coloured in northern Italy and in a second area of which the Balkans and Greece are possibilities. The source of the cobalt colourant used in many Elateia glasses was probably an Egyptian

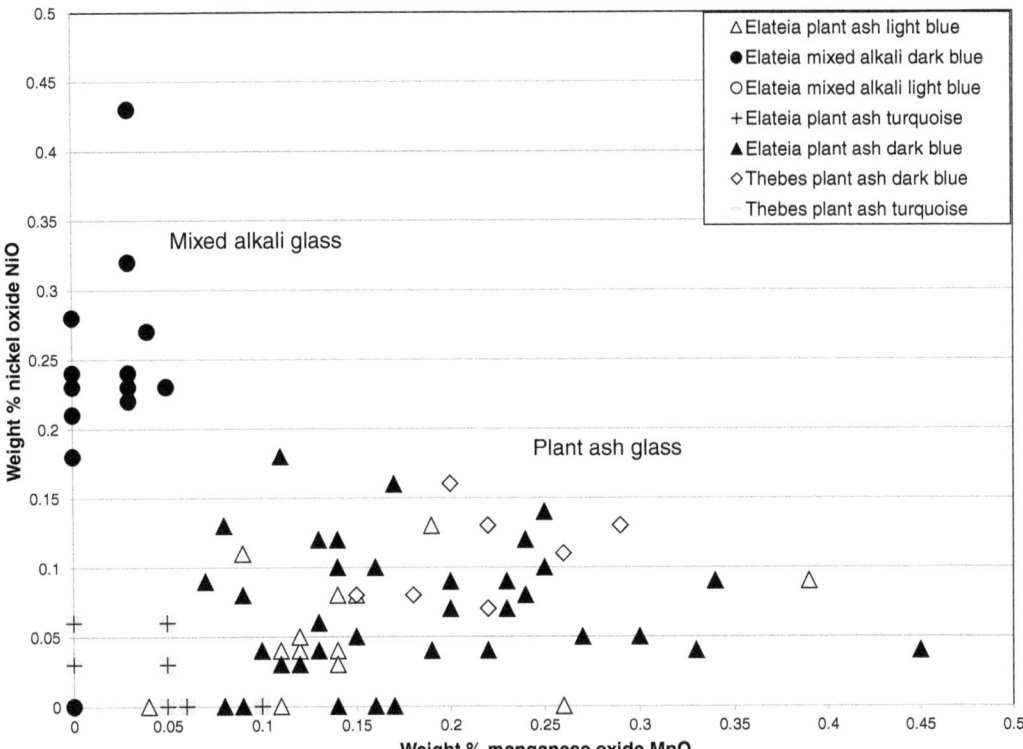

6.5. A bi-plot of weight % nickel oxide versus manganese oxide in glasses from Thebes and Elateia, Greece. Data from Nikita and Henderson 2006.

one, whereas that used in Frattesina blue glasses was probably European, such as the Black Forest deposit in Germany.

A comparison between the levels of alumina and iron oxide in Greek mixed-alkali cobalt blue glasses and those from Frattesina and Mariconda in Italy and turuoise glass from Rathgall in Ireland shows that a group of eleven Elateia glass samples were made from a cobalt source characterised by higher levels of iron oxide than found in Frattesina glass. Most of the rest of the glasses contain similar levels of both alumina and ferric oxide. Some glasses from Frattesina, Mariconda and Rathgall contain alumina levels above 3.5%. This indicates that a second source of alumina is present, perhaps from interaction with the crucible fabric in glasses: there is a clear distinction between the naturally correlated glass, a reflection of the impurities in the sands used and many other glasses with alumina levels above 3.5% (Fig. 6.6).

6.4.1 An Isotopic Approach

The use of isotopes in ancient glass research has recently been reviewed (Degryse, Henderson and Hodgins 2009). It is discussed in detail in Chapter 11.

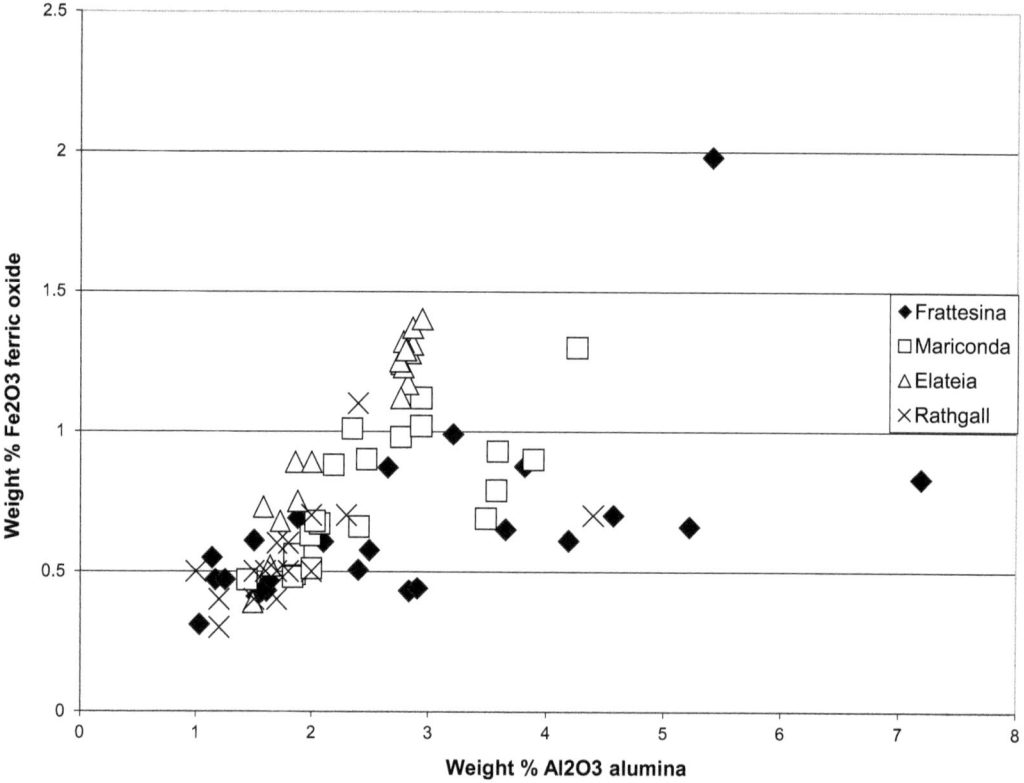

6.6. A bi-plot of weight % alumina versus ferric oxide in mixed-alkali glasses from eleventh-century B.C. Frattesina and Mariconda (Italy), eleventh-century Elateia (Greece) and eighth-century Rathgall, Ireland. Data from Towle *et al.* 2001, Nikita and Henderson 2006 and Henderson 1988a.

Henderson *et al.* (2010) published neodymium and strontium isotope results for a series of late Bronze Age glasses: ten Greek, six Egyptian and eight Mesopotamian. To date, this is the largest data set published for late Bronze Age glass.

The relatively wide variation in the strontium isotope values for Mesopotamian glasses – all samples from Tell Brak (see Chapter 5) – is in stark contrast to the much more constrained variation in Egyptian glasses from Tell el-Amarna and in almost all Mycenaean glass (Fig. 6.7). The level of strontium isotope variation in the Brak glasses is somewhat wider than found in plants growing in northern Syria (Henderson *et al.* 2009) and, on present evidence, suggests that glass was made in more than one centre in late Bronze Age Mesopotamia. Strontium concentrations of up to 1275 ppm in Egyptian and Greek samples (Fig. 6.7) are a reflection of the geochemical contexts in which the plants grew, and there is a distinction from some Mesopotamian glass, which contains less than 449 ppm.

It can be seen in Figure 6.8 that both the concentrations and isotopic ratio values of neodymium provide a distinction between Mesopotamian glasses and

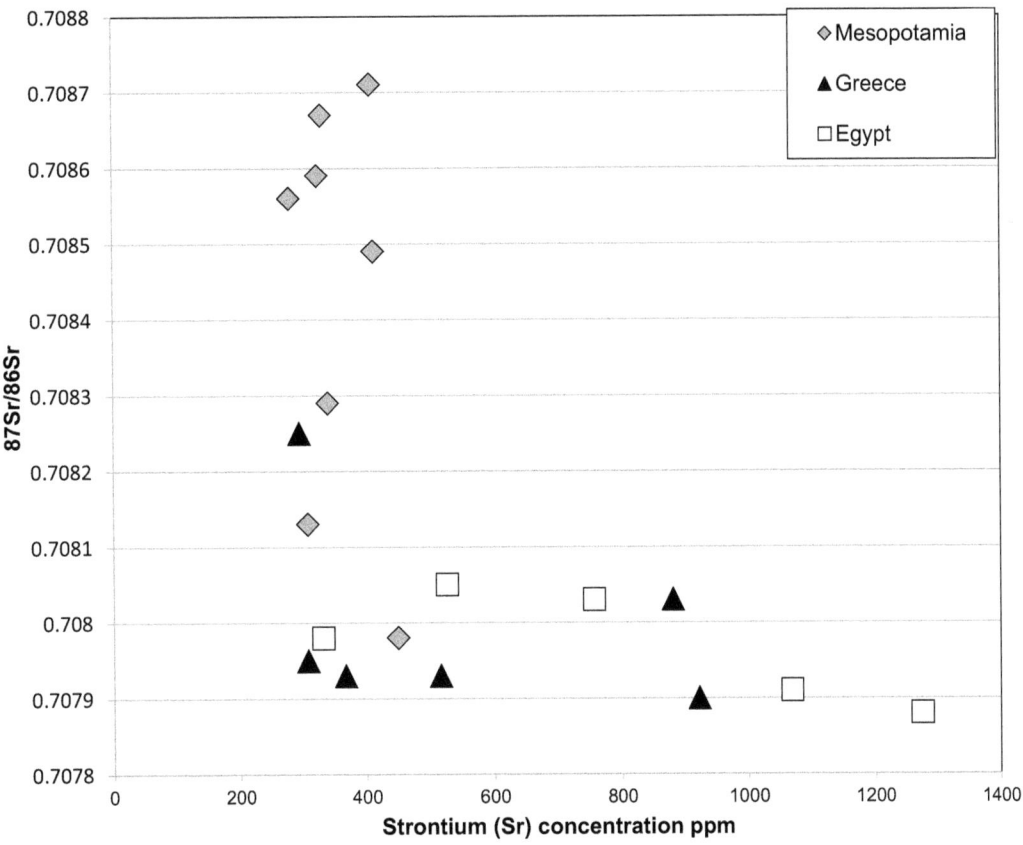

6.7. Bi-plot of $^{87}Sr/^{86}Sr$ versus the concentration of Sr (ppm) in glasses from Mesopotamia, Egypt and Greece.

combined Egyptian and Greek glasses. The Mesopotamian glasses contain very low neodymium impurities (less than 2 ppm) and relatively (geologically) young silica sources. The combined results for Egyptian and most Greek glasses contain higher neodymium impurities (between 2 and 11.7 ppm) and were made with a relatively older source of silica. A plot of the concentrations of neodymium and strontium in these glasses (Fig. 6.9) highlights the evidence for the use of purer raw materials in the manufacture of Mesopotamian glasses, forming a tight group on the plot. The positive correlation of neodymium and strontium for two groups of Egyptian and Greek glasses is not a geologically natural one. An explanation is that both elements were introduced in the glasses in the plant ashes, but that neodymium was introduced in Egyptian glass both in silica and plant ashes (Henderson *et al.* 2010, 13). One positively correlated group is composed of glasses mainly found in Greek Mycenaean contexts. These were made in Egypt – and all are cobalt blue. So this could be tentative evidence that such glass made in Egypt for export was made using slightly different raw material sources and/or that these thirteenth–twelfth century B.C. glasses

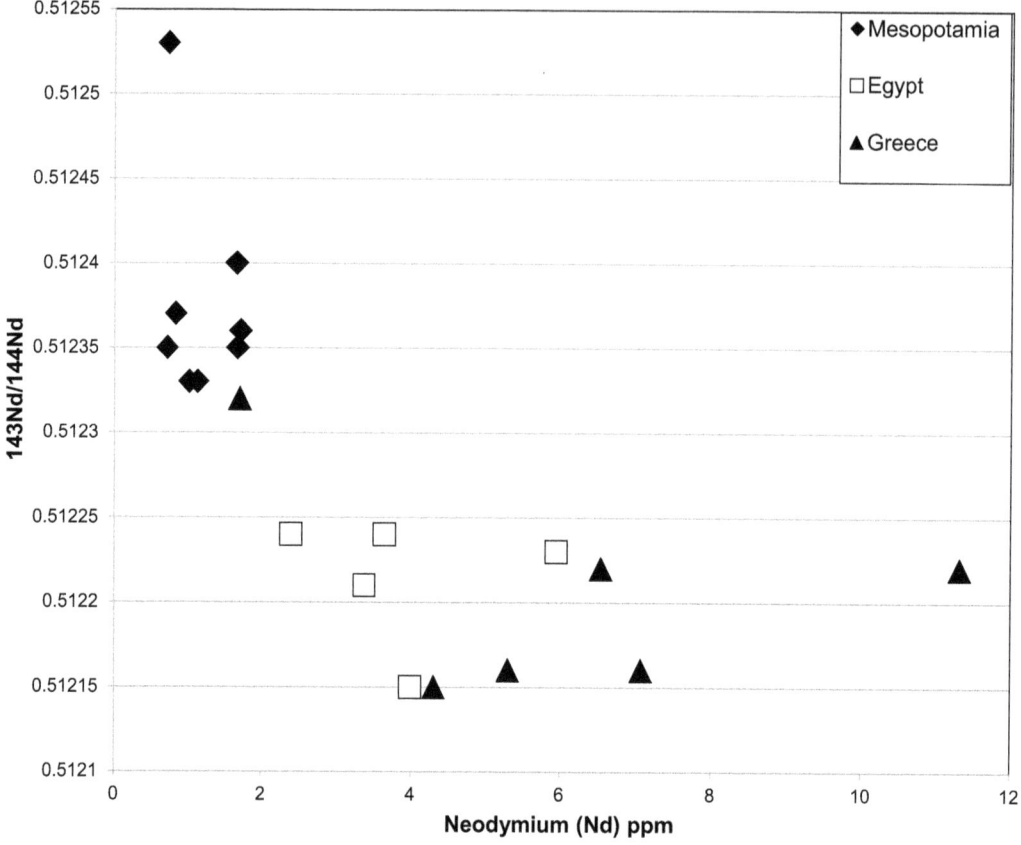

6.8. Bi-plot of ^{143}Nd/^{144}Nd versus the concentration of Nd (ppm) in glasses from Mesopotamia, Egypt and Greece.

(found in Greece) were made at a different location in Egypt but within the same geological zone as defined by isotopes as the Amarna glasses.

The data provide clear evidence for the primary production, provenance and trade in late Bronze Age glass in the Mediterranean. Contrasting neodymium and strontium isotope data distinguish between Mesopotamian glass and combined results for Egyptian and Mycenaean glass (Fig. 6.11). All but one of the Mesopotamian samples are distinguishable from Egyptian ones based on the strontium isotope values. So clearly the use of both isotopes is essential to produce a clear distinction.

If the aim is to provide a provenance for glass by using the local soil geochemistry as reflected in the plant ashes used to make glass (as discussed in detail in Chapters 2 and 11), then it would be hoped that sufficient geological contrasts would provide a fingerprint. Trace elemental analyses provide a distinction between the major production zones of Egypt and Mesopotamia (*a geographical provenance*; Shortland *et al.* 2007) but do not provide a distinction between geological zones (discussed later; see Fig. 6.10). Combining all of the published

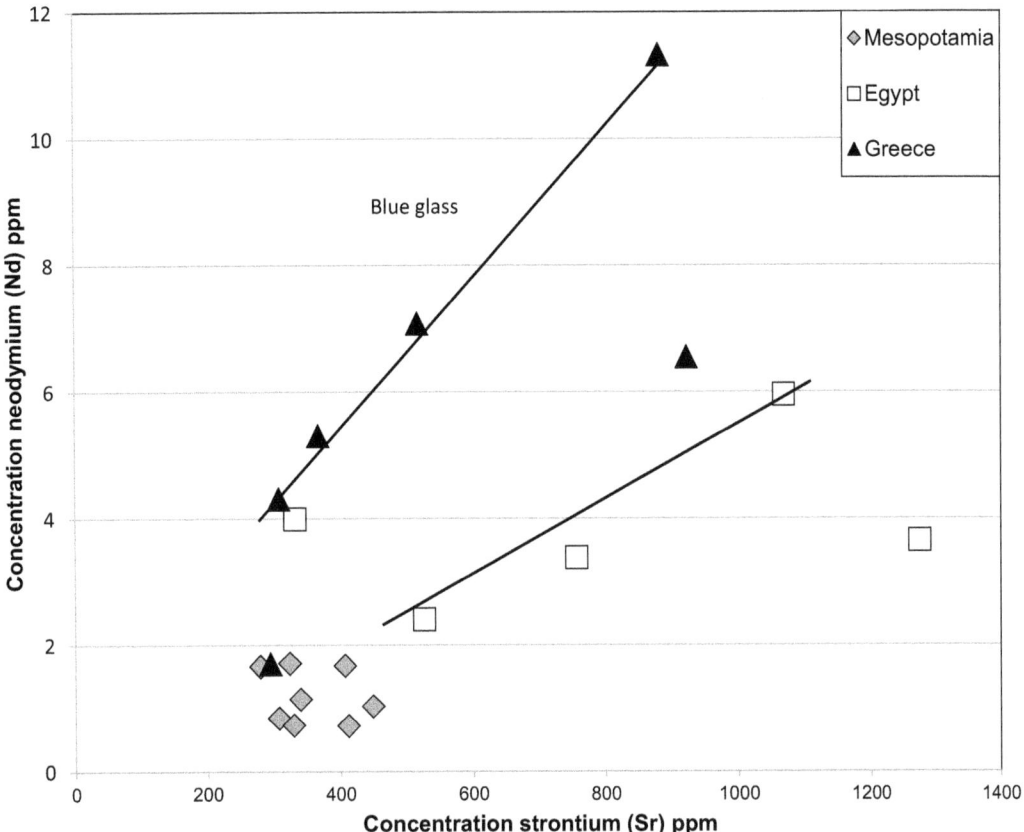

6.9. Bi-plot of the concentration of Sr (ppm) versus the concentration of Nd (ppm) in glasses from Mesopotamia, Egypt and Greece showing two possible correlation lines for some glasses made in Egypt.

isotopic neodymium and strontium isotope results for late Bronze Age glass in the Mediterranean (Degryse *et al.* 2010; Henderson *et al.* 2010) makes a number of things apparent (Fig. 6.11). The most significant is that isotope results provide evidence for the use of raw materials of contrasting geological ages not given by trace elements. When set within the environmental context (Henderson *et al.* 2010) in which strontium isotope variations across the landscape are measured (Henderson *et al.* 2009b), the results are equivalent to *a geological provenance*.

There is a clear contrast between the constrained combined isotopic signatures for glass from the Egyptian sites of Tell el-Amarna and Malkata (with the Greek glass), and the far wider variation found amongst Mesopotamian glass samples (Fig. 6.11). This may indicate that all of the glass was made in one location in Egypt (perhaps at Amarna). In contrast, the wide variation of isotopic data for Mesopotamian glass provides evidence for two (or three) separate production zones. One of these contains the results for Tell Brak ingots: the strontium

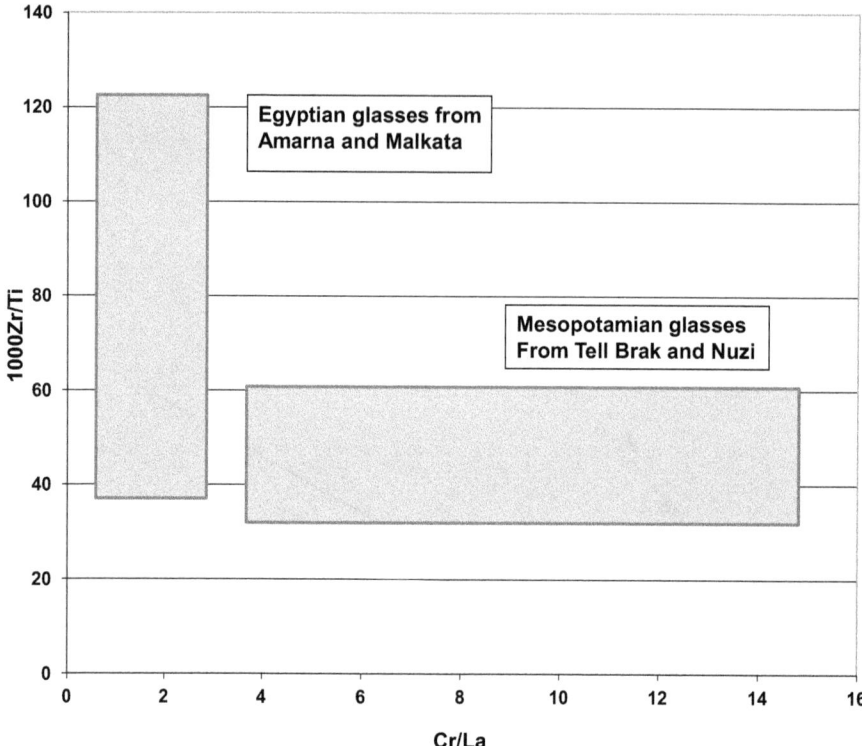

6.10. A bi-plot of chromium/lanthanum versus 1,000×zirconium/titanium in late Bronze Age glasses from Egypt and Mesopotamia (based on Shortland et al. 2007).

isotope values in plant ashes from the region are consistent with a northeastern Syrian production zone, including the site of Tell Brak ('Mesopotamia 1' in Fig. 6.11).

A second Mesopotamian group, formed from glass characterized by a geologically older calcium source (with lower $^{87}Sr/^{86}Sr$ signatures) than the 'Brak' group but with similar $^{143}Nd/^{144}Nd$ signatures can be identified ('Mesopotamia 2' in Fig. 6.11). A third group formed from three Nuzi samples and a single Brak sample ('Mesopotamia 3'?) suggests that a separate production zone existed in northern Iraq; given the large volume of glass found at Nuzi (see Chapter 5), primary glass production may have occurred there. However, the strontium isotope signatures for these glasses are similar to glasses found in northern Syria, falling between the 'Mesopotamia 1' and 'Mesopotamia 2' groups in Figure 6.11: it is the neodymium isotope signatures that distinguish these Nuzi glasses from other Mesopotamian glasses. Indeed, the Nuzi glasses have neodymium signatures that are similar to some Egyptian glasses. In spite of this similarity, if they were made in the same place the strontium isotope results would also be expected to match (Degryse et al. 2010). Nevertheless, a Nuzi-type bead found in Athens falls into the 'Mesopotamia 2' group – and not with the other Nuzi

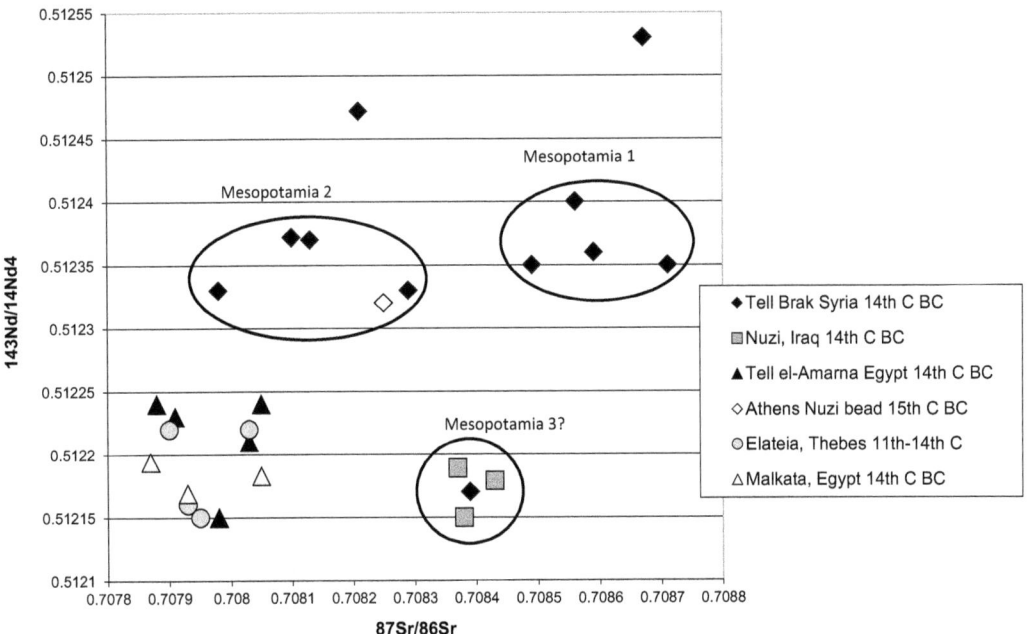

6.11. Bi-plot of $^{87}Sr/^{86}Sr$ versus $^{143}Nd/^{144}Nd$ in glasses from Mesopotamia, Egypt and Greece. Data from Degryse *et al.* 2010 and Henderson *et al.* 2010.

glasses. One result for a Tell Brak glass falls into the 'Mesopotamia 3' group (Degryse *et al.* 2010). The isotope results for Mesopotamia groups 2 and 3 need to be related more precisely to strontium isotope variations across the landscape. This could be achieved by determining Sr isotope variations for plants more intensively, in selected small areas, and over larger geographical areas in northern Mesopotamia. This would create a more statistically rigorous link between isotope signatures in the late Bronze Age glass and the environment.

6.4.2 The Way Forward

Up to c. 2009, there was only a tentative indication that primary glassmaking occurred at Tell el-Amarna. Since then, the publication of the results of radiogenic isotope analyses combined with the identification of lumps of semi-finished glass containing residual quartz with no added colourant and others with added antimony (Smirniou and Rehren 2011) have provided mutually supporting evidence that primary production did take place there. Moreover, Walton *et al.* (2009) published the first set of trace element analyses of Mycenaean glasses. The main result of their research is a compositional distinction between Mesopotamian and Egyptian/Mycenaean glasses, supporting the evidence that separate primary production occurred in the two areas. These results also provided the first evidence of trade in glasses, using trace compositions, from Egypt and Mesopotamia to Mycenaean Greece. Further important trace element results were published by Jackson and Nicholson (2010), who confirmed that cobalt blue glass ingots found on the Ulu Burun shipwreck were of an Egyptian trace element composition agreeing with both the isotope and new archaeological evidence for primary production of glass in Egypt. With only provisional evidence for glass production at Alalakh, Turkey (pers. comm Prof. Dr Yenner) in the earliest glassmaking zone of Mesopotamia, the combined use of trace element and radiogenic isotope analysis to address the question of primary production in different parts of the Bronze Age Mediterranean is one of few ways forward.

Inevitably there are still a number of unanswered questions relating to primary glass production. The logical next step in the scientific investigation of Bronze Age glass in the Middle East is to increase the number of trace element determinations for well-dated glasses from Egypt, Mesopotamia and Greece. Publication of initial suites of trace element analyses for late Bronze Age glasses found in these areas (Shortland *et al.* 2007; Walton *et al.* 2009) have provided clearer compositional distinctions between fourteenth-century B.C. glasses found in Egypt and Mesopotamia. The trace elements associated with silica, in particular, appear to reflect the use of different silica sources in Egypt and Mesopotamia. This could be regarded as a geological provenance (see Chapter 11). However, if the chemical characteristics of possible silica sources (and plants) in Egypt were determined, it would be possible to estimate exactly which elements are associated with Egyptian silica, in what proportions, and which occur in both silica and plant ashes. Once this is has been established, it would be possible to relate these characteristics to Egyptian glasses. Without these results, it might be argued that the trace-element distinctions provide a reasonably convincing scientific case for independent primary production of glass in fourteenth-century B.C. Mesopotamia and Egypt. However, this is not really as convincing as the geological provenance provided by radiogenic isotopes, dividing the Mesopotamian group defined by trace elements (Fig. 6.10) into three separate subgroups.

A key question is whether chemical compositions of late Bronze Age glass are a true reflection of the local soil geochemistry (Shortland 2005) for the areas in which the glasses were made and therefore also for the primary production sites involved. Shortland *et al.* (2007, 788) suggested that the relatively high levels of chromium found in Mesopotamian glasses are due to the use of chromium-bearing *sands* derived from the erosion of ultra-mafic rocks at the headwaters of the Euphrates and Tigris rivers. However, trace element analyses of northern Syrian halophytic plants ashes show that their ashes could be the source of the chromium instead (especially if quartz were used, as described in the cuneiform texts) and would provide one of the compositional characteristics of late Bronze Age glass found in Mesopotamia.

A plot of relative levels of La and Cr in halophytic Syrian plants of a range of genera (Henderson *et al.* 2010) shows a positive correlation with an R^2 value of

0.9376. This correlation is also found in Mesopotamian late Bronze Age glasses (Shortland *et al.* 2007, fig. 4). The range of chromium concentrations detected in northern Syrian samples of the genus *Salsola*, a plant genus thought to have been used for making ancient glass (Ashtor 1992; Barkoudah and Henderson 2006), is between 4 and 14 ppm. These levels fall within the range detected in Mesopotamian glasses, of between 0.83 and 24.8 ppm (Shortland *et al.* 2007) even after dilution with the addition of silica. Plant ashes must have contributed part, if not all, of the Cr in the glass. Walton *et al.* (20009, 1500–1501) have suggested that elevated (and correlated) levels of titanium, zirconium, lanthanum and chromium in Mesopotamian glasses are due to greater contamination with clay or stone, but their contents in plant ashes must also have contributed (and would reflect the compositions of the clays to some extent). Therefore, chromium cannot necessarily be used as a means of characterising the silica source used to make Bronze Age (plant ash) glasses, but it is at least as likely to be a characteristic of the plant ash used. As noted earlier, an apparent lack of published trace element data for Egyptian halophytic plants makes it impossible to contrast chromium contents in Egyptian and Syrian plants in order to test this and also to compare the levels of chromium in Egyptian plant ashes with Egyptian and Mesopotamian Late Bronze Age glasses.

6.5 Scientific Analyses of Mixed-Alkali and Potassium Glasses, and Mixed-Alkali Glassy Faience and Faience from Europe: Evidence of Production Zones?

The site that has yielded the best evidence for the glass industry in prehistoric Europe is eleventh–ninth century B.C. Frattesina in the Po valley, northern Italy (Bietti-Sestierri 1996, 1997): see Chapter 5. Crucibles with glass adhering are a rarity in prehistoric Europe and several have been found at Frattesina. The site has produced the most glass and glassy faience from any site of the period. Vitreous faience has been found on other sites, including earlier ones. Excavations of Mariconda, Citanal Bianco, Poviglio, Prato di Frabulino, Trinitapoli, Chiaromonte, Grotte Vittorio Vecchi and Vicofertile in Italy have produced beads made from both glass and vitreous faience.

The characteristic that unifies most of the glasses and glassy faience found at these sites is their diagnostic chemical compositions. Almost all are of a mixed-alkali type (Biavati 1983) with low levels of magnesia and calcium oxide, and high levels of potassium oxide, the so-called low magnesium, high potassium (LMHK) type (Henderson 1988a, 80–81, fig. 4.1). Two other compositional types which are related to the mixed-alkali glass (because they are characterised by the same low impurity levels) are potassium glasses (also found at Frattesina) and middle Bronze Age (1450–1200 B.C.) Italian soda glasses (Angelini *et al.* 2005, 34, table 1). Mixed-alkali, potassium-rich and soda-rich Italian vitreous materials are compositionally linked by their low levels of magnesium and calcium oxides. The high potassium glasses are especially interesting because they are the earliest recorded in the world. They contain amongst the highest potassium levels of any ancient Western glasses, not to be equalled until the introduction of wood ash as an alkali source in High Medieval northern Europe some 2,000 years later. The discovery of mixed-alkali glasses therefore appeared to provide the first evidence for the independent invention of European Bronze Age glass (Henderson 1988a; Henderson 1988b). The discovery of tenth-century B.C. mixed-alkali and ninth-century B.C. potassium glass beads on the island of Thasos (Henderson 1992) and in the tenth-century phase of the Elatia cemetery in Greece (discussed earlier; Nikita and Henderson 2006) shows that their distribution extends as far east as the Aegean. Although this does not diminish the case for an independent invention of European glass, it raises the possibility that there were other production centres for the glass type. Moreover, given the complex trade links between different parts of the Bronze Age Mediterranean, it must only be a matter of time before examples are found in the Middle East. It is very likely that examples of this glass type will be found in the area bordering the eastern Adriatic.

These mixed-alkali glasses and glassy faience are further characterised by high silica levels of up to c. 80%. Although clearly much of the silica is dissolved to form the glass matrix, the beads and other objects made from it can contain relict silica crystals (Fig. 6.12). The silica levels in some examples fall close to 70%, a level that is not unusual in (fully fused) plant ash (and natron) glasses. Even in these examples there are may still be a small proportion of silica crystals present. The presence of these silica crystals provides a clear link with the earlier technology of faience (see Chapter 1). Examples of faience dating to c. 2000 B.C, with vitreous phases characterised by a mixed-alkali composition from Kerma in the Sudan (Hatcher *et al.*, in preparation), faience from early Bronze Age Poland (Robinson *et al.* 2004) and in what has been described as 'glass' from possible third-millennium B.C. contexts in France (Guilaine *et al.* 1991; Azemare *et al.* 2000), suggests that there was an interesting interlinked development of faience and glass technology in the late third and during the second millennia B.C. over a large area. Without primary evidence for the production of mixed-alkali faience and glass, and the scientific results for more well-dated samples, it is difficult to be more precise about how the technologies were related across this wide geographical area. The earliest examples of mixed-alkali glasses and glassy faience occur in France c. 2000 B.C. or earlier, and somewhat later in Italy (c. 1600 B.C.) and Greece (c. 1000 B.C.).

The microstructures of some 'glasses' can be described as being transitional between faience and glass (Santopadre and Verità 2000) and best defined as glassy faience. Frattesina glassy faience contains tridymite (Verità and Biavati 1989, 296; Santopadre and Verità 2000, 35; Artioli *et al.* 2008a, 244), which indicates that the silica was heated to a minimum of 870°C because (in the right

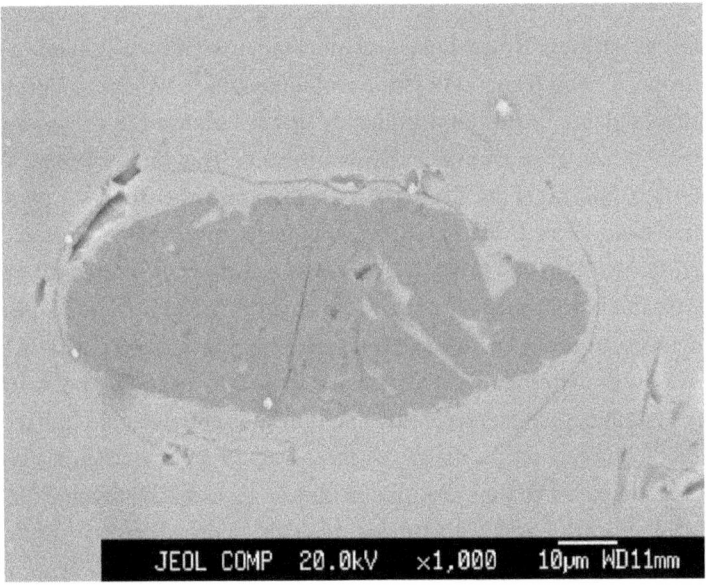

6.12. A crystalline silica inclusion in a mixed-alkali glass from eleventh-century B.C. Frattesina. Note the cracks surrounding the inclusion, caused by the crystal contracting more slowly than the glass matrix ×100.

conditions) the transition from α-quartz to tridymite occurs at that temperature. However, although this gives a minimum temperature, tridymite is stabilised in the presence of an alkali over the range of 870° to 1470°C (Sosman 1965, 45–53), and so crystallisation of tridymite from the glass must have occurred at high temperatures. Middle Bronze Age 1–2 glassy faience from Grotte Vittorio Vecchi and Middle Bronze Age 2 faience from Poviglio and Vicofertile contained traces of cristobalite (Angelini *et al.* 2005, 33). What is more, the occurrence of quartz (tridymite) crystals in glassy faience from Frattesina, and quartz in glassy faience from the earlier sites of Frabulino, Poviglio and Chiaromonte (Santopadre and Verità 2000, table 3), show a link between faience and glassy faience technologies. The presence of a much smaller proportion of silicon crystals in glass produces a more translucent material. Indeed, such a material is best classified as a semi-opaque glass. Although the distinction between glassy faience and glass is an important one, the technological continuum between glassy faience and glass makes it difficult to know when to stop using the term 'faience' and to start using the term 'glass'. However, a far more important point is that it is possible to make a case that an unbroken memory of making a glassy substance (glassy faience and glass) characterised by mixed alkalis was retained over several hundred years in the Mediterranean and probably elsewhere. Therefore, in Europe there is a clear case for technological continuum between faience, glassy faience and glass, even though by the middle and late Bronze Age imported plant ash glass was used at the same time as mixed-alkali glass.

After c. 800 B.C. in Europe, natron glass largely replaced both mixed-alkali and plant ash glasses. Nevertheless, the occurrence of mixed-alkali glasses in a pan-European zone from Italy through Switzerland, Bohemia, Germany and France into England and Ireland indicates that it was used in these zones often alongside plant ash glass from as early as 1400 B.C. (Henderson 1988a, Venclová et al. 2011). Guilaine et al. (1991) have published the analysis of a mixed-alkali glass bead from Gord in France that apparently has a date of 4100 ± 70 BP (calibrated 2895–2420 B.C.). The latest examples of mixed-alkali glasses derive from the late Bronze Age site of Rathgall in Ireland that has been dated radiometrically to between the ninth and eighth centuries B.C. (Raftery 1976). The glass found at Rathgall consists of eighty-eight beads or bead fragments primarily of a turquoise green colour; one is decorated with gold (Raftery 1987).

The alkali source used to make the mixed-alkali and potassium glasses was rather pure compared to that used for plant ash glasses. The levels of magnesia, phosphorus, chlorine and calcium are all consistently lower in mixed alkali than in plant ash glasses. Plants that could potentially provide such mixed-alkali compositions grow in maritime environments. One such is 'verec', also known as kelp. Its ashes contain roughly equal quantities of sodium and potassium (Verità 1985, 21, table 1), and it grows on the northern European coast. However, even if this alkali source was used to make mixed-alkali glasses in northern Italy, it seems unlikely that it was also used to make the mixed-alkali vitreous phase of Sudanese or Polish faience. In any case the use of verec and other maritime plant ashes would introduce elevated levels of impurities, such as magnesia and phosphorus (Dungworth 2005, 256), and such levels are not found in mixed-alkali glasses.

Reduction in the levels of insoluble MgO, 'CaPO$_4$' and CaCO$_3$ can be brought about by purification of plant ashes. This is achieved by solution, filtration and recrystalisation of plant ash, a practice that certainly occurred in medieval Europe and has been suggested for prehistoric Europe (see Chapters 2 and 4). So this needs to be considered as a possible explanation for low MgO and CaO levels in mixed-alkali glasses, glassy faience and faience. However, if this practice did occur, as has been suggested (Brill 1992, Hartmann et al 1997) it would need to have been practiced almost seamlessly for the production of the glassy phase in faience, as early as 2000 B.C. to the manufacture of mixed-alkali glasses in the early first millennium B.C. It is therefore an unlikely interpretation of the evidence (cf. Tite et al. 2006).

A much more probable interpretation is that the alkali raw material(s) used were already pure. Brill (1992) has suggested that potassium-rich saltpetre (KNO$_3$) present in the effloresced salts from latrines and manurial soil might have been used. Its use could account for the low impurities in the mixed-alkali glasses. However, although the saltpetre is accompanied by nitre (Na$_2$CO$_3$), its use does not account for the high soda glasses reported by Angelini et al.

(2005), which in other ways are compositionally very similar to mixed-alkali glasses.

Another possible explanation to consider is that natron (trona), sodium sesquicarbonate, was mixed with a mineral source of potassium to create the mixed-alkali glasses. The earliest natron glasses date to between c. 1000 and 800 B.C. (Sayre and Smith 1961; Sayre and Smith 1967; Henderson 2000; Reade *et al.* 2005; Schlick-Nolte and Werthmann 2003). The combination of a potassium-rich mineral and natron would produce a glass with low levels of chlorine, phosphorus pentoxide and magnesia. Early mixed-alkali glass would therefore have been made from what might have been regarded as crushed rocks. There is no doubt that natron was used in the manufacture of some of the earliest Egyptian Blue in the Fourth and Fifth Egyptian Dynasties and as part of the mummification process in the Fourth Dynasty (David 2000, 373, 384). Most famously, it was later used in the Eighteenth Dynasty mummification of Tut-ankh-Amun (Noble 1969, 436). So for the early production of faience with a mixed-alkali glassy phase in Kerma, the use of natron as a soda source could well be the correct interpretation.

Despite its availability, it is interesting that natron was not used for the manufacture of soda glasses until about 1000 B.C. (see Chapter 4). It would seem that faience, glassy faience and glass, all characterised by their mixed-alkali components, were viewed as parallel and very different branches of the glass industry. Tradition combined with an established *chaîne opératoire* prevented natron being used to make the first natron (soda) glass. The mixed-alkali glass that emerged contained quartz crystals that retained a microstructural link with faience over a period lasting hundreds of years.

One compositional characteristic that runs through the full range of these mixed-alkali faience, glassy faience and glass artefacts is a compositional continuum with a more or less constant (c. 16%) total alkali (in potassium and mixed-alkali specimens; Fig. 6.13). Almost the full range of alkali compositions detected in these potassium and mixed-alkali glasses from a wide range of (other) sites is found in Frattesina glasses. The variation of alkali levels ranges from high potassium oxide-low soda glass to approximately equal contents of both. Given that the alkali compositions form a continuum, we are dealing with a mixing line with high potassium glasses forming one 'end member' and high soda, low potassium glass forming the other one. Whether potassium-rich or mixed-alkali in character, the alkali raw material may well have looked the same.

As for the silica used, the low impurity levels of the alumina alone suggest that a pure source, such as crushed quartz pebbles, was used. Indeed, a pure silica source was evidently used in the manufacture of some of the earliest Bronze Age plant ash glasses, so at least here the two branches of Bronze Age glassmaking – mixed-alkali and plant ash glasses – come together. The use of quartz instead of sand would minimise the possibility of shell fragments entering the glass batch, which would in turn minimise lime levels. Lime occurs in both potassium and mixed-alkali glasses but at low levels compared with most plant ash glasses (Figs. 6.14 and 6.16). The addition of a calcium-rich raw material was therefore essential, as discussed later.

Not a single example of mixed-alkali glass has, so far, been found in late Bronze Age Mesopotamia or Egypt. Even the faience used in Egypt at the time apparently had glassy matrices made from plant ash (Shortland and Tite 1998), but with mixed-alkali matrices in faience from the Sudan. For some reason the technology was not used or even adapted. It is likely that mixed-alkali glass containing unreacted or partially reacted silicon crystals, often used to make beads decorated with a spiral of opaque glass – *Pfahlbautonnchen mit Spirale* (Haevernick 1978, Venclová et al. 2011), had quite a specific cultural meaning. The use of colourants in mixed-alkali glasses has been discussed elsewhere (Henderson 1988a; Brill 1992; Santopadre and Verità 2000; Towle *et al.* 2001; Angelini *et al.* 2004) and so is only summarised here. One characteristic of all of the coloured glasses is that their alumina and iron oxide contents are positively correlated ($R^2 = +0.9417$). This correlation (Fig. 6.6) is a geological one, and so not unexpected, but it may indicate that a similar silica source was used to make them. It indicates that only a minimal contribution of aluminium and iron was made to the glass melt by the colourants and the alkali. The colourants used were Cu II in translucent turquoise glass and Cu I in opaque red glass, in the form of F-centred cubic crystals of metallic copper, sometimes associated with arsenic in the mineral used (Angelini *et al.* 2004, 1180–3); positively correlated copper and tin ($R^2 = +0.752$; Towle *et al.* 2001, table 6), in a ratio of copper to tin of 9:1, shows that scrap bronze (Brill 1992) or (less likely) metal slag was used as a copper source. Santopadre and Verità (2000, 38) have described some copper-rich samples of faience and glassy faience from Prato di Frabulino and Chiaromonte, respectively, that did not contain tin but lead, arsenic in faience and nickel and zinc in glassy faience. On the basis of the compositional variations associated with copper colourants, it can be suggested that a mineral rich in copper and lead and/or arsenic was used (Henderson 1999a).

A blue cobalt colourant rich in nickel, arsenic and iron was used to colour the glasses from Mariconda and Frattesina (Towle *et al.* 2001, 23–4, table 5), with a positive correlation between cobalt and nickel being especially strong ($R^2 = +0.754$). The converse of this is that the cobalt colourant used is not correlated with alumina, lead, manganese, zinc or antimony. Therefore a cobalt-rich ore such as skutterudite (($Co,Ni,Fe)As_3$) from the Black Forest in Germany would have been used. This complex cobalt-rich mineral deposit is characterized by nickel, iron and arsenic (see Chapter 3 and Henderson 1989; Henderson 2000a, 30). Ten of the mixed-alkali cobalt blue glasses from Elateia, northern Greece (Nikita and Henderson 2006, fig. 14), contain higher ferric oxide levels than found in glasses and glassy faience from Thasos in the northern Aegean and

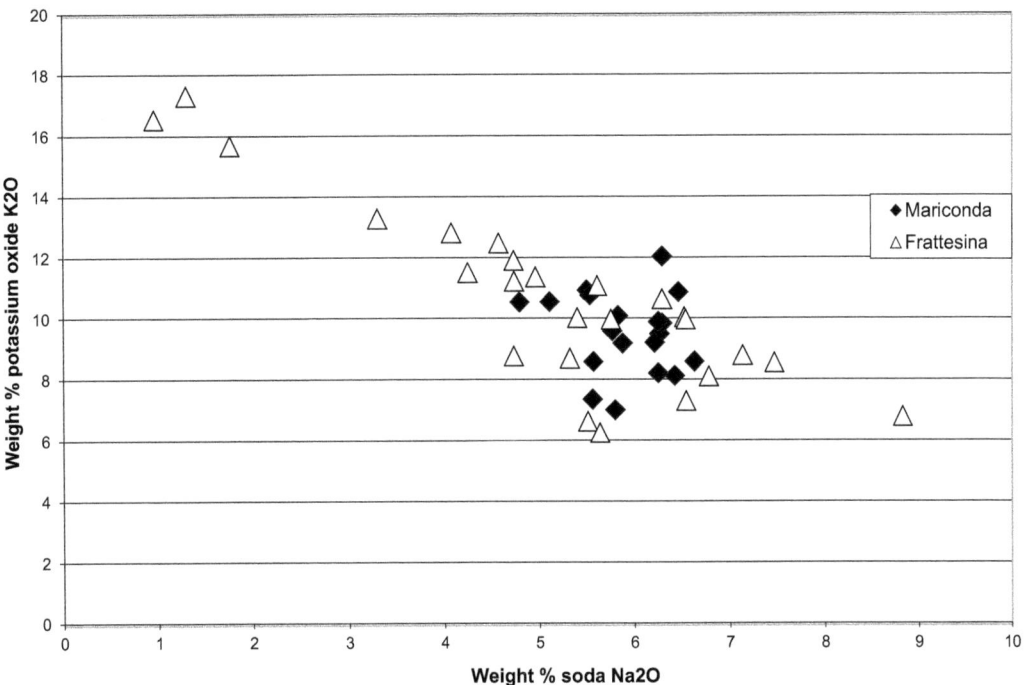

6.13. A bi-plot of weight % soda versus potassium oxide in eleventh-century B.C. glasses from Frattesina and Mariconda, northern Italy. Data from Towle et al. 2001.

the Italian sites of Frattesina, Prato di Frabulino and Poviglio. The higher ferric oxide levels in the Elateia mixed-alkali glasses suggest that they were coloured using a cobalt-rich mineral from a different geological source.

Silica (tridymite) crystals were used for opacification of glassy faience from Frattesina (Verità and Biavati 1989; Santopadre and Verità 2000, table 3). At Hauterive-Champréveyres, silica crystals were used to opacify white glass decoration of mixed-alkali beads (Henderson 1993a). Colourless glasses were of both the mixed-alkali and potassium glass compositions. They were rendered colourless by adjusting the furnace atmosphere: there is no evidence that the glass decolourisers antimony trioxide and manganese oxide were used.

Mixed-alkali glasses are known to have specific physical properties. One of these is that the 'mixed-alkali effect' that increases durability (McVay and Day 1970). Moreover, the higher the atomic number of the alkali present, the greater the absorption of light; sodium has an atomic number of 11 and potassium 19. Thus, relatively high levels of CoO of between 0.17% and 0.24% in mixed-alkali glasses (compared with the levels normally found in plant ash or natron soda glasses) may have been a deliberate attempt to offset the potassium oxide present, which would produce a relatively dark hue.

A comparison of mixed-alkali glass compositions found on different sites provides some interesting insights into regional production for the first time. Although it was suggested that production of 'faience' (we can now add glassy

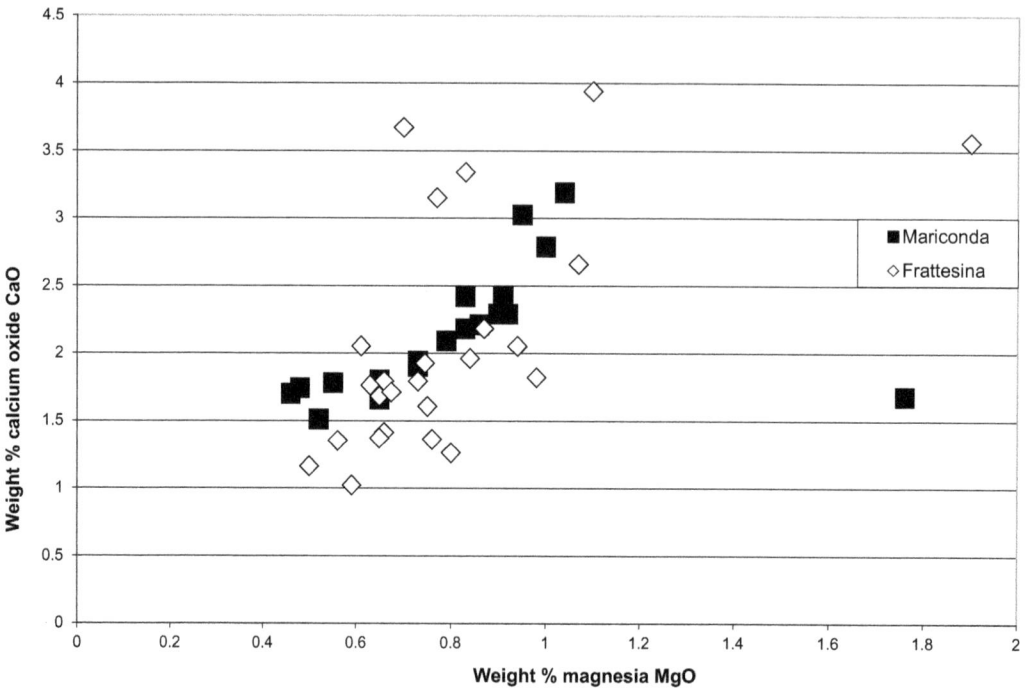

6.14. A bi-plot of weight % magnesia versus calcium oxide in eleventh-century B.C. glasses from Frattesina and Mariconda, northern Italy. Data from Towle *et al.* 2001.

faience) occurred at various locations some time ago (Harding 1971; Harding and Warren 1973; Newton and Renfrew 1970), not many chemical analyses were available at the time. Barfield (1978, 153, fig. 2.1) noticed that a particular 'faience' artefact, the conical button (de Marinis 1975), occurred mainly in northern Italy, and he suggested that it was made locally. By using a scanning electron microscope to investigate the structural characteristics of these buttons, it has been shown that the material used to make them is, in fact, glassy faience. They date to between *c.* 1450 and 1200 B.C. (Bellintani 2000). Like mixed-alkali glasses, they contain low levels of calcium and magnesium oxide but their glassy phase is distinguished from mixed-alkali glasses and vitreous phases in glassy faience in other artefacts by elevated soda and (especially) very low potassium oxide contents (Bellintani 2000, Table 1; Angelini *et al.* 2005, 34, table 1, fig. 2). These diagnostic characteristics support Barfield's suggestion of local production.

In Figure 6.13, the relative alkali levels in glasses from Frattesina and Mariconda are plotted. The data for Mariconda form a cluster whilst those from Frattesina form a mixing line with soda levels ranging between 0.96% and 8.83% and potassium oxide levels between 6.78% and 17.3%. The Frattesina mixing line could be a result of mixing and possibly experimentation. The contrast in the results from the two sites is surprising considering that they are contemporary, lie only 35 kilometres apart and that glassworking occurred in both

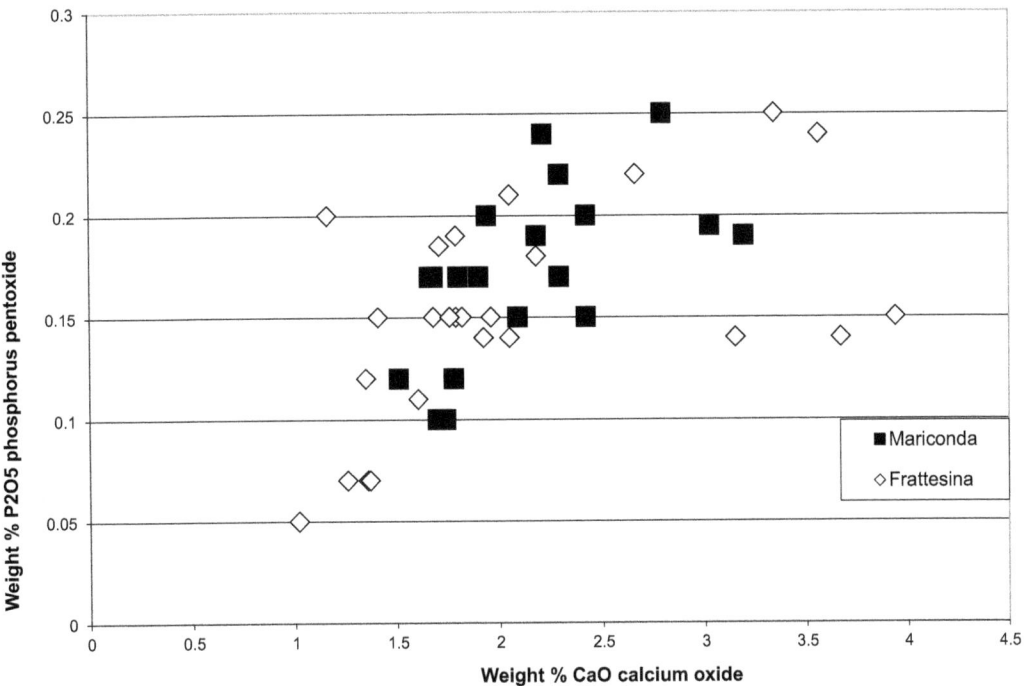

6.15. A bi-plot of weight % phosphorus pentoxide versus calcium oxide in eleventh-century B.C. glasses from Frattesina and Mariconda, northern Italy. Data from Towle *et al.* 2001.

places. One explanation is that Frattesina was an industrial entrepôt and that the glass industry was on a much larger scale. Nevertheless, both glass ingots and crucibles with glass adhering were found at Mariconda (Salzini 1986). As noted earlier, there is also some compositional evidence that the Mariconda glasses derived from a different production centre.

If we now turn to the calcium-rich component in mixed-alkali glasses and the vitreous phase of glassy faience, the plot of phosphorus pentoxide against calcium oxide (Fig. 6.15) suggests that bone ash was a source of calcium used in Frattesina samples. Indeed, the positive correlation between phosphorus pentoxide and calcium oxide in Frattesina glasses containing less than 2.5% calcium oxide is significant. Santopadre and Verità (2000, 38) have reported the presence of calcium phosphate inclusions in one Frattesina sample (presumably because they were not fully dissolved in the glass). They suggest it was introduced as part of the cobalt colourant. However, the use of bone fragments as a calcium source seems a more likely interpretation. In plant ash glasses, the magnesia is introduced in the ashes, producing a well-known positive correlation between potassium and magnesium oxides. In mixed-alkali glasses, no such correlation is found (see Fig. 4.1), but instead there can be a correlation between magnesia and calcium oxide (Fig. 6.14). In this figure, it is notable that although there is a positive correlation between calcium and magnesium

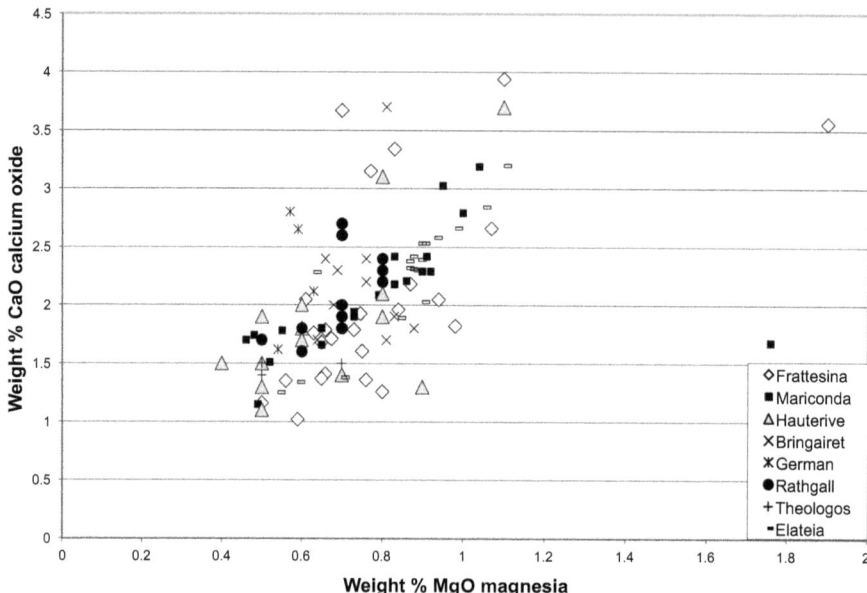

6.16. A bi-plot of weight % calcium oxide versus magnesia in eleventh- to eighth-century B.C. glasses from Italy, Switzerland, France, Germany, Ireland and Greece. Data from Towle *et al.* 2001, Henderson 1992, Henderson 1993a, Henderson 1988a, Nikita and Henderson 2006, Guilaine *et al.* 1991, Hartmann *et al.* 1997.

oxides for many Frattesina samples, the correlation is somewhat stronger for Mariconda results. A mixing line is observable due to the introduction of varying quantities of calcium and magnesium oxides. It is clear from Figure 6.15 that a compositionally more constrained source of calcium was used to manufacture glass found at Mariconda than the glass found at Frattesina. Indeed, very few Frattesina results adhere to the well-defined mixing line for Mariconda. This difference suggests that many of the glassy samples from Frattesina were (surprisingly) manufactured using a somewhat different calcium source and perhaps that they were made in different places. Six Frattesina and three Mariconda samples that contain more than 2.5% calcium oxide fall away from the correlation line, suggesting that a different calcium source was used. However, it is difficult to suggest what that source might have been. Angelini *et al.* (2002, 591–2) have noted a 2:1 ratio of CaO:MgO in mixed-alkali glasses from other Italian sites; they note that the ratio in faience is different at 4:1.

Relative levels of magnesia and calcium oxides in mixed-alkali glasses and vitreous phases in vitreous faience from Switzerland (Hauterive-Champréveyres), Germany, Greece (Theologos and Elateia) and Ireland (Rathgall) provide further insights into the calcium-rich raw materials used to make them (Fig. 6.16). Mariconda samples exhibit amongst the widest compositional variation for positively correlated points in the bi-plot. The data that fit most tightly to this correlation are from Elateia in Greece, forming a group including glass that

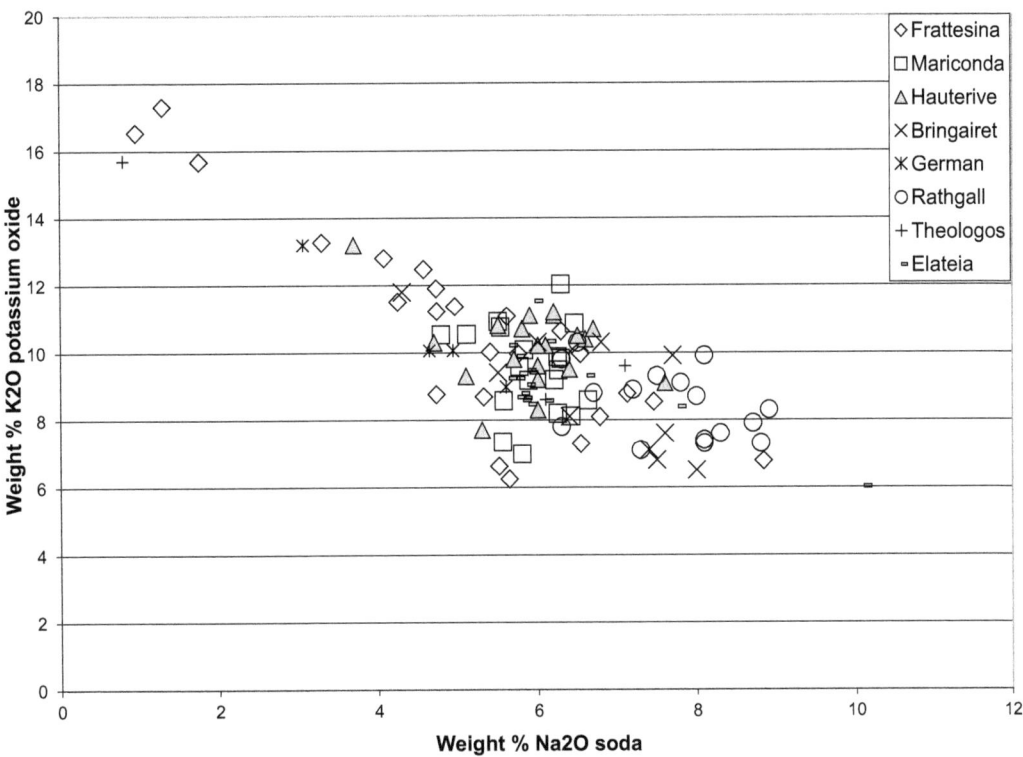

6.17. A bi-plot of weight % soda versus potassium oxide in eleventh- to eighth-century B.C. glasses from Italy, Switzerland, France, Germany, Ireland and Greece. Data from Towle *et al.* 2001, Henderson 1992, Henderson 1993a, Henderson 1988a, Nikita and Henderson 2006, Guilaine *et al.* 1991, Hartmann *et al.* 1997

contains above 3% calcium oxide. However, six Elateia data points do not fit the correlation. Glasses that fall onto the correlation line from widely separated sites were made using a similar calcium-rich raw material. It is perhaps ironic that Frattesina, which is the only site where large-scale evidence of production has been found, has produced a relatively small number of glasses that fit the correlation. The fact that the glass beads from Elateia, a cemetery, tend to have clustered compositions increases the possibility that they were made in batches using the same glass melt. The Hauterive data mainly fall at the lower end of the correlation in Figure 6.16, containing similar relative levels as in some Frattesina glasses. Two Hauterive samples with calcium oxide levels above 3% fall away from the correlation line. Nevertheless, six or seven Hauterive samples fit the correlation formed by the Mariconda and Elateia data, and so these must have been made using the same calcium-rich raw material. The four German samples probably have mixed origins. The latest samples, those from Rathgall of ninth-eighth century B.C. date, mostly fit the correlation in Figure 6.16.

Although it can be seen in Figure 6.17 that Rathgall glasses contain some of the highest soda levels from amongst the mixed-alkali glasses, there are,

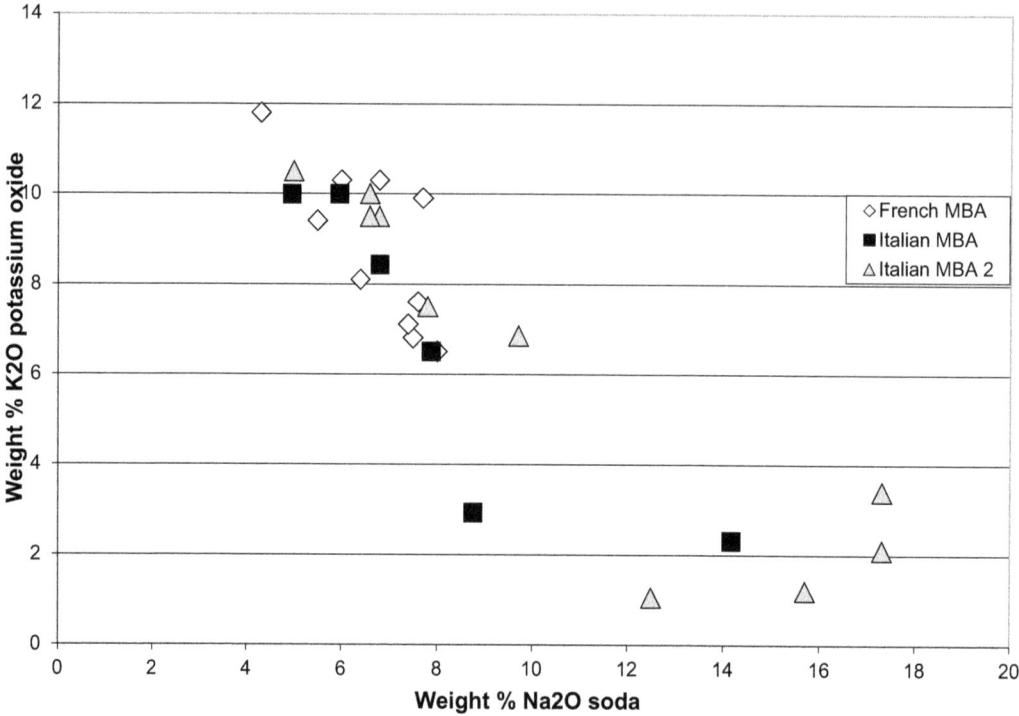

6.18. A bi-plot of weight % soda versus potassium oxide in French and Italian middle Bronze Age mixed-alkali glasses. Data from Guilaine *et al.* 1991, Santopadre and Verità 2000 and Bellintani 2000.

nevertheless, other examples from both Frattesina and Elateia with equally high (and higher) levels. Relatively high soda levels do nevertheless appear to be a distinctive characteristic of most Rathgall glasses. It is possible that the predominantly translucent turquoise-coloured glass beads from the site (Raftery 1987) were heirlooms, but it is more likely that they were made later than the Frattesina glass in a production centre other than Frattesina. A single example of a decorative opaque natron glass was found at Rathgall: its eighth–ninth century date is one of the earliest examples from Europe. It is significant that it was used to decorate a mixed-alkali glass bead, indicating their contemporaneity, at a time when natron glasses were being introduced for the first time. There is probable evidence that glass was worked at Rathgall itself. An unfinished bead with a poorly fired clay core and an applied cracked surface glaze (Raftery 1987, fig. 2, no. 16), the discovery of two fragments of glass shells at Rathgall (Raftery 1987, fig. 2, nos. 14 and 15) apparently also originally attached to a core and a mixed-alkali bead decorated with gold provide the evidence, for which there are no parallels.

It is perhaps fortuitous that the maximum range of potassium and sodium oxide levels in the Frattesina glasses analysed cover most of the compositional range in late Bronze Age and early Iron Age glasses deriving from a much wider

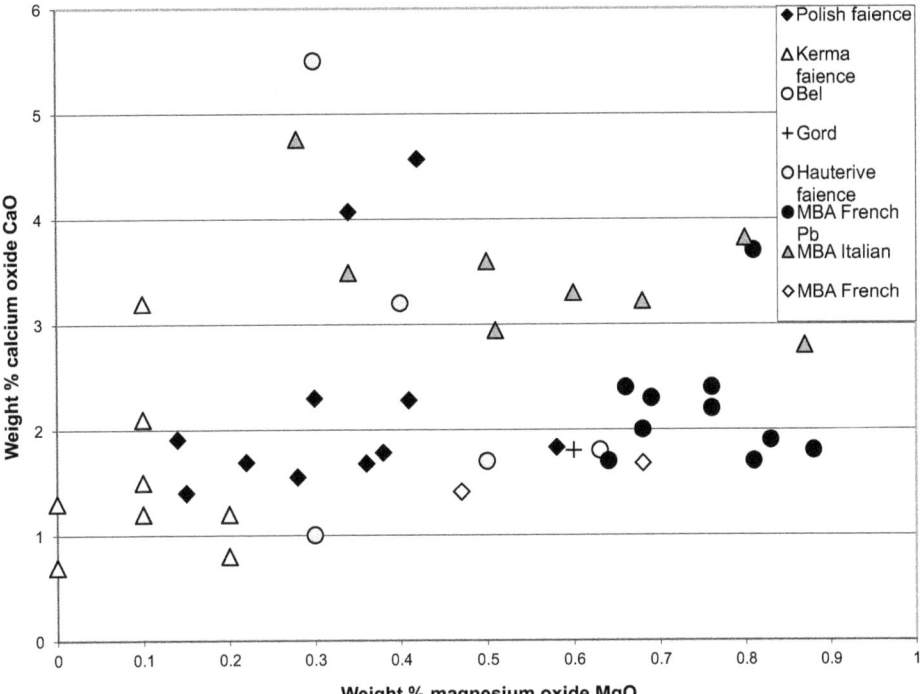

6.19. A bi-plot of weight % calcium oxide versus magnesia in middle Bronze Age French and Italian glass and (vitreous) faience from Switzerland, Poland and Sudan. Data from Henderson 1993a, Robinson *et al.* 2004 and Hatcher *et al*, in prep.

geographical area. (The potassium-rich glasses will be considered in detail in the next subsection.) A separate middle Bronze Age group containing lower potassium oxide levels can be discerned in Figure 6.18. The balance of middle Bronze Age results follows the negative correlation for soda and potassium oxides found in late Bronze Age glasses. Clearly this Middle Bronze Age material is too old to have been made at Frattesina. The mixed-alkali samples with potassium oxide levels above 6% plot precisely on the results in Figures 6.17 and 6.20 (for faience), and this provides evidence for a technological link between middle Bronze Age faience, vitreous faience and glass technologies. The distinctive 'high soda' levels for results plotted in Figure 6.18 are mainly for northern Italian conical buttons made out of glassy faience; the remaining two are for Italian beads.

In Figure 6.19 the French glass samples from Bel and Gord fall close to French middle Bronze Age glass. If the dates are correct for these (c. twentieth century B.C. or earlier), they are some of the earliest mixed-alkali samples in the world, so western Europe (perhaps France) may be where (mixed-alkali) vitreous technology originated, although with such small sample numbers this must remain a tentative suggestion. In this figure a clear distinction between middle Bronze Age glasses from Italy and France can be seen. Most compositions of the vitreous components of faience are different from the glasses containing either lower calcium or magnesium oxide levels.

In Figure 6.20 a relationship can be seen between relative alkali levels in the glassy component of early Bronze Age faience samples and the regions in which they have been found. Figure 6.20 shows the negative correlation in faience from Poland, the Sudan (Kerma) and Switzerland (Hauterive-Champréveyres) and very early glass from France (Bel and Gord). It is significant that five faience samples from Kerma in the Sudan dating to c. 2000 B.C. even conform to this pattern. Although they fall at the low potassium oxide end of the correlation, given the totally contrasting cultural contexts of Sudan and Poland in which the faience was found this is unexpected and demonstrates that a similar alkali source was used to make them. The balance of Kerma samples, also with very low calcium oxide levels, contain higher soda levels than found in Polish samples (Fig. 6.20). The glass samples from Bel and Gord from France, both dating to before 2000 B.C. fall onto the correlation for early Bronze Age samples from Poland. Three of the Hauterive faience samples can be distinguished by their low soda and/or potassium oxide levels.

The adherence of early Bronze Age Polish faience to the established pattern of relative alkali levels for middle and late Bronze Age European mixed-alkali glass samples underlines the clear line of technological development of European faience, glassy faience and glass.

6.5.1 Potassium Glasses

The few potassium glasses from Frattesina are the earliest in the world known to the author. They contain between c. 15% and 17% potassium oxide. Like the mixed-alkali glasses, they are characterised by low levels of calcium and magnesium oxides. Another example was found on the Greek island of Thasos at Theologos (Henderson 1991). Other potassium glasses dating from c. 400 B.C. have been found in India and Southeast Asia (see Chapter 4). Medieval high potassium wood ash glasses are compositionally different from late Bronze Age potassium glasses, with high levels of calcium oxide and elevated phosphorus pentoxide, so wood ash can probably be eliminated as a source of the potassium in prehistoric European high potassium and mixed-alkali glasses.

The Asian and Southeast Asian potassium glasses are, like the Frattesina glasses, characterised further by low levels of magnesium and calcium oxides (see Chapter 4). Because we can be confident that the Asian glasses with low magnesia and calcium were made from a mineral alkali source, this could also suggest that a mineral alkali source was used to make low magnesium–high potassium (LMHK) glass in prehistoric Europe.

It appears that prehistoric Europe and Asia were two distinct production spheres with minimal chronological overlap and that the glasses were made

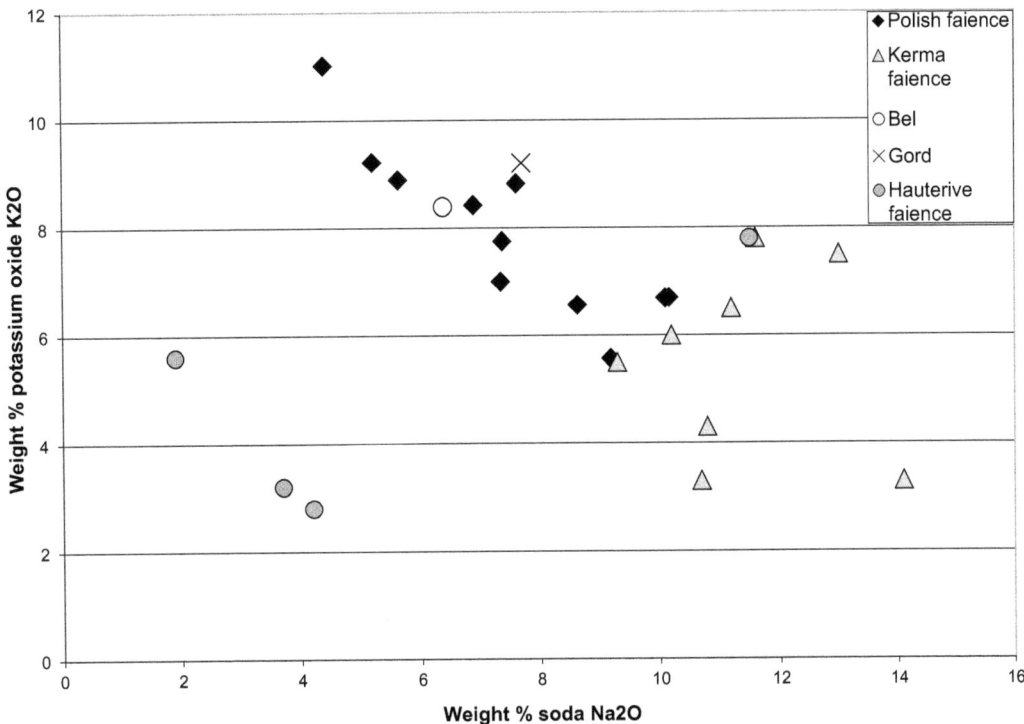

6.20. A bi-plot of weight % soda versus potassium oxide in early Bronze Age French glass and (vitreous) faience from Switzerland, Poland and Sudan. Data from Henderson 1993a, Robinson *et al.* 2004 and Hatcher *et al.*, in prep.

from a similar alkali source. However, given the evidence for the import of plant ash glass into China as early as the 1000 B.C. along the Silk Road, it may be only a matter of time before early high potassium glasses are discovered there. At the moment the evidence suggests that high potassium glasses were made independently in Europe and China, but this may change.

6.6 Local Production Centres in Bronze Age Europe

Early Bronze Age plant ash glass beads have been found in France (Gratuze *et al.* 1998), and the presumption is that they were made in the Middle East, but they are nonetheless some of the earliest glasses in the world. Moreover, Guilaine *et al.* (1991) claim that some of the earliest (Chalcolithic) mixed-alkali glasses have been found in southern France (the Midi), and because of this, they claim that the Midi was a production centre for them (*ibid.*, 258). There are examples of glass beads from secure Chalcolithic contexts. The first, although not analysed, is a bead found next to a copper bead in the Hypogeum of Crottes à Roaix (Vaucluse) with a date of 2150 ± 140 B.C. (calibrated 2400–2300 B.C.). The second is a single mixed-alkali blue (presumably turquoise) bead with an incrustation of ochre from Gord (Compiègne, Oise) with a date of 4100 ± 70 BP (calibrated 2895–2420 B.C.) already discussed. Others were derived from what are claimed as 'Le contexte comporte du campaniforme international et pyrénéen' (ibid., 258–9). A number of other mixed-alkali glass beads derive from Chalcolithic and Bronze Age contexts in the Aveyron (Azemar et al. 2000); some have been rejected from consideration as being later intrusions. However, there is clear evidence here of very early vitreous faience (or possibly glass) dating to at least the first half and certainly the second half of the third millennium B.C. from the Midi.

As already noted, there is middle Bronze Age mixed-alkali glassy faience, such as that from Grotte de Bringairet, Aude, France (Guilaine et al. 1991, 259) and Poviglio and Trinitapoli in Italy (Santopadre and Verità 2000). This shows that there were middle Bronze Age production centres for mixed-alkali glassy materials, probably in Europe. As also noted, a fully integrated study focusing on the production of glass conical buttons in the early Middle Bronze Age in the northern Italian terramare has been carried out (Bellintani et al. 1998; Bellintani 2000), and the scientific evidence supports the distinctive regional character of these artefacts. Barfield (1978) discovered that these buttons had a distinctive form and that their distribution was a regional one (see, for example Fig. 6.18, MBA2). Two French glasses said to be of an early Bronze Age date from Bel and Gord fall close to French examples of middle Bronze Age date. This appears to underline a compositionally regional distinctiveness. Excavations at Frattesina have produced the best glassworking evidence for mixed-alkali and potassium glass dating back to c. 1100 B.C. (see Chapter 5). Using X-ray diffraction Verità and Biavati (1989, 295) determined the presence of quartz, 'alkaline alumino-silicates' and tridymite in a small sample of glass-bearing crucible. Verità and Biavati (1989, 299) suggest that this, together with the fact that the glass is of a diagnostic composition, constitutes evidence for glass-making. However, even without the definite identification of frit and glass-making furnaces, given the scale of the industry, the case is compelling but it remains unproven. Compared with more constrained compositional results for glasses from other sites and regions the results for mixed-alkali and potassium-rich glasses from Frattesina cover the full range of other (late Bronze Age and Iron Age) sites. This reinforces the interpretation that Frattesina was a production site and that perhaps mixed-alkali glass batches were manufactured there from primary raw materials; however, the wide compositional variation may additionally be a result of glass being imported to Frattesina. The consistent ratio of potassium to sodium oxides, for potassium contents above 11% and soda contents above 5%, has an R^2 correlation value of −0.8611. This correlation results from mixing two alkali sources that are probably of a mineral origin. The probable use of either bone ash or dolomitic limestone provides supporting evidence for this interpretation. A lack of any correlation between alumina and potassium means that potassium feldspar was not the source of the potassium.

In the bi-plot of magnesia and calcium oxide (Fig. 6.16) the Greek (sub-Mycenaean) samples fall onto the same tight positive correlation as the glasses from Mariconda and may indicate that they were made in the same place. Moreover, surprisingly many of the Rathgall results also fit the positive correlation, suggesting that a similar calcium source was used to make them. As already noted, a comparison of relative calcium and magnesium oxide levels in late Bronze Age glasses from Frattesina and Mariconda shows that few of the Frattesina data fit the clear positive correlation formed from the Mariconda data. This suggests that two separate production centres are involved.

In sum, during the early and middle bronze Ages, faience, glassy faience and glass were made in different parts of Europe. Building on early work by Newton and Renfrew (1970), Harding (1971) and Harding and Warren (1973), there is now increasing evidence for regional production of these materials in northern and southern Europe. With an increase in the volume of glass produced in the late Bronze Age, there is the intriguing large-scale industrial evidence of glass production at the trading site of Frattesina, but as yet it can not be definitively shown to be a primary production site and, in spite of some interesting compositional differences in calcium-rich raw materials used to make late Bronze Age mixed alkali glass, it is still not entirely clear whether this reflects the existence of more than one production centre.

6.7 Conclusions

Scientific analyses can help to define the technological characteristics of early glasses and can provide links with textual evidence. Moreover, increasingly there is a prospect of being able to provide a provenance for glass using a range of scientific techniques (see Chapter 11). Despite the historical evidence that New Kingdom Egypt relied on Mesopotamia as a source of raw glass, scientific analyses now indicate that separate production spheres for plant ash glasses existed in the two areas. The use of silica and plant ash sources deriving from different geological environments in each region has introduced impurity patterns and isotopes that provide the necessary compositional and isotopic distinctions.

The first set of neodymium and strontium isotope data for late Bronze Age Mesopotamian, Egyptian and Greek glasses has provided a new provenance. When compared to environmental values for strontium isotopes in plant ashes (and neodymium isotope values for sand and quartz samples), the first geological provenance and strong evidence for the independent primary manufacture of glass in fourteenth-century Mesopotamia and Egypt has resulted. As yet there is no isotopic or trace element evidence for primary glass manufacture in Greece

(Henderson et al. 2010). Moreover, the wide variation in strontium and neodymium isotopic values amongst Mesopotamian samples (*ibid.* and Degryse et al. 2010) shows that there was more than one production centre for fourteenth-century Mesopotamian glass, something that trace-element determinations of the glasses have, so far, not revealed (Shortland et al. 2007).

The isotopic data for Egyptian glasses from Tell el Amarna and Malkata of a range of colours, including cobalt blue, fall into the same group (Fig. 6.11). This could indicate that Malkata glass was made at Tell el Amarna or that there is insufficient geological contrast in the raw materials used in the glasses to be able to show an isotopic distinction. The next logical step is to determine the isotopic signatures of glasses from another site where primary glassmaking occurred, thirteenth-century B.C. Qantir, a site that specialised in making red glass (Rehren 1997). Although Qantir is not contemporary with Tell el Amarna, it would at least show whether it is possible to distinguish isotopically between (raw) glasses from the two sites and thereby demonstrate that there is sufficient geological contrast in the raw materials used to make glasses from the two centres.

The strontium isotope results for four cobalt blue specimens of glass found in Greece dating to the thirteenth century B.C. and three dating to the twelfth century (some with neodymium results) show clearly that they fall into the 'Egyptian' isotope group (Henderson et al. 2010, table 1, fig. 3a and 3b). The results are for beads and pendants that can easily survive as heirlooms for generations, so could originally have been made at Amarna. Nevertheless, it is probable that some were made later at locations other than Amarna. This in turn may suggest that the current 'Egyptian' isotope signature represents an area that includes at least two production centres and that there is insufficient geological contrast to be able to detect subregions in which production occurred. The determination of isotopic signatures for glass made at Qantir will be a good way to test this.

Scientific results from ToF-SIMS has suggested that the addition of colourants in Egyptian glasses involved sintering a mixture of colourant and raw glass (Rutten et al. 2009) showing that a minimum of two stages were involved in its manufacture; subsequent ToF-SIMS work has provided more evidence of this (Duckworth et al. 2012). Although already suggested as a separate stage in glass production (Jackson et al. 1998, Rehren 2000) there is also newly recognised archaeological evidence, supported by scientific analysis, showing that a minimum of two stages were involved (Smirniou and Rehren 2011). The evidence from Amarna (Smirniou and Rehren 2011, 62) could indicate that base glass and coloration occurred on the same site in late Bronze Age Egypt, although this may or may not have been the case in Mesopotamia (Henderson 2000a, 59–60). Strontium isotope determinations have already shown that cobalt blue Egyptian glasses, some with a characteristic low potassium level, were made using plant ashes rich in calcium of a consistently similar geological age as non-blue Egyptian glasses (Henderson *et al.* 2010) and therefore probably that an ashed plant was used from the same geological zone to make these differently coloured glasses. This also shows that natron was not used as an alkali to make Egyptian cobalt blue glasses, as has been suggested (see Chapter 3, Section 3.2.1). Therefore, this exciting new chapter in the scientific examination of some of the earliest glasses in the world is only just beginning to bear fruit, and many more fascinating results can be anticipated.

Scientific analysis is also beginning to provide evidence for regional production of mixed-alkali glass and glassy faience in both the middle and late Bronze Ages in southern Europe. Production occurred against a technological continuum amongst faience, glassy faience and glass. Interestingly, this compositional continuum cannot (yet) be demonstrated for the glass that developed in Mesopotamia. It is also too early to be confident that there was a split between primary and secondary glass production centres in middle and late Bronze Age Europe, but it is apparent that there were a number centres where the glass and the vitreous component of glassy faience were fused from slightly different raw materials. The late Bronze Age site of Frattesina in the Po Valley is currently the only site for which there is evidence for large scale industrial activity involving this characteristic mixed-alkali glass: this is the strongest contender for a location of primary glass production. Although even here, using chemical, mineralogical and principal components analysis on a large number of samples, Angelini *et al.* (2009) were unable to confirm that there was more than one production centre involved. The ambiguous discovery of a crucible fragment with plant ash glass rather than mixed-alkali glass adhering to it – one of very few examples from prehistoric Europe – adds to this complex picture.

Given the compositional evidence for regionalised production, the next logical step would be to carry out isotopic determinations of the glass. Neodymium determinations could help to source the silica. It should therefore be possible to state, for example, whether the silica used for the production of mixed-alkali glasses found at Frattesina has a signature that is typical of the silica of the Po Valley dominated by contributions from the Alps. If it is possible to show this, it would provide powerful evidence for primary glass production of European glass in eleventh to ninth centuries B.C. The determination of the strontium isotope signatures should shed light on the types and sources of alkali and calcium used. A combination of glass and environmental isotope results could therefore provide the first crucial scientific evidence for the location of primary glass production. Combined isotopic and trace element analyses of mixed-alkali glasses of a range of dates could potentially provide indisputable evidence for the use of different sources of glass raw materials in different production zones and at different times.

The ritual, social, economic and political contexts in which late Bronze Age glasses were manufactured in Mesopotamia and Egypt are associated with the specialised production of a range of brilliant translucent and opaque glass colours. Production occurred on a relatively small scale, and the glass vessels made were used by only the social and ritual elites. We are reminded of the important ritual value of blue glass in particular because it was clearly used in imitation of lapis lazuli. Sherratt and Sherratt (2001, n. 9) note that blue glass and lapis lazuli were both highly valued in Egypt and up to a point equivalent/interchangeable. The term 'Lapis lazuli', mentioned in Egyptian inscriptions, is used to refer to blue glass, whereas 'royal lapis lazuli' is used for the mineral. Moreover, cuneiform texts distinguish between *lapis lazuli from the mountain* and *lapis lazuli from the kiln* (Oppenheim 1970, 10).

Moorey (1994, 201–2) suggested that there were separate glass production centres in Syria; Nuzi, Assur, Aqar Quf, all in Iraq; and Susa in Iran during the late Bronze Age, all being operated within major royal or temple buildings and serving local needs. It is nevertheless apparent that glass ingots were manufactured for trade. The structure, form and modes of late Bronze Age glass production are therefore a direct reflection of the 'pyramidal' social structure at the time, and further scientific research should be able to illuminate the relationship between social structure and glass production technology.

SEVEN

HELLENISTIC TO EARLY ROMAN GLASS

A CHANGE FROM SMALL- TO LARGE-SCALE PRODUCTION?

This chapter forms a pair with Chapter 8: it provides archaeological and historical information about the production and use of glass in the late Hellenistic and early Roman periods, especially in the Middle East. Chapter 8 provides a discussion of the scientific investigations of the glass.

There is a famous description by the Roman encyclopaedist Pliny the Elder (N.H. I., 65) of the 'invention' of glass by Phoenician traders on the coast of Syria near the River Belus (Nahr Na'man). The translation of it by Trowbridge (1930, 95–96) is as follows:

> According to tradition a ship of natron merchants came to shore and when the men were scattered all along the beach preparing their meal, since there were no stones to support their kettles, they put pieces of natron from the ship under them. When these had caught on fire and the sand of the shore mixed with them, there flowed transparent streams of a new substance, and this was the origin of glass.

With perspicacity, Trowbridge (*ibid.*, 96) emphasises that Pliny was recording a *tradition*. Thus, we should not take it too literally (see Section 2.5). An alternative interpretation could be that a glassy material was indeed formed on the beach, but that it resulted from the fusion of fuel ash (used in the bonfire) and sand rather than natron with sand. Moreover, the Augustan geographer Strabo writing between 26 and 7 B.C., and possibly as late as A.D. 17/19, mentions that glass was made on the Levantine coast.

Nevertheless, the Levantine coast near the Belus River was clearly an important source of sand for Roman glassmaking (Brill 1988, 265–7) and perhaps also in the Hellenistic period. Josephus even describes it as 'a remarkable place no larger than one hundred cubits. [A cubit is about half a metre.] This is round and hollow, and yields such sand as glass is made of'.

Further north on the Levantine coast, Strabo refers to a suitable sand source for glassmaking located near Sidon. Pliny (N.H. XXXVI, 190–3) refers to Sidon as the *artifex vitri* and that it had lost its importance by A.D. 70. The importance of Sidon as a glassworking (and making) centre is further underlined by the large number of glassworkers who 'signed their names' as being from Sidon on mould-blown beakers. When they did this they may still have been resident in the Levant or have migrated to the Italic workshops (Trowbridge 1930, 115, 130; Stern and Schlick 1994, 108).

By the year 2000, no direct evidence for the manufacture of late Hellenistic and early Roman glass vessels in the Levant had been recognised. Excavations in the Souks area of Beirut, however, unearthed large numbers of late Hellenistic and early Roman vessels (Jennings 2000, 41–2). Since then, evidence has been published for both glassworking in Beirut (Jackson-Tal *et al.* 2004; Foy 2005) and now also for glassmaking (Kowatli *et al.* 2008).

Chemical analyses indicate that Hellenistic and early Roman glasses were predominantly natron glasses. However, a small number of plant ash glasses, of a possible Middle Eastern origin, were also made before the Sasanians did between the second to sixth centuries (see Chapter 8). The production of plant ash glass must have occurred very occasionally on a far smaller scale than during the Bronze Age or in the Sasanian or Islamic periods. It is nonetheless intriguing that it seems to have occurred at all, and it would be interesting to attempt to determine using chemical and isotopic analysis where it occurred. A scholiast on Aristophanes' *Clouds* appears to provide historical evidence for the primary manufacture of glass using plant ashes and perhaps even in classical Greece:

> We call *hyalos* that which has just been burned from certain herb and melted by fire for the preparation of certain vessels. (Trowbridge 1930, 99, n. 17)

The Greek word ὑαλος, hyalos, means vitreous. Around 75 to 100 years following the invention of glassblowing on the Levantine coast c. 50 B.C. (Israeli 1991; Israeli 2005) the scale of Roman glass production increased dramatically. Glasshouses were set up in Rome from where glass production spread to the Roman provinces such as in the Ticino where highly coloured glass vessels were produced. By the late first century A.D., highly coloured, including cast, moulded and blown, glass vessels, were supplemented and to a large extent replaced by mass-produced blown pale green glass vessels. By this time such glasses were produced in London. The invention of glass blowing therefore revolutionised glass vessel production, eventually making glass available to all levels of society for the first time over a broad geographical area. Before early

Roman glass production is considered in more detail, the context of the glass industry under Greek control will first be described.

7.1 THE HELLENISTIC AND ROMAN MIDDLE EAST

Alexander the Great won a victory over the Persian army led by Darius III Codomanus in 333 B.C. One result was that the whole of (Greater) Syria was open to Alexander. Syria became a satrapy of Alexander's empire and was viewed by some as being strategically critically important because of its position between the Mediterranean and Mesopotamia, with links to Iran and Central Asia. Ptolemy I controlled the southern region of Syria (which includes Phoenicia). Seleucos I (Nicator), one of Ptolemy's leading generals, was given control of the rest of Syria and Babylon; he later lost these domains, but Ptolemy helped him to regain most of them. The Seleucid period saw urbanisation based on the Greek polis model. It was initiated by Alexander, but Seleucus and his son Antiochus I founded a number of cities. Greek and Macedonian colonists were given land especially in northern Syria: great cities such as Chalcis, Apamea and Antioch were founded (see Fig. 7.1). These cities organised around the polis brought with them the Greek lifestyle in a northern Syrian context, and this would have included dining rituals involving glass vessels. Rulers were also able to control the economy and political structure of these cities and their hinterlands. The Tetrapolis was founded by Seleucus I between 299 and 301 B.C, consisted of Antioch (named after his father Antiochus), Apamea (named after his Persian wife, Apama) on the Orontes Valley, Seleucia (named after himself) and Laodicea (named after his mother Laodice). The latter two are ports on the Mediterranean. The Tetrapolis is thought to have had a total population of between 20,000 and 50,000 (Aphergis 2004, 93). It dominated the production of agricultural produce from the rich agricultural area of northwestern Syria. It is notable that the new cities were given dynastic names, a politically appropriate move.

The Seleucid cities continued to thrive during the Roman period. Rulers that followed Seleucus I fought rivals in the eastern provinces, Palestine and southern Syria, and there was great competition with other Hellenistic kingdoms such as the Antigonids of Macedonia, the Ptolemies of Egypt and Attalids of Pergamon. Sidon, a location which according to Strabo provided sand for making glass was a major city in the Achaemenid and Hellenistic periods. Roman architectural remains on the site suggest that it was still of importance in the Roman period; excavations have shown that the harbour was improved in the first century B.C. (Satre 2005, 164, 260). Another area that was urbanised is referred to as the Decapolis. Although it is unclear which ten cities were included, it was an enclave of highly urbanised city-states, and there was no official alliance between them. Pliny the Elder mentions Damascus, Canatha

ANCIENT GLASS

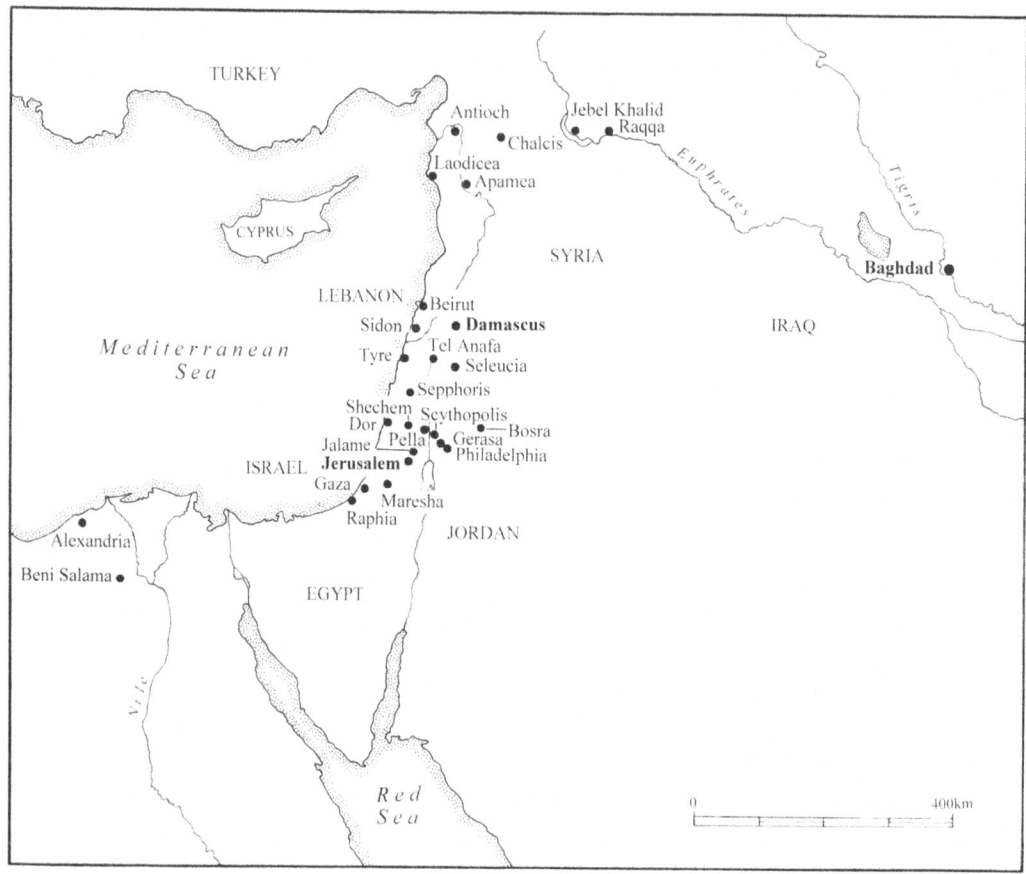

7.1. Map of main locations mentioned in the text.

(Qanawat), Scythopolis (Beth Shean), Pella (Tabaqat el-Fahil), Gadara (Umm Qais), Hippus (Qalaat el-Husn), Dium (possibly Tell Ashari), Philadelphia (Amman), Gerasa (Jerash) and an unknown city, Raphana, possibly Capitolias (Beit Ras). To this list, Butcher (2003, 113) adds Abila, not mentioned by Pliny (see Fig. 7.1).

By the late second and early first centuries B.C. the Romans were extending their territories into the eastern Mediterranean, making alliances with a number of states there. Competition between Roman commanders led to a rather aggressive acquisition of territory. It is claimed that in 64 B.C. 'a Roman general met with a local king at the city of Antioch in northern Syria and told the king that he was out of a job' (Butcher 2003, 19). This meeting between Antiochus XIII Asiaticus (a Seleucid) and Gnaeus Pompeius Magnus (the Roman general, Pompey the Great) marked the start of the Roman province of Syria (Sartre 2005, 389, n. 46). This Pompeian annexation included the Phoenician coast, northern Syria and the inland area of the Decapolis. These were also the most urbanised areas of Syria, where there would have been the greatest demand for glass vessels. There were a number of city-states on the Phoenician coast, the most significant being on an island, the Phoenician city of Aradus. It was a major Hellenistic commercial centre and had gained independence from the Seleucids in 259 B.C. By the time of the annexation, it dominated the nearby coastal cities. Under Augustus, Berytus (Beirut) was founded c. 15 B.C. (Ball 2001, 173). It became part of a powerful city-state, with a large associated territory as far as the Bekaa Valley in Lebanon.

During early Roman rule, the number of cities increased; some settlements were confirmed as cities, others were denied this status. Yet there seems to have been little evidence for a direct imperial involvement in their development, or at least the evidence for this is ambiguous, and in some cases emperors are likely to have been reactive rather than proactive. It would appear that in many cases cities were encouraged to thrive in their own right. In the course of the Roman period, many centres began to lose their Latin character, as the 'Romans' became embedded in the native populations.

By the time Rome annexed the Middle East (Sartre 2005, 39), many of the cities in the southern part of the Levant were under the influence of the Hasmonaeans of Judaea. Although Ptolemy had restored 'freedom' to subject cities, only some, such as Raphia (Tel Rarah), Gaza and Samaria (refounded by Herod as Sabaste) survived after the first century B.C. In 106 A.D. many cities were attached to the new province of Arabia (which included today's Jordan, Syria, Israel, and northwestern Saudi Arabia). By this time glass containers, such as bottles and jars, had become an important means of transporting liquids, such as olive oil and other agricultural produce. By the Nerionian period, glass vessels were described as being cheap in Petronius's *Satyricon* (50-51):

> Pardon me if I say that I personally prefer glassware, since glass vessels do not give off an odour. If they were not breakable, I would prefer them to gold. And now they are very cheap. (Humfrey et al. 1998, 595)

By the second century, most glass vessels were made from 'natural' blue- green glass; luxury vessels were made from colourless glass. By this time, Philadelphia, Gerasa (Jerash) and Canatha formed part of Arabia; the balance of the Decapolis cities formed part of Palaestina Secunda. Kennedy (2007) argues that, although isolated geographically, Gerasa and the Decapolis looked westwards towards the Mediterranean and eastwards to the Arabian desert. Nevertheless, by the first and second centuries A.D. the Romans controlled an increasing number of cities. For example, in the later second century, the formally Parthian-controlled city of Dura Europus came under their control.

Damascus had a large territory that was subject to the Nabataeans, and then the Tigranes in the first century B.C. It is likely that Damascus was one of the cities given to Cleopatra by Anthony because coins minted there carry her head. It was certainly under Roman rule by the time that Nero was emperor (54–68). It acquired the title *metropolis* by the first half of the second century A.D. and *colonia* by the reign of Severus Alexander (222–235).

Palmyra was another important centre. Like many settlements in the Middle East, it was a Hellenistic settlement before it was a Roman one, and it had vast territories extending westward as far as Apamea; to the east and south, its boundaries have not been defined clearly. It was an important stop on the caravan routes across the Middle East, ultimately as far as India (Gagarin and Fantham 2009, 156). There was a Palmyrene military presence as far south as the island of Anatha on the river Euphrates. However, even by the second half of the first century A.D., constitutional characteristics of a Greek city-state can be detected. When Mark Anthony's cavalry raided the city in 41 B.C., it was independent, and it is probable that it was incorporated into the Roman Empire between the reigns of Tiberius and Claudius.

Another region where a number of cities were located is Judaea, with Jerusalem as its administrative capital (Applebaum 1989, 36). Two other principle centres were Sepphoris (where there is evidence for early Roman glassworking) and Shechem, the capital of Samaria. Herod also founded Stato's Tower, which was refounded as Caesarea. Tiberias and Sepphoris were strong Jewish settlements; Petra and Bosra (Bostra) were the two most important centres to the south.

The *Legal Manual* of Constantinus Harmenopolus is a late Byzantine document written in the 530s (Long 2003, 13). It lists the decrees of a prefect, which were recorded by Julianus, known as 'the Architect' of Ascalon (Saliou 1996). This document mentions the suggested citing of industries in relation to the limits of cities in Palestine because of the fire risk and smells produced by them, the smells from cheese makers and the manufacturers of fish sauce (*garum*) being examples. However, makers of quicklime, dyes, olive oil and rush weaving were allowed to set up inside cities, although there were legally defined distances from habitation.

Glassmakers, ironworkers and bronze-casters

> *must not carry out those operations in the cities proper. Or there is a necessary reason that they should inhabit cities and carry on these occupations in them, they must work in remote and sparsely populated parts of the cities. For the danger to the buildings from the fire is considerable; and so likewise is the constant bodily harm to persons.*

It is likely that similar decrees applied to earlier (Roman) periods. During the Roman rule, the population of Syria and the Near East grew considerably, creating a demand for the products from the industries.

7.2 Hellenistic Glass Production

The Hellenistic civilisation in the Mediterranean flourished after the death of Alexander the Great in 323 B.C. and ended when Anthony and Cleopatra defeated the Greeks in 30 B.C. The principle Hellenistic kingdoms were the Macedon and Greece (the Antigonid dynasty), Syria and Mesopotamia (the Seleucid monarchy) and Egypt under Ptolemy and his successors. Amongst the lesser powers was Rhodes, a small island state in its own right. These states competed politically with each other and vied with each other to dominate increasingly large trade networks in western Asia. As Grose (1989, 185) notes succinctly:

> The Hellenistic age witnessed not only the consolidation of political power and the spread of a common culture suffused with Greek ideals and forms but also vast expansion of commercial and manufacturing activity. Long-distance trade increased notably, and new markets were created for both luxury and everyday goods throughout the Mediterranean area and beyond. Thanks to politically and economically favourable conditions, most of the age-old craft industries, including glassmaking, flourished as never before.

Grose (*ibid.*) is of the opinion that the flourishing Hellenistic glass industry constituted a Renaissance. It began in the late third century B.C., occurring in several centres and at different stages and lasted until the Romans adopted it and modified it. However, the question of where exactly Hellenistic glass was made from primary raw materials remains mainly unanswered. The only site where the remains of massive tank furnaces have been discovered, which certainly date to before A.D. 50 and quite possibly to the first century B.C., is Beirut 015 (Kowatli *et al.* 2008), discussed later.

Grose (1981) noted that there were two contemporary Hellenistic glassmaking (forming) traditions: the production of core-formed unguent bottles and cast tablewares including bowls, plates and cups. Grose (*ibid.*, 61) suggests that 'Both industries were probably continuous glassmaking traditions going back to the seventh or sixth centuries B.C.' Core-formed bottles were manufactured from the mid-sixth century B.C. until the end of the Hellenistic era They were all made in imitation of Greek ceramic forms: stamnoi, amphorae, hydriai, amphoriskoi, aryballoi, alabastra and oinochoai. Grose (1979, 55) has suggested that some of the glass 'bowls' may in fact have functioned as drinking cups.

The Hellenistic tradition (which Roman glassmaking grew out of) was dominated in the third and second century B.C. by the manufacture of highly coloured glass vessels produced in moulds using a variety of techniques, such as slumping and casting, to produce mosaic and sandwich gold-glass. The sixth century B.C. core-formed vessels, also highly coloured, continued a tradition that started in the fifteenth century B.C. (see Chapter 5). The colours used in Hellenistic glass vessels were due to a combination of mineral-rich colourants and furnace conditions. Mineral-rich colourants such as cobalt (blue), manganese (purple), copper (emerald green) and opaque red were used to colour the glass; opaque yellow and white colours were produced using antimony-rich materials. Bluish-green, green-brown and colourless glasses were caused by the presence of iron and manganese in the glass melt. 'Water white' colourless glass was produced by using antimony oxide as a clarifier (see Chapter 3).

Vessels could be either monochrome or polychrome, often in imitation of semi-precious stones. Opaque glasses were used especially for the manufacture of polychrome moulded mosaic glass vessels, a technique invented in Mesopotamia (Von Saldern 1970; see Chapter 5). The mosaic technique produced patterns that were sometimes similar to shells or fossils. The vessel forms and decoration vary somewhat according to the region in which they have been found, suggesting that glassworking (vessel production) occurred in separate regional centres, even if the fusion of raw glass from primary raw materials may have occurred only on the coast of the Levant.

As noted earlier, a primary production zone for late Hellenistic (second–early first century B.C.) glass from its raw materials was the Syro-Palestinian coast (Jackson-Tal 2004, 11). Strabo (Geography 16.2.25) refers to Sidon as having been a famous centre for glassmaking using sand taken from between Akko and Tyre; by A.D. 70 it had lost its importance as a glassmaking centre. A second suggested centre for the production of Hellenistic glass vessels is Alexandria (Barag 1985, 59, Grose 1989, 188), and a third is Rhodes. The Roman orator Cicero (106–43 B.C.) refers to glass being imported in Alexandria, but glass could also still have been fused from raw materials there. On Rhodes there is evidence for glassworking and possibly for primary glassmaking, discussed later. Moreover, certain vessel forms appear to be characteristic of the island, and it has been argued that local glass production started as early as the last quarter of the 6th century B.C., with the manufacture of core-formed glass vessels (Triantafyllidis 2003, 131–2). Grose (1981, 61) has suggested that there must have been several different 'centres of manufacture' for Hellenistic glass vessels. However, even if such centres were associated with the manufacture of particular vessel forms, we are still left with the question of how many primary glassmaking centres there were and where they were located.

More than 3,000 fragments of typologically homogeneous cast glass bowls were found during excavations of less than 5% of Tel Anafa, Israel, southeast of Tyre in the Lebanon, associated with an architectural phase beginning c. 125 B.C. (Weinberg 1970). As Grose (1979, 54) has noted, these bowls were the first commonly used glass vessels in antiquity. Previously, any tablewares that existed were truly luxury products, such as found in the rich third century

B.C. tombs at Canosa, southern Italy (Harden 1968a); Strabo, Pliny and Josephus also all refer to the manufacture of glass in southern Italy, as do Jewish sources (Jackson-Tal 2004). The Jewish sources (Mishna), dating to the third century A.D., often reflect late Hellenistic sources: glass is mainly mentioned in relation to purity rights (Jackson-Tal 2004, n. 11). Quite recently, it has been argued that the large scale of production in the late Hellenistic period, especially for faintly tinted green and yellow translucent glass vessels, indicates that on 'consumption' sites, glass was regarded of less of a luxury than has often been postulated.

Evidence of vessel manufacture in the form of glass blanks used for sagging into moulds has been found, along with 120 to 130 conical and hemispherical sagged bowl fragments. This was found with the early glassblowing evidence dating to c. 50 B.C. in Jerusalem (Israeli 2005, 54, 56). Israeli notes that colourless bowls sometimes with a blue-green or yellow tinge predominate. In contrast, at Delos (Nenna 1999, 66–7) and Anafa (Weinberg 1970, 19–21) strongly coloured vessels dominate the assemblages. Jackson-Tal has noted the ubiquity of these monochrome slumped and cast vessels, and the uniformity of the vessel forms found on 'every type' of late Hellenistic settlement. Her discussion mainly focuses on vessels found in Israel, at Maresha, Jerusalem, Dor, 'Akko, Tel Anafa (Weinburg 1970) and HaGoshrim. Large numbers of vessels were also found in the excavations of the Souks in Beirut (Jennings 2000; Sablerolles n.d.). As Jennings noted, if the 12,000 fragments from more than 7,000 glass vessels recovered, including around 200 vessels made either by casting or slumping rather than blowing, were not made in Beirut, then there must have been an active trade in vessels. However, the recent discovery of coloured glassmaking at Bey 015 in Beirut (see Section 7.2.1.1) on a grand scale possibly dating to the first century B.C. appears to address this point. The vessel forms found on a second (domestic) site in Beirut, Bey 006 are mainly those of types 1 through 5 in the following list (Jennings 2000, 42):

1. Conical grooved bowls (Fig. 7.2)
2. Hemispherical grooved bowls
3. Bowls with external groove decoration
4. Ribbed bowls
5. Linear-cut bowls
6. Late plain bowls
7. Cast cups, bowls and dishes

By comparing the Beirut hemispherical bowls to those from Tel Anafa, Jennings (*ibid.*, 56) was able to state that the Beirut examples were made in the same *regional* tradition, but that there were sufficient differences to be able to suggest that they came from a different source. This could therefore mean that raw glass was made centrally and that separate workshops manufactured the two groups.

Jackson-Tal (2004) has published a comprehensive list of the many sites in the Levant where cast Hellenistic vessels have been found. Monochrome tablewares were widely distributed – they have been found in various Greek settlements 'overseas' such as around the Black Sea, Italy, Sicily and Asia Minor. The uniformity in shape of late Hellenistic open sagged bowls means that the 'centres of supply', as O'Hea (2005, 44) refers to them, are difficult to define. Jebel Khalid on the river Euphrates is quite far from the presumed production centre(s) on the Syro-Palestinian coast. O'Hea compared the different proportions of vessel colours found at Jebel Khalid, Jerusalem (Israel) and Pella (Jordan). On the basis of a higher proportion of colourless glass bowls from Jebel Khalid, she has suggested that they were manufactured locally using a clean source of silica, given the Neo-Assyrian and later Sasanian tradition of colourless glass in the region. However, at this time it is likely that antimony trioxide was used for decolourising the glass and that a 'clean sand' was not the (only) way to produce colourless glass (see Chapter 3). If we accept O'Hea's suggestion that the glass was made locally, quartz pebbles from the Euphrates Valley would have provided a relatively pure source of silica. A local source of soda would have been plant ash, yet the glass from Jebel Khalid is liable to be made with natron. The only reliable way of testing O'Hea's suggestion is by chemically and isotopically analysing the glass. Excavations of Maresha, Israel, have produced 106 colourless monochrome cast bowls, which are 50% of the bowls found there (Jackson-Tal 2004, table 1). Such a large number could potentially mean that they were manufactured nearby. However, the best proof of several centres of primary production is by chemically and isotopically analysing raw glass deriving from furnace sites (see Chapter 11) leading to a provenance for glass used to make glass vessels. The 'evidence' of a geographical concentration of vessels of a particular colour may infer that they were imported in bulk from a distant production centre. The mass production of late Hellenistic bowls with simple ornamentation such as horizontal grooves using former moulds has been explained by Barag (1985, 59) as being due to the discovery of 'more economic [manufacturing] methods'. If we examine this suggestion more closely, the controlling factors are likely to be the supply of natron and of fuel, silica being readily available on the Levantine coast. It is possible that natron did become more readily available, but perhaps it was simply that sagging of glass into former moulds was found to be a relatively quick way of producing vessels. Certainly an increase in population would have created a demand by people who could afford glass vessels, fed by an increase in long-distance trade. As noted earlier, the strong city states formed by the Hellenistic Greeks, such as the Tetrapolis, led to the efficient organisation of trade in agricultural produce. This led to the development of more intense long-distance trade, the growth of the 'middle classes' and an increase in demand for glass vessels.

7.2. Conical grooved bowl (after Harden 1968; reproduced with kind permission of the Royal Archaeological Institute).

The big advance in world glass production that occurred c. 50 B.C. in the Levant is described in a somewhat poetic way (with gaps), on a third or fourth century A.D. papyrus translated by Coles (1983, 58) as follows:

First he heated the very point of the iron, then snatched from nearby a lump of bright glass and placed it skilfully within the hollow furnace. And the crystal as it tasted the heat of the fire was softened by the strokes of Hephaestus like.... He blew in from his mouth a quick breath ... like a man essaying the most delightful art of the flute. The glass received the force of his divine breath, for swinging it often like an ox-herd swings his crook, he would breath into.

The earliest evidence for glassblowing was found dumped in a ritual bath in area J of the Jewish quarter in Jerusalem's old city. The dump consisted of large numbers of failed glass vessels, glass rods, tubes, drops and lumps. Although difficult to reconstruct, it is likely that the process involved blowing glass vessels from tubes. This is a process originally suggested by Eisen (1916, 137) as an extension of blown glass bead production. Glass tubes were produced, some of these were stretched with tweezers, and then their ends were closed off by heating them and pinching them together (Israeli 1991, 48–52). The tube tip was then reheated and the whole inflated to produce a vessel, the tube above the blown bulb forming its neck. The evidence from the Jerusalem dump indicates small bottles were the main product. Trailed-on decoration was achieved by winding rods of coloured glass around the tube before inflation.

Associated with these early blown bottles were Hellenistic cast conical and hemispherical bowls as well as flat slabs of glass, presumably for making sagged bowls (Israeli 2005, 54, 56). The coins associated with this immensely important context date to between 104 and 76 B.C.; the bath was sealed by a paved road built by King Herod (37–4 B.C.) in the last years of his reign or soon after. The pottery and (other) glass associated with the glassblowing evidence date to the first half of the first century B.C. (*ibid.*, 47). So the context has a *terminus post quem* of 76 B.C. and a *terminus ante quem* of 4 B.C.

7.2.1 Primary and Secondary Production of Hellenistic Glass

7.2.1.1 Rhodes

Excavations on the island of Rhodes have uncovered evidence that could indicate that primary glass production occurred there. Excavations of the Arfara property in the modern city of Rhodes revealed sealed archaeological layers dating to between the fourth and early third century B.C. that contained lumps of raw glass, large deposits of sand, quartz sand with a 'lime' content and high iron contents, 'burnt' sand and chunks of 'lime' (Triantafyllidis 2000, 32; Triantafyllidis 2003). It is suggested that nearby cisterns may have been used for purifying sand by levigation. The raw glass may have been manufactured in a nearby glass furnace from primary raw materials, but, at present, there is no clear evidence for this. The presence of sand could also be an indication that it was used for glassmaking, but again this has not apparently been proven analytically or isotopically. Chemical compositions of Hellenistic glass indicate that crushed shell fragments in the sand would be the source of the calcium oxide in it. The 'lime' discovered on the site may have been related to the building industry rather than the glassmaking industry. In a different location, 'probably west of the Kaloula property', east of the Lower Gymnasium in Ptolemaion in the ancient city of Rhodes, a complex of rock-cut tanks and cisterns (Triantafyllidis 2000) contained raw glass 'cullet', tile fragments with a layer of glass adhering, glass crucibles, clay containers and a fragment of a bench coated with glass (Triantafyllidis 2000, 31–2, Triantafyllidis 2000a, 193) dating to the end of the fourth to early third century B.C. This evidence could conceivably be the result of glassmaking, but without more direct evidence it seems to show that glass was being worked at or near the location where it was found. Earlier excavations at the Kaloula property to the east of those described above revealed the evidence of a bead workshop probably dating to the late third or perhaps early second century B.C. (Weinberg 1969). Here the evidence includes large numbers of glass tubes especially for the manufacture of opaque yellow glass beads. The evidence was comprehensive, with unusually large dumps of colourless blanks, scrap glass and both malformed vessels and beads. In addition, fragments of luxury glass vessels, including sandwich gold-glass and almost colourless thin-walled plates and bowls, were found. The malformed vessel fragments and the colourless blanks were interpreted by Weinberg as evidence for the manufacture of gold-glass vessels (Weinberg 1983, 37). The house in which the material was found may have been destroyed by the earthquake of 226 B.C.

Evidence for the coloration of glass has been found in a late third–early second century B.C. glass workshop on Rhodes. It has been suggested that colourants were added to imported raw glass there (Triantafyllidis 2000a, 194). Coarse-ware basins (*lekanai*) with layers of single or mixed colourants on their inner faces were found; it is claimed that Egyptian blue, red copper and red lead were found. Lead could have been used to develop opaque colours, such as lead antimonate and copper used to produce opaque red coloured glasses. Lumps of pigments and of lead, rectangular vessels and 'cooking pots', with multiple layers of coloured glass in them, were also found. The evidence found in 'cooking pots' appears to be strong evidence that coloured glasses were being worked or colourants added to raw glass. There is no question that in locations in both the modern and ancient cities of Rhodes, there would have been glass furnaces for reheating the glass to work it. Moreover, Rehren *et al.* (2005) discovered sand grains in raw glass attached to crucible fragments and have suggested that primary glass production occurred on Rhodes. The evidence that might relate to primary glassmaking are lumps of raw glass and sand on one site and glass adhering to crucible fragments on another. The only solid evidence for primary glass production is raw glass attached to *in situ* or dumped furnace fragments, lumps of (overheated) glass frit, or glass frit attached to the vessels in which it was made, but none of these have been found on Rhodes. The presence of raw glass can also be an indication that glass production occurred nearby. Nenna (1999, 160–1) notes the existence of an early first century B.C. workshop on Rhodes, where, amongst other evidence, raw opaque blue glass and tubes were found.

7.2.1.2 Beirut

The archaeological evidence for primary production of Hellenistic glass is rare. The only clear evidence is the discovery of tank furnaces in Beirut, in which highly coloured and nearly colourless raw glass was manufactured.

The site of Bey 015 in Beirut was on the ancient Georges Haddad Street, where the modern George Haddad Avenue is situated. The excavations were initially undertaken by Ibrahim Kowatli within the UNESCO Leb 92/008 framework project. A fairly detailed description of what has been found is provided here because of its rarity and significance for Classical Levantine archaeology. It is also provided to show how difficult it can be to interpret such glass production evidence.

The site is situated in the eastern side of the Seleucid *polis* and the Roman *colonia Julia Augusta Felix*. Three tank furnace complexes were identified at the site, all of which date to before A.D. 50 (Kowatli *et al.* 2008). They were cut into by later pottery kilns and yielded glass vessel fragments that are typical of forms dating to before 50 A.D. and reported on by Sablerolles (Fig. 7.3). The furnaces produced raw glass in very large quantities (Fig. 7.4). A general view of the site is given in Figure 7.5.

7.2.1.2.1 The First Tank Furnace Complex (TFC 1)

The 'First Tank Furnace Complex' is a tentative label given to remains that resemble better-preserved remains in stratigraphically later levels (the second and third tank furnace complexes). It dates to before 50 B.C. The architecture was orientated on a slightly different axis from the later tank furnace complexes (2 and 3). It is located outside the enclosure wall of the Seleucid *polis*.

The southern fragment of the TFC 1 consists of a corner wall, delineating an area below the floor of Tank 2 (TFC 2). A red-brown soil deposit separates the lowest floor of Tank 2 and the top of the wall. An early wall assigned to TFC 1 on the same alignment as the later south wall of Tank 2 (TFC 2) is also separated from the later south wall by a rather thin layer of soil. The orientation of the northern fragment of TFC1 differs slightly from the northern part of the TFC 2, but it seems that the south wall of Tank 2 in the TFC 2, although slightly off alignment, indicates continuity from TFC1 to TFC2. If it is assumed that the southern and northern fragment of TFC1 belong to the same building; the remains could represent the earliest phase of glass production on the site. However, evidence of heated floors and soil deposits were not recognised, so this is a tentative interpretation.

7.2.1.2.2 The Second Tank Furnaces Complex (TFC 2)

A series of plastered floors characterise the second phase at Bey 015. Large deposits of raw glass were attached to these floors, layers of mortar with melted fragments of glass between them and sandstone lumps with melted glass attached to them suggested from the beginning that glass production had occurred there (see Figs. 7.6, 7.7 and 7.8).

The plastered surfaces with glass still stuck to them were found on two levels. The upper ones in Tank 4 were preserved in small patches; stratified floors with glass still attached to them provide evidence that tank furnaces were used more than once in tanks 1 to 4. The walls associated with the surfaces were not found *in situ*. Occasional sandstone blocks may be the remains of original walls or they may have formed part of a rebuild following a glass-melting campaign. Some of the blocks may have been reused for the construction of the later pottery kilns nearby.

7.2.1.2.3 The Third Tank Furnaces Complex (TFC 3)

Plastered floor fragments discovered above the tank 4 floor of TFC 2 have been attributed to a third tank furnace complex (TFC 3) (see Fig. 7.9). A section reveals that the soil below the plastered surface in both tank furnace 3 and 4 does not show the same level of heating as the lower surfaces in tank 4 of the TFC 2. This plastered floor has been interpreted as belonging to a service corridor. It was possible to trace a continuous floor surface over a length of 10 metres.

Hellenistic to Early Roman Glass

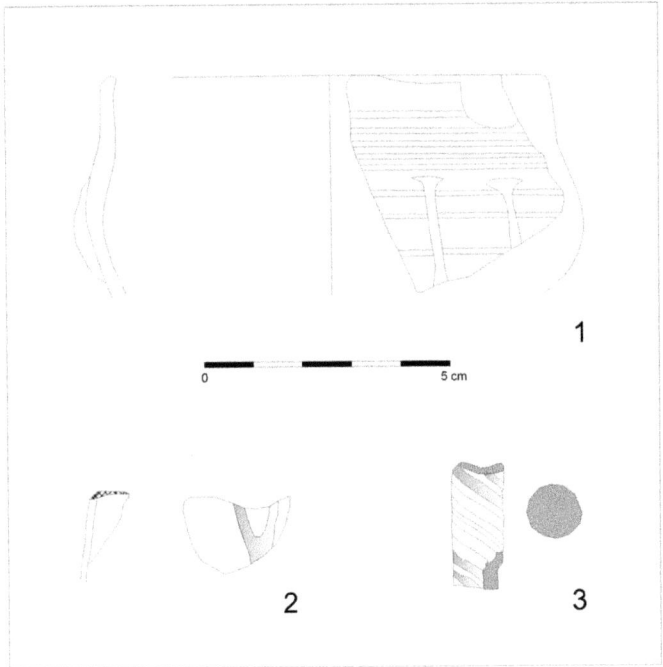

7.3. Glass vessel fragments associated with the Beirut 015 tank furnaces dating to before A.D. 50.

A horizontally placed ceramic tile marked the position of the bottom of the stoke hole in the northeastern wall of the cubicle of pottery kiln 5 (Kowatli *et al.* 2008). This covered the remains of a wall or foundation on top of which the northern wall of the kiln was constructed. This wall was attributed to the construction of TFC 3. The robbed wall to the east of the service corridor was also tentatively assigned to this complex. The remains of a robber trench was found at the same level as the floor of the service corridor. The northern wall of

7.4. Examples of some of the massive amount of raw glass found attached to the tank furnace floors in Beirut 015 dating to before A.D. 50 (reproduced with kind permission of the Director General of Antiquities, Beirut and the Bulletin de Archéologie et d'Architecture Libanaises). The large chunk is 10.2 cm long.

7.5. General view of excavated glass tank furnaces showing TFC 2 in the background (reproduced with kind permission of the Director General of Antiquities, Beirut and the Bulletin de Archéologie et d'Architecture Libanaises).

this area was heavily truncated by the construction of the south walls of pottery kilns 3 and 4. This clearly shows that TFC 3 was constructed and used before the Kiln Complex. The Kiln Complex was built between the second half of the first century and the early second century A.D. atop the remains of the TFCs 2 and 3 (see Fig. 7.9).

7.6. N-S section tank 2 Section (N-S) through Tank 2 (TFC 2) with lower floor remains and still unexposed earlier wall (reproduced with kind permission of the Director General of Antiquities, Beirut and the Bulletin de Archéologie et d'Architecture Libanaises). The maximum height of the section is 67 cm.

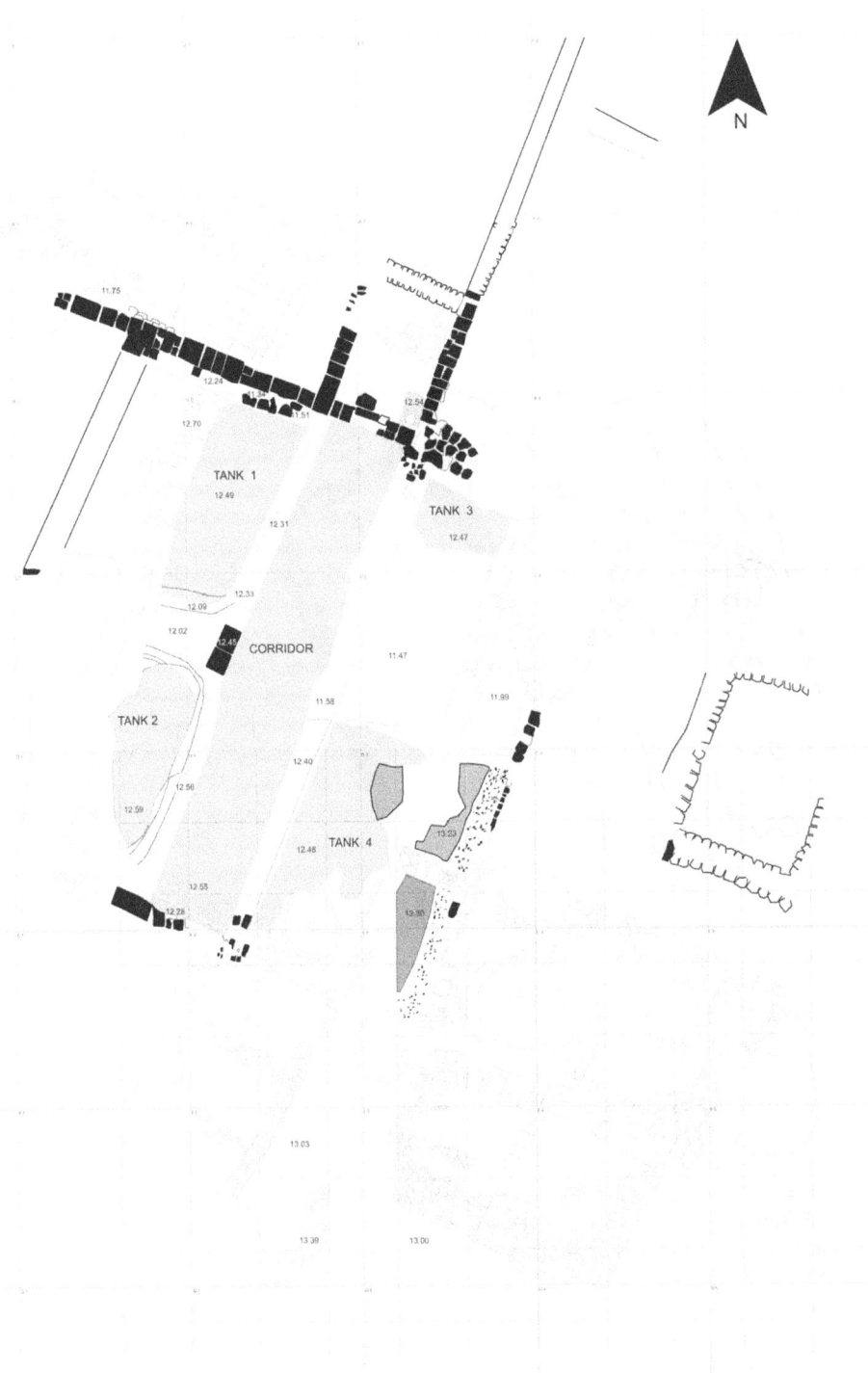

7.7. Plan of Tank Furnaces Complex 2 remains within the context of the excavated remains (reproduced with kind permission of he Director General of Antiquities, Beirut and the Bulletin de Archéologie et d'Architecture Libanaises). Each of the largest squares measures 5 m across.

Ancient Glass

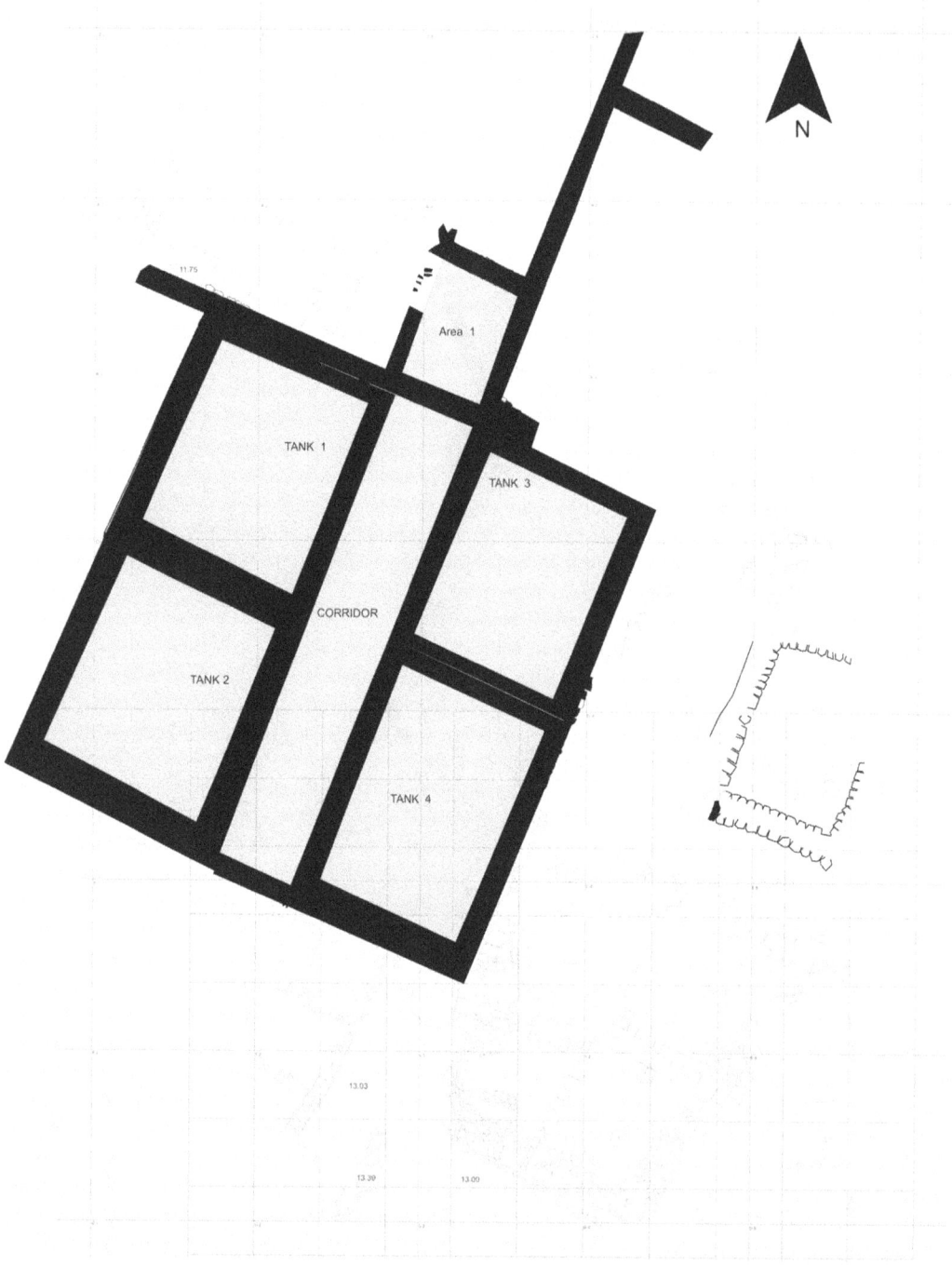

7.8. Schematic plan of Tank Furnaces Complex 2 remains within the context of the excavated remains (reproduced with kind permission of the Director General of Antiquities, Beirut and the Bulletin de Archéologie et d'Architecture Libanaises). Each of the largest squares measures 5 m across.

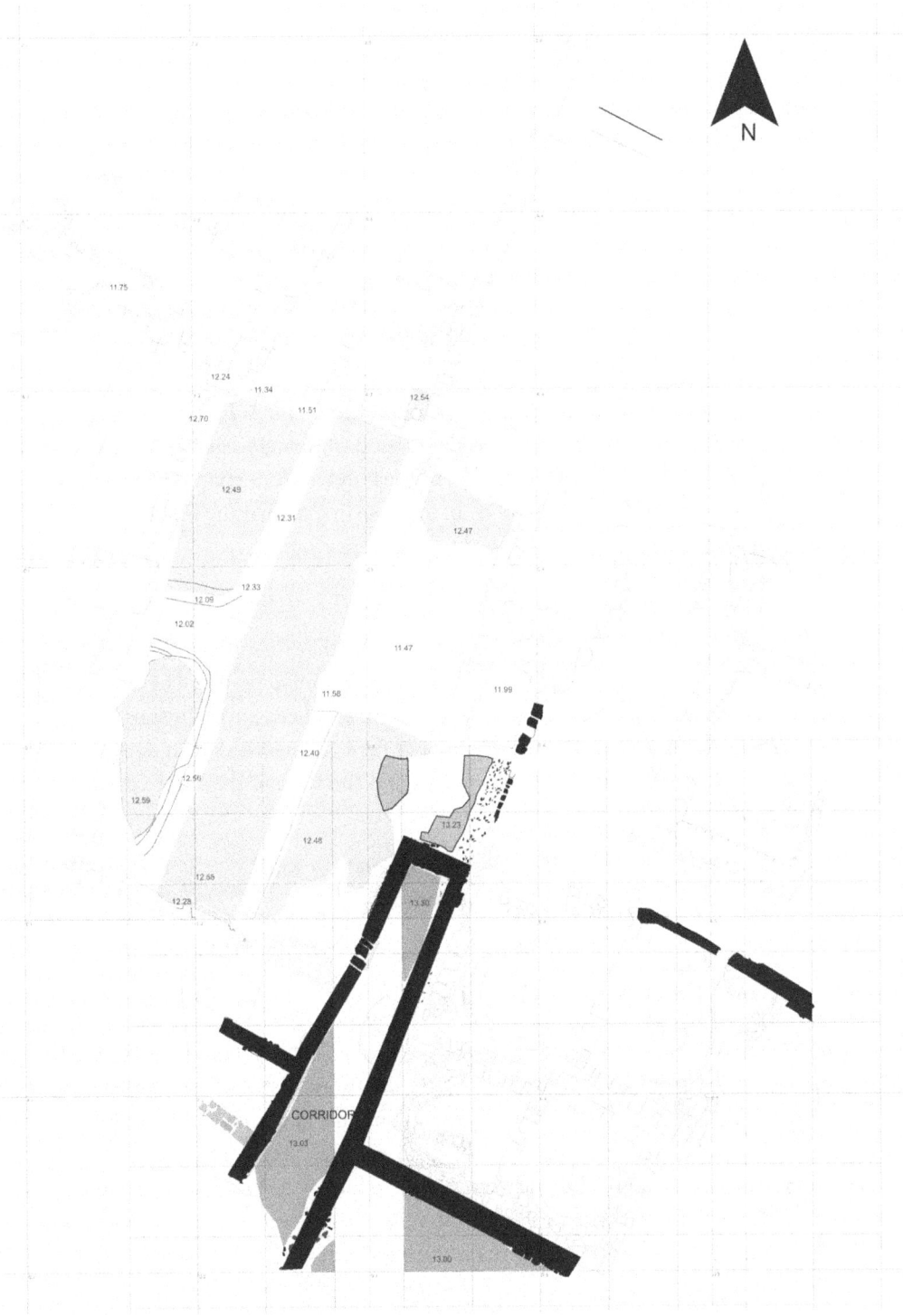

7.9. Schematic plan of Tank Furnaces Complex 3 remains within the context of the excavated remains (reproduced with kind permission of the Director General of Antiquities, Beirut and the Bulletin de Archéologie et d'Architecture Libanaises). Each of the largest squares measures 5 m across.

This is the only definitive evidence, on this scale, for primary glassmaking that includes the evidence of glass-melting furnaces for the late Hellenistic–early Roman period from the Levant. Raw glass has been found in the following colours: green, amber, brown, yellow-brown and nearly colourless with green hues.

At the glassmaking site in Beirut coloured and lightly tinted glass was produced on a massive scale. The manufacture of glass using a combination of natron and sand containing shell fragments referred to by both Strabo and Pliny clearly provides technological continuity between the late Hellenistic and early Roman periods in the Levant. Barag (1985, 59–60) suggests that the account given by Pliny (*Naturalis historia* XXXVI, 191) about the invention of glass on the Levantine coast relates to the introduction of more economic ways of making glass there. A smaller amount of glass would have been produced in the third century B.C. than in the first century B.C., and so it is likely that smaller furnaces would have been involved and a proportionately smaller volume of fuel, the most expensive raw material. The massive tank furnaces from Beirut described earlier would contribute to an increase in glassmaking efficiency, but it is difficult to know if this is what Pliny was referring to.

7.2.1.3 Syria, Greece and North Africa

In southern Syria, Dussart (2000, 92) has noted evidence for glassworking dating to the end of the second century B.C. up to the Augustinian period. The evidence included the remains of a furnace at Bosra, also in southern Syria (Dussart 1998).

Nenna (1999) has summarised the evidence for a glasswork shop at Kerkouase in Carthage dating to the end of the fourth and the start of the third centuries B.C. It included

> un tas de sable fin, des debris de verre fondu vert foncé, . . . de la chaux encore crémeuse et un matière colornate verte.

This evidence is intriguing and could suggest that glassmaking occurred in Carthage. The same house where this evidence was found also had a 'metal workshop'. Nenna *et al.* (2000, 107) have discussed the presence of small temporary workshops for secondary glass production dating to the early third century in the region of Alexandria. These had been installed inside the wall surrounding temples at Tebtynis in the Fayum and at Gumaiyama southwest of Tanis. Nenna is of the opinion that glass plaques were produced at these sites for the manufacture of cartouches and for inlay decoration of furniture and sarcophagi of the type dating to the reign of Ptolemy V (Nenna 1995). A possible third Hellenistic glass furnace dating to the reign of Ptolemy IV (late third century B.C.) has been found in Alexandria itself. Nenna *et al.* (2000, 108) review the strong historical evidence for glass production in the late Hellenistic and Augustinian periods in Alexandria described by Cicero and Strabo. There is mention of glassmelting and a probable reference to cutting glass.

In sum, the principal technological change in glassmaking that occurred between the middle and late Hellenistic period is one of increasing scale. This would have had an impact on the number of artisans involved, their organisation and the quantities of raw materials needed, including the various kinds of clays and stone needed for the construction of furnaces, annealing ovens, crucibles, moulds and other raw materials such as natron, silica and fuel. So although the mass production of cast vessels might be regarded as rather narrow and somewhat conservative in terms of the vessel forms produced, innovation would nevertheless have been involved in restructuring the scale of production. The production of simple vessels in mass produced former moulds that were then decorated with horizontal grooves was a quick and simple way of making bowls (Jackson-Tal 2005, 52).

7.3 Some Hellenistic Vessel Forms

The previous section in this chapter has largely focused on Hellenistic glass production. This section provides examples of Hellenistic vessel forms, but for which there is limited or no evidence for production.

7.3.1 Hellenistic Core-Formed Vessels

Three types of core-formed glass vessels have been defined. These have been found in the Mediterranean and date to between the c. 530 B.C. and the first century A.D. They have been labelled Mediterranean groups I, II and III. The largest, Group I, predates the Hellenistic Groups II and III but is considered here to provide a background for Group II vessels, the first Hellenistic type (see Fig. 7.10).

Group I is the largest and most homogeneous of the three groups. It consists of four main shapes: aryballoi, small jugs (oinochoai), alabastra and amphoriskoi. They were produced in large numbers and have been found across the whole of the Mediterranean area. Both Barag (1970) and Harden (1981) argued that the principal influence on the origin of Group I core-formed vessels was the Iron Age Mesopotamian core-forming industry. The suggestion is that they were either made by locals on Rhodes (Harden 1981, 52–3, Triantafyllidis 2009) or by Mesopotamian craftsmen who moved westward in the seventh century B.C. The main concentration of these vessels is in the Aegean, with a particular concentration in burials on Rhodes and other areas of the (eastern Greek) Mainland. Because the earliest forms of Group I bottles are 'thoroughly Hellenistic', Grose (1989, 110) is of the opinion that they could not have been made in Phoenicia or in Egypt. Even though this may be the case, it is nevertheless important to consider where the raw glass was made: it is possible that it was made on the Levantine coast (Phoenicia) or in Egypt. Core-formed glass vessels

TYPES OF CORE-MADE GLASS OF THE GREEK AND HELLENISTIC PERIODS

	ALABASTRA	AMPHORISKOI	ARYBALLOI	OINOCHOAI	HYDRIAI	UNGUENTARIA
VI-V CENT.						
IV-III CENT.						
II-I CENT.						

7.10. Chronological development of core-formed Hellenistic vessels (after Harden 1968; reproduced with kind permission of the Royal Archaeological Institute).

would have been manufactured in secondary production centres where Greek characteristics would have been introduced. The production of group I vessels ended in the late fifth century.

Mediterranean core-formed vessels of Group II date to the years following the collapse of the Persian Empire (mid-fourth to late third century B.C.). There was a gap of at least two generations before Group II began to be manufactured around 350 B.C. (McClelland 1984, 77–9; Grose 1989, 116). This date more or less coincides with the conquest of the Persian Empire by Alexander the Great, which began in 334 B.C. The 'Canosa group' has a purely Hellenistic character (Fig 7.11). Much has been written about these fine cast tablewares, of which there are about sixty that derive mainly from five hoards of vessels found in tomb groups at Canosa, Italy. They reflect the Hellenised tastes of the citizens of Canosa. Their forms are heavily influenced by the equivalents in metal (skeuomorphism) and represent the first efforts of glassmakers to make entire dinner services. Vickers (1996) has made a strong case for the influence of rock crystal and silver shapes on vessel made from glass. Many Canosa vessels have a faint pale green or yellow tinge, and a small number are highly coloured (Harden 1968a; Oliver 1968; Grose 1981; Grose 1989, 185–9). Harden (1968, 62; 1980) suggests that they were made in Alexandria; others have suggested that

7.11. An example of a vessel from the third-century B.C. 'Canosa group' found in Italy (after Harden 1968; reproduced with kind permission of the Royal Archaeological Institute).

they were made in southern Italy or the eastern Mediterranean. Although the Greeks colonised Italy, there does not appear to be any evidence of either a primary or secondary Hellenistic glass-producing industry there.

The greatest concentration of Group II vessels by far is in Italy, especially in Magna Graecia (southern Italy and Sicily), with few coming from the Levant, the Aegean, Egypt and western Asia (Harden 1981, 53). Some have been found at Sidi Khreish, Benghazi, Libya (Price 1885). Others come from Thessaly, Macedonia, Bulgaria and the Soviet Union. On the basis of the Macedonian finds, McClelland suggested that the vessels may also have been made in Macedonia (1984, 79). Harden (1981) was of the opinion that they were made in Alexandria and Grose (1989, 188) in a southern Italian city. Again, the question of where the raw glass from which the vessels were made also needs to be addressed; the most likely origins are Carthage or the Levant.

Small numbers of core-formed Hellenistic vessels of Mediterranean Group III have been found in Syro-Palestine. This group of vessels emerged in the last half of the third century B.C. The main vessel forms are *alabastra* and *amphoriskoi*. They were mainly translucent blue or green with trailed-on decoration in opaque yellow, white, blue and green colours. These vessels are thought to date between the third and first centuries B.C. (Harden 1981, 53–4; Fossing 1940). However, there is disagreement about the earliest date of their production: Grose (1989, 122) puts it in the mid- to late second century B.C., and McClellan has suggested two production phases, the first during the period between 250 and 150 B.C. and the second between 150 B.C. and the first century B.C. The appearance of new forms such as *alabastra* and *amphoriskoi* in a second phase could suggest that they were made in a new production centre, Cyprus (Harden 1981, McClellan 1984). Another possible production centre is Rhodes (Triantafyllidis 2000, 30). Grose's main argument for the later date for the appearance of Group III is that they do not occur in appreciable numbers until the mid- to late second century. If this is the case, it suggests that Group III vessels appeared about a century after Group II vessels died out.

7.3.2 Hellenistic Glass on Rhodes

The city of Rhodes was founded in 408 B.C. All three types of core-formed glass vessels have been found on Rhodes: the largest numbers of Group I have been found on Rhodes and in Italy. Core-formed Group I juglets with trailed-on decoration, a trefoil mouth and a disc attached to the base of the handle have been ascribed to a Rhodian source. This attribution is due to a concentration of these vessels on the island in contexts mainly dating to between the sixth and fourth centuries B.C. and to the presence of a number of malformed vessels (Harden 1981, 52–3; McClelland 1984; McClelland 1992). One of the products of glassworking on Rhodes was a range of cast and cut glass vessels frequently decorated with a wheel to produce asymmetrical, uneven and irregular designs (Triantafyllidis 2000a, 195, fig. 15).

Seventeen transparent core-formed cast and (relief-) cut luxury glass vessels dating to between the early fourth century B.C. and early third century B.C. have been found and it has been suggested that they were made on Rhodes. However, it has also been suggested that some of these were imported from workshops in the Achaemenid Empire, which stretched from the river Oxus to the west coast of Turkey. Indeed it has been suggested that the Achaemenid glass casting industry influenced the emergence of Hellenistic glass vessels. This is partly because a large collection of high-quality, mainly cast and cut bowls, beakers and jugs in colourless glass (with a green tint) has been found at Persepolis (Barag 1985, 57) and in the western satrapies of the Persian empire (Grose 1989, 80–1). There are various forms of imported vessels such as alabastra, kalikes and mesomphalos phialai. The fourth- and early-third-century green sagged vessels thought to have been made on Rhodes are mainly phialai, skyphoi, bowls and plates. They have unfinished rims and characteristic Rhodian designs, with wheel-cut floral and geometric designs on the walls of the vessels and leaves radiating from the undecorated centre of the base. Plates, in particular, can be undecorated (Triantafyllidis 2000, 31). In the second to first centuries B.C., Syro-Palestinian bowls are also found on Rhodes (Grose 1989, 193–5). There is evidence that a variant of Syro-Palestinian bowls was made there: they are undecorated, unpolished and have rough unfinished rims. The occurrence of a number of deformed bowls and some raw glass with traces of circular cuts on them has been interpreted as evidence of unsuccessful attempts at manufacturing circular glass blanks for the sagging process.

Highly coloured raw glass dating to the third–second centuries B.C. has been found in prehistoric Iron Age Europe. The Hellenistic world is the most likely source for it given the established links between cis-Alpine and trans-Alpine Gaul, especially from the second century B.C. Quantities of raw glass have been found at the third century *oppidum* of Entremont, France (Willaume 1987, 135),

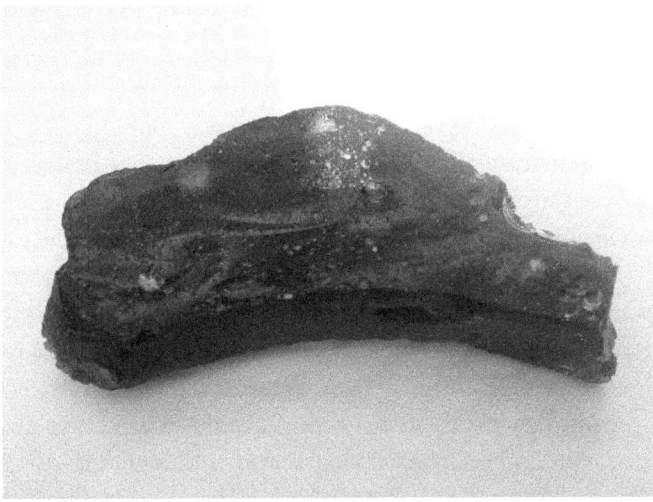

7.12. A large raw piece of glass found on the Sanguinaire shipwreck on the coast of Corsica (length 10.7 cm; photo: J. Henderson). Reproduced with kind permission of Prof. Dr. Rupert Gebhard, ArchäologischenStaatssammlung, München).

at the site of Němčice in Moravia, Czech Republic (Venclová *et al.* 2009) and on the third century B.C. shipwreck of the Sanguinaire, found in the Tyrrhenian Sea off southern Corsica. At least 550 kg of manufactured and raw glass was found on the Sanguinaire (Feugère 1992; see figure 7.12). On the basis of the chemical analysis of some of this glass, a Syro-Palestinian origin has been suggested (Foy *et al.* 2000a).

7.4 Early Roman Glass Production

Hellenistic states were incorporated into the Roman Empire from the first century B.C. Grose (1989, 241) has noted that the early Roman glass industry

> was founded under the auspices of Hellenistic glassmakers from the eastern Mediterranean, it developed rapidly into an independent, purely Roman enterprise. Within two generations... the Roman industry achieved full maturity'.

One result of this was that eventually another increase in the scale of glass production occurred. A second, interrelated feature was the commerce that resulted. For example, several hundred glass bowls, dating to the later half of the first century B.C., were found on the wreck of the *Tradelière* off the southern coast of France (Feugère and Leyge 1989; Foy and Jézégou 1998). The invention of glassblowing under the auspices of Hellenistic glassmakers was enormously important. The Roman glass industry was initially mainly based on a range of techniques that had been used by Hellenistic glassworkers, such as sagging and casting of glass to make bowls and cups. The actual fusion of glass from a probable Egyptian source of mineral alkali (natron) and sand was clearly not a Roman invention (Chapters 2 and 4).

Just like Hellenistic glass, early Roman glass was highly coloured, and this continued to be the case after glassblowing was introduced. Nevertheless the Romans also used nearly colourless, greenish, yellow-greenish and light aqua glass. The evidence of glassworking centres is quite sparse in the early Roman period, but what evidence there is suggests that it occurred on the edges of towns (Foy and Jézégou 1998, 125, Price 2005). Recent excavations in France at Manutention, Lyon (Foy and Nenna 2001) have provided evidence for working highly coloured glass, associated with excavated furnace pits of a keyhole shape, suggesting that the furnaces were probably of a beehive shape. There was no direct evidence of glassmaking there. Although the site in Lyon is dated to c. A.D. 25, the typical container shapes that were introduced when the widespread introduction of mould-blowing and the production of blue-green glass vessels were not yet being made because the first mould-blown containers were not produced in any great quantity until around c. 50 A.D. The reasons why secondary glassworking centres were set up in locations like Lyon must be because of demand for glass vessels by a growing Roman middle class and close links via Marseilles (a thriving port).

The evidence for early Roman glass production under imperial auspices has also been found on sites as widely separated as in the Levant, in Switzerland at Avenches (Morel et al. 1991; Amrein 2001), dating to between 40 and 70 A.D. (Amrein 2001, 12, n. 9) and Reims (Cabart 2003), also dating to the first century A.D. Other first-century furnaces have been found at Bonn (Follmann-Schulz 1991, 36–7, 40). Here three rectangular furnaces dating to 75–100 A.D. have been found. A secondary glassmaking centre has been found at Saintes (La Fenêtre) where there was a furnace with a possible semicircular structure.

Some of the best evidence has been found in Cologne, Germany. Here furnace numbers 235, 236 and 237 are circular (Schäfer 2005; Constanze et al. 2006, figs. 2–5, 76–7, 83); those found at Cologne, Eigelstein 14, which are pre-Claudian–Flavian in date, combined five to six circular structures and one rectangular structure (Doppelfeld 1966, 11–16; Constanze et al. 2006, 77, fig. 6). At Eigelstein 35–7, eleven furnaces were found, of which six were rectangular and five circular in shape. These were originally given a tentative date of pre-Claudian, but have now been re-dated to the second century A.D. (Fremersdorf 1965–1966, 40; Doppelfeld 1966, 16; Amrein 2001; Höpken 2005; Constanze et al. 2006, figs. 7–11).

Amrein (2001, fig. 94, annexe 1) has published a useful list of excavated Roman glass furnaces together with an inventory of Roman glass workshops and by doing so has highlighted the variation in the (excavated) furnace plans. Clearly one should be careful in attributing functions to different shapes of furnaces without the evidence of above-ground structures (Henderson 2000a, 44; Amrein 2006, 58). Amrein has distinguished amongst circular, rectangular, 'semi-circular' and combined structures.

The variation in furnace form at Eigelstein 35–37 is striking, and especially the number of rectangular examples. It is possible that the circular ones were used for blowing vessels and the rectangular ones for annealing them. However, it is claimed by Constanze *et al.* (2006, 78) that raw blue-green glass was produced (i.e. fused) at the site. Without secure scientific and archaeological evidence for this, it is difficult to be certain. In London, glass was definitely worked, and there is a possibility that it was fused from primary raw materials (Shepherd and Heyworth 1991): vitrified bricks and a large number of tank furnace fragments together with lots of evidence for glassworking have been found at 55–61 Moorgate in early-second-century A.D. contexts (*ibid.*, 14). This does not, however, prove that glassmaking occurred.

Beehive shaped furnaces are typical evidence for glassblowing. It is presumed that in many cases they consisted of a bottom chamber in which the fuel was burnt, a middle one containing one or more crucibles containing fluid glass and a top one for annealing blown vessels. They had a key-hole shaped plan. By the first century B.C. some Levantine glass is already a pale green colour; according to the raw glass found, that being worked in first-century Avenches was blue, purple, yellow, emerald green and blue-green–aqua (Amrein 2001, 17–18; Planche 10–18). In her discussion of raw glass, Amrein distinguishes between raw glass that resulted from fusion in a tank furnace, raw glass melted in a crucible and flat plaques of glass destined to be used for tesserae production. Neither glass plaques nor crucibles were found at Avenches. However, the absence of evidence for the use of crucibles does not necessarily mean that they were not used. After all, what else could the glass be heated in and easily extracted from? The evidence for glassworking including glassblowing from Avenches included thousands of moil fragments as well as dribbles, drops and pulls of glass. Amrein (*ibid.*, 17) describes the glass from Avenches as good transparent material, with an absence of impurities and with elongated bubbles (considered to be evidence of blowing). The glass that was worked at Manutention, Lyon, was essentially of the same range of bright colours. It is likely that there was early Roman primary glass production in Italy, although no archaeological evidence has yet been found. Strabo mentions that glass production in Rome involved 'innovations' in glass colour technology and inventions that increased the efficiency of glass production, reducing its cost so that 'a drinking cup could be purchased for a copper coin' (Strabo, *Geography* 16.2.25). However, chemical analysis of early Roman glass has, so far, basically shown that there was continuity in the use of raw materials and colourants used to make late Hellenistic and early Roman glass. This seems to be the case even for glass vessels thought to have been made in Italy, although there is some evidence that raw glass from an origin other than the Levant was used to make vessels in Italy (see Chapter 8).

One of the most famous early Roman glass vessels is the Portland vase. It dates to the last half of the first century B.C. and is thought to have been made for Augustus (Painter and Whitehouse 1990). It is coloured in the Hellenistic 'tradition' with cobalt and decorated with a white cameo design coloured and opacified with calcium antimonate crystals (Bimson and Freestone 1983). Although an exceptional *tour de force* in its own right, there are other cameo vessels using this combination of colours, or in deep translucent purple instead of blue. The cameo technique, of cutting the outer layer of white glass away to produce a carefully executed design, was a Roman invention and reflects the resourcefulness of early Roman glassworkers.

Perhaps the increase in efficiency of glass production referred to by Strabo relates to the introduction of glass blowing in Italy. He notes with confidence that glass was made in Italy from raw materials and that the knowledge spread from there to Spain and Gaul. However, until recently, limited evidence for late Hellenistic and early Roman primary making has been found. The best evidence from Beirut, Lebanon, is described earlier. The raw glass produced there in tank furnaces was of three main colours: aqua, amber (golden brown) and blue. These colours are typical of vessel colours of the late Hellenistic and early Roman period (Grose 1989: 195). One furnace complex may date to before A.D. 50, although their precise dating is difficult. The evidence for the production of predominantly amber linear-cut and ribbed bowls from Bey 002 (Foy 2005) seems to date mainly to the late first century B.C. and early first century A.D. (Jennings 2006, 54), and it may well be that (some of) this glass was manufactured at Bey 015, Beirut.

In the *Periplus of the Eritrean Sea*, written between A.D. 40 and 70, a Greek merchant living in Alexandria states that several colours of glass 'stones' were manufactured at Diospolis, the Greek site of Thebes (Stern 1991a; Stern 1991b). This may be a reference to the addition of colourants to imported ready-made glass, so, if true, this may be evidence for glassworking (the addition of colourants to the glass melt) at Thebes.

However, possible evidence for glassmaking has been found at Autun, Saône-et-Loire. Rebourg (1989a; 1989b) has reported the discovery of four rectangular tank furnaces with 'identical' structures, the best preserved (number 1), orientated east-west, measured 1.9 × 0.88 m, with a semicircular brick structure at one end. Furnace 2 was also well preserved but little remained of furnaces 3 and 4. Number 2 was orientated north-south and measured 1.46 × 1 m. There was evidence that this also had a semicircular brick structure (Rebourg 1989b, fig. 5); Furnace 3, orientated east-west, was partially destroyed by the construction of furnace 1. They are dated to the end of the period we are concerned with here (and later) to c. A.D. 150–250 (Rebourg 1989b, 252–3, fig. 2) but are considered here because they are unusual in a northern European context. Rebourg (*ibid.*) and others (Foy and Jézégou 1998, 125, n. 20) have suggested that these furnaces provide unusual evidence for primary glass production in the occident, although Amrein (2001, 17) is more cautious about labelling the site a primary glass production centre. Raw blue-green glass was still attached to the floor of furnace no. 1 to a depth of c. 5 centimetres and 'frit' was discovered (Rebourg 1989b, 254). Pliny the Elder (*Naturalis historia* XXXVI, 199) refers to *fragmenta temporata adglutinantur*, attesting to the use of recycled ingots. Perhaps glass ingots were reheated in furnaces at Autun but it is likely the bulk of the melt would have been made up of the extensively traded scrap glass. The Autun furnaces are quite small compared with the massive glassmaking tank furnaces found in Beirut and Beni Salama in Egypt (Nenna *et al.* 2005), so this may mean that they were used for glass recycling.

Nenna *et al.* 2000 studied primary glassmaking centres in Egypt, on the coast at Taposiris Magna, southwest of Alexandria, part of an antique industrial complex with evidence for pottery and metal production, although of a probable late Roman date, and at inland locations at two groups of ateliers at Beni Salama and Zakik in the Wadi el Natrun depression 80 kilometres northwest of Cairo. The Wadi el Natrun sites are difficult to date because of the scarcity of finds, but they are considered to be of a late Roman date. For this reason they have not been described in detail here

Like at Taposiris Magna, the Autun furnaces were located between two pottery workshops, and bronze production occurred nearby. Although it is difficult to prove that Autun was a primary glassmaking centre, to assume that all primary glass production occurred in the Middle East at this time would be rash, and scientific evidence is accumulating for there being multiple primary making zones, including isotopic evidence for Roman glass found in north-western Europe with a non-Levantine signature (Degryse and Schneider 2008, Degryse *et al.* 2009, 68). In contrast, the fourth-century Roman glass furnace that was excavated in the Levant, at Jalame, was apparently a secondary glass production centre, at which aqua and yellow-green glass was reworked (Weinberg 1988; Brill 1988). Indeed, aqua-coloured natron glass was found to have been worked in the Byzantine I phase (363–451) glass from Sepphoris (Fischer and McCray 1999, 894–96).

If the primary manufacture of highly coloured glass in Beirut went on as late as A.D. 50, which seems likely, it would not only have fed the emerging blown vessel industry but also the increasingly high demand for glass mosaic tesserae. In later periods raw glass with pale tints was manufactured in large quantities and exported to secondary centres where colourants were added. In Beirut, colourant-rich raw materials, such as cobalt-rich ones, were added to the glass batch within the tank furnaces. Nevertheless, Pliny the Elder (*Naturalis historia* 36) noted, that:

'the [raw glass] masses were melted again in the workshops and coloured then some of the glass is shaped by blowing.'

However, it is probable that opacifying raw materials, such as antimony oxide and other antimony compounds used in early Roman glass mosaics, would have been added in a secondary manufacturing process. Antimony was used almost universally in early Roman opaque tesserae to produce opaque yellow and white; copper would have been added to produce, under reducing conditions (in the presence of iron), an opaque red colour (see Chapter 3).

One of the reasons glass production became increasingly separated into primary and secondary processes was probably because of the introduction of specialised centres for glassmaking and glassblowing after the latter was invented in the first century B.C. Blowing a vessel was more skilled than slumping a blank into a mould. Driven by an expanding scale of production, a second reason for this separation may have been because glass colouration became a separated process. From about the mid-first century A.D., primary production centres focused on melting greenish glass from raw materials. Such centres were in the Levant, perhaps in Egypt, and probably Italy with other possible centres in cis-Alpine and trans-Alpine Europe. Secondary production centres concentrated on importing and reheating glass so as to blow glass vessels, and in some cases to colour it. By the late Roman period, highly coloured glass was still being made with, for example, a flourishing industry producing highly coloured late Roman glass vessels in Germany, especially in the Rhineland (e.g. Harden 1987, 263 *f*). At 4th century Jalame in Israel where green cullet was reheated (Weinberg 1998), there was no evidence for its deliberate coloration to produce deep cobalt blue or purple glass.

In sum, by the late Roman period glassmaking in tank furnaces, the reheating of ingot and scrap glass (cullet), and the addition of colourants, perhaps in smaller furnaces, and the blowing and cutting/engraving of glass vessels all occurred. Compared with early Roman glass production, proportionately less highly coloured glass was being used. The use of highly coloured glass predominated in late Hellenistic and early Roman glass production up to c. A.D. 70, with limited archaeological evidence for the addition of colourants to the glass batch. Further excavations may eventually provide the evidence of a site where mineral-rich colourants were added to lightly tinted scrap and raw glass.

7.5 The Use of Hellenistic and Roman Glass

Hellenistic core-formed vessels were highly decorated with opaque glass, and their manufacture was labour intensive. The core-formed vessels were used for holding small volumes of liquid, like perfumes, the same function that can probably be ascribed to the first core-formed vessels made in the fifteenth century B.C. (see Chapter 5). However, such Hellenistic and late Bronze Age core-formed vessels were obviously produced and used in very different social contexts. Although cast vessels were also made in the early Hellenistic period, they were rare and valuable (Grose 1989, 197). By the middle Hellenistic period, the vessels that had become more common were still used in elite social contexts. One of the first documented pieces of evidence for the manufacture of a complete dinner service is that of the late third to early second century Canosa group of cast, ground and polished vessels from southern Italy and Sicily (*Magna Graecia*; Harden 1968a, Oliver 1968). Hellenistic moulded glass vessels were frequently highly coloured, especially the Middle Hellenistic ones, and by the second century B.C. they were produced on a large scale: witness the 6,000 fluted bowl fragments found at Tel Anafa (Section 7.2). The latter would seem to have been manufactured in the Levant. Before such a discovery it would have been easy to suggest that such bowls were used by the social elites. The discovery of such large vessel numbers from one site suggests that other sectors of society used them.

In the middle and late Hellenistic periods, the range of vessel forms was quite restricted (with some exceptions such as the Canosa group), and the production technology can be defined as somewhat conservative with limited evidence of innovation. Indeed, Finley (1985, 146) notes that Aristotle and Vitruvius saw little virtue or possibility 'in the continued progress of technology through sustained, systematic inquiry'. Furthermore, several classical writers refer to a story that a Roman inventor of unbreakable glass (who had retained the secret) was decapitated by Tiberius, lest the value of 'gold be reduced to that of mud' (*ibid.*, 147).

Urbanised Hellenistic democratic society was quite conservative with an extensive international trade network. Had Hellenistic glass vessels been used in entirely new social/ritual contexts, then it is possible that a greater range of vessel forms would have been produced, potentially at a larger number of production centres using a wider range of raw material sources, leading to greater variability in the chemical compositions. The vessels were basically small open forms that were used for drinking. The occurrence of late Hellenistic coloured or completely colourless cups and bowls and mosaic cane vessels at the military site of Jebel Khalid on the Euphrates in northern Syria shows that a Hellenised military elite had access to such luxury glass. Mosaic cane glassware has not even been found in Hellenistic Jerusalem or in the well-to-do occupants of town houses in Pella (O'Hea 2005, 46). The use of successfully decolourised vessels at Jebel Khalid may conceivably reflect a regional tradition for a taste in colourless glass. The occurrence of such glass in a military establishment appears to reflect the high social standing of some military groups. This is an interesting inference: similar research in the Middle East is essential so as to investigate in greater detail the find and social contexts of Hellenistic and early Roman glass.

The growth of the Roman middle class and the use of glass as luxurious decoration in private and public contexts would itself have created a demand. For example, such vessels were displayed near the entrance to Roman villas, so that visitors could see them. It is only really in the Roman period that dinner sets become common, and middle-class urban life would have involved showing off the glass (just as it does today). Metal dinner sets were replaced with glass ones because the former were said to stink, especially in the summertime.

Wall paintings adorning the walls of late first century Roman villas in Pompeii and Herculaneum sometimes show glass vessels forming part of set meals.

Cool's detailed analysis of the distribution of glass pillar-moulded bowls and cups from contexts dated to A.D. 44 to 60 at Colchester has revealed that there may well have been an association of officers with wine and perhaps 'men' with beer (Cool 2006, 178, table 17.4). A nice insight into the gender-specific use of glass vessels is afforded by detailed research into glass vessels associated with buried individuals, perhaps of Danubian origin (Cool and Baxter 2005). In a third-century A.D. cemetery at Brougham in the military northern zone of the United Kingdom, drinking cups were associated with the burial of 'gentlemen' from the upper echelons of society, but not 'ladies'.

It is clear that other similarly insightful research using Roman glass vessels from secure archaeological contexts needs to be carried out from other regions. This may reveal both similar and contrasting patterns of use in different parts of the empire. If a comprehensive scientific analysis of these same vessels is also carried out, it might be possible to determine the primary production zones for glass used to make such vessels too, and this could be linked to their social and ritual value.

EIGHT

LATE HELLENISTIC AND EARLY ROMAN GLASS

SCIENTIFIC STUDIES

This chapter follows on from Chapter 7: it links the scientific analysis of late Hellenistic and early Roman glass to the archaeological, historical, social and economic aspects that were described in Chapter 7.

Cast glass vessels were manufactured in both the late Hellenistic and early Roman periods. However, a fundamental technological innovation occurred in the first century B.C., glassblowing. This was set to transform the technology of vessel production. Because glass is made to behave in different ways when it is cast or blown, it is important to determine whether glassmakers produced glasses of different chemical compositions, with different working properties, even though the same raw materials were used to make late Hellenistic and early Roman glass (sand and natron). Although natron glass dominated, plant ash glass was used to make a small proportion of vessels. Examples of plant ash glasses have been found at Augusta Pretoria in Italy (Mirti *et al*. 1993), Fishbourne in the United Kingdom (Henderson 1996), Avenches in Switzerland (Amrein 2001), Beirut and Mariana, Golfe de Fos and Ruscino in France (Thirion-Merle 2005) and in early Roman mosaic glass vessels (Nenna and Gratuze 2009). Such glasses were made from coastal or from semi-desert plants; the relatively low contents of magnesium and/or potassium oxides in them (for plant ash glasses) could be diagnostic.

As far as we know, the main production centres for glass vessels in this period are the Levant, north Africa and Italy. Although the archaeological evidence for primary glass production in the Levant is clear, and there is also evidence for this dating to the second century A.D. and possibly earlier in north Africa, (see Chapter 7), there is currently no clear archaeological evidence from Italy or in northern Europe. Early Roman cast and blown vessels of Italian types such as *zarte Rippenschale* bowls, highly coloured cast open vessels in a range of shapes, gold-band decorated vessels, laced mosaic bowls and mosaic ribbed bowls (Haevernick 1967; Harden 1987, 39, 41, 42, 47; Rossi and Chiaravalle 1998; Invernizzi and Vecchi 1998) are often found in graves. It is important to establish whether the raw glass used in these vessels was also made in Italy or whether it was imported. As will be noted later, compositional groups have been defined for specific colourless early Roman vessel forms, suggesting that more that one production zone for them existed, one probably being located in western Europe.

8.1 Levantine Glass and a Comparison with Other Glasses

One key area for the manufacture of raw glass during the first century B.C. and first century A.D. is the Levantine coast. Foy's (2005) timely review of the evidence for the production of moulded bowls in Beirut (see Chapter 7) has been published with compositional data for these and other contemporary glasses (Thirion-Merle 2005).

On the basis of her scientific results produced by using inductively coupled plasma mass spectrometry, Thirion-Merle (2005, 41) has claimed that there is little compositional difference between the raw glass and the samples of contemporary blown and cast glass vessels from Beirut. Furthermore, by using cluster analysis to investigate the structure of the data, she noted that the Beirut glasses were very similar to contemporary glasses from Gaulish sites (Ruscino, Golfe de Fos, Avignon, Mariana, Peyrestortes, St Jean de Garguier and Carpentras) and to Manutention, Lyon (discussed in Chapter 7), including raw glass from Manutention. Thirion-Merle singles out one group that consists mainly of slightly more than 50% of the linear-cut bowls analysed, the majority (8 out of 10) 'first-generation' moulded bowls, some moulded ribbed bowls and a single raw glass specimen. The diagnostic compositional characteristics of this group are high soda and lower levels of silica and alumina than other analysed glasses. However, the group is formed from such a mix of vessel types that it does not appear to have archaeological significance in terms of the production technologies involved.

Further consideration of the data published by Thirion-Merle in 2005 reveals some interesting patterns. A bi-plot of calcium against aluminium oxides (see Fig. 8.1), reflects the use of different sand deposits with different purities. In most cases the clusters are relatable to the vessel types and production technologies used. Eight out of nine of the blown vessels from Beirut form a negatively correlated group on the right-hand side of the figure. A cluster of five vessels made in *terra sigillata* shapes (including those from Ruscino) contain amongst the lowest alumina levels (c. 2.2%–2.3%). In between these two clusters are (1) a negatively correlated group of moulded vessel glass from Ruscino that also includes a moulded Beirut vessel and a raw glass sample from Lyon, and

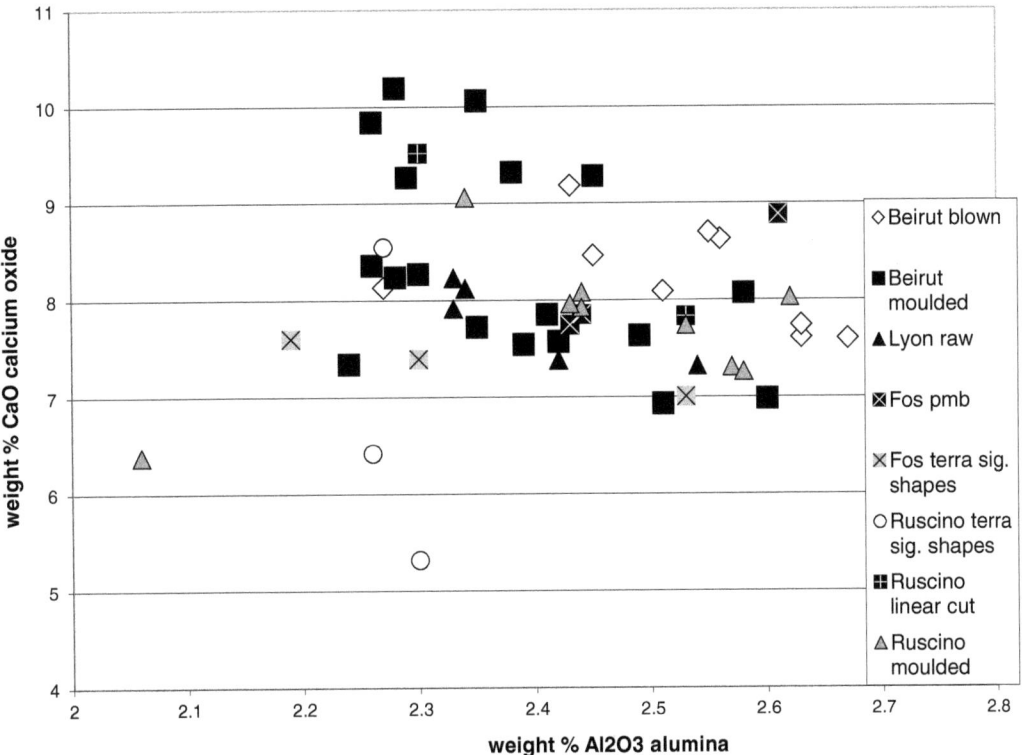

8.1. A bi-plot of the weight % calcium oxide versus alumina in Italian, French and Levantine glasses c. first century B.C.–first century A.D (Pmb = pillar-moulded bowl; Fos = Golfe de Fos, France, terra sig. = *terra sigilata*). Data from Thirion-Merle 2005.

(2) a separate negatively correlated group of Beirut moulded vessels, raw glass from Lyon and a single vessel made in a *terra sigillata* shape. Six Beirut moulded vessels contain greater than 9% calcium oxide. A Beirut blown vessel, a Ruscino moulded vessel and a Ruscino linear cut vessel do as well. Therefore there is a distinction between most Beirut blown vessels, most moulded Ruscino vessels and most vessels made in *terra sigillata* shapes. The three results plotted for pillar moulded bowls appear to fall on a positively correlated line. If the *terra sigillata*–shaped glass vessels are Italian products, then relatively consistent 'low' alumina levels of between c. 2.2% and 2.3% appears to indicate that a relatively pure sand was used to make them (see Chapter 3). However, four of the six *terra sigillata*–shaped vessels are emerald green plant ash glasses (VRR 675, 676, 677 and 678). Three of the plant ash glasses contain between 2.19% and 2.3% Al_2O_3; a fourth contains 2.53% Al_2O_3. The two *terra sigillata* natron glasses contain similar Al_2O_3 levels. The emerald green glasses are relatively unusual examples of classical glasses made from sand and plant ash, possibly in a coastal (?Levantine) location. The relatively low magnesium oxide levels for a plant ash glass could be an indication that coastal plants were used (as seen in

later Islamic plant ash glass: see Chapter 10). If the vessels were made in Italy the raw glass used to make them with would probably have been imported from the eastern Mediterranean.

There are therefore three negatively correlated distributions in Figure 8.1. These are not statistically separable but they consist of (1) most blown Beirut vessels, (2) Ruscino moulded vessels and (3) most Beirut moulded vessels. The correlations between alumina and calcium oxide reflect the natural compositional variations of shell fragments and feldspars in the sand used. Similarly correlated, although less constrained, are relative levels of alumina and calcium oxide in later furnace glasses from Bet Eli'ezer and Ramla dating to between the sixth and ninth centuries (Freestone *et al*. 2000, fig. 7). Given that the data for moulded vessels are based on samples from separate origins, it appears that the glassmakers mainly stuck with the same sand source. From these results it is difficult to be confident about a provenance of the raw glasses or that the results necessarily suggest that a single source was involved.

The zirconium and strontium impurities in late Hellenistic and early Roman glasses provide further information about the raw materials used (see Figure 8.2). Zirconium is present as zircons in sands, and calcium associated with strontium is present either in the mineral aragonite in sea shell fragments in sands or in limestone (see Chapter 3 and 11). In Figure 8.2 a clear distinction can be seen between a group of five *terra sigillata*–shaped cast vessels from Ruscino and Golfe de Fos, with variable and higher zirconium contents of between c. 70 and 100 ppm and the rest of the samples that have with a constrained lower range of zirconium contents of between c. 30 and 40 ppm, including one *terra sigillata*–shaped cast vessel sample from Ruscino. The two groups were clearly made using two different sand sources. Further observations are (1) that in the group containing lower zirconium oxide levels, the cast Beirut vessels and the Lyon raw glass contain virtually the full range of strontium values; (2) most Beirut blown vessels contain less than 450 ppm strontium oxide; (3) a group of Ruscino moulded vessels has the lowest strontium contents and (4) two pillar moulded bowls from Golfe de Fos have amongst the highest zirconium contents in the group.

Although the different classes of glasses are not entirely separable, there are, again, intriguing indications that we are seeing compositional differences perhaps amongst glass made at different furnace sites, presumably all in the Levant. The low level of zirconium in these glasses is a reflection of the low zircon content in Levantine coastal sands (Pomerancblum 1966). Freestone *et al*. (2000, table 3) have shown that raw furnace glass manufactured in the southern Levant during the sixth to eighth centuries using coastal sands tend to contain c. 60 ppm Zr (between 45 and 73 ppm), with a single sample containing 90 ppm and between 320 and 450 ppm Sr. With two exceptions, the strontium results for the southern Levant coincide with the maximum cut-off value of 450 ppm for

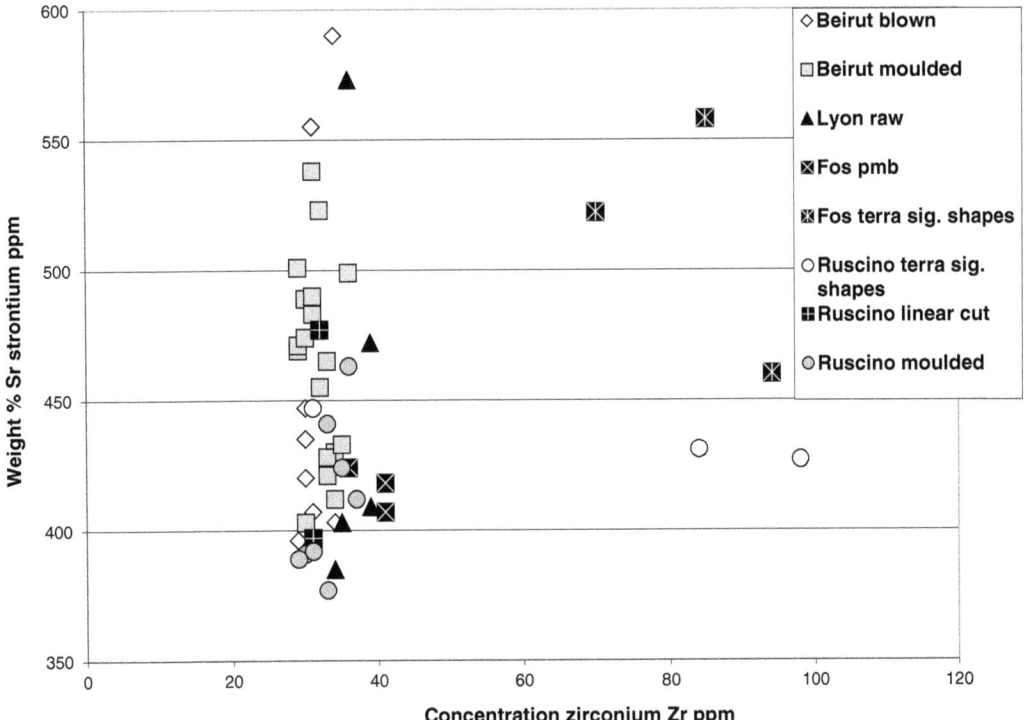

8.2. A bi-plot of the concentrations of zirconium versus strontium in Italian, French and Levantine glasses c. first century B.C.–first century A.D. (Pmb = pillar-moulded bowl; Fos = Golfe de Fos, France, terra sig. = *terra sigillata*). Data from Thirion-Merle 2005.

blown Beirut vessels, the two exceptions containing more than 550 ppm, so this may be significant. The zirconium results for later (furnace) glasses published by Freestone *et al.* (2000) occupy the area between those plotted in Figure 8.2, reflecting the use of different sand sources and presumably different production zones in the Levant. Freestone *et al.*'s results were for the southern Levantine coast of a later (sixth–eighth century) date. Therefore, the sands used to make earlier glass found in Beirut have different impurity levels (especially Zr). This is significant for the provenance of raw and vessel glass and is an indication that glasses made from different Levantine sands can be distinguished in this way. Although four *terra sigillata* cast vessel samples are plant ash glasses and two are natron glasses, except for one, they contain high zirconium levels. A single natron *terra sigillata* vessel result falls into the low zirconium group in Figure 8.1. It can therefore sit at the high calcium end of the group, with the Lyon raw glass and Beirut moulded glass.

It can therefore be stated that these *terra sigillata*–shaped vessels were made using two different sand types, one with higher zirconium and barium (see Fig. 8.3) and lower alumina levels than the other. Furthermore, a sand source with the higher zirconium impurity was used to make both plant ash and natron

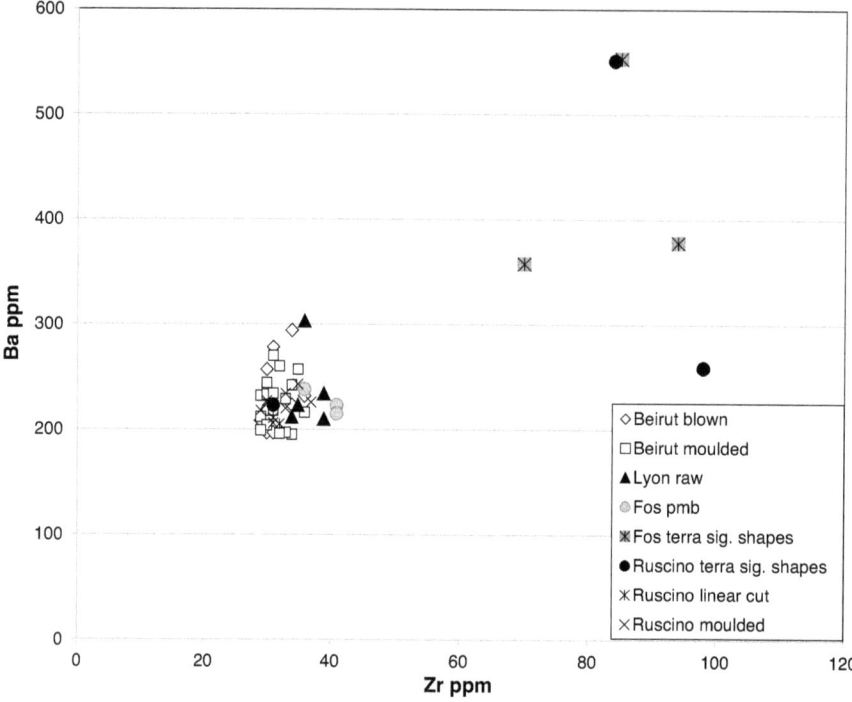

8.3. A bi-plot of the concentrations of barium versus strontium in Italian, French and Levantine glasses c. first century B.C.–first century A.D (Pmb = pillar-moulded bowl; Fos = Golfe de Fos, France, terra sig. = *terra sigillata*). Data from Thirion-Merle 2005.

glasses. Although based on a relatively small data set, this could indicate the same (coastal) production centre made them and that it exploited both natron and plant ash sources of alkali. If it was an Italian (or North African) centre, it is interesting to speculate whether plant ashes were imported from the Middle East or were harvested elsewhere, such as in Sicily. It is equally interesting to note that 'Levantine' sand was probably used to make one Ruscino *terra sigillata* vessel even though it is considered to be of an Italian form. So whichever compositional type of glass was used, it would have been exported to Italy where the Italian vessel form was produced.

The results plotted in Figure 8.4 show that different sand deposits were used for glass production in two periods. There are clearly differences in the types or proportions (or both) of the 'impurities' in the sands used (such as feldspars and shell fragments), giving different compositional signatures in the glasses made from them. The question is what has led to the different sand compositions. One explanation is that processes that deposited and homogenised the sand on the Levantine coast would have led to slightly different proportions of heavy minerals at each location. If a compositional and an isotopic distinction between Belus and Sidon sands could be demonstrated, this would provide an interesting means of distinguishing glasses melted in the two areas.

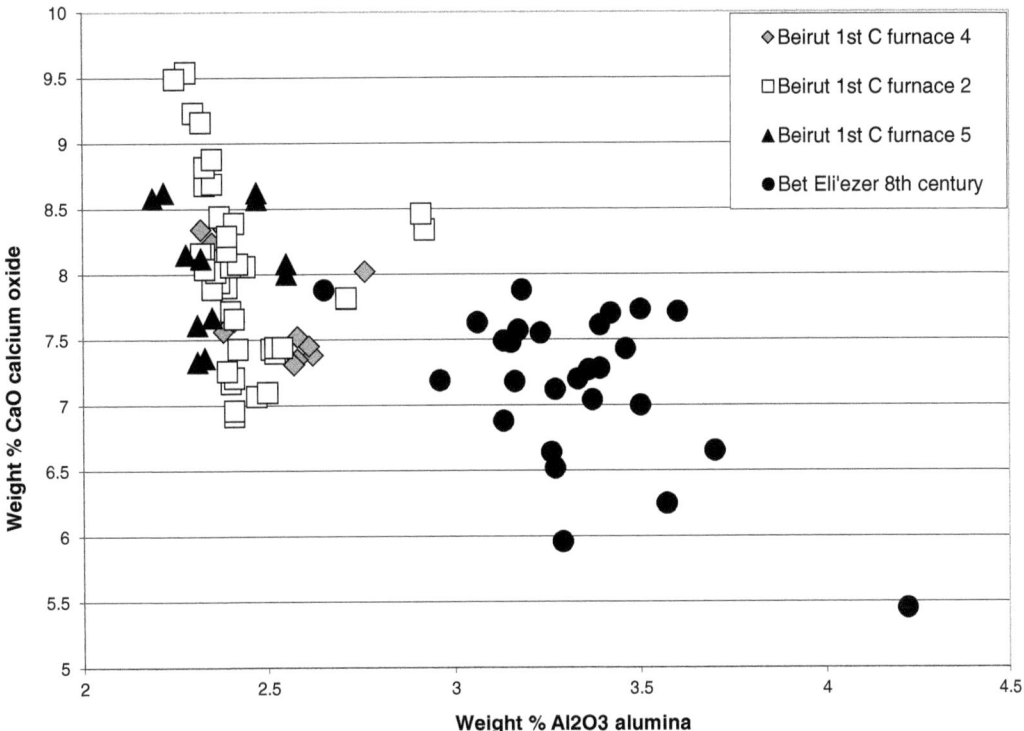

8.4. A bi-plot of the weight % calcium oxide versus alumina in Beirut glass tank furnaces and the eighth-century Bet Eli'ezer glass furnace c. 50 B.C.–A.D. 800. Data from Freestone et al. 2000 (Bet Eli'ezer) and the author (Beirut furnace glass).

The analysis of eleven glasses from Pompeii that date to before A.D. 79 using X-ray fluorescence and laser ablation inductively coupled plasma mass spectrometry appear to provide evidence that a southern Levantine sand source was used to make them. These natron glasses contain strontium oxide levels that fit exactly the range plotted in Figure 8.2. However, five contain between 40 and 50 ppm Zr, and three contain 53.9, 60.3 and 65.4% ppm Zr (De Francesco et al. 2010) and therefore contain Zr levels comparable to (later) furnace glasses made from southern Levantine sands. The remaining three samples have Zr levels that are characteristic of a Levantine sand source from around Beirut. This is an interesting set of data because it appears to provide evidence for primary glassmaking in the southern Levant in the early Roman period where no primary glassmaking sites of this period have been found. Egyptian natron glasses from ell Ashmunein were presumed to have been made from an inland source of Egyptian sand which introduced significantly higher zirconium levels than found in Levantine glasses (c. 160 ppm) and lower strontium levels (c. 100–200 ppm) due to the differences in the silica and lime sources used; tertiary limestone was suggested as the calcium source (Freestone et al. 2000, 73, fig. 9).

In considering the differences in zirconium levels in Hellenistic mosaic vessels, Nenna and Gratuze (2009, 203–4, fig. 4) have suggested two production zones, one in Egypt and a 'Syro-Palestinian' one on the Levantine coast. A number of samples from third–second century B.C. Tebtynis in Egypt, for example, contain between c. 70 and 150 ppm of ZrO, suggesting that they were made using an Egyptian sand source in Egypt. Therefore, either a North African or an Italian origin for the raw glass used to make vessels of 'Italian' type such as *zarte Rippenschale* bowls, highly coloured cast open vessels in a range of shapes, gold-band decorated vessels, laced mosaic bowls and mosaic ribbed bowls can be suggested but this needs to be investigated scientifically.

The next step would be to determine the zirconium levels found in glasses manufactured in inland locations in the Middle East. Moreover, it would be extremely interesting to know whether there were significant compositional and isotopic differences between the coastal sands of North Africa, the Levant and Italy.

8.2 The Relevance of Furnace Glass

If we examine the relative levels of calcium and alumina in raw glasses from three of the first century B.C.–first century A.D. furnaces in Beirut described in Chapter 7 and compare these with the analyses of eighth-century raw glasses from the tank furnaces at Bet Eli'ezer, we can see that, apart from one, all Bet Eli'ezer glasses contain higher alumina levels, so this clearly shows that sand with a greater proportion of alumina-bearing minerals was used to make Bet Eli'ezer glass (Fig. 8.4). This increase in alumina levels in Levantine natron glasses over time has been noted before (Henderson 2002). Bet Eli'ezer glass has been classified as a Levantine II natron composition. If we compare the alumina and calcium oxide levels in Beirut furnace glasses with other Levantine glasses, we find that it fits the Levantine I grouping as defined by Freestone *et al.* (2000). However, some contain lower alumina levels (see Fig. 8.4) and fall into the Egypt II group. Egypt II glasses contain alumina levels ranging between c. 1.5% and 2.4% Freestone *et al.* (2000). These authors defined their compositional types by using raw glass from furnace sites dating to between the sixth and eighth centuries; fourth-century Jalame glasses, a late Roman glassworking site, have chemical compositions that are consistent with this glass type (Brill 1988). Therefore the evidence of first century B.C.–first century A.D. furnace glass from Beirut provides the earliest evidence for the primary manufacture of Levantine I natron glass.

As with any primary glassmaking site, the chemical and isotopic compositions of the raw glass found in the late Hellenistic–early Roman Beirut tank furnaces (Kowatli *et al.* 2008) provide a benchmark against which to compare contemporary raw and vessel glasses found in Beirut and elsewhere. It can be stated with confidence that glass vessels were made in Italy at this time, so a compositional comparison between them and raw furnace glass from Beirut has the potential to provide provenance information, or at least provide supporting evidence that they were not made in Beirut.

A comparison between the levels of alumina and calcium oxide in raw glasses from three Beirut furnaces and the results for glass vessels published by Thirion-Merle (2005) shows several things (Fig. 8.5). First, almost all of the data conform to the broad negatively correlated group already noted, and all contain lower alumina levels than found in the eighth century natron glasses from Bet Eli'ezer. Indeed, if we use these oxides alone, we can state that the data for moulded vessels found in Beirut conform far more closely to the data for the Beirut furnaces than do data for Beirut blown vessels. The blown vessels mainly contain higher alumina levels than the bulk of the furnace glass data. This is not a clear distinction, but it does nevertheless suggest that slightly different glass batches (or glass melts) were involved in making them – and possibly different production sites.

The results for raw glass from Lyon and Golfe de Fos pillar-moulded bowls conform to the Beirut furnace data closely. Two results for *terra sigillata*-shaped vessels from Ruscino with low calcium oxide levels show that they were not made using the glasses fused in the Beirut furnaces. The results for Golfe de Fos fall on the low alumina side of the data distribution. The relative levels of zirconium and barium discussed earlier bear out these compositional differences.

A comparison of the relative levels of calcium oxide and alumina in Beirut raw glasses, with some unpublished analyses for mosaic glass vessels (produced by the author) dating to between the second century B.C. and the first century A.D. shows that some of the second century B.C. vessel glass samples (Grose 1989, 204–5 vessel 210 and 194, vessel 194) contain less than 7.3% CaO, whereas six first-century A.D. mosaic glass vessel fragments from the Kunsthistorisches Museum in Vienna all contain more than 7.5%. This provides tentative evidence for the use of slightly different raw materials, and possibly production centres, in the middle Hellenistic and early Roman periods, but there are insufficient data to be sure. The data for the mosaic glass samples all also fit the Levantine II composition. Stapleton (2003) found little evidence of technological innovation amongst the samples of mosaic glass vessels that she analysed.

If we compare these data with the analyses of Roman glass fragments found in first- to fourth-century contexts from Switzerland, for example, we find that two *terra sigillata*-shaped vessels from Ruscino match a rather different compositional type, labelled European Roman by Arletti *et al.* (2008, 616, fig. 2). Thus, there is clear evidence for both the very early manufacture of Levantine glass and for a separate European compositional type.

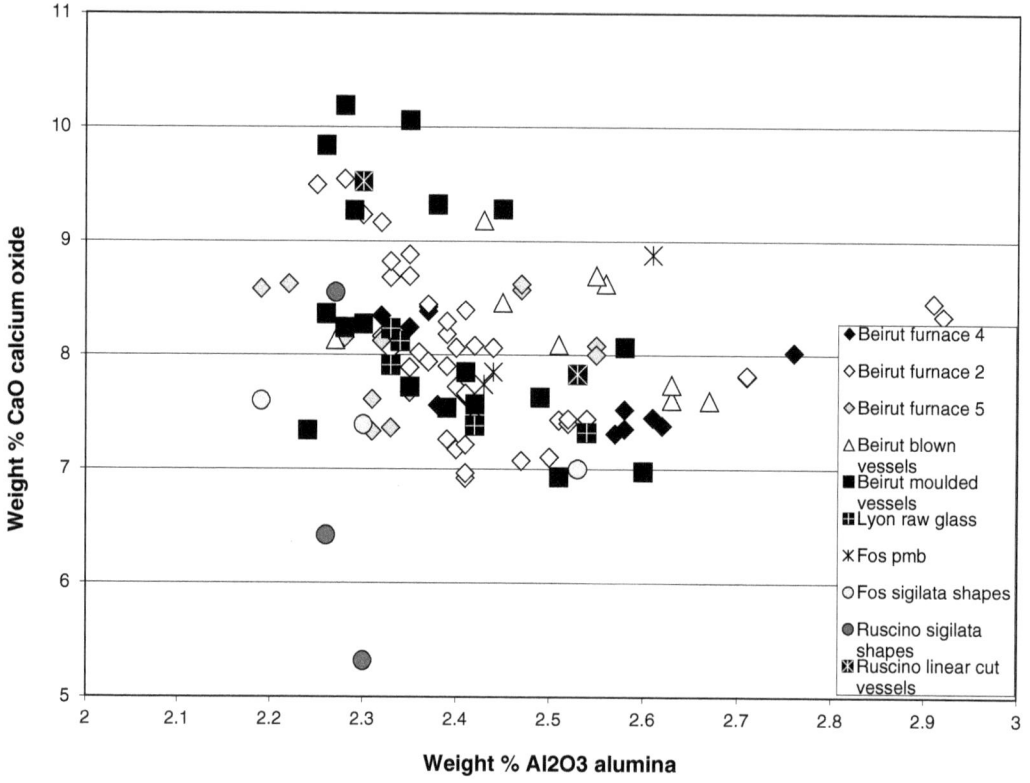

8.5. A bi-plot of the weight % calcium oxide versus alumina in Beirut glass tank furnaces compared with late Hellenistic–early Roman vessels. (Pmb = pillar-moulded bowl; Fos = Golfe de Fos, France, terra sig = *terra sigillata*). Data from Thirion-Merle 2005 and the author (Beirut furnace glass).

Examination of the relationship between chemical composition and glass colour in Beirut furnace glasses has shown that colourless glasses contain less than 17% Na_2O and that samples with greater than 17% Na_2O are frequently strongly coloured. Twelve out of fourteen raw glasses with less than 17% are shades of green or they are colourless; the remaining two are brown and blue colours. The glasses that contain above 17% Na_2O are brown (7), blue (3), yellow-green (4), pale green (3) and aqua (4). There are no aqua-coloured glasses amongst those that contain less than 17% soda. These compositional characteristics are difficult to explain. As can be seen in the Figure 8.6, all of the cast (moulded) vessels contain soda levels greater than c. 17%, whereas half of the blown vessels contain between 16% and 17%. Given that the soda level is especially significant in determining glass melting and working properties (viscosity), the soda level is apparently more critical for moulding glass vessels than for blowing them. Four blown vessels contain the lowest soda levels. Fischer and McCray (1999) noted that levels of calcium and

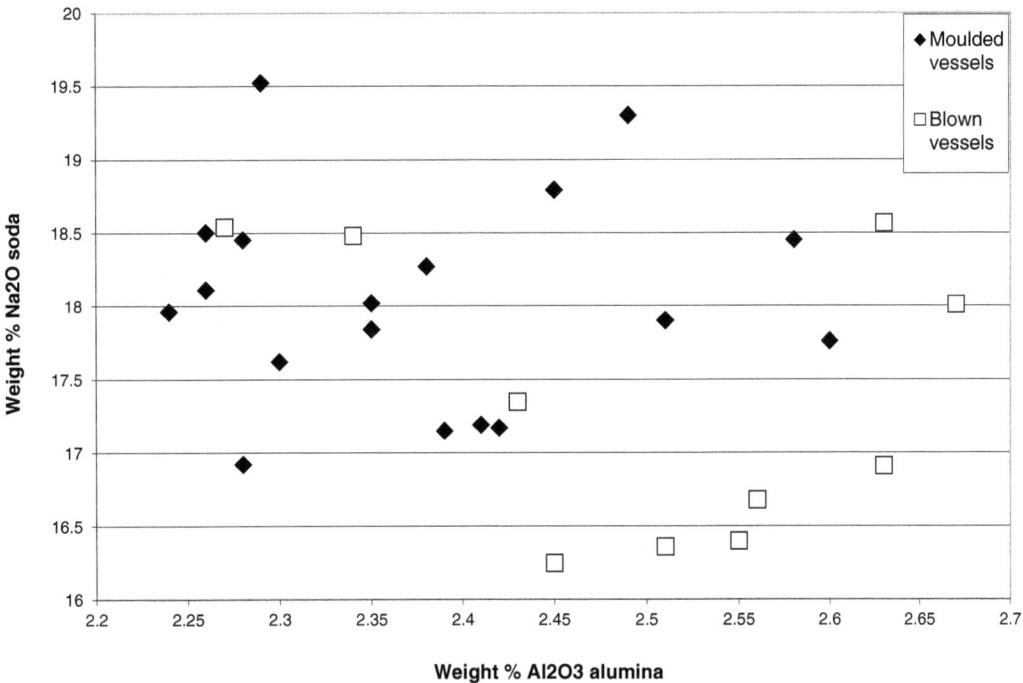

8.6. A bi-plot of the weight % soda versus alumina in contemporary blown and moulded glass vessels found in Beirut (data from Thirion-Merle 2005).

magnesium oxides are 'correspondingly' lower in cast glass than in early Roman blown glass from Sepphoris. A dump of glassworking evidence was found in a cistern there, so they suggested that this provided tentative evidence that glass vessels were made there. Chemical analysis using scanning electron microscopy (SEM), with energy dispersive and wavelength dispersive X-ray spectroscopy were performed on both cast (7 samples) and blown glass (2 samples) vessel fragments. The cast glass contained, on average, higher soda levels than the blown glass, the blown vessel glass being derived from both early and later Roman contexts. Furthermore, the cast glass contained lower levels of alumina, silica, calcium oxide and magnesia. Higher soda would improve the flowing and slumping properties of the glass and at lower temperatures. Brill (1988, 280) noted that a suitable viscosity for sagging hot natron glass into a mould is 5.3 (at a temperature of 810°C). This fluidity could be obtained for only the (average) blown glass composition at c. 100°C. higher (Fischer and McCray 1999, 899, fig. 4). Although this is borne out by the Beirut data published by Thirion-Merle (2005), there are several exceptions. Therefore, although there is some indication that raw glass was selected for its working properties, depending on whether it was destined to be moulded or blown, this is not universally the case.

8.3 GREECE RHODES AND TRANSALPINE EUROPE

The chemical analyses of Hellenistic glass from Vergina, Olympia, Tell Anafa and Morgantina indicate that it is all natron glass. Elevated alumina levels indicate that sand was used to make it. The use of antimony, a decolourant used between c. seventh century B.C. and the first century B.C. or earlier instead, was recognised some time ago (Sayre and Smith 1961; Sayre and Smith 1963), although the change to the predominant use of manganese oxide as a decolourant is now known to have occurred in the second century B.C. instead of the first century B.C. (Henderson 1989). Colourless glass was used to make a range of vessel forms that have been found at the eastern end of the Mediterranean and from the Euphrates Valley eastward (Sayre and Smith 1963, 269, 272, tables III and IV; Sayre 1964, table VII; Sayre 1967, 149). An antimony-rich raw material, perhaps stibnite, was clearly added deliberately to the glass in some form and was used for the manufacture of a range of Hellenistic vessels.

The chemical analyses of Hellenistic glass found on Rhodes (Rehren *et al.* 2005, table 1; Triantafyllidis 2000a, table 2) show that they have a natron glass composition. The iron levels are not unusually high for such glass, and it is clear that antimony trioxide (Sb_2O_3) was used as a decolourant (Triantafyllidis 2000, Table 2). Just as found in Levantine glass, the alumina and calcium oxide levels in the raw Rhodian glass (Rehren *et al.* 2005) fall on the overlap between Levantine I and Egypt II glass, so this could indicate a Levantine origin for it. In the absence of raw glass attached to furnace fragments or lumps of semi-fused glass (frit), it is impossible to prove archaeologically that glass was fused from primary raw materials on Rhodes. However, Rehren *et al.* (2005, 40–2) suggested the presence of raw glass found on the Kakoula property lacking any deliberately added colourants or decolourants, but containing well-rounded grains of quartz, which would have rendered them unworkable, were made locally. This same kind of glass also contained unusually low soda levels and high silica levels for natron glass. Although these glasses may indeed have been fused on Rhodes, another interpretation is that they formed a part of an imported load of raw glass. The 'unusable' glass may also have been used as ship's ballast.

Turning to glass that has been found in southern France, it is fascinating to consider the cultural implications of the chemical analysis of raw blue glass from the third century B.C. Iron Age *oppidum* of Entremont, France. Its compositional characteristics place it in a group with glass found at first-century Manutention, Lyon, the second- to third-century Bourse site in Marseilles, third-century glass from Signoret, Aix-en-Provence, fourth-century St-Martin at Vienne, sixth-century Apollonia in the Levant and some glass from Beirut (Foy *et al.* 2000, 426, fig. 1). The cultural implication is that Hellenistic raw glass was traded with the Iron Age tribe based at Entremont in France. In addition, it appears that the glass that forms Foy *et al.*'s group 3 was made from the same basic combination of raw materials for some 900 years. This indicates a very high level of conservatism in the use of raw materials.

In the absence of evidence for primary glass production in the early and middle Hellenistic Levant, a potential way forward is to determine the Sr and Nd isotope values for well-dated Hellenistic glass vessels from the Levant. The results could then be compared with those for Levantine furnace glass and provide a geological provenance for the glass used to make the early and middle Hellenistic vessels. Such an approach would also answer the question about where the raw Rhodian glass was made.

8.4 Links with Roman Glass

Chemical analyses of Roman glass dating to as early as the first century B.C. apparently show that there is little compositional difference from the contemporary late Hellenistic glass. This is hardly surprising given that the Romans were influenced so heavily by the Greeks with a probable continuity in glass production on the Levantine coast. The possible primary production centres for early Roman glass are the Levant, North Africa and Italy.

In an important article, Nenna et al. (1997) discuss the origin of the glass found at the first century A.D. secondary glassworking site of Manutention in Lyon on the banks of the river Saône (Leyge 1989). Blocks of colourless and coloured glass were analysed using X-ray fluorescence and atomic absorption spectroscopy. Nenna et al. observed the tightly constrained compositions of the glasses, with consistent levels of magnesia, alumina and calcium oxide, suggesting that a single production centre was providing the glass. They then compared twenty analyses from Manutention with fifty-nine from second to third century A.D. Sainte, Charente-Maritime; fifty-eight from Augustinian to fourth-century A.D. Aosta, Italy; forty-two from Rome; nine from Carthage of fourth to seventh century date and fifty-three from fourth-century Jalame. They found a compositional consistency that they suggested results from the glass being manufactured in the Middle East, and specifically using Belus River sand. Nenna et al. (2000) showed that the raw furnace glasses from Taposiris and Wadi el Natrun mentioned in Chapter 7 had different chemical compositions from 'Roman' glasses. Glasses made from the 'Roman' sand source contain a mean and standard deviation of $7.48 \pm 1.18\%$ calcium oxide (CaO), the glass from the Wadi el Natrun lake was found to contain $3.44 \pm 1.68\%$ CaO and that from Taposiris $5.29 \pm 1.59\%$. Nenna et al. (ibid., 85) suggest that these differences are due to the use of the local sands in the three locations containing varying amounts of lime. Nevertheless, some Beirut raw glass contains c. 5% CaO. Other compositional differences between Taposiris glass and 'Roman' glasses were that the former contained higher alumina, iron oxide and magnesia. The reason this publication is important is that it is the earliest to demonstrate that different

Roman glass factories produced glass of distinctive chemical compositions and that could therefore possibly be used as the basis for glass provenance. Nenna *et al.* (2000, 102–4; 107*ff.*) then went on to define the glass made at Maréa-Philoxénité between Alexandria and Taposiris Magna on the coast. This site was occupied between the fifth and the start of the eighth centuries.

In a later publication (Nenna *et al.* 2005), the compositional results for glasses found at three furnace sites at Wadi el Natrun, Zakik, Bir Hooker and Beni Salama, are discussed, but we are given means and standard deviations of six compositional groups that they have defined using cluster analysis rather than the individual compositional results (Nenna *et al.*, 2005, fig. 4). It is not clear why there are six groups and three glass furnaces. Nevertheless, it is clear that even using only calcium and aluminium oxide levels alone, five of the compositional types (wna, wnb, wnc, wnd and wne) contain lower calcium oxide levels than Levantine I and II types with types wna (sixteen samples) and wne (three samples) containing mean levels of 1.87% and 1.94%, respectively (Nenna *et al.* 2005, table 1). As they suggest, these are diagnostic compositions to this part of Egypt due to the use of local sand.

Aerts *et al.* (1999, 2000) published a summary of the chemical analyses of early Roman (4 B.C.–A.D. 68) glass vessel fragments from Qumrân, Israel. The vessel forms analysed were goblets, cups, flasks, bottles, 'chalices' and 'biconical vessels'. They used an electron microprobe to analyse seventy-five samples and synchrotron-induced X-ray fluorescence to analyse sixty-four samples. They therefore determined major, minor and trace components in the glass samples. The three compositional groups that they defined differed according to the heavy metal impurities of copper, tin, antimony and lead, all of which are significantly lower in group III (5 samples) than in group I (45 samples). Group III also contains lower manganese oxide levels than found in groups I or II (nine samples). Heavy metal impurities are found at higher levels in later Roman and early medieval glasses and have been interpreted as being an indication of glass recycling (Henderson 1993; Henderson and Holand 1992). The low levels in early Roman glasses could, nevertheless, be an indicator of recycling, but their presence may simply be interpreted as having derived from mineralogical impurities in the sand, especially in group III, in which the heavy-metal impurities have mean values of (for example) 83 ppm CuO (vs. 209 ppm in group I), 42 ppm of SnO_2 (vs. 113 ppm in group I) and 13 ppm PbO (vs. 156 ppm in group I). When the magnesia levels they found are compared with other published levels for natron glasses, there appears to be a discrepancy. If accurate the result indicates that a distinctive composition of natron was used to make them. Group II contains lower calcium oxide levels than Groups I and III.

Moreover, Aerts *et al.* (2000, 118) found some distinctive differences in the zirconium levels between early Roman Qumrân glass and (later) Roman samples analysed by Caroline Jackson from Mancetter and Leicester in the

United Kingdom (Jackson 1992). This could indicate that different sand sources were used although glass recycling might also have contributed zirconium to the melt. Aerts *et al.* (*ibid.*, 119) make the point that a major change in the late Roman glass industry, which others have observed, is reflected in the chemical characteristics of the glass. They suggest that in the fourth century, a greater proportion of glass was recycled than in the earlier Roman periods.

Mirti *et al.* (1993) conducted inductively coupled plasma optical emission spectrometry and atomic absorption spectroscopy analyses of ten glass samples (including three from a polychrome vessel) dating to the first century A.D. from Augusta Pretoria (modern Aosta) in northern Italy. The vessel forms were bowls and a patera. In common with other early Roman glass, they were highly coloured: blue, amber, yellow, purple and blue-green. It can be assumed that some of these vessels were produced by moulding rather than by blowing, but this information is not provided. Nevertheless, they found that by using sodium, calcium, alumina, magnesia, potassium oxide, titania, strontium oxide and barium oxide as discriminators, they were able to distinguish between first–second and third–fourth century glasses (*ibid.*, 236). For the present purposes, although this may indicate that the same raw glass was used to make cast and blown first-century glasses, one might nevertheless expect there to be subgroups amongst the first-century A.D. glasses.

By focusing on the chemical analyses of specific colourless early Roman vessel forms from the United Kingdom, Baxter *et al.* (1995; 2005) demonstrated that there was greater compositional variation than hitherto suspected. For example, they found that facet-cut beakers are generally compositionally distinct from cast bowls, wheel-cut beakers and cylindrical cups. Moreover, Baxter *et al.* (2005) found that there were at least two subgroups of facet-cut beakers caused by different relative levels of iron, aluminium, antimony and lead oxides in the glasses. The variations in the levels of aluminium, iron and possibly lead oxides suggest that distinct silica sources were used to make the glasses.

The organisation of Roman glass production changed over time, starting with the changes associated with the introduction of glassblowing and spreading north of the Alps after the first century (Follmann-Schultz 1991, 36; Foy 1991, 58–9). Although trade in raw glass ingots and cullet certainly occurred in the Roman period, as indicated by the discovery of shipwrecks such as the *Tradelière* off the coast of southern France (Foy and Jézégou 1998), the distinctive compositional differences found amongst colourless facet-cut beakers nevertheless suggests that different primary production centres were involved in making the raw glasses.

A reasonable assumption here is that primary glass production often occurred close to sources of silica. When this is the case, it can be anticipated that tank furnaces for the primary manufacture of glass destined to be used to make colourless glass may be found in the future; the addition of an antimony-rich material would have occurred in a second stage of production. The compositional distinction that has been found between colourless glasses may partly be explained because it was important to avoid mixing glasses: this could produce unpredictable working properties and potentially make the glass less easy to work. Such selectivity would have been important to Roman glassworkers (Jackson 1992, 198), and so it is likely that (some) colourless glass would remain separated for remelting. Indeed, this is supported by the discovery that one group of second- to third-century colourless Roman glasses found in the northwest provinces (often used to make the more elaborate or higher status vessels such as cage cups and facet-cut beakers) was produced using high-quality sand, and there is little evidence of glass recycling (Jackson 2005; Sayre and Smith 1963, 279). Interestingly, a second (fourth-century) group of colourless glass was made from a base glass with a similar composition to that used to make coloured glass. Paynter's (2006) analysis of Roman colourless glass from sites in the United Kingdom also suggests that the glass characterised with low barium and high antimony was used for cast and facet cut vessels of unusual forms, and that glass with high calcium and low soda matches Foy *et al.*'s (1998) Palestinian group 3. The production of facet cut vessels north of the Alps, outside the region of Mediterranean trade in raw glass (Foy *et al.* 2000a; Foy and Nenna 2001), is therefore likely to have involved relatively small production centres compared with those in the Middle East (Baxter *et al.* 2005, 48, 62). Millett (1990, 159–64). Peacock and Williams (1986, 57–63) have noted that the trade that occurred in the first and second centuries tended to be specifically for the provisioning of the state, Rome and the army. Pliny (*Naturalis historia* 36.193–4) mentions multiple primary production centres for glass.

8.5 Glass and Society

Although there is somewhat limited evidence for the manufacture of Hellenistic glass and for its compositional characteristics, it is interesting to speculate how its production might be related to its use in society. With an increase in the population in the late Hellenistic Levant, demand for glass also increased, along with the expansion of an increasingly active trade network. It is only relatively recently that it has been pointed out that cast bowls were produced in large numbers (see Chapter 7). This was an early example of mass production, so the discovery of the very large tank furnaces in Beirut dating to first century B.C.–first century A.D. should not come as a surprise. The high demand for glass by the second to first centuries B.C. in the Levant must have helped to create the conditions in which glassblowing was invented with the potential to produce glass vessels more quickly on a much larger scale. However, the kinds of vessels produced at the inception of the technique, blown from tubes, were small bottles – in Jerusalem (Israeli 2005) –and it was not until c. A.D. 25 that mould blowing was introduced; by c. A.D. 50, the technique of free blowing was fully adopted, and vessels were mass produced on a far larger scale than had been the case up to that time. So there is no question that the invention had a long-lasting impact on society. Indeed the *nouveau riche* Trimalchio in the *Satyricon* by Petronius dated to the early A.D. 60s states that he would have preferred glass vessels to gold and silver ones but that glass had by then become so common. Mould-blowing led to the mass production of containers, such as square bottles, many of which were used for trading their contents. There is some evidence that by the first century A.D., Roman glass workers selected glass that behaved in a way that most suited to glassblowing rather than moulding – this would have been determined by its chemical composition, particularly the soda levels. Even though early Roman blown glass does tend to contain higher soda levels than, for example, that found at the fourth century site of Jalame (18% rather than mainly between c. 14% and 16%), the evidence is by no means clear-cut.

The spread of glassblowing in the first century A.D. was matched by an increase in the number of glassblowing centres. Glass was still produced in the first- and second-century Levant and Egypt, but increasingly compositional (and isotopic) evidence now suggests that the primary manufacture of glass also occurred in the West. Work on colourless glass especially is showing that its chemical compositions can be linked to vessel forms. Therefore, by the first-second century A.D., especially in post-Neronian contexts, a complex range of vessel forms manufactured in sophisticated ways was used on a variety of site types and in a variety of social contexts (Cool and Baxter 1999; Cool 2003). This seems to be related to the spread of primary glassmaking centres away from the Levant to the West. It is clear that production sequences at the time create a series of fascinating links amongst glass production, provenance (using chemical/isotopic analysis) and the use of specific glass vessel types in a range of social and religious contexts.

The form of glass production in the late Hellenistic and early Roman periods changed from being relatively conservative to one in which a high level of experimentation occurred. This ultimately led to glass becoming a common material in both urban and rural environments and used in increasingly wider social contexts by wider sections of society. Changes in society led to changes in the ways that Roman glass was produced, involving both the range of vessel forms and glass chemical compositions.

NINE

ISLAMIC GLASS

TECHNOLOGICAL CONTINUITY AND INNOVATION

This chapter provides the archaeological, social and political background to Islamic glass, especially in the Middle East. Both Chapters 9 and 10 are designed to be read in conjunction with Chapters 2 and 3; the latter provide information about the raw materials used to make glass. Scientific research of Islamic glass is discussed in Chapters 10 and 11.

9.1 The Social, Political and Religious Contexts of Islamic Glass Production

Islam means 'submission to the will of god'. Mohammad, an Arab trader from Mecca, the last of the prophets, was born in Mecca in 570. He began to receive his first revelations from the Angel Gabriel c. 610 and in 622 moved with his followers to Medina, where he died in 632. He was both a political activist and Messenger of God (Allah). Muslim power was consolidated under the successors to Muhammad, the Khalifahs or Rightly Guided Caliphs (A.D. 632–61). The spread of the Muslim faith was rapid. Between 633 and 650, a period of only seventeen years, the Muslim armies conquered Palestine, Iraq, Syria, large parts of Iran and Egypt. By the eleventh century, the Muslim world covered an enormous area stretching from southern Spain to central Asia. For the locations of the principal sites mentioned, see Figure 9.1.

From 661 onwards, the Muslim world was ruled by a series of dynasties that controlled different areas at different times. The first of these was the Umayyad dynasty that extended the area of Islamic domain beyond that conquered before 650 to cover the area between Spain and Iran. The geographical extent of the caliphates and the centres of their control changed frequently. For example, between 661 and 749, Damascus formed the first capital of the Umayyad

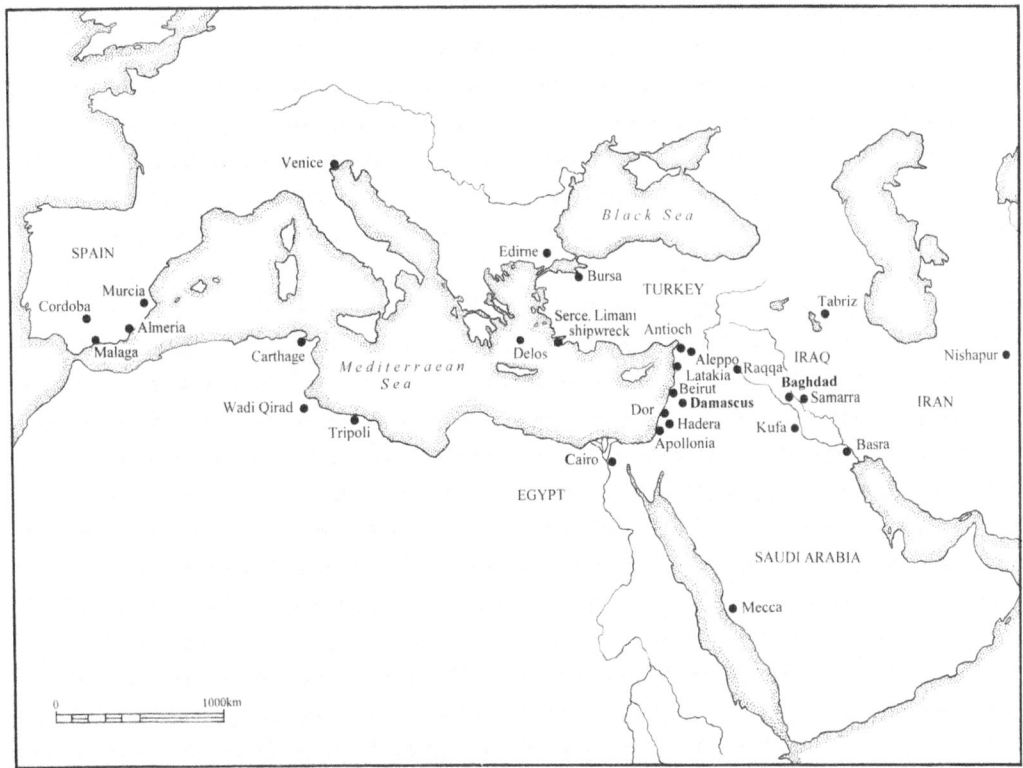

9.1. The principal sites mentioned in the text.

caliphate. At this time the hub of Greater Syria was formed, consisting of modern Syria, Lebanon, Israel, Jordan and Palestine. It is worth tracing the political changes in one particular area, Syria, to exemplify how these changes occurred. The Umayyad caliphate was followed by the 'Abbāsid caliphate. This lasted from 749 until c. 850 in Syria but continued uninterrupted in Baghdad until 1258. Between 868 and 905, the centre of control shifted to Cairo under the Tulunids, followed by the Fatimid caliphate, also based in Cairo between c. 920 and 1171. The Ayyūbid caliphate, again based in Cairo, held sway between 1171 and 1250. The Mamluk caliphate that followed was based in both Cairo and Damascus between 1250 and 1517. There were also smaller areas of control. For example, northern Syria was controlled by the Zengids between 1127 and 1222 from their base in Aleppo. There were three long-lived dynasties: the 'Abbāsids occupied an area known as Greater Iraq between 749 and 1258; the Fatimids occupied Egypt and Greater Syria between 969 and 1171; and the Seljuqs occupied Iran and Anatolia between 1040 and 1194.

The control that the caliphs had over their domain varied according to the period concerned. The 'Abbāsid caliphate, especially under the caliph Hārūn al-Rashid, has been regarded as the most successful in terms of political centralisation, the coordination of the administration and his armies. From his capital,

Baghdad, the caliph supported each aspect of society, including scholarly learning, poetry, religion, medical research, alchemy and astronomy. Some aspects of scholarship were even tied to the palaces. During the time when Hārūn was caliph and under his son al-Ma'mun, until the mid-ninth century, every aspect of the arts was supported, including the industries that fed them. There were clear developments in architectural style and decoration and the production of objects made from metal, (glazed) ceramics and glass. The caliph controlled a vast area linking the Mediterranean Sea to the Indian Ocean, and it was at this time that a 'boom' in the economy occurred (Ashtor 1976, 70 *ff.*, Hodges and Whitehouse 1983, 130). As a result of the high degree of coordination and centralisation, movement of all kinds of things across this vast area, including merchants, craftsmen, pilgrims, scholars and armies became easier. Some of this movement would have occurred along the Silk Road. The artisans could be moved over thousands of kilometres and brought their techniques (Carboni 2001, 4), styles and ideas with them and may have moved with the caliph if he moved his centre of control (Henderson 2003), together with the paraphernalia of entire workshops. This presence of artisans from a variety of origins must have led to a consequent fusion of technical and technological ideas in the workshops controlled by the caliphs and governors, although individual culturally distinct contributions may sometimes be difficult to identify in the results. For example, even the kinds of fuels used during a firing cycle or the ways in which fuels were stored to be used in metal smelting and glass melting or the firing of pottery kilns may have been influenced by the geographical origins of the artisans involved.

There are several historically attested uses of foreign artisans and 'foreign' materials from the early Islamic period. Ya'qubi (c. 278/891) describes how the caliph al-Mutasim was able to bring artists and workmen to Samārrā from Antioch, Basra, Kufa and Misr. Another reference is provided by the works of the tenth-century geographer al-Muqaddasī (d. 381/991). According to him, the Umayyad caliph al-Walīd (705–715) gathered artisans from India, Persia, the Maghreb, and Byzantium for the construction of the mosque in Damascus between 705 and 712. Both glass tesserae and Byzantine mosaicists were imported (see Fig. 9.2); when travelling overland, the mosaic glass tesserae would have been transported by mule or camel (Hamidullah 1960, 285). Eighteen shiploads of gold and silver were brought from Cyprus; 'The implements and mosaics for the mosque were sent by the king of the Greeks'. Al-Walīd was so desperate that he blackmailed the emperor by threatening to destroy Christian churches in Muslim lands if his request were not met. Even though the caliph's conviction was that first-rate work could come only from Byzantium, the Muslims saw such a favour as the emperor's subservience to them. In contrast, the Byzantines interpreted the granting of the 'privileged use of highly technical training' to the 'barbarians' as an imperial favour and something that enhanced the prestige

9.2. Mosaics composed of glass tesserae in the Umayyad mosque in Damascus, Syria, constructed in the early eighth century (photo: J. Henderson).

of the emperor (Graber 2004, 82–83). Graber (*ibid.*) interpreted al-Walīd's 'call for workers from Byzantium as being 'a sign of the Muslim prince's accession to universal power'.

Byzantine mosaicists were used for the decoration of the eighth-century mosque in Medina and the tenth-century Umayyad mosque of al-Ḥakim II in Cordova (Levi-Provencal 1963, 393; Graber 2004, 264; El Cheikh 2004, 99). Even by the tenth/eleventh centuries, al-Marwazī (who died after 514/1120) notes that the Rūm (Byzantines) were indisputable masters in the applied arts, although they were considered inferior to the Chinese. There was a consensus that they were masters in all fields irrespective of medium. Arabs genuinely appreciated Byzantine skills in architecture, crafts and fine arts. Another important historical insight into Islamic glass technology is provided by al-Ṭabarī (d. A.D. 923), *Tārīkh* tertia series 2:1194 (Gibb 2004, 52):

> We began to pull down the mosque of the Prophet in Safar 88 (January 707). Al-Walīd had sent word to inform the Byzantine emperor (*Sahib al-Rum*) that he had ordered the demolition of the mosque of the Prophet and that he should aid him in this work. The latter sent him 100,000 *mithqāls* of gold, and sent also 100 workmen, and 40 loads of mosaic cubes. He gave orders also to search for mosaic cubes in ruined cities and sent them to al-Walīd, who sent them to [his governor in Medina] Omar b. Abd al-Aziz.

The use of glass tesserae, which derived from ransacked Byzantine monuments, for the glass industry also occurred in other parts of the world. Tesserae have been found in some numbers on early ninth-century Viking-age emporia in northern Europe (Näsman 1979), and the practice of recycling 'Roman' (presumably Byzantine) glass tesserae was still continuing in twelfth-century northern Europe, according to the monk Theophilus Presbiter (1979).

There was a reliance on the models provided by Byzantine economic and administrative life during the Umayyad caliphate and there were strong diplomatic links between Byzantium and Islam between A.D. 639 and 750. Moreover, there was no compunction about imitating Byzantine forms and practices (El Cheikh 2004, 54). An important contribution to Islamic glass technology continued to be made by the Byzantines in later periods, although not in the construction of 'Abbāsid imperial monuments. There were direct diplomatic links between 'Abbāsid and Byzantine courts, especially during the reign of Hārūn al-Rashīd (170–193/786–809). Hārūn maintained diplomatic links with Empress Irene and the Emperors Constantine VI and Nicephorus I. Indeed, it has long been recognised that Byzantium exerted considerable direct and indirect influence on Islamic art in Umayyad and early 'Abbāsid eras (Graber 2004).

Even though the Islamic caliphate and the Byzantines were at war almost perpetually, commercial and courtly relations continued uninterrupted. In the ninth century, for instance, al-Tabarī described how in 832 John the Grammarian was sent by Theophilus to the caliph al-Ma'mun, who was in Damascus at the time, with a letter urging peace, the benefits being peace and an exchange of prisoners (Kennedy 2004, 139), although the caliph still attacked. Later Ibn Jubair noted that there was a flourishing trade between Syrian cities and the ports of Tyre and Acre during the Crusades, 'which was almost completely unaffected by the military operations between the opposing princes and their armies' (Gibb 2004, 57–8, n. 15).

In the late Islamic period from about 1419, the Ottoman state invited the 'Masters of Tabriz' from Iran to oversee the insertion of glazed tiles in mosques and minarets. The Masters of Tabriz not only oversaw the production and construction of the glazed tiles but also influenced the design of the glazed patterns on the tiles. One of their first projects, the mosque complex of Sultan Mehmed I, shows clear Iranian aesthetic (Atasoy and Raby 1989, 83). They also oversaw the tiling of the mosque complex of Sultan Mehmed I, constructed between 1419 and 1424, and the mosque of Murad II in 1425, both in Bursa. They tiled mosques in or near Edirne, another early capital of the Ottoman state: Şah Malek Paşa Camii, completed in 1429; the mosque of Murad II near Edirne, completed in 1436; and Murad II's mosque, Üç Serefeli, datable to between 1437–8 and 1447–8. Their last work was the so-called tomb of Cem Sultan in Bursa that was originally built in 1479.

Large groups of artisans with particular skills were brought from hundreds or even thousand of kilometres away for the construction and decoration of large buildings. Examples are the enormous palatial desert castles, such as eighth-century Ukhadir in Syria, and those associated with urban complexes such as Samārrā in Iraq. Meinecke (1996, 24–5) describes how labour was transferred to al-Rāfiqa, Syria, in the eighth century for its construction. Muqaddasī mentions that workers derived from all over Syria, the Jazīra and from Iraq, including Baghdad, for the construction of Madīnat al-Salām (Heidemann 2006, n. 50). In addition, 'job lots' of polished blocks of glass would undoubtedly have been manufactured for decorative purposes. At al-Raqqa, polished translucent aqua green blocks were found *in situ* flooring the audience chamber of a ninth century 'Abbāsid palace thought to be that of Hārūn al-Rashid's son, the Caliph al-Mu'taṣim (r. 833–42) (Salibi 1954–5; Catalogue 1976, fig. 79; Baker 2004, 6). At Samārrā, polished millefiori blocks and tiles were discovered during excavations of Jawsaq al-Khāqānī, another palace built by the Caliph al-Mu'taṣim between 836 and 842 (Lamm 1928, 109–10, nos. 304–12; Hasson 1979, 34, no. 64; Carboni and Whitehouse 2001, 148, catalogue no. 61; see Fig. 9.6). These could have been used for decorating floors or walls, although wall decoration seems most likely. These structurally complex palaces needed to be planned. They were highly decorated using a range of other materials, such as stone (including marble), plaster, brick, and, in some, superb relief stucco decoration. It was at this time that the caliphate was organised most effectively and when it had greatest control. There would also have been a peak in the interaction between artisans, the exchange of ideas and the sharing of knowledge in the workshop environments. These examples serve to underline the virtual certainty that when glass production occurred on a large scale in the Islamic world, it involved artisans from a wide range of cultural backgrounds, bringing with them their own technological traditions.

9.2 ISLAMIC GLASS TECHNOLOGY

As in other periods, there are several different and interrelated aspects of glass technology. These include the selection and use of raw materials, the use and construction of furnace types, practical aspects combining raw materials, melting glass and producing a range of vessel forms, the ethnicity of the artisans and the relationship between glass technology and other industries. Clearly Islamic vessels, windows and beads made from glass were a reflection of the period in which they were made. Starting in the ninth century, the vessel forms were easily identifiable as being typically 'Islamic'. For example, in the ninth, tenth and eleventh centuries, vessels with relief-cut decoration (see Fig. 9.3), possibly manufactured in Egypt, Iran or Syria, were clearly Islamic in inspiration. Ninth century lustre-ware vessels were probably manufactured in various Islamic centres (Fig. 9.4). Later, during the time of the Ayyūbid (1169–1250) and Mamluk (1250–1517) caliphates in Syria and Egypt, highly decorated gilded and enamelled decoration was applied to vessels, the enamels involving an exotic range of raw materials (including lapis lazuli) (see Fig. 9.5). These enamelled vessels were typically mosque lamps, although other forms such as bottles, beakers and basins were also made. Some of these vessels clearly had a high

9.3. Cut glass bottle (reproduced with kind permission of Dr. Stephan Weber and the Staatliche Museen zu Berlin – Museum für Islamische Kunst).

value: products from Syria were given as gifts or commissioned by Ayyūbid rulers, Jaziran atabags and Anatolian Seljuq sultans. These easily identifiable features were a product and a reflection of the cultural norms of Islamic society, and therefore wherever they were used there was an implicit understanding

9.4. Lustre painted (reproduced with kind permission of Dr. Stephan Weber and the Staatliche Museen zu Berlin – Museum für Islamische Kunst).

9.5. Enamelled horseman on a twelfth- to fourteenth-century colourless bottle (photo: J. Henderson; produced with kind permission of Dr. Stephan Weber and the Staatliche Museen zu Berlin – Museum für Islamische Kunst).

on the part of the beholder that they were derived from an Islamic context and everything that it stood for. Whether all of those who took part in the *chaîne opératoire* to produce the glass, the natron and silica collectors, the plant ashers, the fuel collectors and the colourant providers were Islamic is another matter, but the results were nevertheless perceived as a symbol of the Islamic world.

Because the Islamic domain varied in its extent over the centuries, it incorporated vast areas with a wide range of raw material sources and artisans from different origins. As is discussed later, a major change in Islamic glass technology occurred in the early ninth century, when plant ash glass replaced natron glass fully, as reflected in the chemical compositions of dated glass weights (Matson 1948; Gratuze and Barrandon 1990). It is certainly no coincidence that at about the same time the vessel forms became clearly Islamic in character and that a number of other technological innovations occurred. Glass used in Muslim society until about the early ninth century was mainly made from natron (see Chapter 3). Although control over natron production was in the hands of the Muslims, because the most likely source of natron was at Wadi el-Natrun in Egypt, the production of natron glass was essentially a continuation of a well-established Roman and (early) Byzantine industry. Along with the use of these 'Roman/Christian' raw materials, the earliest vessel forms used by the Muslims were Byzantine/Christian in character, mainly because they were probably made in the Byzantine workshops on the Levantine coast. As noted earlier, Byzantine mosaicists and glass tesserae, made using Byzantine technology, were brought to Damascus to create the mosaics of the great early eighth century Umayyad mosque there. At this stage, there appears to have been little Umayyad patronage or control over the primary production of glass. The glass vessels may therefore have been perceived as having 'Christian' origins or attributes. Carboni and Whitehouse (2001, 4) note that in the seventh and eighth centuries, a lack of glass objects with moulded Arabic inscriptions shows that an official status was not conferred on the objects by new Umayyad rulers. Moreover, some glassmakers in the Levantine area were certainly Jewish, as noted in the Cairo Geniza (Goitein 1968, Vol. 1, 94), and they must have been making vessels for Islamic, Byzantine/Christian and Jewish markets. Even though some Jewish glass artisans would have converted to the Islamic faith in the seventh century, the range of blown glass vessel forms had become restricted, and innovation in the art of vessel production had slowed down. By the ninth century, plant ash glass had virtually replaced natron glass, at which point the technology may have been viewed as 'Islamic'. In technological terms, the introduction of plant ash glass as part of a new *chaîne opératoire* presumably involved Islamic artisans in most of the crucial decision-making stages, although this is by no means certain.

At this time production of raw glass and glass vessels with distinctive Islamic designs occurred in Islamic centres. It was a contrast to glass production under 'Byzantine' auspices using techniques 'as received' from the Roman world. Carboni (2001, 3–4) has noted that before the ninth century there was a minimal artistic contribution by Islamic glassmakers and that at no point did a member of a powerful dynasty have his or her name carved or moulded on a glass vessel. The latter, in particular, appears to signify that the dynastic rulers did not regard glass as being of great importance. Carboni (2001, 15–17) has suggested that before the ninth century, two groups of 'proto-Islamic' vessels can be identified, one influenced strongly by late Roman vessel forms, the other influenced by Sasanian decorative traditions, as discussed later.

9.2.1 Raw Materials and Historical References to Their Use

Most Islamic glass is composed of three major components – silica, soda and lime – (see Chapters 2 and 3). Scientific analyses of Islamic glasses have revealed that two quite distinct types of soda-lime-silica glasses were made between the seventh and fourteenth centuries in the Middle East. The (natron) glass used and produced from the seventh to the late eighth centuries was primarily a mixture of Egyptian natron (trona), a mineral source of alkali, and sand. The sand also contained shell fragments, which provided the 'lime'. However, from about A.D. 800 or slightly later, the main source of the soda-rich alkaline flux changed from natron to plant ashes (see Chapters 2 and 4), a 'reintroduction' of the most ancient technique to make glass (see Chapter 6). Although plant ash glass was produced sporadically in the Middle East during the period between 800 B.C. and A.D. 800, notably by the Sasanians (Brill 1999; Mirti *et al.* 2008), and to a

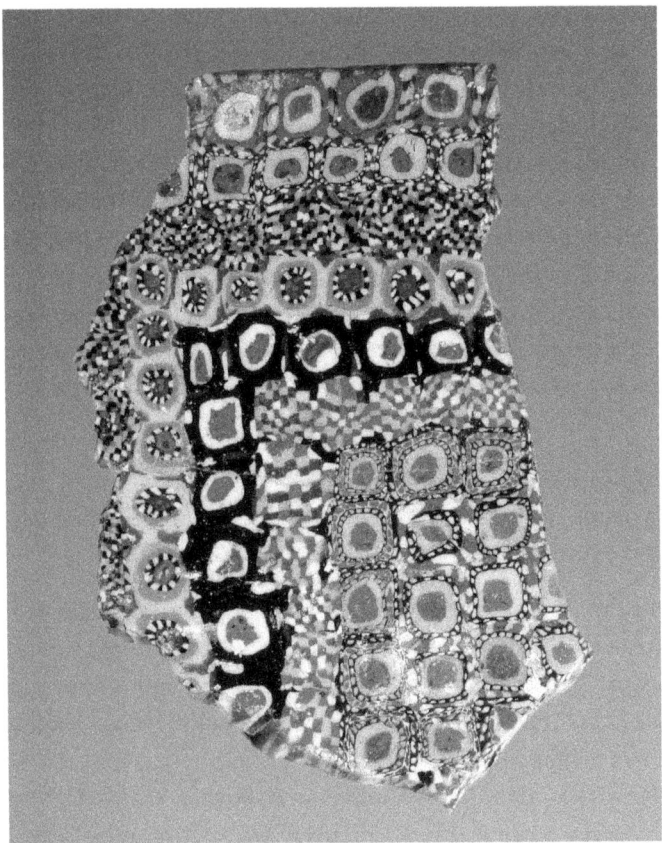

9.6. Millefiori wall plaque from Samārrā no. 200 (reproduced with kind permission of Dr. Stephan Weber and the Staatliche Museen zu Berlin – Museum für Islamische Kunst).

degree during the Umayyad caliphate, natron glass technology dominated until c. A.D. 800. The compositional characteristics of ashes of halophytic plants of the Chenopodiaceae family, the most likely to have been used, are discussed in detail in Chapter 2. In the context of Islamic glass production, the word *al(-)kali* is clearly of an Arabic origin, meaning 'the alkali' and is the same as the word for the halophytic plant 'kali'. However, a slight confusion arises because *kalium*, the Latin word for potassium (with the chemical symbol K), is not the principal fluxing component of many of these halophytic plants; *natrium* (soda) is.

If a two-component recipe was used, the third major ingredient of plant ash glasses, calcium oxide, would have been present in the plant ash. However, the levels of calcium can vary considerably for plants of the same species growing on different geological bedrocks (see Chapter 2). There is some evidence based on scientific analyses, from ninth-century al-Raqqa, that a third primary raw material was used, perhaps a feldspar (see Chapter 10). It is probable that plants of the genus *Salsola* were used as a source of sodium, and, as noted in Chapter 2, the possible species used may have been *S. kali, S. jordanicola, S. vermiculata* and *S. sueda*. *Salsola soda*, which grows in maritime environments, may also have been used. As early as 985, the Arabic geographer al-Muqaddasī mentions ūshnan (kali) as an export from the province of Aleppo (Ashtor 1992, 482). A further reference to the use of plant ash as an alkali source was made by Abū'l Qāsim bin Ali bin Muhammad bin Abu Tahir, a historian in the Mongol court who was a member of the Abu Tahir family of potters. He was a member of a leading family of potters and was writing in c. 1300. The family of potters lived in Kashan and produced luxury pottery (Allan 1973). He wrote a treatise titled 'The Virtues of Jewels and the Delicacies of Perfumes' in which there is a chapter on 'a kind of alchemy', the art of ceramics. His reference to plant ashes as a source of alkali in glazes (essentially the same as glass) is as follows:

> The fifth (substance) is shakhar, which they call qali, which is made by burning pure, fully-grown ushnan plants. . . . The best shakhar is that which has a red-coloured centre when broken, with a strong smell.

His reference to the selection of fully grown plants clearly shows that plants with the greatest biomass and therefore the highest potential amount of sodium carbonate were harvested in late summer; Smedley and Jackson (2006) have demonstrated that the greatest biomass of fern (used to make medieval glass) is also in the late summer.

Haloxylon articulatum, known today as *Hammada scorparia*, and *Anabasis* were used widely when Ashtor (1992) was writing. Moreover, he notes that in more 'modern' times, the Bedouin come to Aleppo ten times a year in great caravans known as the 'caravans of the keli'.

The full reintroduction of plant ash glass production from the late eighth-early ninth centuries provided the foundations and the technology for what in the West is regarded by some as the finest glass in the world, the products of the Venetian glass industry and its later European inspirations. Indeed, the link between Islamic and Venetian glass industries was clear from as early as the thirteenth century, when documents dating to the last quarter of that century establish beyond doubt that plant ash was exported from the Levant to Venice (Zecchin 1973). By the mid-fourteenth century, Levantine ashes were exported from Aleppo, Sarmin, Beirut, Latakia, Tripoli and Ramla (Ashtor 1992, 508). Although plant ash was also imported from Egypt, the greater part came from Syria. Given that both soap and glass were made from plant ashes, it is significant that the soap and glassmaking industries expanded at the same rate in fifteenth-century Venice (McCray 1999, 103). The enormous demand that was created for both glass and soap appears to have driven the production and supply of plant ash at a time when glass production in the Middle East was on the wane.

The Venetian glass industry continued to be fed by Middle Eastern suppliers of plant ashes into the seventeenth century, by which time the evidence for a

Middle Eastern glass industry is minimal. We have an interesting comment by the seventeenth century humanist and traveller George Sandys (1578–1644) on the supply of plant ash to the Venetian glass industry from Alexandria. In 1610, travelling through the desert between Alexandria and Rosetta, Sandys (1670, 90) comments as follows:

> here and there a few unhusbanded Palmes, Capers and a weed called Kall [probably meaning kali] by the Arabs. This they use for fuel and then collect the ashes which crush together they sell in great quantity to the Venetians; who equally mixing the same with the stones that are brought them from Pavia, by the river of Ticimum, make thereof their crystalline glasses.

If Syrian plant ashes were regarded as being of the finest quality (compared with that from Alexandria and Spain), perhaps it is no coincidence that northern Syria also appears to have been part of the area – northern Mesopotamia – where the earliest (plant ash) glass in the world was made. The quality of plant ash was clearly of great importance to glassmakers. As noted in Chapter 2, the quality can be defined by the proportion of sodium and calcium carbonate in the ash. Plants that were ashed in Syria were referred to as the *Chinane* plant there. They are still being ashed today because they form one of the main ingredients of soap – the ash is mixed with oils to produce it. The Muslims are credited with the invention of hard soap made from soda-rich plant ashes; in the High medieval West, potassium-rich soaps made from tree ashes were gelatinous (Ashtor, 1992, 481). The Middle Eastern residuary ash, known as 'keli' today, is still sold (as soap) in the markets of Aleppo and Damascus. It is no coincidence that the soapmaking and glassmaking industries ran in parallel. In the tenth century, al-Muqaddasī described al-Raqqa in the following way: 'It is a source of fine soap and olives' (Heidemann 2006, 41). It is, however, sobering to realise that archaeological excavations have revealed the evidence for glass manufacture on a large scale there, but this is not referred to by al-Muqaddasī.

Al-Biruni, in his description of glaze (and glass) production, notes that natron was used as the alkali 'and those types of borax and borax salt which are related to it and determine its speed of melting'. It is clear that al-Biruni is discussing the fluxing effect of alkalis. However, there is no current evidence of borax having been used to make Islamic glass and glazes, so the precise identity of what he refers to is obscure. The text does, however, remind us that it is easy to make assumptions about the alkali (and other raw material) sources used in glaze and glass production and that we should not be surprised if we find evidence in future research for the use of other mineralogical alkali sources.

The sources of silica in the Islamic world could vary quite widely. The possible sources are quartz pebbles found in riverbeds, and various sources of sand:

from beaches, inland geological deposits, riverine deposits and evaporite deposits. The purity of the silica used varied according to the nature of the deposits exploited (see Chapter 3). Quartz pebbles can be the purest form of silica, especially white ones, which contain the lowest levels of iron-rich minerals, followed in descending order of purity, by geological deposits, evaporite deposits, beach sands and riverine deposits. The purity of the silica used would have been significant for glassmakers because iron-rich minerals produce a natural green colour in glass. However, quartz pebbles with minimal iron impurities would not have necessitated as much furnace control. An example of a silica source is the river Euphrates. Its origins are in the mountainous area of Anatolia. It then flows through northern Syria and into Iraq, ultimately flowing out into the Persian Gulf. The river brings with it a wide range of pebbles and mineralogical inclusions that become mixed with the sand and deposited at different points along its course. If used to make glass, some of these inclusions would therefore be introduced.

Al-Biruni's description of lead-rich glaze (dated 990) refers to the melting together of a mixture of sand and keli by heating them together for a few days. He mentions that flint (*marwat*) could be used as a source of silica 'which are brilliant white stones from which fire can be struck'. However, a brilliant white colour would suggest that he is referring to quartz. This impression is supported by his mention of their occurrence in ravines and wadis. The account therefore appears to be slightly contradictory.

Al-Biruni goes on to say that various colours of glass were obtained by adding different minerals, such as iron oxide for yellow, copper for green, manganese for a wine colour, and lead for white. If lead was used in combination with tin oxide, as is regularly found in Islamic opaque white enamels and decorative opaque white glass, an opaque white colour was produced – even though lead was not essential to produce the colour. It is possible that the use of lead being referred to is to make lead oxide-silica glass, which would have an added brilliance to a colourless glass due to a greater refractive index than found in plant ash glasses, but this would not have been an opaque 'white' colour but transparent 'water white'. In the same document (p. 79), al-Biruni distinguishes between mineral glass, which is mined like rocks (presumably rock crystal), and plant ash glass, with the addition of magnesia to make it 'flexible'. The importance of magnesia is also mentioned by Almass'oudi (957) in his book مروج الذهب Meadows of Gold (p. 161), where he writes that glass is formed when sand, keli, and magnesia are heated to high temperatures for a long time. However, scientific analyses of Islamic (plant ash) glasses (see Chapter 10) reveal magnesia is *generally* associated with potassium (Henderson *et al.* 2004, fig. 4) and is therefore thought to derive from plant ashes. Moreover, magnesia actually has the effect of increasing the melting temperatures of glass melts and, if anything, would reduce the time during which glass remained 'flexible' or workable. On the other hand, the addition of lead would increase the 'flexibility' of glass, increasing its working period, but this is clearly not what is being referred to. The only other major component in Islamic glass that could be a contender is calcium, but again this increases its melting and working temperatures, and it is difficult to envisage how it could increase its flexibility. Experimental archaeology clearly has a role to play here in testing the effect of increasing concentrations of magnesia on the flexibility of plant ash glass melts.

As noted earlier (and in Chapter 3), the green coloration of glass from iron can be removed by the use of a combination of furnace atmosphere and the use of glass decolourant, such as manganese oxide. Although it can produce a purple colour under the appropriate furnace conditions, it is also known as 'glassmakers soap', because it can produce a colourless glass. Moreover manganese oxide is almost universally present (as an 'impurity') in Islamic glass of all colours. Al-Jabir, the eighth- and ninth-century alchemist, specifically mentions the use of manganese in glass technology (Sarton 1927, 532). However, because manganese is present even in glasses that are neither colourless nor purple, it appears to have been added routinely whatever colour was made. Deep purple and brown glasses contain much higher manganese oxide levels (Henderson *et al.* 2004). In different furnace atmospheres the addition of manganese oxide would therefore have allowed the glassmaker to produce a purple or colourless glass. Nevertheless, it was not new to glass technology: the Romans certainly deliberately used it. Examples can be found amongst the fourth-century glass found at Jalame (Brill 1988, 259).

9.2.2 The Change from Natron to Plant Ashes

This topic is also discussed in Chapter 4. Although we can show that the levels of soda dropped in the glasses being manufactured on the Levantine coast during the first millennium (See Fig. 4.2 and Henderson 2002) and that the principal period when this was felt keenly by glassmakers was in the sixth and seventh centuries, this would also appear to form part of a general 'malaise' in glass production during this period. Perhaps the increasing difficulty of being able to blow glass as the soda levels dropped contributed to this 'malaise'. The range of classically inspired Byzantine vessel forms was relatively narrow, and, despite archaeological evidence for the manufacture of glass on a large scale (Gorin-Rosen 2000), there is little evidence of innovation in glass vessel technology. Moreover, one of the kinds of 'proto-Islamic' vessels produced in the seventh and eighth centuries, facet-cut bowls, was inspired by and based on Sasanian vessel forms. Such vessels were mainly manufactured using plant ash glass technology (Brill 1999). The raw materials and knowledge for producing plant ash glass were available in the Sasanian area of Iran where it had been produced between the second and sixth centuries. It is unlikely this knowledge would have been lost, because Sasanian decorative techniques provided a strong contribution to some early Islamic wares (Carboni 2001). Large parts of Iran formed parts of the Islamic domains; they were included even in its initial phase of expansion before 650.

So it might be regarded as somewhat surprising that apparently there was little motivation to adopt the plant ash glass technology earlier than the 9th century. Had this occurred centres such as Damascus could have produced its own glass using locally available plant ash and silica. In so doing, the glass would have become truly 'Islamic', without having to rely on the Byzantine glass production centres on the Levantine coast. One explanation for this is that until the mid-eighth to ninth centuries, the Caliphate and its retinue was not especially interested in developing Islamic industries, arts and crafts that could be easily perceived as Islamic as opposed to Roman, Christian, or Jewish. Indeed, as Hourani (2002, 36) noted 'More systematically than the Umayyads, the 'Abbāsids tried to justify their rule in Islamic terms'. Arts and crafts would have formed an innate part of every aspect of life under the 'Abbāsids. The Islamic religion and culture, as expressed in arts and crafts, would therefore have formed a means of belonging to a Muslim society. So it seems likely that a technological explanation of natron sources drying up was not the only contributing factor to a 'delay' in the introduction of 'Islamic' plant ash glass technology: the political impetus was simply lacking. It is no coincidence that enormous strides in the development and innovation of clearly Islamic industries, arts and crafts occurred during the reign of one of the most effective caliphs, Hārūn al-Rashid – and that of his son al-Ma'mun – in the context of a centralised state. These developments constituted an industrial revolution and occurred in centres such as Baghdad, Samārrā and al-Raqqa.

9.3 Production Centres

Principal urban centres tended to be located on rivers so as to facilitate transport and water supplies. For example, Aleppo and al-Raqqa were located on the Euphrates, Samārrā and Baghdad on the Tigris and Damascus on the river Barada. Halophytic plants grow abundantly in and around these areas. In contrast to natron glass, plant ash glass production could occur without recourse to importing one or other primary raw materials, sometimes over vast distances. Glass production centres can either be primary, where glass was fused from primary raw materials, or secondary, where raw glass was imported, reheated and worked by cold-working, moulding, casting and blowing the glass into vessels and objects. Both primary and secondary production can occur in the same location, but this is not always the case (see section 11.2). Thus, without archaeological evidence, historical sources need to be 'interrogated' to be sure that they are referring to primary glass production, with the associated range of activities. In other words, when 'glass production' is referred to, it could be primary or secondary glass production, or both.

Lamm (1929–30; Lamm 1941) suggested that there were specialised Middle Eastern glass production centres for enamelled vessels in Damascus, Aleppo and al-Raqqa. Moreover, historical evidence indicates that the Qadisiyya quarter of Baghdad was a glass production area (Lamm 1929–30, 498). During the Ayyūbid caliphate, Aleppo was a primary glassmaking centre: Yāqūt (d. 1229; Kitāb mu'jam al-buldān (Dictionary of Countries), I, 631) , who lived in Aleppo, has described the use of fine white sands for glassmaking there. The cosmographer al-Qazwīnī (d. 1283) mentions the sūq al-marzuqīn, or market of enamellers, in Aleppo, which sold amazing decorated objects (Irwin 1998, 25). The source of the sand mentioned by Yāqūt was Jabal Bishri on the south side of the Euphrates between al-Rusāfa (to the west of al-Raqqa) and Tadmur (Palmyra). The sand is described by Yāqūt as being 'like lead white (*kal-isfidāj*)' and used as a colourant in glass (Irwin 1998, 25; Heidemann 2006, 40–1). Because a mixture of lead and sand by themselves do not produce a colour in glass (unless it is the green colour imparted by iron-rich minerals in sand), this may be an elusive reference to the production of enamels using lead, given that coloured Islamic enamels often contain lead (Henderson and Allan 1990; Henderson 1998; Freestone and Stapleton 1998). Islamic opaque white enamels are produced with tin oxide in a lead-rich matrix and opaque yellow enamels with lead-tin oxide crystallites. Taken together with historical reference to the market of the enamellers, this evidence therefore lends a degree of support to Lamm's suggestion that Aleppo was a production centre for Ayyūbid (11th century) enamelled vessels. Aleppo was even involved in exporting plant ash for glass production to Europe from at least as early as the thirteenth century.

An alternative interpretation is that the combination of lead oxide and silica was used for the manufacture of translucent lead oxide-silica glass, which appears to have been manufactured by the Muslims from around the ninth century (see Chapter 4) . For example, lead oxide-silica glass was used to make windows of the twelfth century Qasr al-Banāt in al-Raqqa (Henderson *et al.* 2004). The presence of lead in the glass produces some rather distinct colours, such as emerald green. In the thirteenth century the best quality lead glass (*minā'*) was said to have been made in Syria, Egypt and the Maghreb (Irwin 1998, 25). This is mentioned in Naṣīr al-Dīn al-Ṭūsī's *Tansūkh-nāma-yi īlkhāni* (Soucek 1960).

The historical sources for the production of glass in Tyre, which was part of the Crusader Kingdom of Jerusalem in the twelfth and thirteenth centuries, is particularly rich (Irwin 1998; Carboni *et al.* 2003; Barkoudah and Henderson 2006). In the tenth century, al-Muqaddasī reported that cut and blown glasses were produced at Tyre. In 1163, the traveller Benjamin of Tudela (Navarre), a Spanish Rabbi, reported that about 400 Jews were engaged in glass-production and ship owning there (Wright 1848, 80). 'Long before that', the Jews apparently had a 'monopoly' of glass made with lead (Dillon 1907, 148). In the twelfth century, al-Idrīsī, a geographer, wrote that Jews located in the suburbs of Tyre produced both glass and ceramics (Chebab 1979, 429). William of Tyre commented on the sand from Tyre being suitable for glassmaking, and James of Vitry noted that the region of Tyre had high-quality 'potash' (a generic term for plant ash; it is unlikely to be potassium oxide, its literal translation) used both for glass- and soapmaking (James of Vitry 1611, 1098).

Carboni and Whitehouse (2001, 4) suggested that glass was produced in the area of royal Samārrā in iraq. A survey of the site by Northedge and Faulkner (1987) revealed a centre for glass production close to the Samārrā palaces and on the north bank of the river Tigris in an area known locally as m'mal al-Zujāj (Zujāj is an Arabic word for glass). Given its urban context, it is likely that primary glass production occurred there. In the ninth century, glass vessels similar to second and third-century forms were apparently copied in the region. In association with the range of other industrial activities that were probably carried out there, the large amount of glass discovered there (Lamm 1928) provides a degree of support for the evidence of glass production.

Equally, the large amount of glass found at Nishapur, especially the nearly colourless ware with shallow-cut decoration, has been used as evidence that at least some of the glass was made locally. The fact that Nishapur was a stop on the Silk Road, however, does increase the range of possible geographical sources for such glass. Excavations in the city of Hama, Syria, have also yielded a vast amount of twelfth- to fourteenth-century glass (Riis and Poulsen 1937), which could suggest local production. Carboni (2001, 184) has suggested that Hama, Damascus, Aleppo and al-Raqqa were all glass 'production' centres during the Ayyūbid period (1171–1260).

Lamm's consideration of Ayyūbid and Mamluk enamelled vessels (1929–30) suggested to him that they had been made in one of three Syrian cities: Damascus, Aleppo or al-Raqqa. However, Cairo (Fustat) was certainly a Mamluk centre for the production of enamelled vessels, and this he does not mention (Scanlon and Pinder-Wilson 2001). Moreover, glass vessels inscribed in Kufic can provide us with evidence for their place of manufacture, though not necessarily for the glass itself. For example, the inscriptions on three 'stained glass' vessels, two cups and a bowl, indicate that Damascus was one production centre and that Egypt, perhaps Fustat, was another production centre in a period spanning most of the eighth into the early ninth centuries (Carboni 2001, 52). Although this relates to glassworking, there is also a possibility that primary glass production occurred in these centres.

Archaeological evidence for glass production in al-Raqqa dating to the eighth–ninth, eleventh and twelfth centuries (discussed later) indicates that the centre had a long tradition, so it may have been instrumental in producing some of the first enamelled and gilded vessels. No archaeological evidence for enamelling has been found, but this is hardly surprising given the minimal evidence it is likely to leave. Carboni and Whitehouse have noted that glass, which was first gilded and then enamelled, was made for rulers in areas adjacent to the Jazīra in which al-Raqqa is located: modern northeastern Syria, southeastern Turkey and northern Iraq. Al-Raqqa was closer geographically and culturally to Anatolia and Iran than Aleppo and Damascus (Carboni and Whitehouse 2001, 204–5). Indeed, as noted elsewhere (Henderson *et al.* 2005), the potential multicultural input to glass production can easily be envisaged in al-Raqqa. In general, practitioners of 'Arabic science', including glassmakers and glassworkers, are likely to have been Muslims, Christians, Jews or 'pagans' and involved Iranians and Armenians (Cooperson 2005, 83). Moreover, Heidemann (2006, 41, n. 54) has noted that there may be indirect historical evidence for glass production in al-Raqqa: as noted earlier, al-Muqaddasī describes al-Raqqa as being 'a source of fine soap (ma'dan al-sābūn al-jayyid) and olives'. Because the same halophytic plant ashes were used in glass and soap production, the two were clearly interconnected. By the time of the Mamluk caliphate (beginning c. 1250), glass was apparently manufactured in Damascus when it was under state supervision by a low-grade soldier (al-Qalquashandī 1913–20, 4:188), and there is evidence that it was worked in Beirut (Jennings 2006, 223). The evidence for primary glass production in Fustat is less strong (Irwin 1998, 25).

There is historical and archaeological evidence for the manufacture of glass in Islamic Spain. The geographer Ibn Said al-Maghribi (1213–86) describes Murcia, Málaga and Almería in al-Andalus as being important glass production centres (De Gayangos 1984; Jiménez Castillo 2000, 2006). Excavations of glass production sites have been carried out in Murcia and Almerí (Castillo and Martínez 2000). The glass workshop at Puxmarina Street in Murcia, dated to the twelfth century, was located near the mosque and the castle inside the medina. Excavations revealed three rectangular and two circular furnaces, with the remnants of three others. Raw glass adhering to a rectangular (tank) furnace, crucible fragments and large dumps of debris were found (Jiménez Castillo 2000; Jiménez Castillo *et al.* 2005).

Excavations in the Crimea at Solkhat have also produced evidence for what appears to be a fourteenth-century Islamic secondary glass production centre (Kramarovsky 1998, 99, fig. 22.7). The evidence consists of two (? single pot) furnaces with an inner diameter of c. 1.05 metres that faced each other and a third one adjacent to the northern one, perhaps an annealing oven. A 'container' was found that still contained ash. Glass waste, colourless glass fragments and what was probably a Mamluk enamelled glass fragment were found. Excavations elsewhere in Solkhat have produced a range of Mamluk glassware, including parts of an enamelled mosque lamp.

By the end of the thirteenth century and into the fourteenth century enamelled vessels, especially mosque lamps, were commissioned by Mamluk rulers in Cairo. By this time, Cairo took over as the centre of the Mamluk caliphate, and it is most likely that large numbers of vessels, and especially those commissioned by wealthy rulers, were made there. Nāṣir Muḥammad ibn Qulāūn, who was responsible for commissioning some of these vessels, had a rather orthodox attitude toward the arts, including their decoration. This often resulted in a sequence of nondescript quadrupeds as decoration and an avoidance of the human form. Royal patronage of the production of mosque lamps continued until the late fourteenth century (Carboni and Whitehouse 2001, 206–7).

The existence of centres of specialisation for different glass vessel forms and for objects between the eighth and fifteenth centuries has been documented by Shatzmiller (1994, 201, 224–226). Documentary evidence shows that the Islamic glassmakers distinguished between thirteen 'types of manufacture' with specific production centres being known for individual specialisations such as window glass, lamps, drinking glasses, bottles, flasks and beads.

Trade in raw and scrap glass occurred with production centres. Excavations of an early-eleventh-century shipwreck off the Turkish coast at Serçe Limani have provided evidence for the export of glass on an enormous scale (Bass and Van Doorninck 1978; Lledó 1997; Bass et al. 2009). The cargo included approximately 2 metric tons of raw glass blocks of up to 30 centimetres across (Bass 1984, 65), which would have been collected directly from glass factories. As many as 11,000 objects described as 'rough-rimmed glass discs' by Bass (1984, 67, fig. 4), were grouped together in the hold. Judging from the published photo, these appear to be lid moils, a by-product from the manufacture of glass bowls and beakers. They would therefore have been gathered directly from a glass-blowing workshop. The cargo also included broken vessels and more than eighty intact glass vessels, including many typical Islamic pattern-moulded examples.

9.4 Archaeological Evidence for Production Sites and Furnaces

'Islamic' glass first started being manufactured in the seventh century. The archaeological evidence shows that the natron glass used at the time was mainly being melted in large tank furnaces on the Levantine coast at sites such as Apollonia, Dor and Hadera/Bet Eli'Ezer that were then using local sand and imported natron (Goren-Rosen 2000). Furnaces at the Byzantine-early Islamic site of Hadera were up to 4.5 metres long and 2.5 metres wide and could therefore produce enormous quantities of raw glass in a single firing. It has been estimated that 8 tons of raw glass was produced in each furnace. The foundations of seventeen tank furnaces were found at Hadera: if all were fired at the same time, 136 tons of raw glass would have been produced (Gorin-Rosen 1995, 42–3; Gorin-Rosen 2000, 52–3; Freestone *et al.* 2000, 67). Whether we refer to the site of Hadera as Umayyad or Byzantine is irrelevant, because the natron glass made in the Levant was presumably used by Jews, Muslims and the Byzantines. Production at Hadera may have extended into the eighth century (Freestone *et al.* 2000, 67, 72).

After c. A.D. 800, the Muslims used tank furnaces to make plant ash glass. The best evidence for the *inland* source of plant ash glass is at al-Raqqa in northern Syria. Here the furnaces were located in an industrial complex of 2 kilometres in length in an area known as Mishlab, or Maktalta, where glass manufacture occurred alongside glazed and unglazed pottery production (Henderson 1999b; Henderson *et al.* 2005). Some 10 metric tons of destroyed tank furnace fragments with raw glass attached were found during excavations on sites dating to the late eighth–ninth (Tell Zujaj), eleventh (Tell Fukhkhar) and twelfth centuries (Tell Bellor), so the manufacture of raw glass occurred over a period of c. 400 years in al-Raqqa.

Rescue excavations in Mishlab in 2010 (Khalil and Henderson 2011) have revealed the first *in situ* tank furnace, at Tell Abu Ali, together with glazed pottery kilns. The site is at the eastern end of the industrial complex and formed part of a strip of industrial sites along the northern edge of Mishlab. The new glass furnace is of a probable ninth-century date. Although sectioned by a machine, its approximate dimensions can be suggested from what remains as being c. 3.7 × 2 metres; the surviving southern wall was 28 centimetres high. Raw purple and green glass was still attached to the furnace floor. The first evidence for the use of slabs on both the floor and walls of the furnace was also found (Fig. 9.7). The furnace was used at least twice (Fig. 9.8). At the time of writing, the phase II furnace appears to be the first example of an ancient downdraft furnace, with a large heating chamber located underneath the floor of the furnace (Fig. 9.9) and a gap between the double southern wall providing a flu for the hot air to rise into the upper chamber. The double thickness of the supporting walls at the southern side of the furnace is c. 1 metre wide (Figs. 9.9 and 9.10). Unusually, the furnace floor slopes slightly from north to south (Fig. 9.9). The two floors associated with phases I and II of the tank furnace had glass adhering to their surfaces (Fig. 9.8) so the upper floor had raw glass attached to both its top and bottom surfaces. Large numbers of such furnace floor fragments have been found at Tell Zujaj, another ninth century furnace site on the same industrial complex. The Tell Abu Ali excavations have therefore provided the *in situ* evidence for the occurrence of such evidence. The phase I furnace appears to have been heated by the use of a brick structure built against its western wall. Presumably this incorporated a forced draft system, although a heating chamber for the air appears to be absent; perhaps it was destroyed in the phase II construction.

9.7. The southern wall of the al-Raqqa tank furnace showing a rill of vitrified cement that divided the slabs attached to the wall. Photo: J. Henderson.

The rescue excavations also revealed part of a curving brick furnace wall (Fig. 9.10) associated with a floor. This was located above the tank furnace with a layer of fill between them. The furnace was floored with slabs to which raw glass was attached (Fig. 9.11). As is often the case, it is difficult to be sure what form the complete furnace would have had. The remains could be the end of an apsidal furnace or of a circular one. If circular, then the received wisdom would suggest that it was used for glassworking, the expectation being that it had been a beehive-shaped three-chambered furnace (Charleston 1978). However, the excavated floor, presumably the lowest one, had no signs of

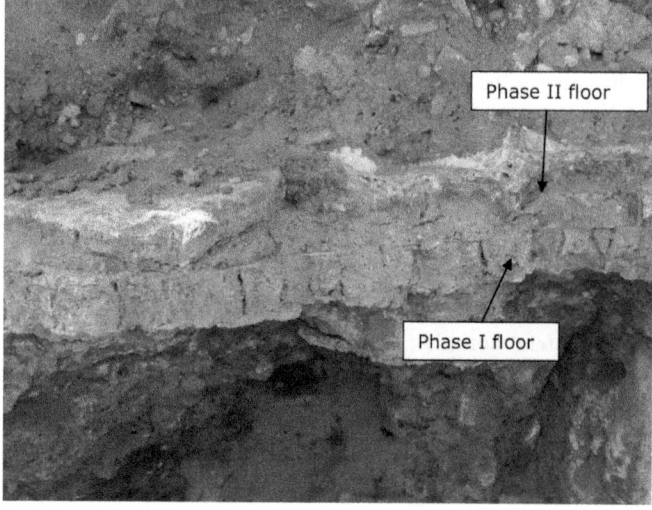

9.8. The two-phase floor of the glass furnace. Photo: J. Henderson.

ISLAMIC GLASS

9.9. A section through the furnace showing the heating chamber located below the furnace floor. Photo: J. Henderson.

having been used for burning fuel: instead it had raw glass attached to it. Therefore, it can be suggested tentatively that this might be the remains of a circular furnace in which glass was heated or potentially even fused from raw materials. Additional excavations should help to address questions about the way in which the furnaces operated. The discovery of rectangular pottery kilns at both Tell Abu Ali and Tell Aswad to the east shows that this structural shape

9.10. A view of the furnaces from above facing east, showing the sectioned circular tank furnace, a sectioned 'circular' furnace and the mud-brick structure built against the western wall of the tank furnace. Photo: J. Henderson.

was used contemporaneously in both glass and pottery technologies: they may well have been constructed by the same groups of people.

Excavations at Tell Zujaj, southwest of Tell abu Ali on the same industrial complex, revealed the bases of beehive-shaped glass furnaces (Henderson 2000a, fig. 3.43) and the dumping of tank furnaces on a massive scale. Thus far, this industrial complex at al-Raqqa is the only Islamic inland site that has provided evidence for primary and secondary glass production. This may be because it is the only inland Islamic glassmaking site that has been excavated intensively: it may well be that the discovery and excavation of others will confirm that this is a model for inland (plant ash) glass manufacture in the Islamic world (Henderson 2003).

The coastal site of Tyre in Lebanon has produced the most complete evidence of tank furnaces used for plant ash glass production (Aldsworth *et al.* 2002). These date to between the tenth and twelfth centuries A.D. and possibly later (*pers. comm.*, Dr. David Whitehouse) The furnaces were even larger than those excavated at Hadera: at Tyre it has been estimated that 32 tons of raw glass was produced each time a furnace was used. If the raw materials were fritted first, each firing would probably have produced an even larger volume of glass. The reason why the Tyre furnaces were larger than those at Hadera is probably because a mixture of silica and plant ashes would have occupied a far larger volume than silica and natron would.

However, as already noted, had a process called *fritting* been carried out, it would have reduced the volume needed for glass production. The manufacture of frit was described by Abū'l Qāsim in his treatise dated 1301. It was carried out in a kiln called a *biraz*. The process takes eight hours, during which time the mixture is mixed well (Allan 1973, 114). This process allows the silica and the plant ashes to interact at lower temperatures than during glass melting, and it is arrested before glass starts to form. The result is a grey-coloured friable crystalline material (Fig. 1.5), which would occupy a far smaller volume in the furnace than a mixture of sand and plant ash would, however much the plant ash was crushed. Apart from reducing the volume occupied by the glass batch, the reason glassmakers fritted their raw materials was to purify them. Whether natron or plant ashes were used as the alkaline source, salts such as carbonates, sulphides and sulphates would break down during the fritting process, producing gaseous carbon dioxide, carbon monoxide and sulphur dioxide. The release of gases from the melt at this initial stage of glass production would lead to fewer gas bubbles being present in the fully fused glass; these could potentially produce opacity and a lower-quality glass.

The difficulty for archaeologists is that because frit is friable and can be prone to weathering, it tends not to survive. Indeed, it appears that it is only when the process has been allowed to go too far and a glassy network starts

9.11. The rills of glass and cement that divided the slabs attached to the "circular" furnace floor. Photo: J. Henderson.

to develop, producing semi-fused glass, that 'frit' survives (Henderson 1995; Paynter and Dungworth 2011, 137). A third reason for fritting a glass batch is that its overall melting temperature is reduced, especially if scrap glass is later added. Furthermore, if sufficiently high temperatures are reached in the furnace, any remaining gases can be driven out of the glass melt, producing translucent raw glass. The tank furnaces at al-Raqqa were heated to a minimum temperature of 1250°C (McLoughlin 2003). Because fuel would have been the most expensive raw material, it would seem logical that fritting would have been carried out, although this is dependent on whether potentially ridged glassmaking traditions were adhered to or not. The addition of scrap glass (cullet) to aid glass melting is supported by the ethnographic evidence of tank furnaces found in small glass workshops in villages northeast of Agra in the province of Uttar Pradesh, northern India (Sode and Kock 2001). Here only the top half of the melt in a tank furnace was used; the lower half of the melt was considered unusable and added to the succeeding melt as cullet (see Chapter 3). Moreover, although 'molten frit' was referred to, the process of fritting *sensu stricto* did not form a part of the procedure of glass melting in the tank furnaces reported by Sode and Kock (*ibid.*, 165). At twelfth century Tell Bellor in the al-Raqqa industrial complex, a pit associated with fritting, perhaps the remains of a fritting oven, was discovered. Another departure from the Indian ethnographic evidence is the frequent occurrence in early-ninth-century and twelfth-century contexts of high-quality translucent glass still attached to the remains of glass furnace floors, presumably showing that most of the glass melted in the furnace was usable and that only the very last layer of glass was left behind. So, not surprisingly the *chaîne opératoire* in Islamic al-Raqqa appears to have been different from that in late-nineteenth-century India, and it illustrates the dangers of extrapolating too literally from ethnographic evidence.

9.5 Archaeological Considerations for Scientific Analysis

As mentioned earlier, excavations of Islamic primary glassmaking sites can produce raw glass attached to tank furnaces or furnace fragments. The raw furnace glass can then be used as a basis for defining the compositional types of glass manufactured on the production site. Raw glass found attached to furnaces can be produced as a result of fusing the primary raw materials (plant ash and sand or crushed quartz) or frit, and this is the ideal situation for glass provenance. However, the raw glass can also be produced by mixing cullet with the primary raw materials or frit, aiding the fusion of the raw materials. From the point of view of provenance, the question is whether the added cullet was locally made or imported from another production centre. If the latter it may have been made using ashed plants derived from a different geological zone and silica of a different geological age and purity. If this is the case, the different impurity patterns associated with using different raw material sources could produce a misleading compositional picture. However, when dealing with the large-scale production of plant ash glass especially, the chances of this happening are low. Once a number of large tank furnaces were installed, and the procedures for making the glass were established, it would be unnecessary to use glass cullet other than that which was made locally. Moreover, if such mixing did occur, and a sufficiently comprehensive chemical characterisation of furnace glass is carried out, it should be possible to observe compositional evidence of mixing. An even more effective means of demonstrating the mixing of raw materials of different geological origins is by using isotopic analysis (see Chapter 11).

The archaeological context in which the raw glass is found is important. Raw glass attached to furnaces is potentially the 'purest' and most useful in representing local glass production. Deposits of raw glass chunks disassociated from the production site in which they were made are less useful. Even though they may represent the product(s) of discrete production centres, the location of the production centres is obviously unknown. The raw glass found at Banias of tenth- to thirteenth-century date (Tsaferis and Israeli 1994; Gorin-Rosen 2000) and on the shipwreck of the Serçe Limani dating to the early eleventh century are examples of the latter. The raw glass found at Banias may indeed be the product of a single glass batch (Freestone *et al.* 2000), but the locational information provided by raw glass attached to *in situ* furnaces has been lost, and this reduces its significance for provenance studies.

Trade in a wide range of things occurred over great distances in the ancient Islamic world, over water and land. The material traded would have included glass vessels, and even glass cullet. Evidence of trade in scrap vessel and raw glass is provided by the Serçe Limani shipwreck with a *terminus* date of 1025. Here the tons of glass, including raw glass, probably exported from the Levant, would have been sufficient to make large numbers of vessels or glass mosaic tesserae. In addition, there is textual evidence for the export of glass from Antioch to Venice in the second half of the thirteenth century (Carboni 1998, 101–2, and n. 12). The French prince Bohemond of Antioch taxed the Venetians for importing *verre brizé*, broken glass (cullet, *vitrum des laboratum*), plant ash (*alumen album*) and beach sand. This suggests strongly that Antioch was a primary glassmaking centre. One of the results of such trade is that vessels made in the Venetian style at the time would have the same chemical and isotopic composition as glass made from Syrian raw materials in Syria. A number of key texts have been published highlighting the important contribution that the Muslims made to the manufacture of glass vessels, amongst which are Lamm (1928), Kröger (1995), Carboni (2001), Carboni and Whitehouse (2001), Scanlon and Pinder-Wilson (2001) and Goldstein (2005). However, a number of questions about where Islamic raw glass was made and where specific vessel forms, some a unique contribution to glass technology, were made remain to be answered. Amongst the questions that could be investigated with scientific analysis are the following:

1. In the ninth century, glass vessels and architectural decoration were made in an innovative Islamic style and in a style influenced by Roman glass production techniques. The latter included the use of *millefiori* as floor or wall decoration (Carboni and Whitehouse, 2001, 18–19; Fig. 9.6). Although scientific analysis of glass excavated at al-Raqqa has shown that there is a relationship between some specific vessel forms and chemical composition during the ninth-century period of technological transition, it would also be of interest to investigate whether the 'Roman' influence further manifests itself in the production of *millefiori* glass in the use of natron glass rather than 'Islamic' plant ash glass.
2. The cold-cut glass vessels of the ninth and tenth centuries were probably made in several locations, including Egypt (Scanlon and Pinder-Wilson 2001), Greater Syria or Iran, where there is a concentration of such glass at Nishapur (Kröger 1995), but even this is not certain. Currently there is a compositional distinction between Nishapur and al-Raqqa plant ash glass. However, more chemical and isotopic analyses should provide clearer indications as to where these cut glass vessels were made and at the very least indicate whether they were made in more than one centre.
3. Ayyūbid and Mamluk enamelled and gilded vessels were almost certainly made in several centres. Many (mosque lamps) are known to have been royal commissions, with production centres in Damascus, Cairo and

al-Raqqa being likely. With a sufficiently comprehensive scientific investigation it would be possible to test the suggestion by Lamm (1929–30) that they were made primarily in Aleppo, Damascus and al-Raqqa. Therefore, chemical and especially isotopic analyses could provide clear evidence for different glass production zones.

TEN

ISLAMIC GLASS

SCIENTIFIC RESEARCH

This chapter provides discussion of the scientific research that relates to the social, economic and archaeological aspects of Islamic glass discussed in Chapter 9.

Islamic glass production occurred over a wide geographical area. During the first 100 years or so, the current evidence suggests that glass was probably made for the Muslims by Jewish and Byzantine glassmakers (see Chapter 9). By the time the Muslims did manufacture glass from primary raw materials, the wide variety of chemical compositions, together with archaeological and historical evidence, indicate that it was manufactured in a range of locations. The principal raw materials used for making Islamic glasses, silica and plant ashes, have been described in Chapters 2 and 3. The chemical and mineralogical compositions of plant ashes, including major, minor and trace components, vary significantly according to a number of factors. These are discussed in detail in Chapter 2; the principle factors appear to be the genus and species of the plant, the mineralogical/chemical compositions of the drift and bedrock geology and the purification of the ashes that removed insoluble magnesium and calcium carbonate.

Figure 10.1 indicates that there is a relatively wide variation in alumina and calcium oxide levels in ninth- and tenth-century Islamic glasses from the Gulf, the Middle East, North Africa and Iran (a smaller number of twelfth to fifteenth century samples have also been plotted). The figure shows that alumina and calcium oxide levels, which reflect mainly the varying compositions of silica and plant ashes used, respectively, form clusters for some decorative vessel types and glass origins that are suggestive of separate production zones. There are well-defined groups of colourless and coloured Nishapur glasses; clear subgroups of glasses from Fustat, largely due to wide variations in calcium oxide levels (including very low and very high values); glass from twelfth- to fifteenth-century Qasr es Segir (Qasr al-Seghir) in northern Morocco, examples of lustre- and cameo-decorated vessels with high alumina levels; and two separate groups of lustre-decorated vessels with low and high alumina levels.

Results for glass from Qasr es Segir (Redman *et al.* 1979) have been published by Brill (1999b, VII G). Most contain a minimum alumina content of 3%, with a majority containing more than 4%. Although a small number of earlier al-Raqqa vessel glasses contain similar relative levels of alumina and also magnesia, there are no compositional parallels for the balance of these later glasses. It is not unreasonable to suggest that these Moroccan glasses were melted from primary raw materials in North Africa, though such high alumina levels could also suggest an origin for the raw glass further east, perhaps in India.

As we saw in Chapter 6, a new and promising picture for the chemical and isotopic analyses of Bronze Age plant ash glasses in the Mediterranean is emerging. It is very likely that the same genera of halophytic plants were used to make both Bronze Age and Islamic glass. As we shall see here, with a wider range of Islamic plant ash glass vessel forms and more material to scientifically analyse, it appears that not only is it possible to detect compositional (sub-)types of Islamic plant ash glass, but patterns of regional glass production based on chemical compositions (Henderson 2003a).

Compositional analyses of Islamic glasses have been published by a number of research groups around the world. Amongst these are those of lustre-ware vessels by Brill (1970a), Islamic glass weights and stamps by Gratuze and Barrandon (1990), Ayyūbid and Mamluk enamelled vessels (Henderson and Allan 1990; Henderson 1998; Freestone an Stapleton 1998), analytical data for a variety of vessels from a range of locations by Brill (1999b, section VII), marvered glass vessels (Henderson 1995a), unpublished data for Mamluk mosque lamps (discussed in Henderson 2003b), the great glass slab at Beth She'arim (Brill and Wosinski 1965; Freestone and Gorin-Rosen 1999) and the glass from the glass factory and palaces at al-Raqqa, Syria (Henderson *et al.* 2004) and the factory at Tyre (Freestone 2002), African beads (Robertshaw *et al.* 2003; Robertshaw *et al.* 2010), central Asian glass (Adburazskov 1972; Adburazskov 2009; Brill 1972; Brill 1989), and others. The discussion that follows considers these data in a broad context.

10.1 Umayyad Glass

Any published chemical analyses of glasses of seventh-century date (or references to them) from the Levant (Dussart 1995; Fischer and McCray 1999) have demonstrated the persistence of 'Byzantine' low magnesia natron glass. In Israel, natron glass continued to be used, even in later centuries (Fischer and

ISLAMIC GLASS

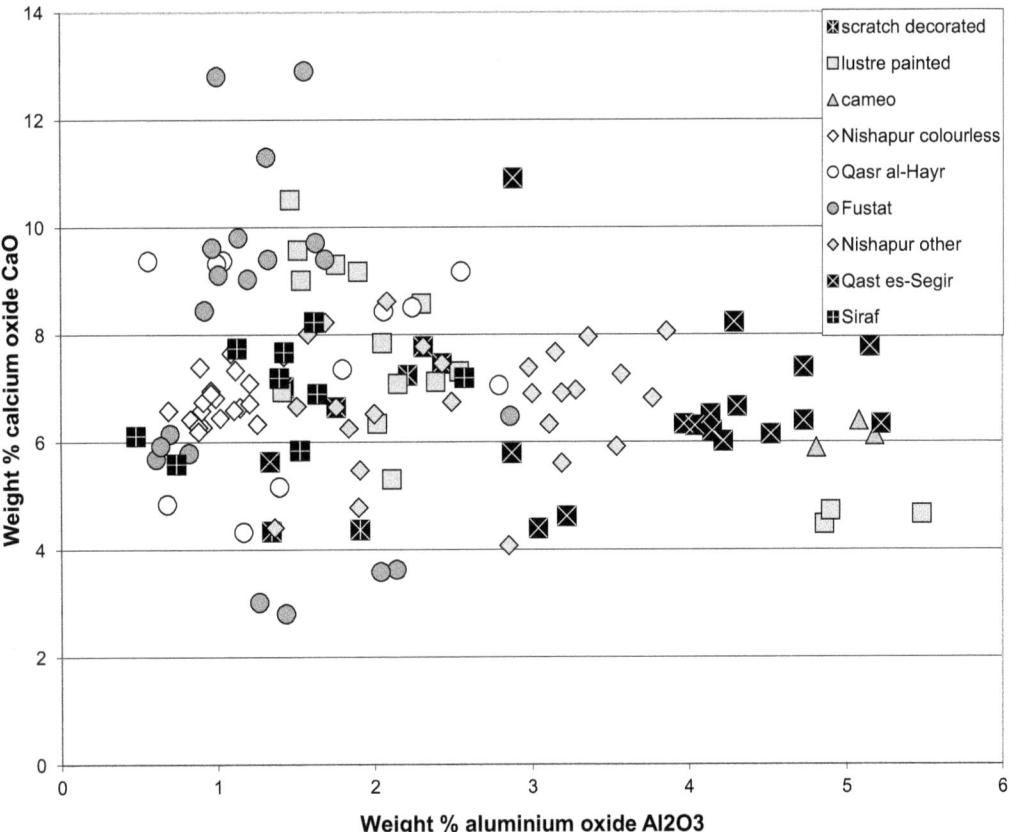

10.1. A bi-plot of weight % alumina versus calcium oxide in specific types of mainly ninth- and tenth-century Islamic vessel decorative types (lustre, scratch and cameo) and glasses from Fustat, Nishapur, Qasr es Segir and Siraf (data from Brill 1999b).

McCray 1999). Much Umayyad period glass appears to have been made on the Levantine coast using an established (classical) natron glass technology. Freestone *et al.* (2002, 173) have suggested that Levantine II natron glass was actually an 'Islamic' natron glass, replacing Levantine I (Byzantine) compositions, the latter being typical of the Levant between the fourth and seventh centuries. Levantine I glasses are characterised by relatively high calcium oxide levels of between 8% and 9% and alumina levels of between 2.5% and 3%, both of which are slightly higher than found in the typical natron glasses found in Roman Europe. Byzantine/Umayyad natron Levantine II glasses tend to contain higher alumina levels of between 3% and 3.5%, lower levels of calcium oxide (6%–8%) lower levels of soda (10%–13%) and silica levels rising to between 73% and 76% (Freestone *et al.* 2000, Table 2). The furnaces at Hadera (Bet Eli'ezer), where seventh- to eighth-century Levantine II glass was manufactured on an enormous scale, are described in Chapter 9.

The chemical analyses of eighth-century glasses from Qasr al-Hayr al-Sharqi (Brill 1999b, table VII D) are presumably of an Umayyad (eighth-century) date, even though there was a twelfth- to fourteenth-century A.D. town there (Graber et al. 1978; Genequand 2005). The contextual information for the samples analysed is lacking; they are mainly of a plant ash composition, with a single sample having a composition approaching that of natron glass. When Qasr al-Hayr al-Sharqi data are compared with that from ninth-century al-Raqqa, a comparably wide range of magnesia levels is found in glass from both sites (Fig. 10.2), but there is no evidence in Qasr al-Hayr al-Sharqi glass for the far higher levels of alumina found in the plant ash glasses and the natron glasses from al-Raqqa containing the lowest magnesia. The results for Qasr al-Hayr al-Sharqi could probably indicate that plant ash glasses were produced in the eighth century in inland locations. However, there is also a possibility that such glasses represent recycled Sasanian glass. Until primary glass production sites of this date are found and the glass characterised scientifically, it will always be difficult to know exactly where and when such glasses were made. The well-dated eigth-century primary glass production sites are located on the Levantine coast where natron glass, but apparently not plant ash glass, was made (Freestone et al. 2000). Moreover, Foy (2003, 87) has reported the existence of seventh- and eighth-century glass from Tunisia of a natron (Levantine 2) composition. Chemical analyses of glass, including samples of a probable Umayyad date from Qal'at Sem'an in northern Syria (Dussart et al. 2004), showed that some (dating to before A.D. 750) had slightly elevated magnesia and potassium oxide levels. On the basis of this, the authors suggested that the mainly Early Christian and Umayyad vessel forms were made from a mixture of natron and plant ash glass (they call it a 'near Roman' composition). There is also some evidence for the use of such glass at al-Raqqa from late-eighth- to early-ninth-century contexts (Henderson et al. 2004, fig. 4).

10.2 'Abbāsid Glass

As mentioned in Chapter 9, al-Raqqa is the only scientifically excavated inland Islamic glassmaking complex where the primary production of raw glass and secondary production by blowing occurred, certainly in the early ninth century and probably in the late eighth century. Given the scale of the extramural industrial complex, located between two urban centres, it is likely that glass decoration such as carving cold glass and lustre decoration also occurred there. However, archaeological evidence for these activities would not survive. Nevertheless, in the writings of al-Jābir, an alchemist/early scientist, who was a contemporary of Hārūn al-Rashīd, there is a detailed description of lustre decoration and the colouring of glass in the eighth century (al-Hassan 2009). As a probable resident of al-Raqqa (Heidemann 2006), al-Jābir probably oversaw lustre decoration there.

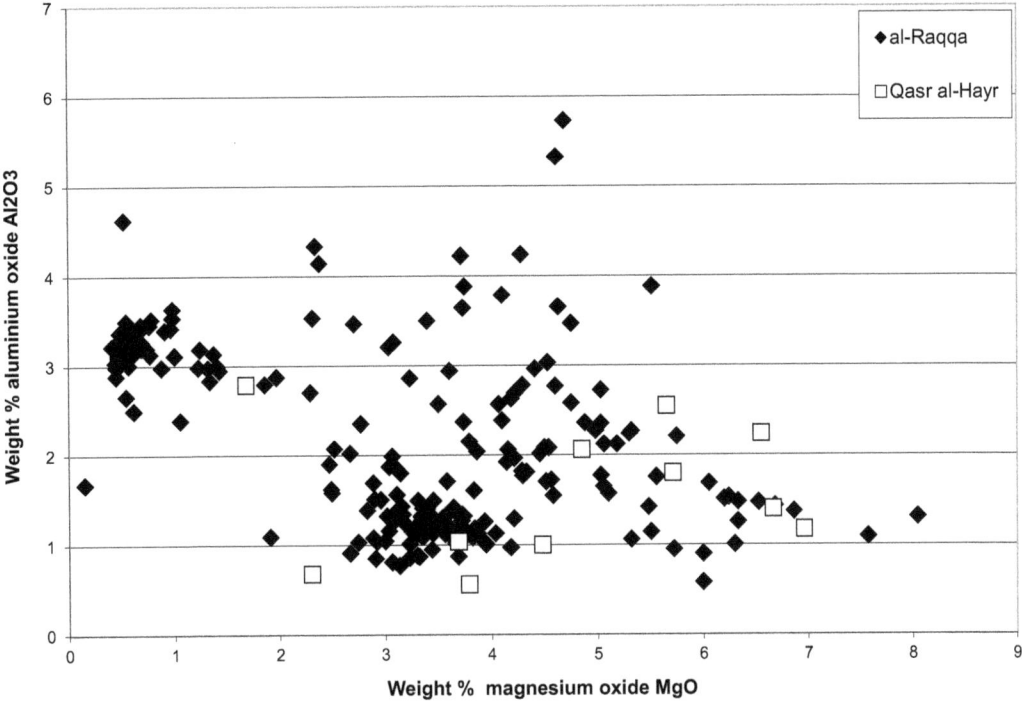

10.2. A bi-plot of weight % magnesia versus alumina in glass from ninth- and eleventh-century contexts in al-Raqqa and probably seventh- to eighth-century Qasr al-Hayr al Sharqi, both in Syria. Data from Brill 1999b and Henderson *et al.* 2004.

Glass production at al-Raqqa is characterised by its massive scale, the presence of the caliph, the fact that al-Raqqa was not just a regional centre but a cosmopolitan one, where Christians and Muslims lived (Heidemann 2006), and the possible presence of al-Jābir. Chemical analyses of glass found there show that both natron and plant ash glasses were being worked there. The change to a domination of plant ash over natron glass occurred in the ninth century (see Section 4.5), and we can see that change in al-Raqqa: it was one of several important transformations in Islamic industry, crafts and arts which occurred at this time. This change in glass technology may have occurred in several places, including al-Raqqa. As an (inland) capital of the 'Abbāsid' empire, it is likely to have occurred there.

The chemical analyses of al-Raqqa glasses using electron probe microanalysis (Henderson *et al.* 2004) appear to reveal the widest compositional range from any ancient glassmaking site in the Middle East. The compositional range includes three principle types of glass, two made from plant ashes (types 1 and 4) and one from natron (type 3), with a third plant ash glass type (2) represented by a smaller number of analyses (Fig. 10.3).

It is clear from the variation of magnesia and alumina impurity levels associated with plant ashes and (partly with the) silica, respectively, that type 1 plant ash glass and the natron glass have the most consistent and constrained chemical compositions. In turn, this indicates that raw materials of essentially the same types were used to make each respective type, and if purification of plant ashes occurred (which seems unlikely), the same techniques were used repeatedly to make the glass. As mentioned in Chapter 3, natron, a mineral, is a considerably purer source of alkali than plant ash and has a very restricted number of sources compared with plant ashes. The variation in alumina levels in the natron glasses is somewhat wider than found in type 1 plant ash glasses because the alumina-rich feldspar content in Levantine sands probably used to make the glass is correspondingly more variable than in the range of silica sources used in al-Raqqa glasses (Chapter 11).

Unlike type 1, type 4 plant ash glass exhibits a much wider compositional variation. First, the MgO levels in type 4 vary between 2.5% and 8.0% with 'end members' probably representing the use of different plant genera and/or the use of plants growing on different bedrock geology. The mixing indicated by the negatively correlated distribution of data for al-Raqqa type 4 in Figure 10.3 probably results from mixing varying proportions of raw glasses containing maximum and minimum contents of both components. This has led to the suggestion that the mixing resulted from experimental glass production. If the ashes had been washed, reduced levels of magnesia and calcium oxide would enter the glass. However, al-Raqqa plant ash glasses contain either low magnesia or calcium oxide, but no samples contain low levels of both, and so ash purification did not occur. The wide variation in alumina levels of between 1% and 5.5% could indicate that silica of a range of purities was used and certainly isotopic results show that silica of different geological ages was used to make the glass type (see Chapter 11). Relatively pure silica sources that would have introduced c. 0.5% alumina in some type 4 glasses are quartz sand or crushed quartz; feldspars in sands are the likely source of the higher alumina levels of c. 2.5% and higher found in other examples of type 4 glasses (see Chapter 3). However, Figure 4.10 reveals that the alumina present in some experimental (type 4) glass is positively correlated with calcium oxide, suggesting that a third major raw material was used to make the glass (see the right-hand data in the al-Raqqa results plotted in Fig. 4.10). Plant ashes have been considered to be the normal source of calcium in plant ash glasses, but given that some halophytic plant ashes have been found to contain insufficient levels of calcium to make Islamic glass of known compositions (Barkoudah and Henderson 2006), the need for a separate calcium-rich raw material is not a surprise. This third primary raw material rich in both alumina and calcium oxide could be a feldspar. Moreover, the positive correlation between alumina and calcium oxide has been found in plant ash Sasanian glass (Mirti *et al.*, 2008), suggesting that a third primary raw material was also used to make it. Nevertheless, when Figure 4.10 is compared with Figure 10.4, it can be seen that the lower alumina levels are to be found in

ISLAMIC GLASS

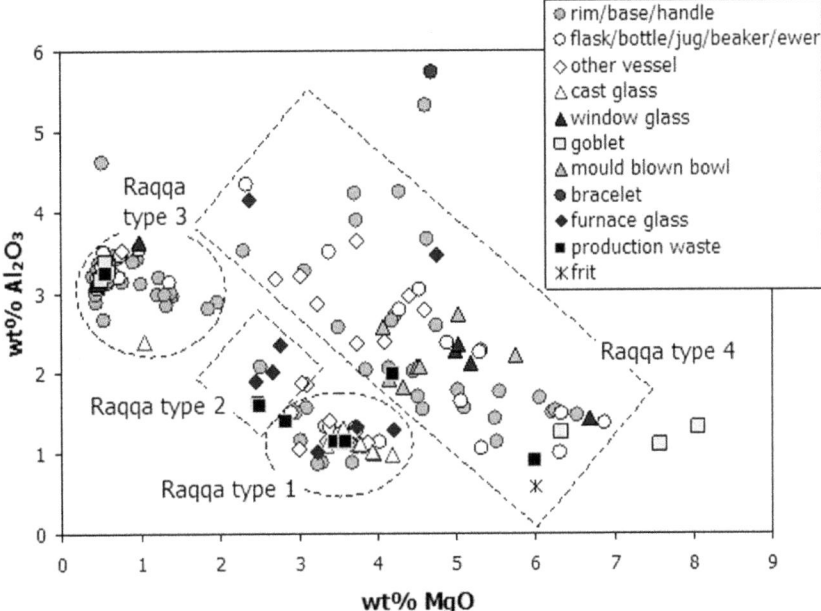

10.3. A bi-plot of weight % magnesia versus alumina in ninth- and eleventh-century glass from al-Raqqa according to glass and artefact type (data from Henderson *et al.* 2004 where glass types are defined) and artefact type.

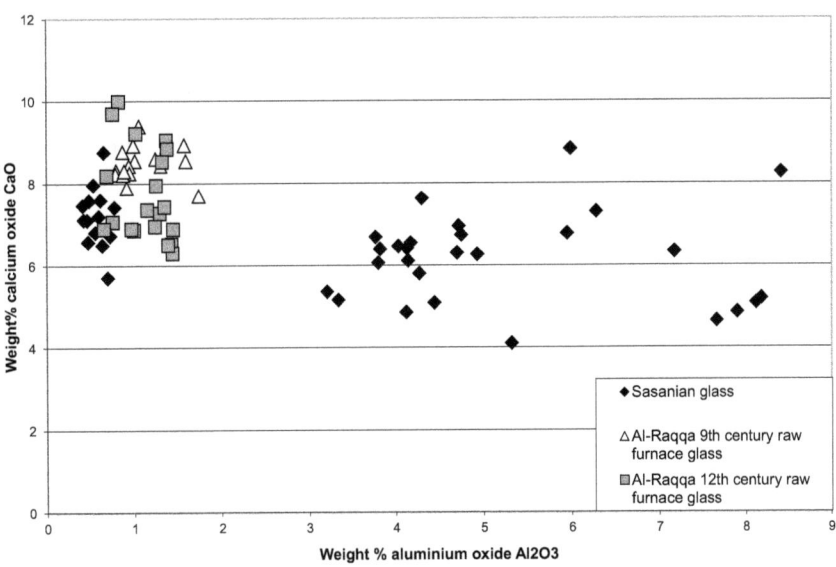

10.4. A bi-plot of weight % alumina versus calcium oxide in raw ninth-century (TZ) (data from Henderson *et al.* 2004) and twelfth-century (TB) furnace glass (data from Meek 2005) from al-Raqqa Syria and second- and sixth-century Sasanian glasses (data from Mirti *et al.* 2008).

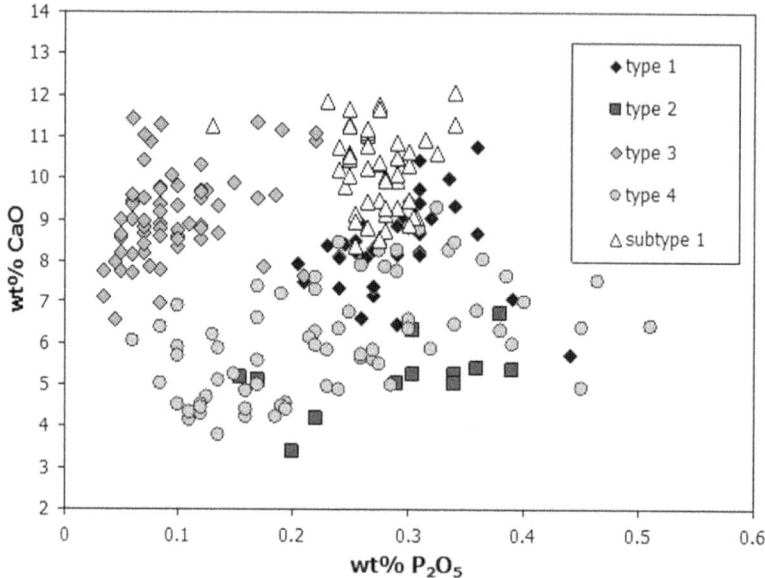

10.5. A bi-plot of weight % calcium oxide versus phosphorus pentoxide versus calcium oxide in ninth- and eleventh-century al-Raqqa glasses type (data from Henderson et al. 2004 where glass types are defined).

ninth- and twelfth-century raw furnace glasses from al-Raqqa. The elevated levels in al-Raqqa glasses of up to 5% alumina therefore appear to derive from 'foreign' glasses. Very high alumina levels of up to 8.6% are to be found in Sasanian glasses (Fig. 10.4); a source for such al-Raqqa glasses may therefore be further to the east as discussed later.

Although an aluminium contribution from sand is likely, the combination of sand containing high alumina levels with plant ash is unusual amongst Bronze Age and Islamic glasses (Henderson et al. 2004; Nikita and Henderson 2006). Therefore, the use of an alumina-rich primary raw material is likely in some cases. Figure 10.5 is a plot of calcium oxide against phosphorus pentoxide in al-Raqqa glasses, and this provides a fairly clear distinction between natron (type 3) and plant ash glasses. The natron glasses contain a consistent combination of some of the lowest phosphorus levels (below 0.1%) combined with high calcium levels (up to nearly 12%). Such low phosphorus levels of c. 0.1% are to be expected from a relatively pure mineral source of alkali. The range of phosphorus pentoxide contents in experimental (type 4) plant ash glasses extends from close to the lowest in the figure (0.05%) to a maximum of 0.52%. Elevated levels are associated with the use of plant ashes. It is also notable that there is a weak positive correlation between the oxides of calcium and phosphorus in types 2 and 4; they also contain some of the lowest calcium oxide levels down to c. 3.4%. We have suggested elsewhere that bone ash could provide a calcium source, especially in type 2. However, the evidence for the use of feldspar as a

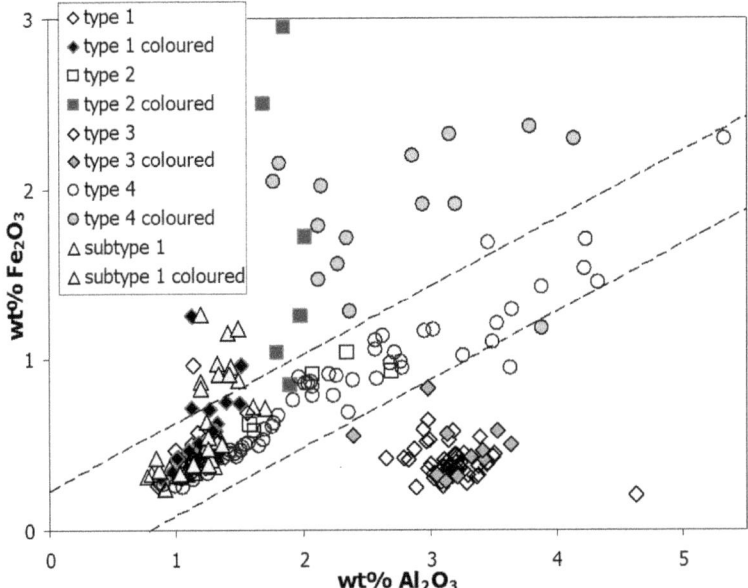

10.6. A bi-plot of weight % alumina versus ferric oxide in ninth- and eleventh-century al-Raqqa glasses showing natural positive correlation between broken lines and modifications of relative levels due to additive colorants. Data from Henderson et al. 2004.

calcium source in type 4 glass is stronger (Fig. 4.10). With the proviso that some calcium could be introduced in shells in river sand, Figure 10.5 also shows that plant ashes of type 1 and subtype 1 contain comparable calcium oxide levels to the levels provided by seashells found in natron glasses (type 3) at al-Raqqa. Type 1 plant ash glasses contain consistently higher calcium oxide levels than found in the experimental plant ash glass, hence the need for an alternative source in the experimental glass. It can be demonstrated that there was a slight difference in the composition of the plant ash glass used over time at al-Raqqa. Many of the eleventh-century plant ash glasses (type 1a in Fig. 10.5) are compositionally distinct from ninth century type 1 (Henderson et al. 2004, 454).

The plot in Figure 10.6 sheds light on the glass-colouring technology in ninth- and eleventh-century al-Raqqa glasses. No distinction can be made between highly coloured and 'natural green' colours of both natron and type 1 plant ash glasses with respect to levels of iron and aluminium. In contrast, there is evidence for a divergence from the positive correlation found in 'naturally coloured' green glasses (defined in Fig. 10.6) in highly coloured types 2, 4 and some subtype 1 plant ash glasses. This shows that an iron-rich raw material with relatively higher iron was added to these glasses – perhaps a relatively impure sand. The net resultant colour would also be due to the furnace atmosphere.

The relative levels of manganese and iron oxides in al-Raqqa glasses are, to a limited extent, related to their respective compositional types (Fig. 10.7). The level of manganese oxide is consistently lower in the mainly aqua-coloured

Ancient Glass

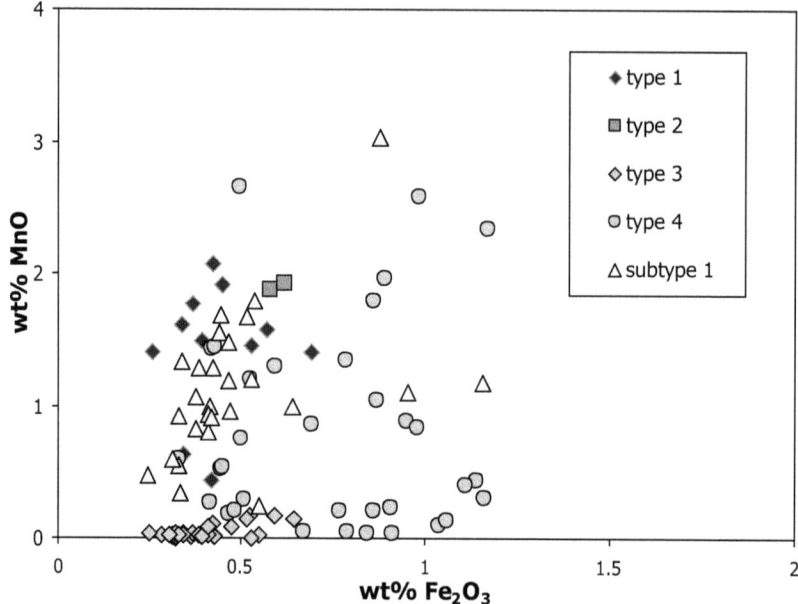

10.7. A bi-plot of weight % manganese oxide versus ferric oxide in al-Raqqa glasses by compositional type (data from Henderson et al. 2004 where glass types are defined).

natron glasses than in most plant ash glasses, reflecting the deliberate addition of manganese oxide at levels above c. 0.5% in many plant ash glasses (Fig. 10.7). The oxide levels in plant ash glasses of type 1 and many subtype 1 glasses are far more constrained than those for type 4, the latter exhibiting the widest variation in both oxides.

The relationship between levels of iron and manganese oxides and specific glass colours is shown in Figure 10.8. Evidently manganese was added deliberately to create purple plant ash glasses (made with a sand with relatively low iron levels). An iron-rich raw material was added deliberately to create olive green glasses. Blue glasses with elevated manganese oxide levels are also compositionally distinct but mainly because they also contain high iron levels. In this case it is difficult to attribute either the manganese or the iron to the use of a particular cobalt source (see Chapter 3). Although manganese was added deliberately in some cases, production of all glass colours would have been strongly affected by the furnace atmosphere.

Mixing of raw glasses at al-Raqqa in the ninth century would have affected the working properties of the glasses. Mixing of (nonlocal) raw and scrap glass seems to have occurred on a limited scale even though large quantities of raw and scrap glass were traded in the Islamic period (see Chapters 9 and 11). However, the political, economic and social contexts in which glass production at al-Raqqa occurred were of regional and international significance, and this is why experimentation is likely to have occurred there.

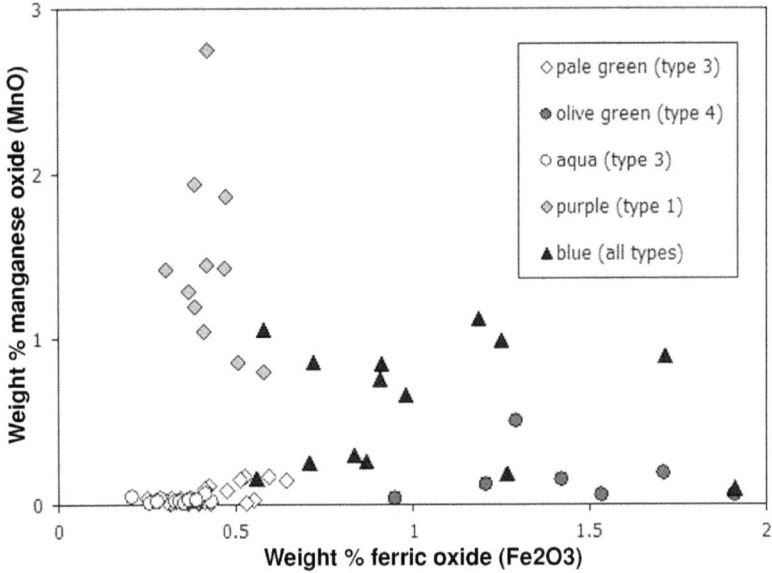

10.8. A bi-plot of weight % manganese oxide versus ferric oxide in specific glass colours of al-Raqqa glasses. Data from Henderson *et al.* 2004.

The evidence for experimentation with glass technology at al-Raqqa, including innovation with furnace design (see Chapter 9) forms part of the ninth-century Islamic industrial revolution, which led to a recognisably Islamic material culture. The rebirth of plant ash glass technology in the ninth century, which invariably contained manganese oxide (Sayre and Smith 1961), the substance that Jābir, an early scientist, specifically mentions (Sarton 1927, 532), may have had an ethnic significance, being regarded as quintessentially Islamic. Moreover, Jabir who resided in Baghdad and al-Raqqa, for example, may have contributed to its rebirth in Greater Syria and Iraq using locally available raw materials. However, evidence of seventh- and eighth-century primary glass production in the inland areas of Greater Syria has not been found (yet). There was probably a relatively short chronology for the ninth-century glassmaking site of Tell Zujaj in al-Raqqa, of perhaps thirty years, between c. A.D. 800 and 830 (Henderson *et al.* 2005). However, on the basis of evidence provided by the chemical analyses of Egyptian dated glass weights (Matson 1948; Sayre 1967; Gratuze and Barrandon 1990), the earliest time that the (re-)introduction of plant ash glass could have occurred was c. 845, even though suitable plants were available in Egypt. Therefore, al-Raqqa, an important centre of the caliphate for a short time, may be one of the earliest sites where the transition occurred; the change may well have occurred in several other inland centres at the same time. Perhaps Egyptian glassmakers used Egyptian natron for longer than glassmakers outside Egypt, such as in Greater Syria.

A small number of glasses from al-Raqqa defined as type 2 (see Fig. 10.3 and Henderson *et al.* 2004) are partly characterised by a relatively low magnesium oxide level of less than c. 3% compared with most other plant ash glasses from al-Raqqa. This characteristic is also found in raw chunk glass from Banias (Freestone *et al.* 2000), which dates to between the eighth–ninth and eleventh centuries (Freestone *et al.* 2000, table 2), in later glass from the tenth- to twelfth-century coastal site of Tyre (Freestone 2002, table 1), in raw glass found on the eleventh-century Serçe Limani shipwreck (Brill 1999b, 178–87; 2009; Henderson 2003a) and unpublished data for vessel glasses samples found in Beirut and Damascus (see Fig. 10.10). All of these glasses were found or possibly made in coastal locations, so it is possible that they were made with halophytic plants that grew in coastal locations characterised by lower magnesium oxide levels than found in glasses made in inland locations such as al-Raqqa and glasses from Iran, as discussed later. Isotopic analyses have confirmed that, in some cases, a likely coastal source of silica was used to make them (Henderson *et al.* in preparation).

The scientific results from al-Raqqa have provided us with baseline data for an inland primary glassmaking site against which to compare glasses from other primary glassmaking sites. When one or more comparable inland sites are discovered, and the same degree of scientific investigation is carried out, it will be possible to place the results from al-Raqqa in a broader context (Henderson *et al.* 2005).

10.3 Glass Compositions and Vessel Types from the Middle East

Chemical analyses of specific glass vessel types from al-Raqqa show that ninth-century glassworkers sometimes used the experimental plant ash glass (Henderson *et al.* 2004, type 4) rather than the contemporary plant ash glass that would become the established type for some 600 years to come (type 1). 'Abbāsid goblets were similar in shape to Byzantine ones and may therefore have been viewed as symbolising the Byzantine sphere of production and way of life. It is interesting to note, therefore, that of the five analysed from al-Raqqa two were made with 'Byzantine' natron glass, and the other three with the experimental plant ash glass. Prior to full publication of the data, Freestone et al. (2000, 72) suggested that such natron glass may have been imported from the 'Byzantine' factory at Hadera (Bet Eli'ezer) on the Levantine coast to be used at al-Raqqa. However, the al-Raqqa data (Henderson *et al.* 2004) shows that although some contain the same high silica levels as in Hadera (Bet Eli'ezer) raw glass, al-Raqqa glass contains higher calcium and somewhat higher soda levels. It was therefore not made at Hadera (Bet Eli'ezer). None of the goblets analysed were made using the plant ash composition that came to dominate the Islamic world: plant ash glass of type 1. Instead, the experimental plant ash glass was used. This may reflect the manufacture at al-Raqqa of what was regarded as a 'foreign', or non-Muslim, vessel form, perhaps by Christian artisans living in al-Raqqa using the experimental recipe. The main thing is that the goblets found were made using both natron and experimental plant ash recipes at this time of transition. In contrast, vessels found at al-Raqqa with cut 'honeycomb' facets, a decorative technique which is clearly an echo of earlier Sasanian vessel forms (see Chapter 9), were almost exclusively (seven of ten samples) made with the experimental (type 4) plant ash glass with high magnesia levels. Of the remaining samples, one was made from natron glass and two with the 'standard' plant ash (type 1) glass. No raw 'experimental' glass containing high magnesia and low calcium oxide levels has been found at al-Raqqa, presumably because it was being manufactured further east, perhaps in Iran. A Mesopotamian label for such glass (Kato *et al.* 2010) is probably too broad because the evidence suggests that parts of northern Mesopotamia (including northern Syria) did not make glass of this chemical composition. Nevertheless, what has been labelled an 'experimental' glass composition falls on the high magnesia end of a magnesia-alumina mixing line (Figs. 10.2 and 10.3), suggesting that Iraqi/'Iranian' glass with high magnesia contents was mixed with raw glass made in al-Raqqa. Indeed, there is some isotopic evidence for this, as described later, but it is not conclusive proof.

The idea that Sasanian glass technology made a direct contribution to early Islamic glass technology is therefore substantiated chemically. There is here an interesting fusion of environmental, technological, historical and geographical factors involved. The late-eighth- to early-ninth-century palace complexes at al-Raqqa were glazed with small green, turquoise, blue, purple and brown windows. The buildings were clearly deeply symbolic of a highly successful caliph at the height of his powers: again, only glass of the experimental plant ash type was used to make them (Henderson *et al.* 2004), even though raw natron glass and plant ash glass of the 'standard' composition was available, and isotopic results show that the glass was imported to al-Raqqa (see Chapter 11).

As noted earlier, a large amount of Islamic glass was found at tenth-century Nishapur, which could suggest that it was made there. The area which Nishapur was located in specialised in the production of specific types of decoration (lustre or painted, incised and wheel cut), reaching an apogee in the ninth and tenth centuries (Kröger 1995, 37). Amongst these glasses are some colourless carved vessels. When their compositions are compared with the earlier al-Raqqa glass and vessels (Fig. 10.1), they are tightly constrained with consistently low levels of alumina, indicating the use of a relatively pure silica source and elevated soda (Henderson 2003, 119, fig. 8). This indicates first that a consistent combination of raw materials was used to make them, second that the compositional range may represent the product of a single melt and third that this formed part of the specialised production of colourless glass with carved decoration. The production of these carved vessels (and perhaps the raw glass) may have occurred independently in Iran and Iraq (Kröger 2010, 29).

A further consideration of the relationship between Islamic vessel form and chemical composition amongst ninth- and tenth-century material is provided by the data plotted in Figure 10.1. It shows that three specimens of plant ash cameo glass vessels all contain very high alumina levels. Although the sample numbers are small, the result suggests that a consistent use of a relatively impure sand was used to make the glass. On the other hand, a small number of lustre painted vessels, using a silver- or copper-rich paint, was also made using an impure silica source. Exactly where such vessels were made is still unknown (Carboni and Whitehouse 2001; Scanlon and Pinder-Wilson 2001, 11). Egyptian glassworkers are thought to have been the earliest to have used this decorative technique (Lamm 1941, 18–19), and Fustat has produced the earliest Islamic examples, dating to c. 780 (Scanlon and Pinder-Wilson 2001, 110; Carboni 2001, 51–2); the other probable production area is Syria. Lustre-painted wares were made both with the experimental glass composition and the plant ash glass that was set to become the mainstream Islamic glass type. It is possible that a wide compositional range for lustre-painted vessels is a reflection of several production centres. Those containing the highest alumina levels are clearly compositionally distinct from al-Raqqa glasses.

The relative levels of calcium and alumina in scratch-decorated (sgraffito) vessels is far more constrained (Henderson *et al*. 2009b, fig. 7), with a series of six negatively correlated data fall into the al-Raqqa experimental glass cluster (Fig. 4.10) and a single sample into the al-Raqqa 'standard' plant ash compositional group; the six vessels may well have been the products of a single melt. There are other examples that were made from natron glass, and so raw glass, in all probability made on the Mediterranean coast, may have been exported to and blown in Islamic glassworking centres. A production centre in northern Syria (perhaps al-Raqqa) or Samārra, Iraq, has been suggested (Carboni 2001, 80–1). A tight negative correlation between alumina and magnesia for six scratch decorated vessels falls precisely on the correlation (mixing) line observed in the al-Raqqa 'experimental' type 4 (Fig. 10.3). Several of these vessel samples derive from al-Raqqa, so they could have been made there in the ninth century. Nevertheless, this cannot be regarded as a definitive means of provenancing them (see Chapter 11).

For the Ayyūbid and Mamluk periods (eleventh – fourteenth centuries), there is historical evidence for glass production in both Damascus and Aleppo as noted earlier, but only Tyre and al-Raqqa have currently produced archaeological evidence for primary glass production in the ninth to twelfth and tenth to twelfth centuries respectively (see Chapter 9). A plot of calcium versus alumina for selected groups of glass shows several things (Fig. 10.9). First, almost all the glasses contain less than 1.5% alumina and many of the results for

ISLAMIC GLASS

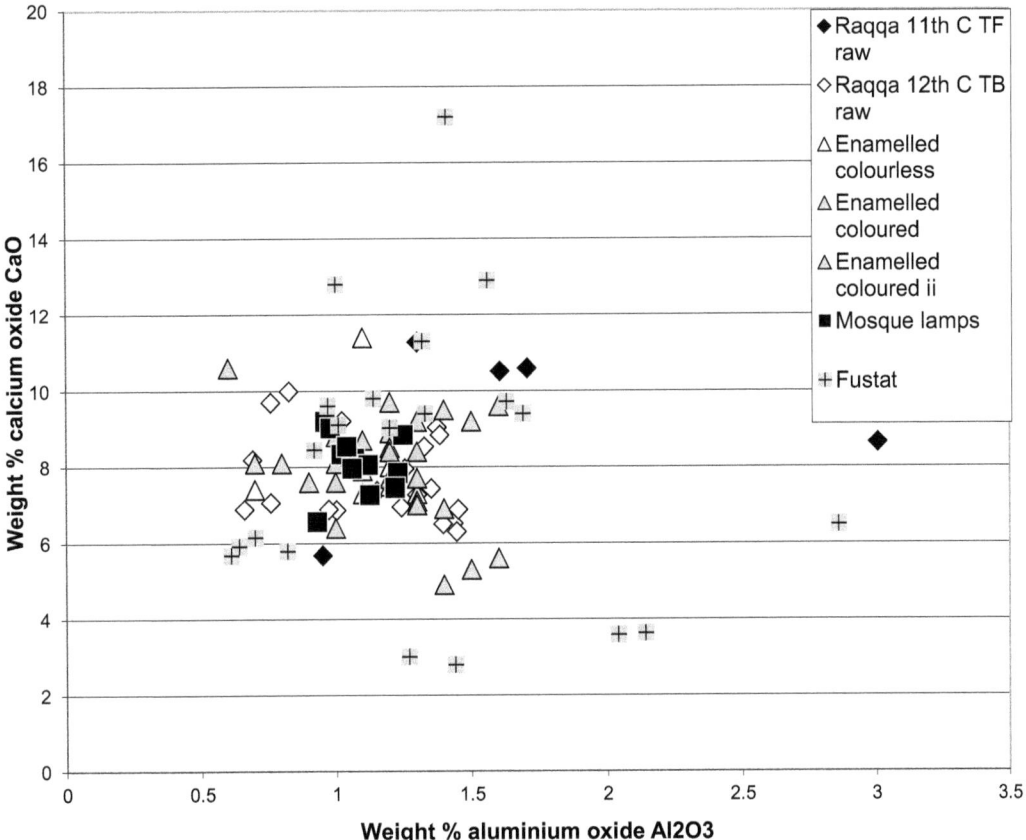

10.9. A bi-plot of weight % alumina versus calcium oxide in tenth- to fourteenth-century glasses from al-Raqqa, Fustat, enamelled vessels and enamelled mosque lamps. Data from Henderson *et al.* 2004, Brill 1999b, Henderson and Allan 1991 and the author (mosque lamps).

tenth-century Fustat glasses with low calcium oxide levels fall away from the main cluster and probably represent the products of a separate production zone. The compositional variation of twelfth-century furnace glass from Tell Bellor (TB), one of the sites in the al-Raqqa industrial complex (Meek 2005), encompasses a field that is as wide as the majority of the enamelled mosque lamps and bottles tested. These results alone suggest that the enamelled glasses were made in a relatively restricted area of the Islamic world or, as we shall see, that these oxides are not an especially good means of characterising these glasses.

Colourless bodies of enamelled Ayyubid and Mamluk vessels labelled 'enamelled colourless' in Figure 10.9 using data published in Henderson and Allan (1990) have quite tightly constrained alumina and calcium oxide levels. In Figure 10.10, these 'twelfth- to fifteenth-century enamelled' vessels also show a constrained composition within the 'Mainly N. Syrian/Egyptian' area. This suggests that they may be products of a single production centre or zone. The results for the bodies of 'Enamelled ii' group using data published by Freestone and Stapleton (1998) and 'enamelled coloured' using data published by Henderson (1995) in Figure 10.9 exhibit a relatively wide compositional range. However, the data plotted in Figure 10.9 for the translucent green and brown bodies of mosque lamps are distinctive: they contain a restricted range of between 0.93% and 1.25% aluminium oxide but variable magnesia levels of between 2.5% and 5.9%.

These results are likely to reflect the use of a characteristic Egyptian sand source; some of these vessels are dedicated to important Egyptian emirs and were certainly blown there. It is likely that the glasses were also melted from primary raw materials in Cairo. The data for 'mosque lamps' plotted in Figure 10.9 includes the oxide concentrations for a blue bottle, of possible Syrian origin. This falls away from the rest in Figure 10.9, containing the lowest calcium oxide level of 6.56%. Moreover, raw eleventh-century glasses from Tell Fukhar, al-Raqqa (probably imported from the Levant, see Fig. 10.10) can be distinguished from Mamluk glasses and group with probable Levantine products. A comparison with the chemical analyses of eight 'raw' glasses from two tank furnaces in Tyre (Freestone 2002, table 1) show that two compositions from furnace 2 group with the raw glasses from eleventh-century Tell Fukhar, al-Raqqa, whereas those from furnace 1, with characteristically high calcium oxide contents, are closest compositionally to some Fustat glasses. Indeed, the Tyre furnaces may be the source for these particular Fustat glasses. Although based on small numbers of analyses, these relatively constricted results for Mamluk enamelled mosque lamps, in particular, provide grounds for suggesting that it may be possible to distinguish compositionally between Syrian and Egyptian Mamluk products.

The plot of magnesia versus calcium oxide contents in ninth- and eleventh-century furnace glass from al-Raqqa compared to Sasanian glass and both colourless and coloured glass from Nishapur, the eleventh century Serçe Limani shipwreck, the bodies of twelfth- to fourteenth-century enamelled vessels and glasses from Beirut, Damascus and Bahrain (Fig. 10.10) reveals several interesting things:

1. Results for eleventh-century Serçe Limani raw glass, eleventh-century raw glass from al-Raqqa, twelfth-century Beirut and eleventh- and twelfth-century Damascus are characterised by relatively low magnesia levels (less than c. 3.5%) and most by calcium oxide levels of about 9% and above. The high calcium levels suggest that shell fragments in beach sand was the calcium source. These glasses were probably manufactured on the Levantine coast, although their strontium isotope results are slightly lower than would be expected (Brill and Fullagar 2009, fig. 2, 554) and suggest that an older source of lime provided by plant ashes has been mixed with the sand containing 'modern' shells.

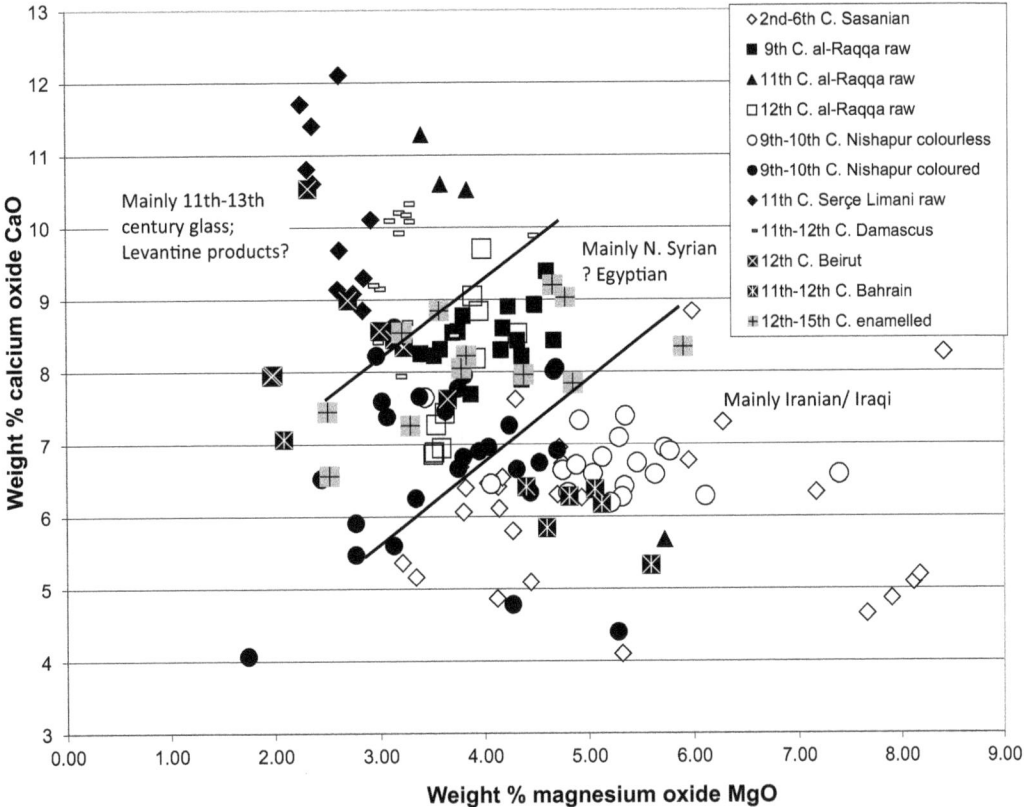

10.10. A bi-plot of weight % magnesia versus calcium oxide in Sasanian, ninth- to twelfth-century raw (furnace) glass from al-Raqqa and glass from twelfth- to fifteenth-century Nishapur, the Serçe Limani shipwreck, Damascus, Beirut and Bahrain (data from Brill 1999b; Henderson *et al*. 2004; Meek 2005; Mirti *et al*. 2008; Damascus, Beirut and Bahrain data from the author).

2. Sasanian and Nishapur colourless (and some coloured) glasses are characterised by relatively low calcium oxide and high magnesia levels, as are some of the glasses excavated from Bahrain. These may have been made in Iran or Iraq.
3. The tenth-century 'coloured' glasses from Nishapur are characterised by relatively low calcium oxide and separable from the ninth century raw glass from al-Raqqa, suggesting that they were made in a different production zone. Some also fall in or close to the higher magnesia group.

Figure 10.11 has results for specific types of vessel decoration using the same suggested production zones as in Figure 10.10. Here we can see several things:

1. There is a clear relationship between composition and the eastern production zone where cut and carved vessels found at Raya and al-Raqqa were made.

2. The glass used to make the bodies of lustre-painted vessels were probably made in the Levant, Syria (perhaps in al-Raqqa) and perhaps Egypt.
3. Data for scratch-decorated and most cameo decorated vessels fall into the 'Syrian' zone.

Amongst the chemical analyses of al-Raqqa glass (Henderson *et al.* 2004), the colourless glass samples containing more than 5% magnesia also contain low calcium oxide levels of less than 6.4%, suggestive of an 'Eastern' origin.

These results therefore show that by pursuing a well-defined sampling and analytical strategy to include closely dated diagnostic vessel forms, greater definition of production spheres and provenance is likely to result. However, although we can recognise compositionally tight groups formed from vessels of the same shape (e.g. mosque lamps) or decorated in the same way (e.g. scratch decorated), this is not the same thing as providing a true provenance for the glass used to make them – it simply shows that the same or a very similar combination of raw materials was used to make the vessels within that group. This may have occurred in the same place and perhaps at the same time using glass from the same melt. However, if we bring the results for ninth-, eleventh- and twelfth-century raw glass from al-Raqqa into the equation, we can at least be quite certain when a vessel was *not* made there using the data available to us and in some cases when it probably was. This is still not an entirely satisfactory means of provenancing glass, although it represents an essential first step; the determination of radiogenic isotopes in the glasses makes an essential contribution (see Chapter 11).

10.4 Other Islamic Glass Compositions

Brill is one of few scholars who has investigated Islamic glass from Afghanistan scientifically (Brill 1972; Brill 1989). He has suggested that the glasses that contain potassium oxide levels of greater than c. 4.0% are characteristic Afghan products (Brill 2001, 28). These potassium oxide levels are also found in other central Asian centres such as glass from near Tashkent, Bukhara, Samarkand, Fergana and Dzhambul (Adburazskov 1972). The higher levels of potassium oxide alone would suggest that a separate production zone for plant ash Islamic glass existed in central Asia. Brill has suggested that a different plant ash composition is the explanation. However, we know that this is only one possible factor, others being the geochemistry of the soil in which the plants grow and the season in which they are harvested (Chapter 2). Moreover, if there were several primary production centres that used local halophytic plants introducing elevated potassium oxide levels, the characteristic is apparently a regional geological-botanical one.

A comparably high potassium oxide level (5.98%) has recently been found in a Mamluk armlet from Jordan (Boulogne and Henderson 2009, 72). Being easily

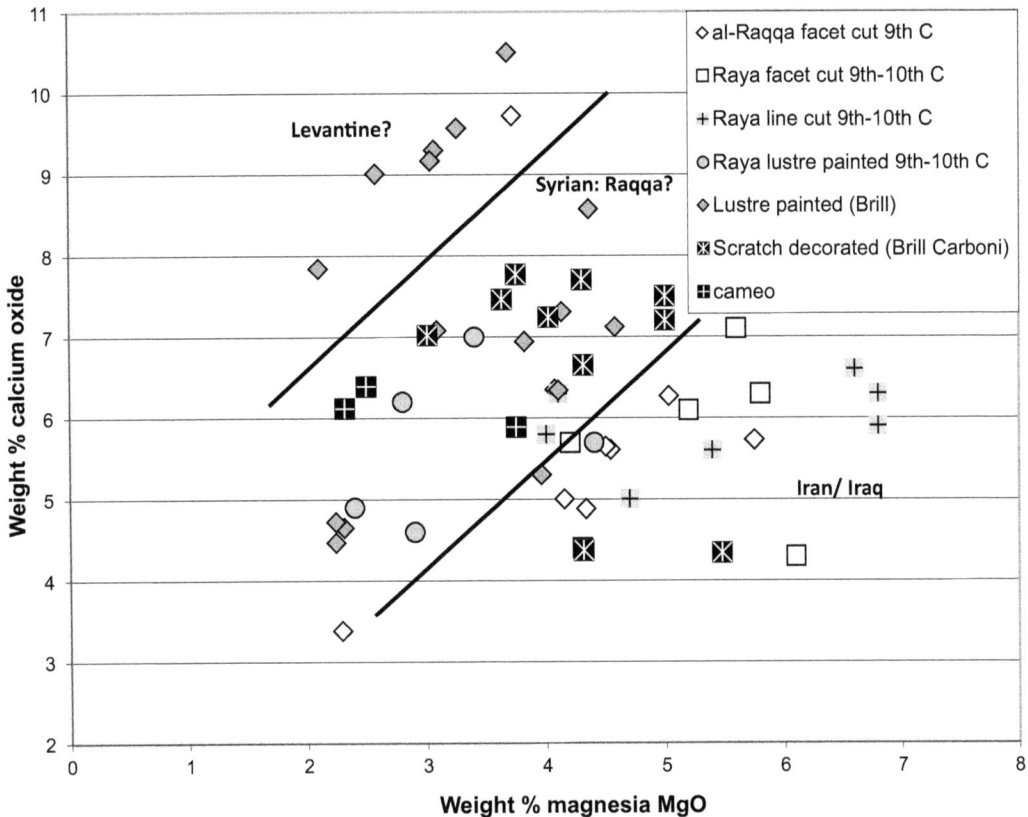

10.11. A bi-plot of weight % magnesia versus calcium oxide in ninth- to tenth-century glass vessels of specific types, including examples from al-Raqqa and Raya (data from Brill 1999b; Carboni 2001; Henderson et al. 2004; Kato et al. 2010).

transported, armlets can potentially have a wide range of origins and chemical compositions. Boulogne and Henderson (2009) discovered the following compositional types amongst the thirty-six armlet samples from Mamluk and Ottoman Tell abu Sarbut and Khirbat Faris, Jordan that they published: (1) plant-ash glass, (2) high-alumina glasses, (3) mineral glass made with an alkali-rich mineral other than natron, (4) a single natron glass.

The high alumina glass is discussed here in detail because it illustrates the complexities of trade and recycling of glass in the Middle East. These high-alumina glasses are the first to have been published in detail for glasses found in the Middle East. The alumina levels detected range from 5.95% to 10.25%. This is far higher than found in typical Islamic plant ash glasses, which normally contain a maximum of c. 4% (Figs. 10.3 and 10.9). They also contain higher potassium oxide levels than found in typical Middle Eastern Islamic glasses which contain a maximum of c. 4% (Fig. 10.12); consistently higher phosphorus pentoxide levels, with a maximum of 0.8%; and relatively low calcium oxide levels.

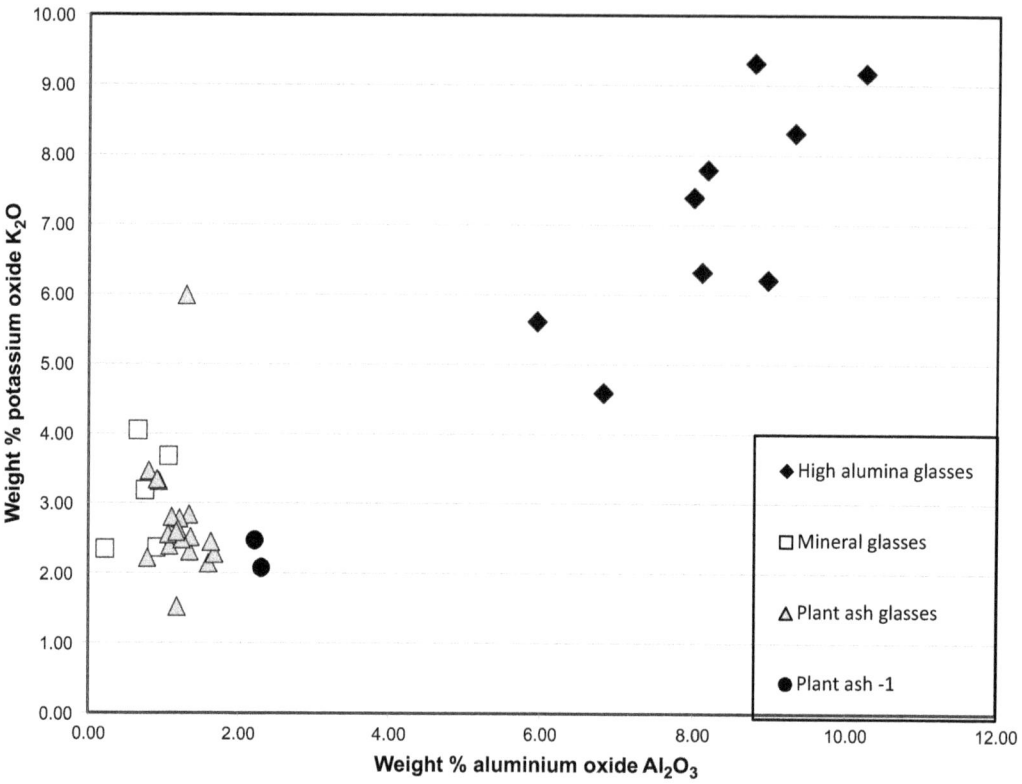

10.12. A bi-plot of weight % alumina versus potassium oxide in twelfth-century and Ottoman glass armlets from Jordan showing the wide range of glass compositions. Data from Boulogne and Henderson 2009.

All of the high alumina armlet samples contain high titania (0.76%–1.29%) an iron oxide (1.4%–12.93%), the latter generally causing the glass to have a deep translucent brown or green (apparently 'black') colour. As noted in Chapter 4, high alumina glasses have been found in Africa, India, Sri Lanka and southeastern Asia (Brill 1987; Lankton and Dussubieux 2006; Dussubieux et al. 2008). The Indian ones were probably made from a granitic source of sand and the alkali reh. However, the compositions of the high alumina glasses from Tell Abu Sarbut and Khirbat Faris in Jordan were unlike those published from these regions and their very high soda levels, amongst other characteristics, have been interpreted as resulting from a mixture of Middle Eastern plant ash glass and high alumina glass. The mixing of these glasses may have occurred in the Middle East using imported high alumina glasses from India, or possibly Afghanistan (Boulogne and Henderson 2009). The characteristically high potassium oxide levels may have been produced by mixing 'central Asian' high potassium glasses with high alumina glasses. Therefore the Mamluk bangle glasses were probably made from glass imported along the Silk Road.

ISLAMIC GLASS

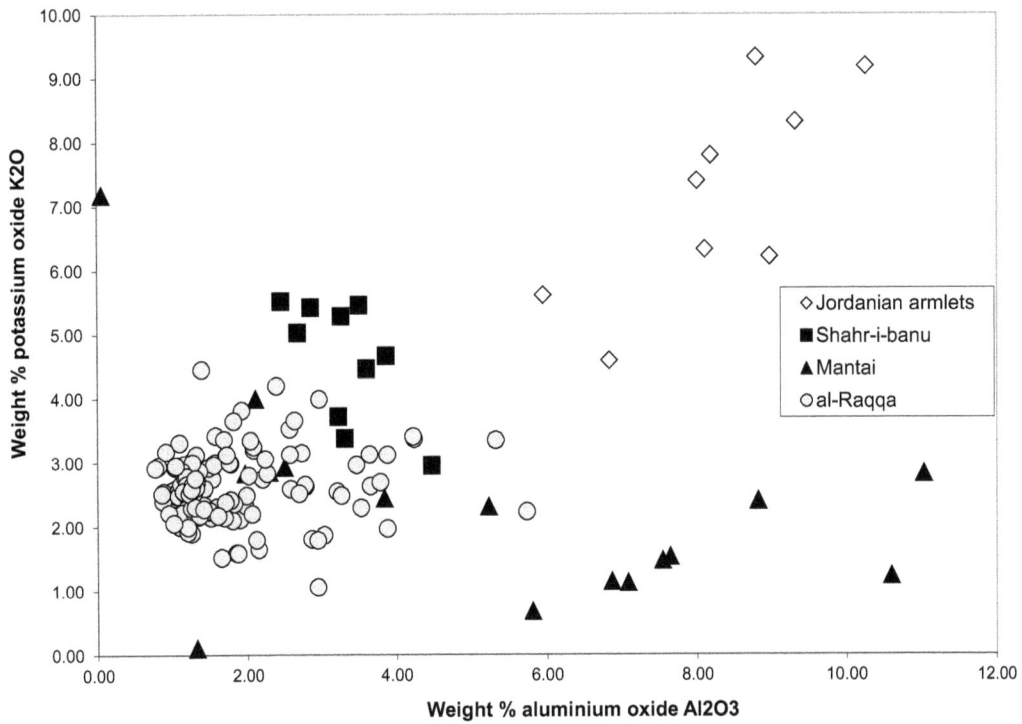

10.13. A bivariate plot of weight % alumina versus potassium oxide in four groups of Islamic glass: Mamluk and Ottoman armlets from Tell Abu Sarbut and Khirbat Faris, Jordan (data from Boulogne and Henderson 2009); glass vessel fragments from seventh- to the thirteenth-century Shahr-i-banu, Afghanistan (data from Brill 1999b, XIVA, 342–3); raw bead and vessel glass fragments of a mainly eleventh-century date from Mantai, Sri Lanka (data from Henderson, in press); ninth- and eleventh-century al-Raqqa, Syria (data from Henderson et al. 2004).

Figure 10.13 shows that in comparison to plant ash glasses from the Middle East, exemplified by ninth- and eleventh-century glasses excavated from al-Raqqa, Syria, the Jordanian bangles and ninth-century glasses from Mantai, Sri Lanka, contain significantly higher alumina levels. The bangles contain higher levels of both potassium and aluminium oxides. The al-Raqqa glasses that contain more than 4% potassium oxide must be of a nonlocal origin. Some of the Islamic glass from Shahr-i-banu, Afghanistan is distinguishable from Middle Eastern glass by its higher potassium oxide levels. In technological terms, the variations in potassium and aluminium oxides shown in Figure 10.13 indicate that the use of both silica and alkali raw materials vary considerably across the Islamic world from central Asia to the Middle East.

'Middle Eastern' plant ash glass has also been found in Afghanistan and Sri Lanka (see section 4.10). Although it is what might be described as a 'Middle Eastern' type, its occurrence in these places could potentially mean that it

299

was nevertheless also made there. The corollary is that high alumina and high potassium glasses are a rarity in the Middle East. Compositional types not found at al-Raqqa plotted above are rarely found in the Middle East. This suggests that the raw materials used to make them were not available in the Middle East. Raw glass of the 'central Asian' composition was probably imported to make the armlets discussed earlier.

10.4.1 Islamic Glass from North Africa and the Persian Gulf

Three recently published articles, two on glass from the Persian Gulf (Kato *et al.* 2009; Kato *et al.* 2010) and one of North African glass by Robertshaw *et al.* (2010), have provided further intriguing evidence for the variety of chemical compositions used for the manufacture of Islamic glasses.

Chemical analysis of natron glass from eighth-century contexts in Raya, an important Red Sea port, and ninth-century contexts at a monastery of Wadi el al-Tur, near Raya at the southern end of the Sinai Peninsula, using non-destructive portable X-ray fluorescence spectroscopy has revealed what appear to be four different glass types, one of a probable Levantine origin and three of an Egyptian origin (Kato *et al.* 2009). These suggested provenances are based on the work by Sayre and Smith (1974), Gratuze and Barrandon (1990), Freestone *et al.* (2000) and Foy *et al.* (2003) and relate to relative quantities of aluminium and calcium oxides. Their eighth-century N1 type was made using a lime source that introduced relatively high levels of aluminium, calcium and strontium oxides (probably maritime shells) from a probable Levantine production area. Their eighth-century N2-a1 type is characterised by low calcium and strontium oxide levels and has a probable Egyptian source. The third type, N2-a2, has low calcium oxide but much higher levels of aluminium, titanium and iron oxides. The distinction between N2-a1 and N2-a2 was found by Foy *et al.* (2003); Nenna *et al.* (2005) have even published data for the primary glassmaking site of Wadi el Natrun. However, as Kato *et al.* (2009, 1705) have pointed out, there is a 400-year difference in date between Wadi el Natrun and Raya. Because it is unlikely that the Raya glass is recycled, it is apparent that this same type of glass was manufactured by the Romans and the Muslims, but that the production site for the later glass is still to be found.

Their mainly ninth-century N2-b natron glass type is characterised by low alumina and high calcium oxides and is almost certainly of an Egyptian origin. The researchers analysed vessels decorated in a variety of ways. For example, three scratch-decorated and two impressed decorated were made from natron glasses of their Egyptian eighth-century N2-a1 type with low calcium oxide levels (3.49 ± 0.89% CaO); this might suggest that glass with low levels of lime were selected to successfully incise and impress the decoration. However, their ninth-century N2-b type, which contains $11.0 \pm 1.03\%$ CaO (and elevated titanium and iron oxides) was also used for the manufacture of impressed, scratch-decorated and stamp-decorated vessels. It is likely that the glasses with higher lime levels would have been 'shorter' (with a shorter working period). It is therefore clear that Islamic glass vessels with the same decoration were manufactured using glass of varying compositions and that would therefore have behaved in different ways at high temperatures. This work underlines the complexity of ancient glass technology. It is clearly an oversimplification to assume that glasses with specific working properties were necessarily selected for the manufacture of specific kinds of vessel decoration. Moreover, this interesting research suggests that scratch-decorated and impressed natron vessels were possibly made in more than one region.

The compositions of plant ash glasses from Raya and the monastery of Wadi al-Tur (Kato *et al*. 2010) have provided more evidence for separable compositional types of 'Mesopotamian', Levantine and Egyptian/Mediterranean plant ash glasses. The often colourless 'Mesopotamian' glasses are referred to by Kato *et al*. as PA (Plant ash)-2. They are characterised by high silica, with a relatively low iron content and, in some cases, a higher magnesium-to-calcium oxide ratio compared to Syria-Palestinian glass (Henderson 2003a; Brill 2005; Kato *et al*. 2010).

The glass with high magnesia (and low calcium oxide) contents from al-Raqqa, of the 'experimental' type 4, for which there are strontium and neodymium isotope results, are vessel glass fragments – significantly no raw furnace glass of this composition was found at al-Raqqa suggesting a 'foreign' source for it (as discussed earlier and in Chapter 11). If we compare the strontium isotope results for these non-Raqqa vessel glasses with the summary diagram published by Brill and Fullagar (2009, fig. 2) the results for both of them fall just off the lowest end of the results for Nishapur and are quite distinct from the values for Serçe Limani glass and Tyre (of later date) found in more western locations. Their chemical compositions could be diagnostic to glasses made in Iran, for example. With a strong Mesopotamian influence on the manufacture of vessel glasses in ninth-century Baghdad and Sāmarrā, Iraq (the latter was the 'Abbāsid capital between 836 and 892), an influence that would have extended westwards to al-Raqqa, it may be that the social and cultural meaning of the glass was extended also to glass artisans at al-Raqqa. Similar raw materials introduced the characteristic magnesia:calcium oxide ratio in these glasses. If 'magnesium-rich alluvium from the Euphrates and Tigris Rivers' (by which Kato *et al*. [2010, 1392] presumably mean the chemical composition of the plant ashes used) have produced this characteristic Mg:Ca ratio as Kato *et al*. have suggested, it is necessary to define how far up these rivers to the north such a contribution could have been derived. The Euphrates, for example, flows from

Turkey across northern Syria. Additional isotopic characterisation (see Chapter 11) and determination of a greater range of trace elements in the glasses may provide greater clarity. Glasses with a high MgO:CaO ratio may have distinctive isotopic signatures, and this will help to confirm that they were manufactured in a different (Iranian/Iraqi) production zone.

Colourless stamped, impressed and moulded plant ash glass vessels from Raya fort and the Wadi al-Tur monastery dating to between the tenth and twelfth centuries with a low MgO:CaO ratio (Kato *et al.* 2010, 1393) have a similar composition to Ayyūbid and Mamluk enamelled glass bodies with a possible Syrian provenance (Henderson and Allan 1990). In contrast, unpublished strontium isotope results for Beirut glasses show that both the probable coastal sand source and the source of calcium from the plant ashes used are geologically young and therefore that the plants probably grew in a coastal location. This suggests that coastal halophytes were used to make such glass. Additional trace and isotope results may be able to distinguish between glasses of these compositions made at various Mediterranean coastal locations.

Ninth- and tenth-century vessels with cut, stamped and threaded decoration were produced using both of the main types of plant ash glasses identified at the Raya fort and Wadi al-Tur monastery (Kato *et al.* 2010, fig. 4). As found in natron glasses, plant ash glasses with contrasting levels of calcium oxide (and therefore with differing working properties) were used to make vessels of the same decorative type. The working properties are dependent on the balance of other components that would affect the glassworker's ability to blow, mould and cut the glass, but calcium oxide levels would nevertheless have been important.

Robertshaw *et al.* (2010) provide compositional results for glass beads. As with the armlets discussed earlier, we should be aware that beads can travel over long distances very easily, and therefore their find spots could be a long way from where the glass was fused from primary raw materials or where the bead was made. Nevertheless, Islamic glassmakers clearly used a wide range of raw materials in the manufacture of glass, and the evidence mainly from ninth- to eleventh-century al-Basra, Morocco, adds to this impression. In addition to plant ash (soda-lime-silica) glass (Robertshaw *et al.* 2010, 371–4), the investigation of thirty glass beads revealed the presence of three kinds of lead-rich glass: lead-silica, lead-soda-silica and lead-silica with high calcium and phosphorus contents. They also found soda-lime glasses with high alumina levels and a single example of a potash-calcium-silica bead. (The latter may well be of a much later production date.) The other lead-rich glasses are of interest because there are few other examples from Islamic sites. Other examples are four early-eleventh-century emerald green vessel fragments from the Serçe Limani shipwreck (Brill 2009) and a single emerald green window fragment from the twelfth-century princess's palace in al-Raqqa (Henderson *et al.* 2004, table 1, sample 47). Two lead-soda-silica beads from al-Basra were probably produced by mixing lead-silica and plant ash glasses. The high levels of calcium and phosphorus in the third type of lead-rich glass appear to be the result of adding bone ash or possibly apatitic sand.

10.5 Conclusions

Islamic glass can be characterised in a number of ways scientifically. The emergence of a true cultural identity is partly reflected in the appearance of new vessel forms and decoration in the early nineth century, both of which involved experimentation. Mass production of glass occurred in urban contexts; both specialised and unspecialised production of vessels for diplomatic, religious, alchemical and functional purposes occurred.

Islamic glass artisans were influenced by the Sasanids who, until the sixth century, occupied a large area, including Iran. Decorative techniques such as facet cutting were adopted by Islamic glassworkers. The other main influence was a classical Roman/Byzantine one, as revealed by the beautiful polychrome *millefiori* glass plaque found at Samarra (see Fig. 9.6 and book cover). Over and above these early influences, Islamic glassmakers and workers innovated with lustre-ware glass and glazes, other forms of cut glass technology and plant ash glass technology. They invented the use of tin as an opacifier in glazes and produced the first lead-silica glass in the 'West'.

It is worth considering the extent to which the 'new' glass technology, that of plant ash glass, formed part of the growing awareness of what it meant to be a Muslim as opposed to a member of 'the other', whether a Christian or a Jew. In the late eighth and ninth centuries, the 'Abbāsid caliphs, especially Hārūn al-Rashīd and his son al Ma'mun, encouraged the translation movement based in Baghdad (Gutas 1998, 7) and the development of what would be called today 'the sciences'. So it is no coincidence that the technological innovations in ninth-century glass formed part of this wave of interest in the sciences. It could be that the caliph and his retinue regarded such developments in technology as representing the Islamic state and its people. However, this might be overstating the case, partly because such activities formed part of everyday life and generally would normally not impinge significantly on courtly and religious life. The exceptions might be the manufacture of the occasional vessels regarded as prestigious and the specially coloured window glass and wall/floor glass used to decorate palaces.

Plant ash glass had been made from the same basic raw materials by both Sasanians and Islamic glassmakers. Indeed, because the Sasanians inhabited 'inland' locations they naturally chose the locally occurring raw materials, plant ash and sand. In contrast, their contemporaries, the Romans followed by the Byzantines, largely used coastal sand and natron instead. The main difference between

Sasanian and Islamic (plant ash) glassmaking was the scale: the Islamic glass industry occurred on a far larger scale. The other main influence, especially in the early stages of Islamic glass technology under the Umayyads, was the Byzantines. Until the late eighth century, the Byzantines were regarded as the masters of technology. They were invited by the Umayyad caliph to manufacture the thousands of glass mosaic tesserae (presumably out of natron glass) for the early eighth-century mosque in Damascus, an enormous imperial building that rivalled equivalent Christian structures. The caliph even invited Byzantine artisans to insert the mosaics. However, partly under the auspices of the caliph Hārūn al-Rashīd, the change to plant ash glass became synonymous with Islamic glass technology. It is no coincidence that its development occurred at a time that translations of Greek and Syriac 'scientific' texts into Arabic occurred, providing a stimulus for experimentation.

Such experimentation can be investigated by using chemical and isotopic techniques. It becomes possible to identify characteristic strands that make up the development of Islamic glass technology. One is the determination of its provenance, including raw furnace glasses, and eventually glass vessels. Tight groupings formed from chemical compositions of glass vessels with diagnostic types of decoration, such as scratch-decorated, or from diagnostic vessel forms, such as mosque lamps, when compared with trace element and isotope results and the scientific analyses of raw factory glasses can provide evidence for production centres or zones. This is especially surprising given the potentially very wide range of plant ash compositions that could have been used (Barkoudah and Henderson 2006) and adds weight to the suggestion that some compositional groups are the result of single melts. An alternative interpretation is that the glassmakers controlled their batch compositions as tightly as possible by always using sand from the same location and ashed plants of the same species, growing in the same location and harvested at the same time of the year.

Whereas chemical compositions provide a useful means of suggesting the use of particular raw material types, the use of isotopes can not only bolster these suggestions but can also provide a refined geological provenance for the glasses when set in an environmental context. Although the two approaches are interdependent, clear patterns of provenance for large numbers of samples should eventually revolutionise the study of Islamic glass. Isotopic determination of carefully chosen glass samples from centres for which we have historical references to manufacture, but no current archaeological evidence, such as Aleppo, Damascus and Cairo, could provide the first way of proving scientifically what kind of glass was made there. Moreover, it will be possible to prove or disprove the current theories about the production of diagnostic vessels types such as the proposed production of colourless cut vessels in Nishapur in Iran and/or Fustat in Egypt and the range of Ayyūbid and Mamluk enamelled wares.

In sum, in the complex and sophisticated world of the Islamic caliphate, glass was certainly produced in urban and semiurban environments. The wide range of glass chemical compositions across the Islamic world provides great potential for establishing glass production zones. The use of chemical analysis, combined with stable and radiogenic isotope determinations in an environmental context, has the potential to provide a provenance for glasses even in the absence of primary production sites.

ELEVEN

THE PROVENANCE OF ANCIENT GLASS

To be able to determine the movements and provenance of people (Price and Gestsdóttir 2006; Montgomery *et al.* 2007) and objects is important because it can provide a crucial strand of evidence for the interpretation of a wide range of important aspects of the ancient world. The most obvious of these is being able to trace migrations and diasporas of peoples and being able to show that trade in objects of a range of social, political, ritual and economic values has occurred.

Obsidian is a natural glass that has lent itself to provenance in a way that is impossible for man-made glass. Its chemical composition unquestionably reflects the relatively unique geological context in which it was formed. Given a sufficient number of chemical analyses of obsidian sources, we can therefore be certain of its provenance (Renfrew *et al.* 1966; Tykot 2002; Rosen *et al.* 2005; Henderson 2000, 305*ff.*). Therefore, it has been possible to determine the sources of obsidian tools found on sites hundreds of kilometres away from their origins, which led to a step change in modelling exchange and trade patterns in prehistory (Renfrew 1975). The same has been possible for other geological materials used in antiquity, such as carnelian (Insoll *et al.* 2004), turquoise (Weigand *et al.* 1977), limestone (Holmes and Harbottle 1994), soapstone (Ige and Swanson 2008) and jade (Hsiao-Chun Hung *et al.* 2007). In contrast, the involvement of humans in the production of glass, the selection of a range of raw material types from a variety of origins and their subsequent purification and modifications at high temperatures has had a direct impact on its chemical composition, thereby making its provenance more of a challenge.

Nevertheless, this challenge is well worth addressing because it can lead to a range of fascinating insights into several possible aspects of the ancient society and economy. These include contributions to the study of

- specialisation in glass vessel and glass colour production in different locations;
- trade in raw glass and finished objects from different sources;
- the occurrence of provenanced glass in specific archaeological context types, such as burials, destruction contexts and occupation layers, reflecting variously ritual, political or economic factors;
- the use of local raw materials to make glass in association with their study in a broad environmental context and
- the degree of glass recycling.

If the glass comes directly from a furnace, then it may have been produced without the addition of scrap glass (cullet) from other sources. It is often assumed that such raw glass is 'pure' and that its scientific characterisation provides a provenance for it. Associated questions are whether the raw materials used carry the local geological signature into the glass and whether the isotopic ratio remains unaffected by the production process. Before we consider whether raw glass can indeed be considered 'pure' in this context, it is first necessary to consider models for glass production and then what compositional and mineralogical characteristics of raw materials are transferred directly into the glass and remain unaffected by production procedures. The use of stable and radiogenic isotopes is then discussed as a way of provenancing ancient glass.

11.1 Models for Glass Production and Glass Provenance

As noted in Chapters 4 and 6, most of the earliest glass in the west and Middle East was of a plant ash composition until about 800 B.C.; from c. 800 B.C. until c. A.D. 800 natron glasses were used; from c. A.D. 800, plant ash glasses started to be (re-)introduced in the Mediterranean and wood ash glasses north of the Alps. Models for glass production are worth considering, especially given that the determination of radiogenic isotopes is a means of suggesting where glasses were made even in the absence of industrial evidence. Indeed, the results of isotope analysis can potentially be used to test models for glass production, as discussed later. Primary production centres can be defined as locations where the glass was fused from primary raw materials; secondary production centres imported raw glass, its colour may have been modified and then beads, windows and vessels manufactured from it. Brill and Wosinski (1988, 283–4) noted that

> *Glassmaking can be divided into two operations: an engineering stage, producing the material from its ingredients [primary production], and a handcrafting stage [secondary production], fashioning the material into objects.*

The suggested model for the production of raw (soda-lime) glass based on what was regarded as being a great similarity in their chemical compositions led to the following suggestions.

Henderson (1989, 39) provided the following possible production models:

1. 'Strict adherence to the SLS (soda-lime-silica) glass recipe was achieved because a *limited number of centres* produced the glass. The glass was traded or exchanged with other centres where glass artefacts characteristic of specific regions were produced.'

or

2. 'The same SLS recipe was used due to close communication between glass artisans enabling technological knowledge to be shared over a broad geographical area, leading to widely dispersed [primary] manufacturing centres.'

In sum, model (1) consists of a few primary production centres and a much larger number of secondary production centres; model (2) consists of multiple primary production centres, interspersed with secondary production centres, increasing the likelihood that primary and secondary glass production occurred on the same site.

Freestone *et al.* (2000) applied these models to the manufacture of glass in the Levant. Lemke (1998) discussed the 'traditions' involved in (Roman) glass production and discussed (chemical) compositional variations in relation to such a tradition. Lemke took his definition of a tradition from Hobsbawm and Ranger (1994, 1):

a set of practices, normally governed by overtly or tacitly accepted rules . . . which seek to inculcate certain values and norms of behaviour by repetition, which automatically implies continuity with the past.

Lemke (*ibid.*, 279) suggests that material culture provides 'fossilized remnants of traditions'. This is true, but the statement underplays the sequence of decisions involved, which would potentially have been made by a range of different groups. Both of the discussions by Lemke and Freestone *et al.* (2000) dealt with the production of natron glass, where the sources of the alkali are constricted and in all likelihood derived from Egypt (Chapter 3). Although, as Pliny noted, if natron were traded, it would seem that the coast of the Levant was one area of primary glass production, probably from as early as the sixth century B.C. and until the ninth century A.D., by which time *plant ash glass* was made there. The focus on natron glass production lent itself to the claim that model (1) was the dominant mode of production (Freestone *et al.* 2000). However, this was partly determined by the proximity of the primary production sites to the source of beach sand and an established way of obtaining Egyptian natron.

Halophytic plants that provide a suitable alkali source occur widely, so they would have been easier to obtain than natron. It is therefore likely that there were many more primary glassmaking centres for plant ash glasses from c. A.D. 800 dispersed over a wider area. Indeed, the primary production of plant ash glass (Mirti *et al.* 2008) by the Sasanians (second–sixth centuries A.D.) may have been quite widely dispersed across their territories for the same reasons. Although the archaeological evidence for primary glass production of Bronze Age Mesopotamian glass is lacking, isotopic results (especially for strontium) are starting to indicate that the primary manufacture of glass was not just concentrated in one place (Chapter 6). This therefore increases the likelihood that primary and secondary glass production occurred on the same site and that decentralised production occurred in the Bronze Age, Sasanian and Islamic periods; it is likely that evidence for this will eventually be found. In medieval and postmedieval Europe (eleventh to seventeenth centuries), primary and secondary production of wood ash glass definitely occurred on the same sites.

11.2 Glass Provenance and the Chemical Composition of Raw Materials

The chemical compositions of glass raw materials were described in Chapters 2 and 3; the chemical compositions of ancient glasses were described in Chapter 4. Here the extent to which the compositional characteristics of glass raw materials can be used as a means of provenancing the glasses made from them is addressed.

11.2.1 Silica

The chemical composition of silica used in glassmaking mainly reflects its geological origins and formation process (Highley 1977). As noted in Chapter 3, silica sand can be accompanied by a range of mineral inclusions (e.g. feldspars, zircons, monazites, chromite and titanite) and by shell fragments. The presence of these inclusions is related to the age and formation of the sand. The chemical compositions of these inclusions will contribute to the chemical composition of the glass made from the sand. Possible silica sources include river sand, beach sand, geological deposits and crushed quartz pebbles. If a silica source has a diagnostic mineralogical and chemical composition that reflects its geological source and it can be proven to have been used in glass production, this provides locational provenance for the silica used, and it helps to chemically characterise the glass. The wide distribution of silica of a range of ages subjected to a range of formation processes provides the potential for contributing to glass provenance. However, if, for example, the silica source is quartz, as may be indicated by low levels of aluminium and neodymium, these chemical characteristics may not help with provenance: different quartz sources may have similar impurity patterns. However, such characteristics do provide a means of establishing the purity of the silica deposits used in glassmaking and the type of silica source used. Overall, differences in the mineralogical characteristics as reflected in rare earth elements such as lanthanum and neodymium in coastal and inland sand deposits, in particular, can assist in characterising it. If sand sources local to the glassmaking centre were used (e.g. in the Levant), characteristic mineral inclusions will help to confirm a provenance for the glasses made from them (see Chapter 8).

11.2.2 Natron, Trona, Nepheline, and Reh

There are a limited number of mineral evaporate sources (Chapter 2). It is generally accepted that most natron used in the manufacture of soda-lime Hellenistic and Roman glasses was derived from Wadi el Natrun (see Chapters 2, 7 and 8). In most cases, this must be true; the mineral would have been transported to glassmaking sites on the Levantine coast. However, there are other possible soda-rich evaporite sources, such as in Macedonia, Turkey and northern Syria (see Chapter 2). Therefore it is necessary to keep an open mind as to whether these other sources were exploited. If, for example, Turkish nepheline was used, although also a relatively 'pure' mineral, it may have a different impurity pattern from that introduced with natron from Wadi el Natrun. Therefore, as with sand, if a characteristic mineral alkali was used for glassmaking, some of its chemical characteristics may be carried into the glass made from it, and this will help to source the alkali used. The impurities associated with the soda allow the glass to be placed within a tradition of glassmaking (Chapter 4). Unlike sand, natron does not contain variable suits of heavy minerals. The relative proportions of different salts that evaporite minerals are formed from vary across the deposits.

However, reh, used in India as an alkali source (Chapter 2), does provide distinctive chemical characteristics, but as for natron, its relatively limited occurrence may also limit its potential as for provenancing glass using its isotopic and chemical signatures.

11.2.3 Plant Ashes

Unlike silica and natron, the use of plant ash as a source of alkali has a greater potential to provide provenance information. The various parameters that effect the chemical compositions of plant ashes are discussed at length in Chapter 2. To summarise, they are as follows:

1. soil geochemistry in which the plants grow;
2. the plant species and genera, and therefore its physiognomy (certain plant genera appear to retain constant ratios between components irrespective of the geology in which they are growing);
3. the way in which it is ashed;
4. whether plant species were mixed before ashing and
5. the season in which the plants are harvested.

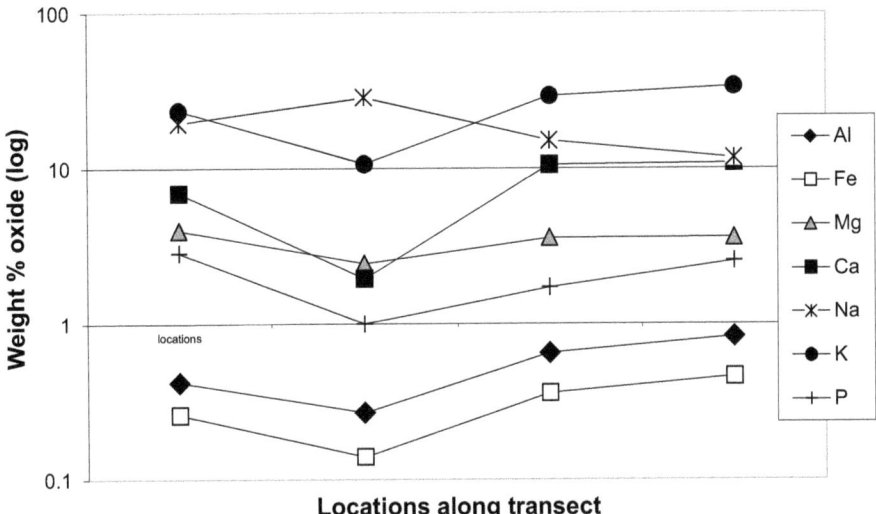

11.1. Log % oxides of aluminium (Al), iron (Fe), magnesium (Mg), calcium (Ca), sodium (Na), potassium (K) and phosphorus (P) in four ashed *Salsola* plant samples from four locations along the river Balikh, northern Syria. Data from Barkoudah and Henderson 2006.

Despite this list of parameters that can affect the chemical compositions of plant ashes and the glasses made from them, it is nevertheless possible to define differences in plant ash glass compositions, which could be helpful in provenancing glasses made from them. Moreover, had the technique of ashing or the mixing of ashed plant species caused significant variations in the chemical compositions of the glasses made from them, the glassworker would have had difficulty predicting the behaviour of the glass at high temperatures; ash mixing could make it more difficult to define composition groups of plant ash glasses made from them (c.f. Fig. 10.10). Although Rehren's (2000) model for glassmaking, in which the lowest temperature eutectic trough would produce glass of limited compositions, there is, nevertheless, evidence for distinctive compositional groups of plant ash glasses which show that some chemical characteristics of the plant ashes, partly determined by the geochemistry of the soils in which they grew, are carried through into the glass melt.

- By chemically analysing samples of halophytic plants along a transect and chemically analysing their ashes, the ways in which the plants respond to the geochemical variations in the soil in which they grow can be demonstrated. Plants of the genus *Salsola* were collected from four locations along a 29-kilometre transect of the Balikh Valley in northern Syria (Barkoudah and Henderson 2006, table 2, samples 14, 15b, 16 and 17).

The results of major and minor chemical analyis are shown in Figure 11.1 The most southerly sampling location was the eastern end of the Islamic industrial complex at al-Raqqa. One sample is *S. jordanicola*, the rest are samples of

S. sp. *(sueda?).* The variation in the levels of aluminium, iron, magnesium, calcium, potassium and phosphorus oxides across the landscape seen in Figure 11.1 follow the same pattern of relative increases and decreases in all four locations. The relative contents of sodium oxide display the opposite pattern. For the levels of alumina and iron oxide (occurring at below 1%), this positive correlation is to be expected (see Fig. 2.13). The levels of Mg, Ca, K and P, all at levels above 1% oxide, also all follow the same trends.

The same pattern of variations of trace levels of barium, copper and strontium in the same samples of *Salsola* plants can be seen in Figure 11.2. Since calcium and strontium are chemically linked, so it is surprising to see that their relative variations do not match precisely. Variations in barium and copper across the landscape are somewhat flatter and comparable to magnesium.

The reason soda levels follow the inverse pattern to the other oxides is because the total sodium and potassium oxide contents need to be retained at levels at c. 40% or below so that the osmotic pressure across plant cell walls, controlled by the balance of alkali ions, can function effectively. It appears that plants belonging to the genus *Salsola* are unable to tolerate total alkali levels above 40%, so the ratio is obviously important. The other plant genus that was sampled at two locations along the same transect (at al-Raqqa and at 25 kilometres to the north) was *Girgensonia* sp., a potassium-rich plant. The pattern of relative change for soda, alumina, iron oxide and calcium oxide, is the inverse of that found for *Salsola* ashes. However, for magnesia, phosphorus pentoxide and potassium oxide, it is the same. Thus, the plant physiognomy of *Girgensonia* apparently dictates a somewhat different balance of components. As noted in Chapter 2, there are clear compositional distinctions between plants of the genera *Salsola* and *Girgensonia*.

Overall, results from this set of ash analyses from halophytic plants taken along a transect show that variations in the underlying geology produce variations in the chemical compositions of ashed samples of *Salsola* sp. *(sueda?)* and *S. jordanicola*. Furthermore, the plant physiognomy can be shown to dictate the relative levels of sodium and potassium. Therefore, at the stage of ashing, it can be assumed that the chemical composition of plants of the genus *Salsola* growing in the same location should be fairly predictable (Barkoudah and Henderson 2006). The levels of components unaffected by the glass production process should therefore offer a means of provenancing the glass made from the ash, assuming that the plants used grew close to the production centre. Relative levels of calcium, strontium and phosphorus in plant ashes appear to be most sensitive indicator of variations in the underlying geology, showing greatest relative variations in Figures 11.1 and 11.2. Although the parameters listed earlier, which could affect the ash compositions, need to be considered, this result is a starting point that when combined with the experimental synthesis of glass in the long term could contribute to the provenance of plant ash glasses.

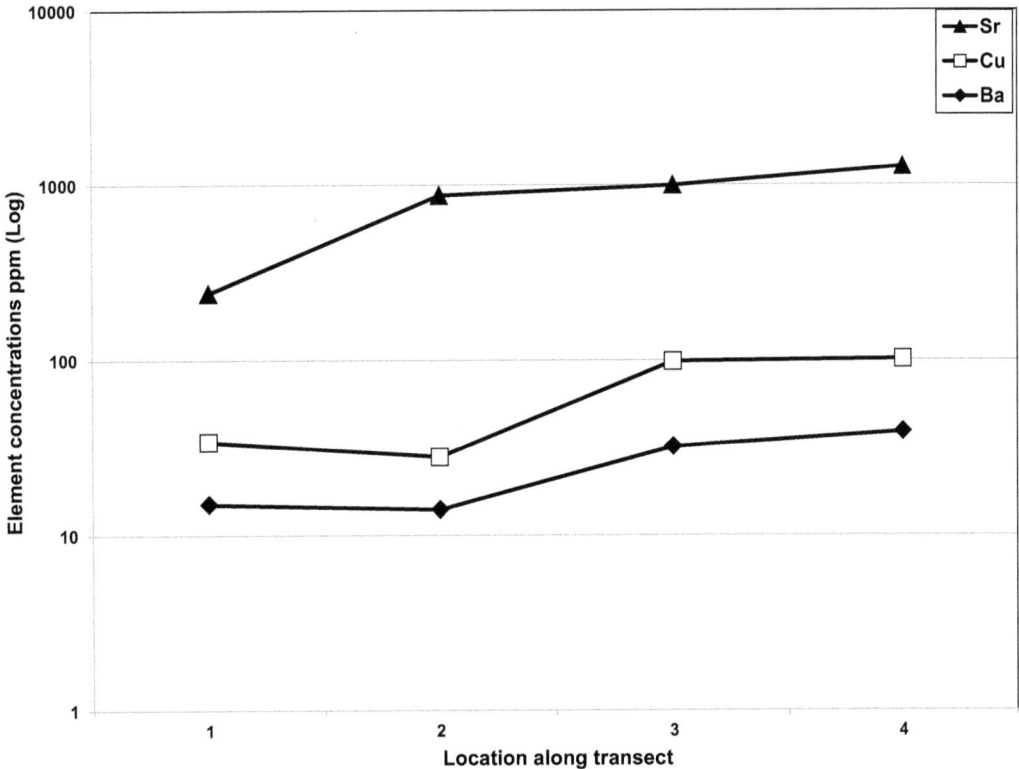

11.2. Log % oxides of barium (Ba), copper (Cu) and strontium (Sr) in four ashed *Salsola* plant samples from four locations along the river Balikh, northern Syria. Data from Barkoudah and Henderson 2006.

For glass provenance, an environmental approach is important. Halophytic plants rich in sodium carbonate growing close to the production site may have been used. To test this, the strontium isotope signatures of such plants need to be defined. Large amounts of ash would have been required for a tank furnace batch, and if some were imported from a geologically contrasting area this would be apparent in the resultant strontium isotope signature. An assumption is that if the halophytic plant genera that still grow in that environment have not changed significantly over time. For the plants in al-Raqqa this assumption appears to be valid. Environmental coring has revealed that halophytes were the dominant plant species in the primary period of glass manufacture there in the ninth century (O'Hara *et al.*, in preparation). Given that halophytes grow in semi-desert environments and on the fringes of deserts, it can be stated that today's environment at al-Raqqa is similar to that of 1,200 years ago.

To test whether the halophytic plants growing around a glassmaking site were used to make the glass there using chemical analysis, the plants should be harvested intensively in the late summer, when they reach their greatest biomass – the time of year when glass was probably made. Harvesting of plants should occur along a number of transects across the zone so as to highlight, in a sensitive way, any variations in the soil geochemistry that could affect the plant chemical compositions. All species of halophytic plants present (see Chapter 1) should be harvested to determine, in detail, whether the physiognomies of particular species have an influence on the chemical composition of the ashes and glasses. If an additional source of calcium were added to the glass batch in the furnace, this would obviously change the elemental ratios. The elemental ratios of potassium: magnesium, calcium:magnesium, and alumina:iron for each species from different locations in the zone sampled should be compared with the glass fused on the glassmaking site. This will make it possible to test whether there is a direct relationship between the ash composition and the glass made from it. At the same time, chemical analyses of the glass will reveal whether plants of contrasting genera, with very different elemental ratios in their tissues, were used to make them. In the past, it has been suggested that the use of different plant species produced correspondingly different compositional groups of glass (Lilyquist *et al.* 1993; Rehren 2001; Šmit *et al.* 2004). However, as demonstrated earlier, variations in geology have a significant effect on different samples of plants belonging to the genus, *Salsola*, and so this interpretation needs to be reconsidered carefully. To place the findings of this exercise in a broad context, the same experiments should be repeated for a second plant ash glassmaking site located on a contrasting geology. The results from the second site will provide a direct comparison and help to verify the methodology. Where possible, the same species of halophytic plants should be harvested on and around the two glassmaking sites.

Therefore, in the long run, chemical analyses of both plant ashes and plant ash glasses from production sites could provide compositional fingerprints that constitute a kind of geological provenance. By themselves, separate compositional groups of glasses can provide us with a general idea about where they were made. The compositional characteristics of plants are partly determined by the geological characteristics of the soil in which they grow. If these characteristics can be identified in the glass, there may eventually be the prospect of claiming a geological provenance for the glass this way. As with isotopic studies, discussed later, the success of such a provenance study relies on the existence of production sites being located in areas of contrasting geology, and this may not always be possible. However, when the chemical characteristics of the plant ashes are not sufficiently diagnostic to the production site, those for the silica used may be.

11.3 Glass Provenance and Batch Formation

The extent to which the original chemical compositions of glass raw materials can be modified during glassmaking must also be considered. Some of the procedures that can affect the chemical compositions of raw materials (and therefore the glasses made from them) are listed in Section 11.4.

11.3.1 Alkali Sources

Natron glass is manufactured from materials that are relatively pure, and its chemical composition is more likely to reflect directly the chemical composition of the glass raw materials and, more completely, the glass *melt*. Nevertheless, the conditions of glass melting and the possibility of glass mixing should be considered (see Chapter 3). In contrast, as already discussed, there are more possible sources of compositional variation that can affect the manufacture of plant ash glasses. A discussion of the factors that can affect plant ash compositions was summarised earlier in the chapter and presented more fully in Chapter 2.

Turner (1956) noted that the only components readily reactive with the silica are carbonates, bicarbonates, sulphites and sulphides. Sodium chloride and potassium chloride are so stable that they melt and volatilize without decomposition; sulphates react very slowly unless they are reduced by carbonaceous or other agents in the melt. Smedley *et al.* (1998, table III) provide a useful list of the melting points of individual oxides often found in glass batches. Bateson and Turner (1939) showed that the solubility limit of sodium chloride in a soda-lime-silica glass is 1.4%. Therefore if a plant ash contains a mixture of alkali salts, including sodium carbonate, sodium chloride, sodium sulphate, potassium carbonate and potassium chloride, the glass made from it will not contain the same proportion of the sodium and potassium oxides as in the ashes (Barkoudah and Henderson 2006). Nor can it be claimed, as Tite *et al.* (2006) have done, that because the plant ash has a mixed alkali composition, a mixed-alkali glass will necessarily result if used as a glass flux. The reasons for this are (1) potassium and sodium chlorides are minimally reactive; (2) it is impossible to establish how much chlorine in the melt will attach itself to sodium and potassium; (3) high levels of chlorine in plant ashes reduce the level of potassium (Tanimoto 2007). Admittedly, relatively low levels of potassium in ashes cannot possibly produce a glass with more than a low level even once unreactive KCl is accounted for. Theoretically it should be possible to establish the proportion of each compound in a plant ash by using X-ray diffraction spectrometry. However, as noted in Chapter 2, the complex X-ray diffraction patterns produced by plant ashes are of limited assistance partly because ionic substitutions can occur when the ashes are heated.

Moreover, if a plant ash is purified by dissolving it in water, the levels of insoluble components, calcium carbonate, calcium phosphate, magnesium carbonate and carbon, will be reduced. The levels of soluble components, sodium carbonate, sodium hydroxide, sodium chloride, potassium chloride, potassium sulphate, silicates and phosphates will be retained if the liquid is recrystallised. Purification would have involved solution, filtration and evaporation of the ashes. Evidence from glass chemical analyses suggests that ash purification occurred when Venetian *cristallo* glasses were manufactured, producing both low calcium and magnesium oxides and resulting in crizzling (Frank 1982, 76; Newton and Davison 1989, 141; Freestone 2001, 620; Henderson 2006, 306–7). However, chemical analyses of Bronze Age and Islamic plant ash glasses show that it is very unlikely that plant ash purification was carried out when they were made, whether the glass was made in crucibles or in tank furnaces; low calcium levels are generally not associated with low magnesium levels in these glasses. Although the leaching of woody ashes is possible during the manufacture of mixed-alkali glasses (Brill 1992), there are reasons for rejecting this (Chapter 6).

11.3.2 Ashing and Fritting

Another way that plant ash compositions can be modified is the maximum temperature at which they were ashed. Research by Misra *et al.* (1993) is discussed in detail in Chapter 4. They have shown that the levels of soda are reduced by increasing temperatures, and above 800°C, potassium levels are reduced in glass melts where wood ash is used as a source of alkali. This has implications for melting of all types of ancient glasses, although Misra *et al.* (1993) found that the extent of alkali reduction (caused by heating the ash) was reduced by fritting. The process of fritting appears to be described in Bronze Age cuneiform texts (Oppenheim *et al.* 1970., 19, 75) to produce semi-fused (sintered) materials called *tuzkû* and *zukû*, so it appears to have been carried out from a very early period of glass production.

Fritting is a solid-state reaction. It allows gases, such as carbon monoxide, carbon dioxide and sulphur dioxide produced from the breakdown of carbonates and sulphates, to be driven off, and it thereby reduces the proportion of bubbles in the glasses produced. It can also help to reduce the overall melting temperature of the batch. Without carrying out a series of experiments to test the effect of fritting on the relative levels of components in plant ashes and in the glass made from them, it is difficult to estimate its effect. However, it is likely that the main difference is that a (soda-lime) glass batch heated to its melting temperature of c. 1150–1250°C. will contain bubbles of CO, CO_2 and SO_3; these should be absent; if fritting has occurred, their number should be reduced. Also, if no fritting was carried out, a maximum loss of ash components (such as the alkalies) would occur as the temperatures rose above 1000°C in the primary glassmaking furnace. The relic stones (incompletely melted glass raw materials) that can be excluded from glass melts and could form part of the scum that formed on the surface can provide useful clues about the raw materials used (Basso *et al.* 2008).

11.3.3 Bronze Age Glass

Rehren and coworkers have focused on glasses made in crucibles, especially in the context of thirteenth-century Bronze Age Egypt at the Ramaside site of Qantir (Rehren 1997; Rehren and Pusch 2005; Pusch and Rehren 2007). The conclusions they have reached therefore relate to the small-scale production of plant ash glass in crucibles. One significant aspect of the research is the identification and study of a parting layer on the inner face of a crucible and whether it modifies the chemical composition of the glass made in the crucible. It is claimed that the interaction of the parting layer with the glass can have an overriding effect on primary elemental signatures in the glass (Rehren 2008, 1348). However, evidence from isotopic analysis appears to show that this is not (always) the case (see Chapter 6).

From the point of view of provenance, Rehren (2008) hypothesised that the chemical differences observed amongst Bronze Age glasses found in Egypt, Mesopotamia and Greece (Nikita and Henderson 2006; Shortland and Eremin 2006) relate more to different manufacturing techniques than to geological signatures. However, Rehren (2008, 1349) suggests that a lack of correlation between sodium and magnesium and between sodium and calcium is proof that not all magnesium and calcium was introduced in the plant ash. This does not take into account the fact that the balance of sodium and potassium in plant ashes is controlled by the *plant physiognomy*, whereas the magnesium and calcium levels in plants are controlled (largely) by geology (see Chapter 2). Indeed, the balance of alkalis follows an opposite trend to that for magnesium and calcium in plants (see Section 11.3). Furthermore, the presence of unreactive alkali salts will change the proportions of alkalis that end up in the glass. Therefore, it is not surprising that sodium and magnesium and sodium and calcium are not correlated.

Another feature of the glass production process that can effect the composition of the glass is the temperature of the melt. It has been noted that the percentage of calcium oxide can increase as the temperature of the melt increases (Shahid and Glasser 1972; Shugar and Rehren 2002) with a rise of 4% when the temperature of the melt rises from 900° to 1000°C. If there is a parting layer on the inside of the crucible, as the temperature rises it is increasingly absorbed into the melt, and this can introduce a proportion of lime. Furthermore, it has been noted that in the context of crucible melting that magnesium can be precipitated as diopside crystals once the magnesia level in the glass melt exceeds 4%, and this process therefore reduces the concentration of magnesia in the glass melt (Rehren 1997).

Some of these factors could be significant in the context of the relatively small-scale manufacture of plant ash glasses in Bronze Age crucibles. One of the key factors which could determine how significant they are is the size of crucibles involved. When much larger crucibles are used to produce plant ash glass, such as in sixteenth- and seventeenth-century A.D. Europe, there would be a far larger volume of glass relative to the surface area of the inner face of the crucible. If calcium enrichment occurred, it would only minimally modify the gross chemical composition of the glass. A similarly minimal modification at the edges of glass melts would have occurred in tank furnaces. A 'reduction' of magnesia levels could still have occurred, but again only in a minimal way at the interface between the glass and the furnace wall. Indeed, the layer of glass attached to tank furnace floors, where diopside and wollastonite crystals formed (McLoughlin 2003, 165–73), is often discarded. The presence and growth rates of wollastonite crystals in the five ninth-century tank furnace fragments from al-Raqqa, Syria Dr. McLoughlin tested suggest that the heating cycle of the furnaces lasted for days. So do these factors mean that it is impossible to provenance Bronze Age glasses using their chemical compositions? Although an increase in the chlorine levels in plant ashes leads to a decrease in the potassium levels (Tanimoto 2007), the compositional differences in late Bronze Age glasses do nevertheless apparently reflect regional differences in the technology employed (see Chapter 6). However, because the balance of potassium and sodium in plants is controlled by the plant physiognomy, variations in the geochemistry in the soil would not be reflected in their levels in plant ash compositions/glasses. One explanation for the regional differences in potassium levels summarised broadly in Figures 6.2 and 10.13 would therefore seem to be a combination of the plant physiognomy and manufacturing techniques, rather than the use of different plant species or differing salinity in the soils. Halophytes are alkali tolerant up to a maximum level, and all grow in highly alkaline environments, but variations in soil salinity are unlikely to have an impact on the levels of alkali in the plants.

Unlike the impurities associated with plant ashes, the impurities associated with silica are unlikely to be affected to the same extent by raw material preparation and the other production processes. For example, in Bronze Age glasses the lowest alumina levels are found in non-blue Mesopotamian glasses, reflecting the use of a relatively pure silica source to make them, and the highest in non-blue Mycenaean glasses, reflecting the use of a less pure sand (Nikita and Henderson 2006), and this is borne out by trace levels of neodymium (see Fig. 6.9). As noted in Chapter 3, there are other elements that can be associated with silica and that are nonvolatile, such as chromium, lanthanum, titanium and zirconium. Therefore these elements could potentially be a means of characterising the silica source (e.g. Fig. 6.10). Nevertheless, Pusch and Rehren (2007) have shown that impurities introduced by stone tools used to break up and crush quartz pebbles of contrasting lithologies can introduce impurities in the quartz dust. They have demonstrated that the impurities such as lanthanum and chromium associated with the tools can be a reflection of the tool lithologies rather than the geological source of the quartz used to make the glass. However, even here, an assumption that chromium is introduced only with silica appears to be an oversimplification in that a contribution would also have been made by the plant ash or dust attached to the plant (Henderson *et al.* 2010, 12, fig. 4). Elements such as neodymium in quartz are more likely to reflect the use of differing silica sources. Indeed, if neodymium impurities were introduced from both the quartz and the metamorphic or igneous rock tools, a way of showing this would be by determining the neodymium isotope ratio in the glass. If mixing had occurred, an average of the two isotope values would result that is likely to be distinguishable from a quartz isotopic value. The geological age of the silica in the crushing tool and the in quartz pebbles is likely to be different. However, there is no apparent evidence for this from available isotopic data (see Fig. 6.8).

11.4 Glass Provenance and Its Chemical Compositions

11.4.1 Technological and Cultural Factors

The chemical composition of an ancient glass is determined by a number of decisions and actions on the part of its manufacturers. These include the choice of raw materials, the ways they may have been purified and the fusion processes used. The chemical composition is the sum of all of the chemical characteristics of the raw materials used to make glass, including any scrap glass (cullet) and colouring materials added. The following is a list of some of the features of the production process, divided according to whether they are technologically or culturally determined, that contribute to glass chemical compositions:

1. *Technological characteristics* and the chemical compositions of the glass batch/melt

The chemical composition of the glass is unique to the specific furnace melt and will be a reflection of both the time and place where the glass was made. The key here is to define the components in the glass using chemical, isotopic, or possibly 'microstructural' characteristics that provide a *locational* provenance. Even if cullet were introduced, it may also have been made in the same place. If glass were produced on a large scale, the cullet may have been produced in the same furnace, or in a nearby one. It would therefore probably have been made with the same silica and alkali sources, having the same chemical and isotopic signatures. It is also possible to prove when 'foreign' cullet has been mixed in isotopically. There is a chance that the location where the raw glass was formed within the tank furnace may affect its chemical composition. As noted earlier, if close to the furnace wall or floor, it could be enriched with elements such as aluminium, iron and calcium found in furnace bricks.

However, glasses enriched with excessive levels of these elements would make them unworkable. Indeed, there is ethnographic evidence from India which indicates that the lower half of a melt was sometimes recycled in subsequent melts (Sode and Kock 2001). For the production of glass in crucibles, such as occurred in thirteenth-century Egypt (Rehren 1997), the glass batch could be enriched by interaction with the buffering material used to line the crucible, as discussed earlier.

2. *Cultural factors*, the decisions made by those involved and the chemical composition of the glass batch/melt:
   a. The *traditions* adhered to by the glassmakers in different regions would ultimately have been ethnically determined. These may have included the species, genera of plants selected for ashing, perhaps determined by flower colour; the temperatures achieved during ashing; the purification of ashes; the taste of the plant ash used as a guide to its quality; the mixing of plant species/genera; the colour of the silica selected; the purification of sand; the types of fuel selected for the furnace at different stages of melting; the proportions of raw materials used; whether the raw materials were fritted and the clays used to make the crucibles.
   b. The availability of the raw materials, partly determined by supply routes and possible trade. This can change over time.
   c. A decision to exploit local raw materials, if available, or to import them.

A somewhat different series of decisions would be taken if glass were fused in a tank furnace rather than reheating it in a crucible before working it (see Chapters 5, 7 and 9). Decisions taken could have an impact on, for example, the preparation of a mould to make a moulded mosaic glass vessel, the manufacture of a decorated wound glass bead or a blown glass vessel. The reheating of glass before one of these activities occurs is a time when both 'foreign' scrap glass and colourant materials may have been added to the melt.

It can be seen that technological and cultural factors inevitably intermesh and overlap, creating an interesting challenge when investigating the provenance of ancient glass.

11.4.2 The Chemical Compositions of Plant Ashes and Plant Ash Glasses

By comparing the chemical compositions of northern Syrian plant ashes with plant ash glasses found there, it should be possible to reveal (1) when the compositions of the plant ashes and the glasses match, potentially providing provenance information and (2) whether there were similar production practices in the use of plant ashes for Bronze Age and Islamic period glass.

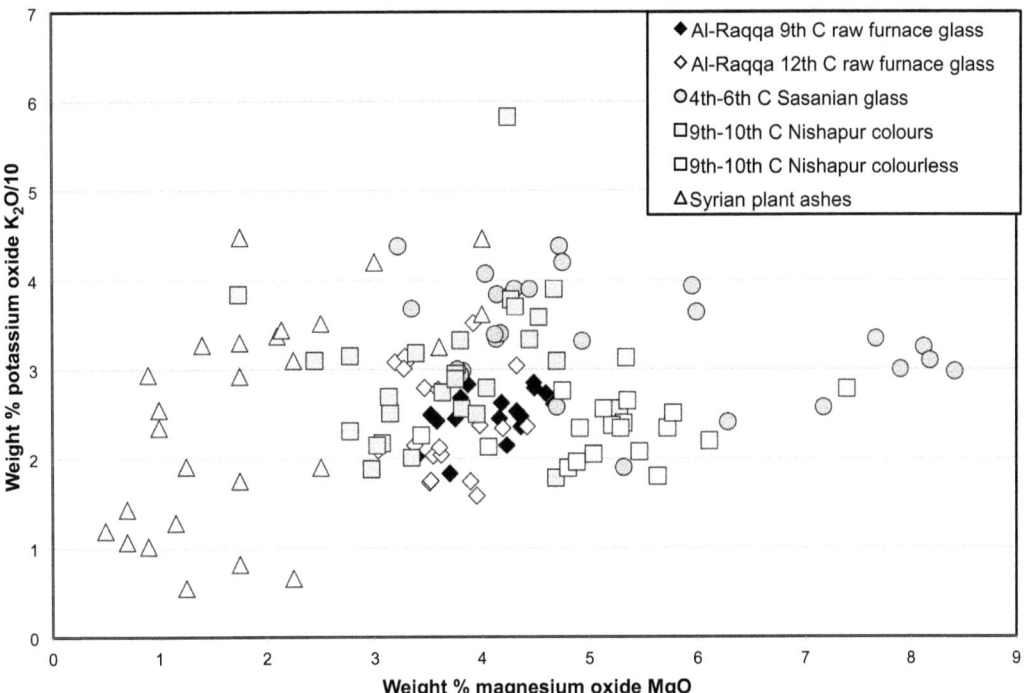

11.3. Weight % magnesium versus potassium oxide (divided by 10) in raw furnace glass from al-Raqqa (ninth century, TZ; and twelfth century, TB), Sasanian glass and Syrian plant ashes. Data from Henderson *et al.* 2004, Meek 2005, Mirti *et al.* 2008 and Barkoudah and Henderson 2006.

It is well known that high magnesia levels characterise Bronze Age, much Islamic, Venetian and northern European sixteenth- and seventeenth-century glasses (see Chapter 4). Furthermore potassium oxide and magnesia levels can be positively correlated in such glasses. The perfectly valid assumption is that high magnesia glasses were made from plant ashes (Sayre and Smith 1967). Therefore, if this is the case, and plants of the genus *Salsola* were used on a regular basis (see Fig. 2.11) and the ratio were not changed by glass production, we might expect there to be a positive correlation between potassium and magnesium oxides in both the glasses and *Salsola* ashes. In Figure 11.3, the results for sodium-rich plant ashes fall onto a general positive correlation line. However, while there is a general positive correlation for glasses that contain between c. 3% and 4% magnesium oxide, this is lost for glasses that contain higher levels than this. The plotted results for potassium oxide in plant ashes are divided by 10. It appears that if indeed plants of the genus *Salsola* were used to make these glasses, various factors would lead to a reduction in potassium levels to fit the concentrations found in glasses (in addition to the diluting effect of silica). These include the combination of potassium with chorine to form KCl (Rehren 2008)

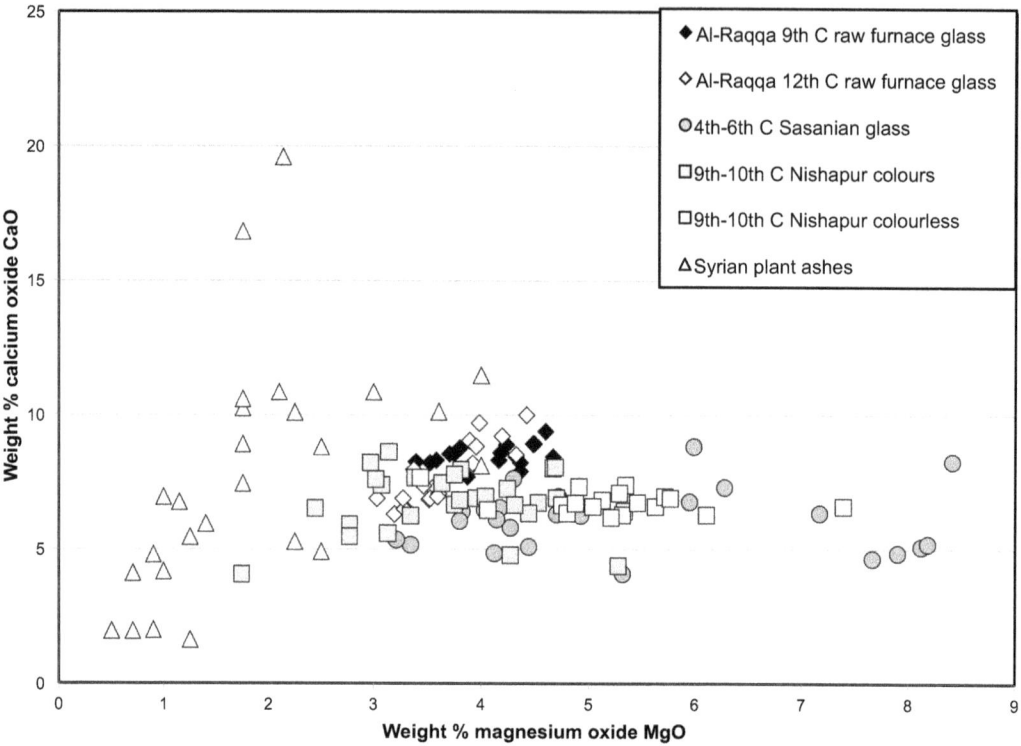

11.4. Weight % magnesium versus calcium oxide in raw furnace glass from al-Raqqa (ninth century, TZ; twelfth century, TB), Sasanian glass and Syrian plant ashes. Data from Henderson et al. 2004, Meek 2005, Mirti et al. 2008 and Barkoudah and Henderson 2006.

and the potential loss of potassium from the ash at high temperatures (Misra et al. 1993). The plant physiognomy partly defines the level of potassium in the ash (see Fig. 11.1).

The relative magnesium and calcium oxide levels in ninth-century al-Raqqa raw furnace glasses (see Fig. 10.10) are positively correlated with a similar ratio to the plant ashes but if these plant ashes were used, an additional magnesium component has been added. Levels in the Sasanian glasses are not correlated in Fig 11.4. The dilution of the oxide levels in the plant ashes by combining them with silica could potentially provide a maximum of c. 4% to 5% CaO and 2% to 3% MgO in the glass produced, which is not quite high enough for many of the glasses seen here. However, the concentration of the oxides in the ashes would increase with ashing at temperatures higher than that used (600°C). Although clearly relating plant ash compositions to ancient glass compositions is complex for the reasons already mentioned, if the same ratio between two components is found in the ash and the glass, a degree of confidence that this was the genus used to make the glass is provided. The corollary is that where the ratios do not match, other explanations need to be offered to account for this.

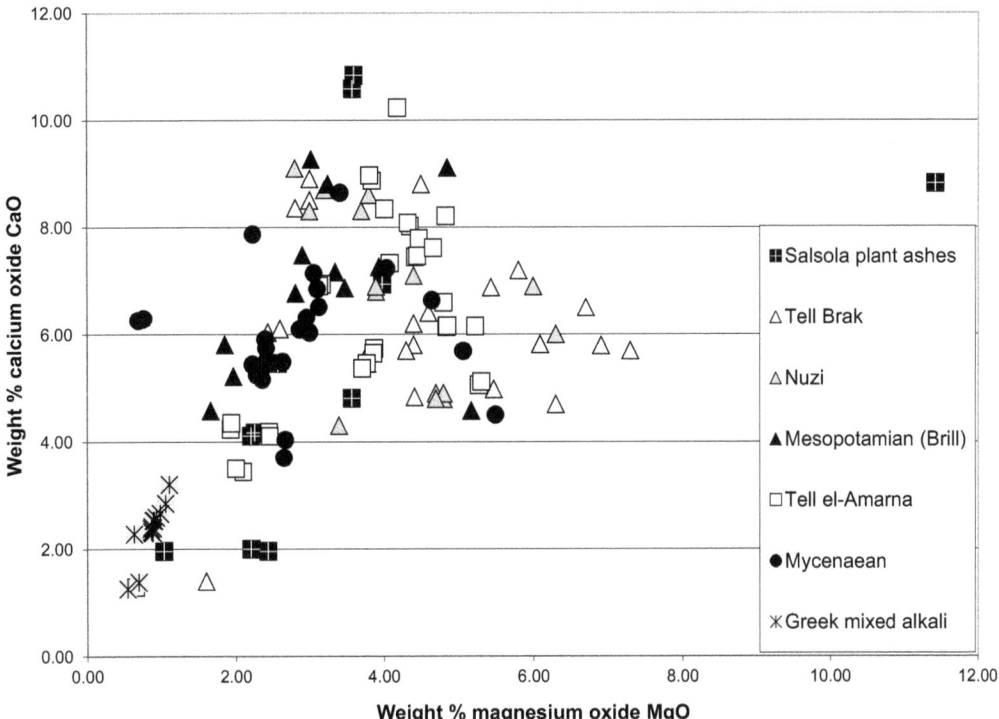

11.5. Weight % magnesium versus calcium oxide in plant ash glasses from Tell Brak, Nuzi, 'Mesopotamia', Tell el-Amarna, Greece (Mycenaean), mixed-alkali Greek glasses and plant ashes. Data from Henderson 1997, Shortland and Eremin 2006, Brill 1999b, Henderson et al. 2010, Nikita and Henderson 2006 and Barkoudah and Henderson 2006.

The data plotted in Figures 11.3 and 11.4 show that plant ashes of the genus *Salsola* sp. from the geological terrain of northern Syria were probably not used to make Sasanian glass. This is not really a surprise, but it does at least show that there is potential in the approach, especially when considering areas of geological contrast.

A plot of magnesium against calcium oxides in selected late Bronze Age glasses, and in *Salsola* plant ashes from northern Syria (Fig. 11.5), shows that Egyptian, many Mycenaean and Mesopotamian glasses fall onto the *Salsola* correlation. Many of the results for Mycenaean and Mesopotamian glass published by Brill (1999b) forming an especially tight correlation. However, many of the glasses from Tell Brak in northern Syria, many of the glasses from Nuzi and a smaller number from Amarna contain higher magnesium oxide levels than detected in any of the plant ashes. As with the Islamic glasses, this again suggests that an additional source of magnesia has been added to these glasses. This probably derived from plants growing in a different geological environment (such as in Iraq) or the use of a different plant genus. Indeed by comparing Figures 11.4 and 11.5 for Islamic, Sasanian and Bronze Age glasses, we find that the higher magnesia glasses with plotted points falling away from the correlation line occur in both. The reasons for this evidently need to be investigated more closely, but because Sasanian glasses fall into this area, and most Bronze Age glasses from Nuzi in northern Iraq do as well, it could indicate that all the samples with elevated magnesia were made in an area to the east of northern Syria, stretching into Iran. The plotted results for mixed-alkali glasses in Figure 11.5 indicate that a very different calcium source was used to make them, probably of a mineralogical origin (see Chapter 6).

These compositional relationships suggest that *Salsola* was likely to have been one of the genera of plants used to make some Bronze Age and Islamic glasses. Data for plant ashes plotted in Figures 2.9 and 2.10 show that ashes of *Salsola* and non-*Salsola* species can be distinguished. Therefore when data for glasses fall away from the correlation, one explanation is that non-*Salsola* species have been used to make the glass; a second is that an additional source of the elements involved has been added. The question that remains is what causes the association of magnesium, potassium and calcium in plant ashes of the *Salsola* genus. One explanation is that feldspars, other minerals and ions rich in calcium and potassium in the soil are drawn up into the plant by transpiration, but there are many complex factors involved in determining the concentrations of chemical elements in plants and plant parts that are not understood completely (Watanabe *et al.* 2007).

The relative levels of alumina and iron oxide in *Salsola* and non-*Salsola* plant ashes that derived from northern Syria, in late Bronze Age glasses from Tell Brak, Nuzi, Tell el-Amarna, and in mixed-alkali glasses are given in Figure 11.6. There is an almost exact match between the correlations for ashes and the glasses from Tell Brak, Tell el-Amarna, Nuzi and three samples of Mesopotamian glasses published by Brill (1999b). This suggests that we are seeing a common geochemical correlation in both Egyptian and Mesopotamian glasses. This would be present in both the silica and the plant ashes used and that glass production processes have not disrupted the correlation. Nevertheless, some non-blue glasses from Nuzi and Tell Brak and several glasses from Mesopotamia published by Brill (1999b) do not fit the correlation. The cobalt alum used to colour blue Mycenaean and Amarna glasses has increased the alumina level in the glasses, shifting them to the right. The two groups of Greek glasses labelled 'Mycenaean' in Figure 11.6 are all cobalt blue. The group with the highest iron oxide and alumina levels are mixed alkali tenth-century glasses; the group with lower levels are plant ash glasses. However, for the alumina levels to increase in non-blue glasses that fall away from the correlation line, an additional factor such as interaction with the crucible lining, or an elevated level of iron (in sand, for example) must be involved.

In contrast, a bi-plot of relative levels of iron and aluminium oxides in Syrian plant ashes and raw plant ash glasses from three production sites in al-Raqqa,

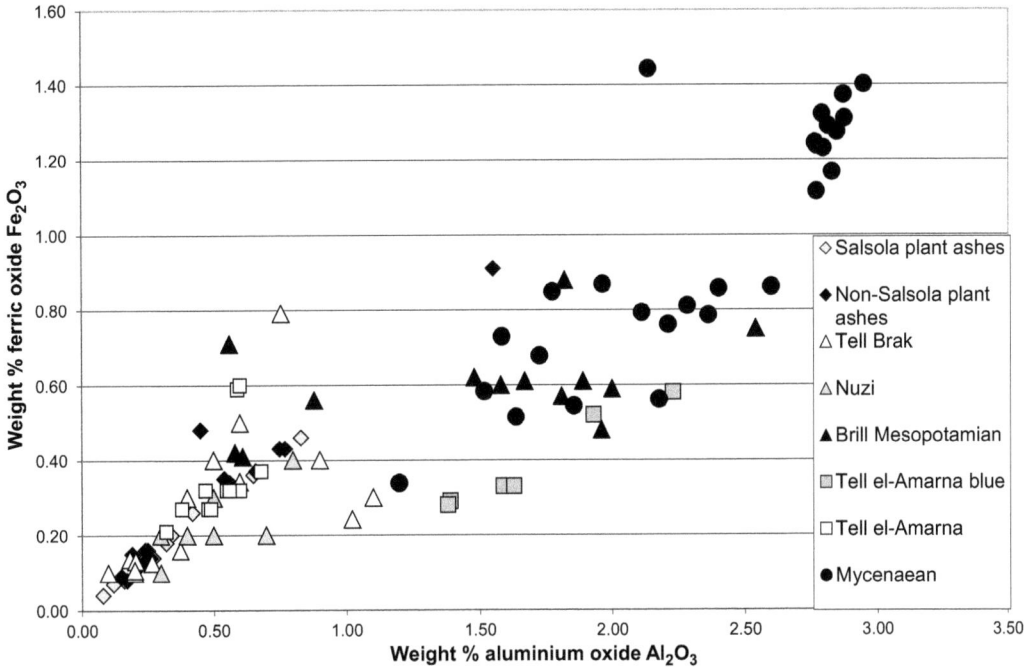

11.6. Weight % alumina versus iron oxide in plant ash glasses from Tell Brak, Nuzi, 'Mesopotamia', Tell el-Amarna, Greece (Mycenaean), mixed-alkali Greek glasses and ashes of *Salsola* and non-*Salsola* plants. Data from Henderson 1997, Shortland and Eremin 2006, Brill 1999b, Henderson *et al.* 2010, Nikita and Henderson 2006 and Barkoudah and Henderson 2006.

northern Syria: ninth-century Tell Zujaj (TZ), eleventh-century Tell Fukhkhar (TF) and twelfth-century Tell Bellor (TB) (Fig. 11.7) show that extra alumina occurs in some even though the iron levels are similar to those in plants for some. Even here the same correlation line is retained; the higher iron and aluminium oxide levels presumably derive from the use of an impure source of silica, containing feldspars.

11.4.3 Summary

1. When the correlations between the ratios of element oxides found in plant ashes and glasses match, it suggests that the ratios in the ashes have been retained during the glass production process.
2. If an additional calcium-rich raw material has been added, it would change the calcium-to-magnesium oxide ratios. For example, for Bronze Age glass, calcium enrichment could occur by interaction with the crucible wall.
3. In some cases, although there may be insufficient levels of, for example, calcium in the plant ashes, if the same ratio is found in the glasses and ashes, it suggests that plants of the *Salsola* genus were used but perhaps from a different geological context.

4. Where the oxide levels found in Bronze Age glasses fall away from the plant ash calcium oxide/magnesia correlation, this may indicate a. that the plants were growing in contrasting geological locations; b. that a different plant genus was used; or c. that a separate magnesium-rich raw material was added.
5. Why is it that the absolute levels of oxides in some plant ashes are too low to account for the levels found in Bronze Age, Sasanian and Islamic glasses, even though they fall on the same correlation line? Approximately 2:1 ash: sand by weight would have been mixed (Smedley *et al.* 1998, 150–1). One consideration is that levels of components in the plant ashes would have been diluted by fusion with the silica, another is that they would have been concentrated when ashed. It can be assumed that, for example, calcium oxide levels in plant ashes will be reduced by two-thirds by mixing with silica. If everything else is equal, a plant ash containing 12% calcium oxide will contribute 4% to the glass batch. However, this accounts for only a small proportion of both the plant ashes analysed and the glasses analysed.

This suggests either

1. that plants with far higher CaO levels were exploited from those for which results are available and that once suitable plant colonies were defined they would have been used consistently, or
2. that the ashes of similar plants tested were used and that an additional calcium-rich raw material was added. However, to fit the correlation line, the additional raw material would have had the same relative levels of calcium and magnesium oxides, and
3. that plant ashing increased the concentration of calcium significantly.

Overall, this approach provides a link between the chemical compositions of soda-rich plant ashes and the chemical compositions of Bronze Age, Sasanian and Islamic glass. The results show constrasts in the geological source of plant ashes and/or, the genera of plants used, and/or, for example the enrichment of calcium during crucible melts or the addition of a separate calcium-rich raw material. A logical next step is to chemically analyse plant ashes deriving from northern Iraq and western Iran and to combine them with experimental glass melting.

11.5 The Use of Isotopes to Provenance Glass: The Importance of an Environmental Approach

A tacit assumption is often made that the chemical compositions of raw furnace plant ash glasses made at different production sites are characteristic to those sites. Chemical compositions are a summed result of traditional approaches, the raw materials used and the ways in which the raw materials may have been prepared and purified. However, as we have seen, chemical compositions can

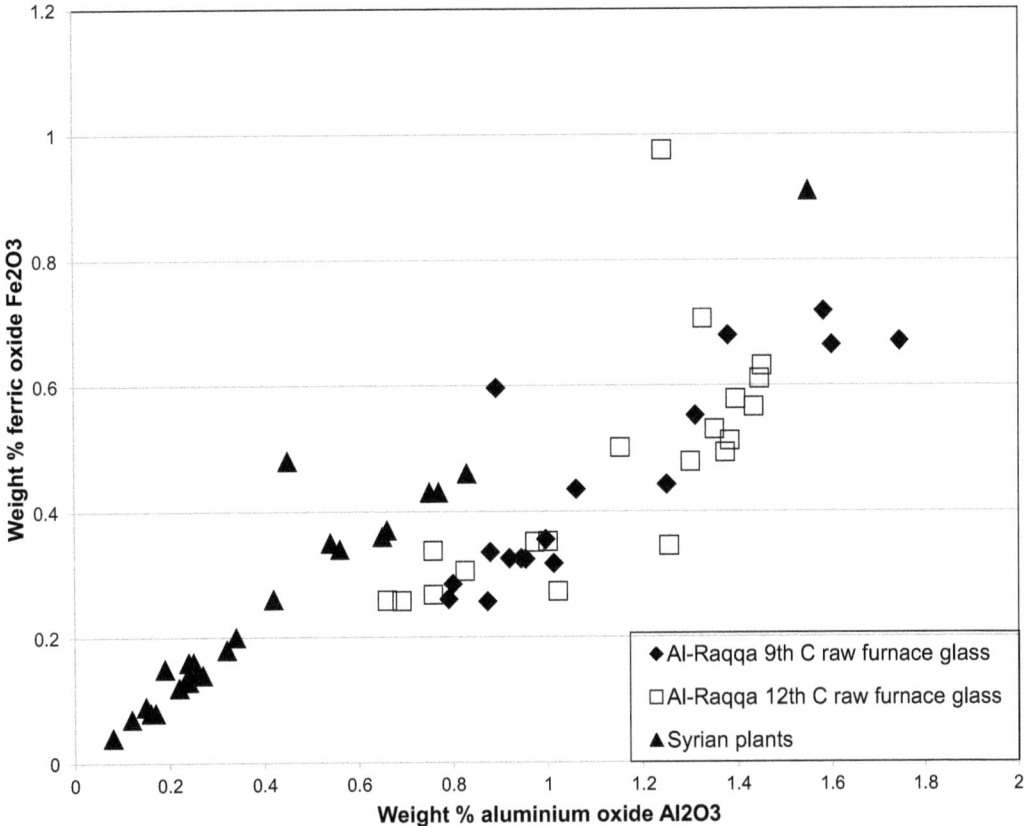

11.7. Weight % alumina versus iron oxide in raw furnace glass from al-Raqqa: ninth century, TZ; eleventh century, TF; twelfth century, TB; and in ashes of Syrian plants. Data from Henderson *et al.* 2004, Meek 2005 and Barkoudah and Henderson 2006.

provide a general *geographical provenance*. Therefore, an independent means of providing a provenance for glass would be a major step forward. As a starting point, the isotopic characteristics of furnace glasses provide the best means of constraining the possible parameters which can affect isotopic compositions of ancient glasses. This is mainly because if the production is carried out on a sufficiently large scale, the chances that 'foreign' scrap glasses were added to the batch are minimised. The addition of colourant-rich raw materials at a secondary stage could also modify the isotopic composition of the glass melt. However, the results for such raw glasses from primary glassmaking sites can be used as baseline data against which the results for highly coloured vessel glasses can be compared. The same might be argued for the results of chemical analyses. However, isotopic studies can also provide reliable *independent* evidence for mixing of glasses made using raw materials deriving from contrasting geological origins, which is more difficult to demonstrate using chemical compositions.

When set in an environmental context, the isotopic results for raw furnace glasses take on a greater significance. Variations in the isotopic characteristics of alkalis and silica sources across broad areas of the landscape surrounding production sites provide a means of demonstrating which silica sources were used and which geological zones the plants derived from (irrespective of species). Therefore, these procedures can provide the best possible means of producing a *geological provenance* for the glasses. By determining strontium isotope values for plants growing along transects, it becomes possible to show contrasts in geology. Using the variation of strontium isotopes in plants, we can define the variations across the landscape and compare the signature with that at individual glass production centres.

Given that such work is relatively new, the principles behind its use are given here, mainly in relation to plant ash glasses. (For a description of the mass spectrometry techniques used to determine isotopes in ancient glasses, see the Appendix).

11.5.1 Calcium Sources, Strontium Isotopes and Glass Provenance

One radiogenic isotope found in plants ashes and glasses is that of strontium ($^{87}Sr/^{86}Sr$). Strontium, and its isotopes, ^{88}Sr, ^{87}Sr, ^{86}Sr and ^{84}Sr, substitute for the calcium that occurs in plants as calcium carbonate. ^{87}Sr is radiogenic and is formed by beta decay of the alkali metal ^{87}Rb over time. ^{87}Rb has a half-life of 48,800,000 years. The remaining isotopes of strontium do not decay. Therefore, because ^{87}Sr begins to form from the moment that the rock is formed, a comparison between it and ^{86}Sr provides a measure of the age of the calcium. Because calcium carbonate dissolved in water containing $^{87}Sr/^{86}Sr$ is taken into the plant from the soil by transpiration and the plant is then ashed and fused with silica to make glass, $^{87}Sr/^{86}Sr$ enters the glass. The chemical and $^{87}Sr/^{86}Sr$ isotope composition of the plant ash used reflect the age and rubidium content of the bedrock geology in which the plants grew (Freestone *et al.* 2003). The isotope ratio is unaffected by exchange reactions or by kinetic effects in the production process because they have low mass differences (Faure 1986). That there is no fractionation has been shown in practice by comparing Sr isotope results for glass from primary glassmaking sites with those for ashed plants ashes deriving from near the site (Henderson *et al.* 2005; Henderson *et al.* 2009). When used in the manufacture of glass, this technique provides important information about location, which is key to an independent geological provenance for glass production.

The specific sensitivity of using $^{87}Sr/^{86}Sr$ determinations to provenance glass is therefore dependent on sufficient contrasts in the geological age of the calcium source, such as found in the Middle East (Beydoun 1977; i.e. it depends on the age and the origin of the calcium-rich rock). At a simple level, old marine limestone-baring soils, common in the Middle Eastern steppes, can be contrasted with the calcium-rich seashells in sands that have formed from 'modern' seawater.

Another source of calcium is feldspars in sands. Moreover, $^{87}Sr/^{86}Sr$ values of plants (irrespective of species) have been found to be the same as in the bedrock and surficial geology in which they grow, for a range of geological environments (Montgomery et al. 2006). However, to define the range of $^{87}Sr/^{86}Sr$ isotope values for a production zone, $^{87}Sr/^{86}Sr$ values for plants within and between major geological zones need to be determined, as noted earlier. By doing this, an index of signatures for the landscape concerned can be established (Evans et al. 2010), and a comparison with isotope signatures for raw furnace glass from primary glassmaking sites and (secondary) vessel glass can be made.

The concentrations of strontium in plant ashes and glasses made from them are independent of the strontium isotope signature. Concentrations can vary according to the nature of the bedrock and the soil geochemistry in which the plant grows as well as the physiognomy of plant genera and species, as described earlier. Both desiccation and ashing will increase strontium concentrations significantly. Once plants are ashed, strontium concentrations are unaffected by the process of glass production.

There are several possible calcium sources. The bedrock geology and its contribution to soil geochemistry is one source of calcium. The range of lime-baring materials include limestone, calcium-bearing feldspars (plagioclase feldspars, Na, Al, Si_3O_8-$CaAl_2Si_2O_8$) in soil and sand and seashell fragments in sand. Seashells are basically composed of the mineral aragonite ($CaCO_3$) with a small contribution from older limestone (Freestone et al. 2003).

When the organism grows its shell, it takes in calcium-containing seawater, which therefore has a 'modern' $^{87}Sr/^{86}Sr$ signature that is c. 0.7090. The $^{87}Sr/^{86}Sr$ signature of young volcanic rocks can be as low as 0.704 (Carter et al. 2011). Their signatures are therefore very different from that of a Holocene sea shell source of lime. Freestone et al. (2003, 27) noted that a maximum calcium oxide contribution from a pure calcium feldspar end member (anorthite, $CaAl_2Si_2O_8$) would be 1.6%. (As noted in Chapter 3, clay minerals, and mafic minerals such as amphibole and pyroxene, amongst others, can be present in sand.) Geological analysis of Israeli beach sands (Emery and Neev 1960) has shown that only 3% of feldspar is present, reducing the calcium oxide contribution to 0.6%. Therefore, a minimal component of c. 10% lime in Israeli sand could be contributed by noncarbonate minerals.

The strontium isotope signature from shell-rich sand (0.7090) is different from the $^{87}Sr/^{86}Sr$ of old marine limestone (e.g. 0.7080) found in northern Syria. Given that shell fragments appear to have provided the principal source of lime in natron glasses, the 'modern' $^{87}Sr/^{86}Sr$ signature of the shells can be expected to dominate in these glasses. However, the presence of feldspars could modify the $^{87}Sr/^{86}Sr$ signature by producing a slightly older signature overall (Leslie et al. 2006), but it is still possible to indicate that a shell source of calcium in sand has been used. Therefore, although $^{87}Sr/^{86}Sr$ signatures in natron glasses provide a means of defining the use of beach sand, we still do not know whether there are distinctive $^{87}Sr/^{86}Sr$ signatures for beach sands that derive from the Levantine coast, northern Africa (e.g. Alexandria), and the sources mentioned by Pliny in Italy (see Chapters 3 and 7; Degryse et al. 2008). A map of strontium (and neodymium) isotope variations across the Mediterranean basin (Weldeab et al. 2002) suggests that such discriminations in beach sands should be possible. Although the 'modern' $^{87}Sr/^{86}Sr$ signature in shell fragments would dominate in all three cases, it may still be possible to show that there are distinctive proportions of feldspars, other calcium-baring minerals or a contribution from volcanic minerals that could produce a diagnostic $^{87}Sr/^{86}Sr$ or neodymium isotope signature, as discussed later. Beach sands have mineralogical contributions from the rivers that flow into the Adriatic and Mediterranean (see Chapter 3), and it is these that could produce a characteristic isotope signature. The massive scale of Byzantine and early Islamic (Umayyad) natron glass production in the Levant from the sixth to the ninth centuries (Gorin-Rosen 1995; Gorin-Rosen 2000; Freestone et al. 2000; Henderson 2003) used a combination of local sands and imported (Egyptian) natron.

The geological formation ages of limestones in the Middle East can be similar. Nonetheless, this may be an advantage when $^{87}Sr/^{86}Sr$ isotopic signatures of Islamic glasses made in regions, such as Syria, Iran, Egypt (and Oman) are compared. If there are diagnostic isotopic signatures for each region, these will also provide insight into the extent of glass trade and exchange between them. For the Bronze Age, contrasting isotopic signatures between Mesopotamian and Egyptian plant ash glasses have already been demonstrated (Henderson et al. 2010).

11.5.2 Silica Sources, Neodymium and Oxygen Isotopes and Glass Provenance

A second radiogenic isotope ratio that helps to provenance ancient glasses is neodymium ($^{143}Nd/^{144}Nd$). This isotope is found in accessory phosphate and silicate minerals and is generally present in sandstones at c. 20 to 80 ppm. Insignificant amounts are taken into halophytic plants from the Middle East used for glass production (Barkoudah and Henderson 2006, table 2). This isotope can therefore be traced through the glassmaking process. The $^{143}Nd/^{144}Nd$ ratio of a rock, such as quartz, is dependent on its formation age and the concentration of the parent element, samarium, in it.

Therefore, the use of sand deposits of contrasting geological ages to make glass can be detected using $^{143}Nd/^{144}Nd$ signatures. The sand deposits used may have derived from beaches, quarries, rivers or evaporite deposits; quartz pebbles may have been collected from river beds. As noted for strontium isotopes, the neodymium isotope values of sand and quartz relate to their age and origin. As noted in Chapter 3, silica is ultimately derived from the erosion of mountains. Rivers deposit the silica in the form of sand on shorelines. For example, it is possible to distinguish isotopically between beach sand on the Lebanese coast ultimately derived from the erosion of the mountains that make up the anti-Lebanon combined with material ultimately derived from the Nile and Alexandrian beach sand deposited by the river Nile. Different trace impurity levels have been noted for Egyptian and Levantine glasses, which accord with the different silica sources used to make them (Freestone *et al.* 2000; Freestone *et al.* 2005; see Chapter 3). The majority of isotope data available for Egyptian sediments comes from studies of the Nile and its sediment load. This is dominated (95%) by basalt-derived, volcanogenic material from the Ethiopian Highlands (Grousset *et al.* 1988; Freydier *et al.* 2001) and has εNd values of c. -3.3 (Goldstein *et al.* 1984). Scrivener *et al.* (2004) cite a εNd range of -1 to -1.5 as being typical of the Nile; Tachikawa *et al.* (2004) document a value of Nd -1.9 from a leached sediment. The best current indication of neodymium sand composition from Egypt is probably the two sand samples from Wadi el Natrun (Degryse and Schneider 2008), which give values of 0.512289 and 0.512199 (εNd $= -6.8$ and -8.6 respectively), although these are unlikely to match those of Alexandrian beach sand. Nevertheless, Levantine sands have values that are too radiogenic (Degryse and Schneider 2008) to have been used for the manufacture of Egyptian Bronze Age glasses (Henderson *et al.* 2010)

Using neodymium isotope determinations, it should be possible to discriminate between sand derived from the erosion of old metamorphic rocks and sand derived from the erosion of a younger rock, such as granite, and this provides provenance information. Moreover, the concentrations of neodymium within glass can help to indicate the type of silica used, because the purer the silica, the lower the concentrations of impurities. Again the isotope ratio is unrelated to the concentration of the element detected; in practical terms, when a pure silica source has been used and the neodymium levels are low (c. 1–2 ppm) a larger glass sample is needed to determine the ^{143}Nd/^{144}Nd accurately.

The determination of ^{143}Nd/^{144}Nd in ancient glasses and in silica samples taken from a range of silica sources (an environmental approach) is therefore a means of relating the silica-rich glass to the specific age (and type) of the silica raw material used to make the glass. Where natron glass was made close to the shoreline of a sea, such as the Mediterranean, it is likely that nearby beach sand was used, and this is supported by historical and archaeological evidence (see Chapters 3 and 7). In addition, (subtle) differences in the geology of Lebanon and Israel could eventually lead to distinctions in the sources of sand used on the Levantine coast, such as between the sand found at the mouth of the river Belus and that found at Sidon (*cf.* Degryse 2008; Henderson *et al.* 2009b). In turn, these isotopic results could be linked to the silica used for the manufacture of glasses at specific primary production sites such as found in Beirut (Kowatli *et al.* 2008), Dor, Bet Eli'ezer and Bet She'an (Goren-Rosen 2000). A gradual increase in alumina levels over time has already been noted for glasses made and found on the Levantine coast (Henderson 2002, 596 and see Chapter 8). This may be due to the exploitation of sands with an increasing proportion of feldspars, possibly from different locations on the coast. The silica component of river-borne sand will have a ^{143}Nd/^{144}Nd signature, which is dominated by the age of the rocks in which the river originated. Quartz pebbles found in the Tigris and Euphrates Valleys are likely to have similar ages, both ultimately being derived from Turkish mountains. However, quartz derived from the Nile or from Italian rivers such as the Tiber or the Po is likely to have diagnostic ^{143}Nd/^{144}Nd signatures. Geological deposits of sand that are known – and possibly still quarried today – are likely to be fairly homogeneous, to have a diagnostic ^{143}Nd/^{144}Nd and would not be expected to contain shell fragments that could affect strontium isotope signatures. It is just as important to determine the variations of ^{143}Nd/^{144}Nd signatures in contrasting silica sources across the landscape as it is to define variations in plant ^{87}Sr/^{86}Sr signatures. Although, as mentioned in Chapters 3 and 8, it is possible to distinguish broadly between the kinds of silica used according to the levels of impurities thought to be associated with silica, such as Nd, Ba and Al, this does not necessarily help and does not provide direct evidence of geographic origin for the silica source. Therefore, ideally sand and quartz pebbles should be sampled from beaches, the shores of rivers, geological deposits and evaporite deposits. Their ^{143}Nd/^{144}Nd signatures provide a means of discarding some silica sources from consideration, and provide evidence for the use of others in glass production. It is somewhat simplistic to assume that a production site would necessarily have used only a silica source, although this may well have occurred.

How do ^{143}Nd/^{144}Nd signatures in silica raw materials and in glasses provide a geological provenance? If the silica used derives from a source that is local to the production centre, then indeed this does provide a geological provenance and this is the case for Levantine beach sands. It becomes slightly more complex in inland situations. Here the use of silica sources is less predictable. Nevertheless, if a geological deposit of sand were consistently supplied to a glassmaking centre, the ^{143}Nd/^{144}Nd signature provides a geographical provenance for the sand used and potentially a diagnostic fingerprint for the glass. It certainly defines the source of the silica and therefore contributes to the *chaîne opératoire*. It could also help to define a traditional and consistent use of a silica source within a particular production zone over time. Moreover, isotopic analysis can provide evidence for glass trade. The location of silica used for making glass at inland locations but not close to the glass production centre is therefore best described as secondary geological provenance. However, if the sand used is local to the production centre it provides a geological provenance for the glass. Contrasts in the ages of silica used would be determined by a regional survey; in the context of Syria the ^{143}Nd/^{144}Nd results for silica sources may in fact provide more localised locational information than for strontium isotopes, because potentially there is a greater contrast in the ages of the silica used, as discussed later. Overall, the combination of isotopic results for strontium and neodymium provides new and important insights into the types and sources of raw materials used in plant ash glass production.

A third isotope, also largely related to the silica sources used to make glass, is oxygen: ^{18}O/^{16}O. Translucent glass is composed of oxides of which silicon oxide dominates. Oxygen isotope ratios in glasses can therefore be regarded as basically being a reflection of the age of the silica used. Some of the earliest investigations of oxygen isotopes in ancient glasses were carried out by Brill (1970). He suggested that oxygen isotopes could be used to detect the source of silica in glass because a natural range of isotopes values occurs in silica that is dependent on the geological conditions of quartz formation. Sand, via the erosion of a particular type of granite, would give low values (typically c. 10‰–12‰ vsmow), whereas low temperature metamorphic quartz, typical of vein quartz, would have higher values, perhaps up to 20‰ (vsmow).

However, ^{18}O/^{16}O determinations appear to be less diagnostic to silica than ^{143}Nd/^{144}Nd signatures (Henderson et al. 2009; Meek 2011). Silvestri et al. (2010) found that they were unable to distinguish between Belus and Campanian sands using oxygen isotope determinations, which alone underlines the limitations of the technique. They found that natron glass had consistently higher δO^{18} values than some plant ash glasses (although not potassium-rich plant ash glasses), though this does not help to provenance glasses. There are a number of possible reasons for the limitations of the technique. First, there will be oxygen contributions from other components of the melt. The first of these is the oxygen in sodium oxide derived from a breakdown of sodium carbonate during the melting process. Another oxygen source is calcium oxide deriving largely from the breakdown of calcium carbonate. Both soda and calcium oxide may be associated with oxygen of different ages from that contributed by the silica. However, having said this, it is possible that in combination these oxygen sources could potentially provide a unique fingerprint for glass made in a specific location. In combination with strontium and neodymium isotope signatures, oxygen isotope results may eventually play a useful role.

11.5.3 Lead Isotopes in Glass

Although not a major constituent of most glass, the isotopes of any lead present can provide a comparison of the average Pb isotope composition of the raw materials used to make the glass. Possible sources of lead are impurities in lead-bearing minerals in sand, contaminants taken up by plants (Chapter 2), lead in heavy-metal impurities introduced by glass recycling (Henderson and Holand 1992; Henderson 2000, 70–1) and lead used as a raw material.

Lead isotopes of a range of archaeological glasses have been determined, mainly by Brill and co-workers (e.g. Barnes *et al.* 1986; Brill *et al.* 1979; Brill *et al.* 1984; Brill *et al.* 1991). These studies have shown that it is possible to distinguish, in general terms, using ^{208}Pb/^{206}Pb vs ^{207}Pb/^{206}Pb between the sources of lead in archaeological objects (including glass) from, for example, China and Egypt. However, leads from Spain, Wales and Sardinia fall into the same grouping as do leads from England, Italy and Turkey. So this has severe limitations as a provenancing tool. However, Wedepohl *et al.* (1995) have been able to link lead isotope characteristics of medieval glass to lead ore deposits in Germany. A much larger number of lead isotope determinations taken from metal artefacts from a relatively small area has underlined the complexities of recycling and other parameters in attempting to source metal used in artefacts, especially once mixing of lead sources occurred (Rohl and Needham 1998). No such study on this scale has yet been attempted for glasses, although recent determinations of lead isotopes in Islamic glazes from Fustat, Egypt, have provided evidence for the potential complexity of lead use in the Islamic world, including an indication that the lead used travelled over considerable distances (Wolf *et al.* 2003). Where glass has been opacified with a lead-rich opacifier, such as lead antimonate, and therefore not an impurity, there is a greater likelihood of being able to suggest a lead source with confidence, as in the case of Bronze Age Egyptian glass (see Chapter 5).

11.5.4 Mixing of Raw Glass and Glass Raw Materials

Glasses made in different workshops with different isotopic characteristics can be mixed, and this might be considered a problem. However, mixing can also be detected using isotopic mixing lines. If at least one 'end member' is defined, then any mixing of 'foreign' glass will modify the isotopic plot, especially if mixing of different proportions of the glasses with these different isotopic signatures has occurred, producing a mixing line (Degryse *et al.* 2006) as in Figure 11.11. In addition, strontium isotope determinations are a useful way of disentangling the complexities of compositional patterns when plants of different genera and glasses made from them have been mixed during glass production. Although different plant genera may produce glasses of different chemical compositions, if they derive from an area dominated by the same bedrock geology, the same isotopic signatures will consistently be obtained for them and in the glasses, irrespective of the genera or preparation technique used.

Export of raw glass and ingots could lead to the mixing of glasses on secondary production sites. However, both the mixing of raw glass and raw materials from contrasting geologies would be revealed with isotopic mixing lines. Mixing can be shown to have occurred even without identified examples of the original unmixed glasses (in geological terms, the 'end members'), because one isotope signature will 'dilute' the other, producing a range of values (Freestone *et al.* 2005; Henderson *et al.* 2005, 669; Degryse *et al.* 2006).

11.6 A Case Study: Strontium, Neodymium, Oxygen and Lead Isotopes of Glass Found at the Glassmaking Site of Al-Raqqa, Syria

Plant ash glass was the dominant glass type in the Islamic period from c. A.D. 800 in the Middle East. The use of isotopic techniques to investigate raw furnace glass from Islamic primary glassmaking sites provides an ideal means of testing the efficacy of this approach for glass provenancing and for testing the principal models for glass production described in Section 11.2. As described in Chapter 9, Islamic glass production occurred in centralised urban contexts: there is both archaeological and historical evidence for its production in locations such as Tyre, Aleppo, Damascus, Cairo and al-Raqqa. Export of raw and scrap glass occurred, with comprehensive evidence of this from the Serçe Limani shipwreck (Bass 1984; Bass *et al.* 2009). There is also some evidence for the export of plant ashes (Chapter 9), but it is unclear the extent to which this occurred within the Islamic world. The evidence for the glass industry at al-Raqqa is described in Chapter 9.

The main aims for using isotopes to examine glass production at al-Raqqa are the following:

1. To investigate whether the radiogenic and stable isotope signatures of glass raw materials, raw furnace glass and artefact glass from al-Raqqa (and the surrounding area) could be used as a means of providing an unambiguous provenance for the glass made there, thereby providing a 'local' signature, or at least a restricted regional signature for the glass.
2. To identify distinctions between the 'al-Raqqa' signature and glass made in other production centres revealed by a 'foreign' isotopic signature.
3. To build up a strontium map of variations of strontium isotope signatures for plants growing in the vicinity of al-Raqqa and across quite large areas of Syria and the Lebanon, thereby placing the al-Raqqa signature into a regional context and allowing the potential of strontium isotope analysis to be assessed as a glass provenance tool.
4. To estimate the degree of mixing of raw glasses and/or cullet using the isotope results.

The selection of samples of raw and artefact plant ash glass was based on the chemical characteristics of the glass fused at al-Raqqa. Most samples derived from the ninth-century site of Tell Zujaj, where the most comprehensive

Table 11.1. The strontium and neodymium isotope results for twenty-five samples of glass from al-Raqqa, Syria

Sample	Sr ppm	$^{87}Sr^{86}Sr_{(n)}$	Nd ppm	$^{143}Nd/^{144}Nd_{(n)}$	$\delta^{18}O$ V-SMOW	Artefact, date	Compositional type
RAQ 33–34c	517	0.708178	5.639	0.511987	13.7	Raw, 9th C	1
RAQ 33–34M	503	0.708090	5.863	0.512056	14.9	Raw, 9th C	1
RAQ 45–46	542	0.708148	6.348	0.512102	13.7	Raw, 9th C	1
RAQ 50	325	0.708208	7.716	0.512204	13.5	Vessel, 9th C	2
RAQ 54	628	0.708121	6.676	0.512091	13.7	Vessel, 9th C	1
RAQ 62	494	0.708374	6.569	0.512071	13.1	Vessel, 9th C	1
RAQQA-26	582	0.708204	5.890	0.512118	13.9	Raw, 9th C	1
RAQQA-268	469	0.708293	4.894	0.512190	13.7	Bangle, 11th C	4 mid Al
RAQQA-269	208	0.708289	5.599	0.512203	13.7	Bangle, 11th C	4 mid Al
RAQQA-35	602	0.708131	5.724	0.512084	13.5	Raw, 9th C	1
RAQQA-38	361	0.708150	4.020	0.512114	16.6	Raw, 9th C	4 hi Al
RAQQA-40	515	0.707595	11.016	0.512340	14.9	Raw, 9th C	2
RAQQA-41	463	0.708448	3.321	0.512148	13.5	Vessel, 9th C	4 mid Al
RAQQA-42	1026	0.708354	4.132	0.512123	13.9	Raw, 9th C	1
RAQQA-47	226	0.708175	4.4314	0.512107	14.0	Vessel, 9th C	4 lo Al
RAQQA-49	726	0.708110	7.1348	0.512023	14.0	Raw, 9th C	1
RAQQA-58	309	0.708152	2.1167	0.512206	13.7	Vessel, 9th C	4 lo Al
RAQQA-60	648	0.708620	5.8618	0.512027	13.2	Raw, 9th C	1
RAQQA-61	170	0.709347	9.3907	0.512084	15.0	Vessel, 12th C	2
RAQQA-66	112	0.708556	5.8347	0.512244	13.7	Window, 9th C	4 mid Al
RAQQA-67	426	0.708520	6.2719	0.512221	13.0	Window, 9th C	4 mid Al
RAQQA-80	70	0.708428	7.4148	0.512095	13.5	Vessel, 9th C	2
RAQQA-81	291	0.708220	3.6019	0.512203	13.5	Vessel, 9th C	1
RAQQA-84	172	0.708150	5.877	0.512035	13.4	Raw, 9th C	4 mid Al
RAQQA-87	663	0.708129	5.8154	0.512118	13.6	Raw, 9th C	4 hi Al

C = century.

evidence for primary glass production in tank furnaces was discovered (see Chapter 9). Thus, a key part of the investigation was to test the isotopic compositions of multiple raw glass samples taken from furnace floor fragments. This would first provide a base-line against which to compare vessel and other artefact glasses and second allow a direct isotopic comparison between the raw glass and strontium isotope values of local plants and neodymium isotope values in local silica. Other samples were ninth-century vessel fragments from Tell Zujaj, two ninth-century window samples (RAQ 66 and 67) from the west palace complex north of the industrial zone in al-Raqqa; two eleventh-century armlets from Tell Fukhar (Tonghini and Henderson 1998); and a single twelfth-century vessel fragment (RAQ 61) from the princess's palace within the walls of al-Rafica, al-Raqqa's 'twin' city, one kilometre to the west of al-Raqqa. The results for twenty-five glass samples from al-Raqqa are given in Table 11.1.

Three plant ash compositions (Henderson *et al.* 2004, types 1, 2 and 4) are noted in the last column of Table 11.1. These chemical compositions are

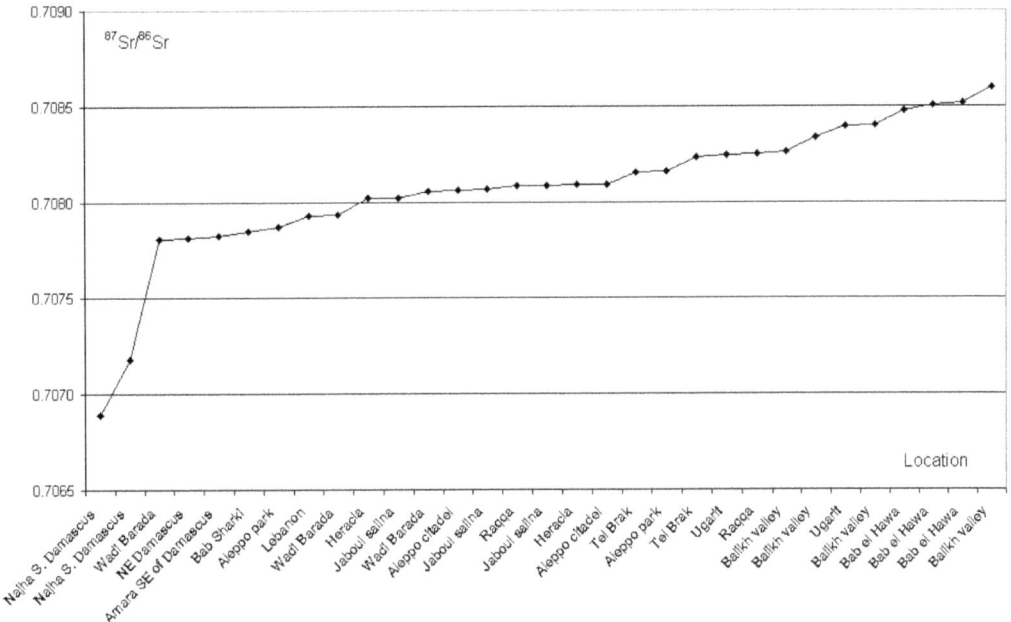

11.8. Strontium isotope variations across the Syrian landscape showing a clear dip in the values for the area in southern Syria around Damascus. Data from Henderson *et al.* 2009.

discussed in detail in Chapter 10 and defined elsewhere (Henderson *et al.* 2004). The reason the levels of alumina in type 4 are noted in the last column is that they relate to the position of the samples on a mixing line formed from relative levels of magnesia and alumina (Fig. 10.3). The 'end members' are high alumina-low magnesia and low alumina-high magnesia. By using isotope analysis it was possible to shed light on whether the raw materials had a local or non-local origin and where they fell on the mixing line. Underlying the interpretation of isotopic results is the crucial geological characteristics of Syria and the surrounding countries. The geology of Syria is dominated by Cretaceous, Eocene and Neogene sedimentary deposits. A survey of $^{87}Sr/^{86}Sr$ variations in plants from different sites in this area (Fig. 11.8) shows that it is not wide (0.70689–0.70860), especially compared with an area with widely contrasting geology such as the United Kingdom, as reflected in mineral waters, where the range was found to be between 0.7059 and 0.7207 (Montgomery *et al.* 2006; Evans *et al.* 2010).

The highest values ($^{87}Sr/^{86}Sr > 0.7086$) in Syria are from Eocene and Neogene geology in northern Syria around Bab el Hawa, west of Aleppo, in the Balikh Valley close to al-Raqqa and a point to the north of it (Fig. 11.8). Cretaceous rocks would produce a minimum $^{87}Sr/^{86}Sr$ value of 0.7072 (MacArthur *et al.* 2001): the lowest $^{87}Sr/^{86}Sr$ values are from the Cenezoic basaltic area southeast of Damascus (Asch 2005). Plant samples SYR 26 and 27 from this area give the lowest values (0.7069 and 0.7072, respectively). Therefore, the restricted range of $^{87}Sr/^{86}Sr$ values found has both advantages and disadvantages: it means that it may not be easy to distinguish easily between some glasses made in different parts of Syria, but on the wider international scale, both within and without the Islamic world, it should provide a quite well-constrained Syrian signature.

The $^{87}Sr/^{86}Sr$ isotope results plotted against strontium concentrations for raw furnace glass, vessel glass, window glass and bangles from al-Raqqa are shown in Figure 11.9. For comparison, our plant ash data and published glass data from the primary production site of Tyre and the secondary production site of Banias, both in the Levant, together with the site of Ra's al-Hadd in Oman (Leslie *et al.* 2006) are also included. These results confirm comprehensively our initial discovery (Henderson *et al.* 2005), that the plants growing in the vicinity of al-Raqqa have a limited range of strontium isotope values, which is also found in raw furnace and some vessel glass from al-Raqqa.

The results demonstrate the following:

1. *local plants* were used to make both of the main chemical types of plant ash glass found at al-Raqqa types 1 and 4,
2. some of the vessels from al-Raqqa were made and blown there,
3. there is no fractionation of the isotopes in the plant ash when glasses are made from them, and
4. this forms the basis of glass provenance using strontium isotopes in the region.

The $^{87}Sr/^{86}Sr$ value for raw glass from Banias in the Levant is low and is consistent with having been produced using plants growing in soil with a basalt bedrock close to Damascus. Indeed the $^{87}Sr/^{86}Sr$ value for Banias ground water of 0.7072 does not match the Banias raw glass (Starinsky *et al.* 1980), providing supporting evidence for Banias being a secondary production site. The isotopic data from Tyre form a tight group, which, somewhat unexpectedly, is similar to the results from al-Raqqa. Given the similarity between plant signatures from some parts of the Levant, including Ugarit and Beirut, it is therefore not necessary to interpret the discovery of such a similarity in strontium isotope signatures in glass from al-Raqqa and Tyre as being evidence for the export of plant ash from al-Raqqa to Tyre (Leslie *et al.* 2006, 262). Single results for vessel glasses from Ra's al-Hadd and al-Raqqa are characterised as having higher $^{87}Sr/^{86}Sr$ results than al-Raqqa plants and raw glass, showing that they were not made in the area. The vessel glass with a strontium isotope value of 0.709347 derived from the twelfth-century princess's palace in al-Rafica has a much younger $^{87}Sr/^{86}Sr$ signature than that of the typical signature for al-Raqqa raw glasses and therefore must have been imported. However, see Figure 11.13.

The Provenance of Ancient Glass

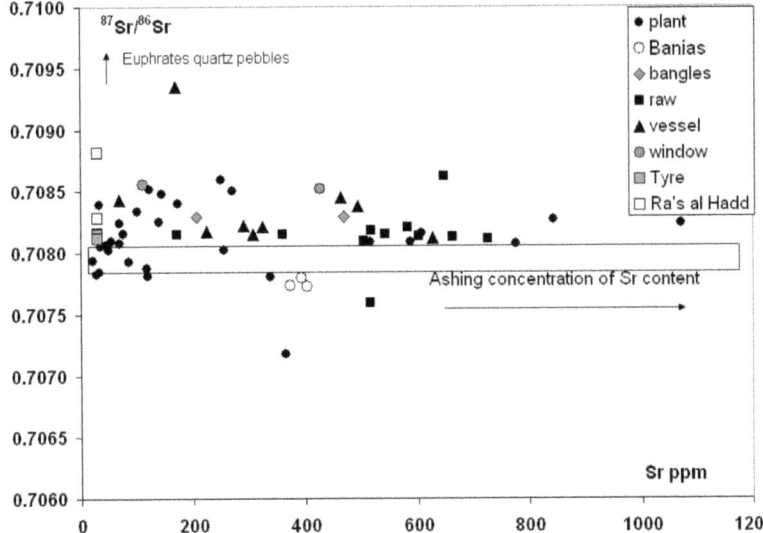

11.9. The concentration of strontium versus strontium isotope ratios in glass from al-Raqqa and in plant ashes from Syria and Lebanon, including ashed plants from al-Raqqa. (The Sr concentrations of samples from Tyre and Ras al-Haad have been given arbitrary values of 30 ppm.) Data from Henderson *et al.* 2009.

The result from Ra's Al-Hadd has been interpreted by Leslie *et al.* (2006) as indicating that the glass was made from older continental sand. The results for Banias raw glass and for some vessel fragments from al-Raqqa and Ra's al-Hadd are clearly distinct from al-Raqqa raw glass and from the plant results. These results show that within this geographical area, $^{87}Sr/^{86}Sr$ results are not always especially good at discriminating between production zones. Greater precision could be obtained by carrying out more detailed surveys of variations of $^{87}Sr/^{86}Sr$ in plants across the landscape, especially in areas of complex geology and in contrasting geological zones outside Syria and Lebanon.

11.6.1 New Information about the Use of Glass Raw Materials

Sand and crushed quartz are the two possible sources of silica used to make glass. Quartz is clean, whereas sand can contain significant levels of other minerals, such as feldspar (see Chapter 3). The latter can affect both the isotopic and the chemical compositions of the glass. Table 11.1 provides $^{143}Nd/^{144}Nd$ isotope results for (Islamic) plant ash glasses.

Figure 11.10 gives the neodymium results plotted for the three chemically distinct types of plant ash glasses found at al-Raqqa (Henderson *et al.* 2004). The first thing to note is that in most cases, the chemical types correlate with the isotopic groupings. Type 2 glasses are characterised by containing the highest concentrations of neodymium, including the outlier glass, al-Raqqa 40 (*ibid.*, table 1, analysis 22). The sand used to make them is therefore the least pure. The neodymium results for type 4 glasses can generally be distinguished from type 1 by being more radiogenic and therefore of a relatively young geological age. The results for type 4 glass, especially raw glass containing high alumina levels, are most radiogenic.

These results therefore provide new information about the silica types used in Islamic glass production according to purity and their ^{143}Nd/^{144}Nd isotope signatures. Type 1 plant ash glass was made with silica of a relatively constrained geological age and type 4 with a wider range of geological ages. The sands used to make types 1 and 4 must have largely remained segregated on the site. This may be because their impurity levels gave the glass specific working properties and possibly because different groups of artisans were involved; the decision to keep them separate clearly formed part of the *chaîne opératoire* at al-Raqqa. The smaller number of type 2 glass samples contain the highest neodymium levels (and also the highest iron levels) and, unlike types 1 and 4, are frequently highly coloured (Henderson *et al.* 2004, fig. 8). The artisans must therefore have used the different qualities of sand to their advantage. Type 2 glasses are compositionally distinct, and higher iron levels must have given the sands a darker colour, even though they are not necessarily isotopically distinct, as discussed later.

Unpublished ^{143}Nd/^{144}Nd isotope signatures in sand deposits across Syria and along the Levantine coast show a precise match for the sand used to make some of the al-Raqqa glass: it is a source lying to the south east of al-Raqqa, near Palmyra.

11.6.2 Evidence for Mixing of Raw Materials

Mixing of glass has the potential to confuse provenance studies. In contrast to the segregation of different types of sand at al-Raqqa, another aspect of glass technology that isotopic studies can be used to investigate is the mixing of silica- and calcium-rich raw materials and glass. The line plotted in Figure 11.11 shows that there was mixing of silica of different geological ages and purities within the compositional types 1 and 4. Points A and B represent the extreme Nd values and the points in between (A+B) mixtures of different proportions of these. The glasses at one end have low Nd concentrations and a relatively high (geologically younger) ratio and those at the other end higher Nd concentrations and lower (geologically older) ^{143}Nd:^{144}Nd ratios.

In Figures 11.10 and 11.11, we can see that 2 type 1 raw furnace glasses fall at the right hand end of the line. Two other type 1 raw furnace glasses fall in the middle; at the left-hand end are two type 4 vessel glasses. The results for the vessel glass therefore suggest that they were made with an admixture of non-Raqqa (younger) silica.

THE PROVENANCE OF ANCIENT GLASS

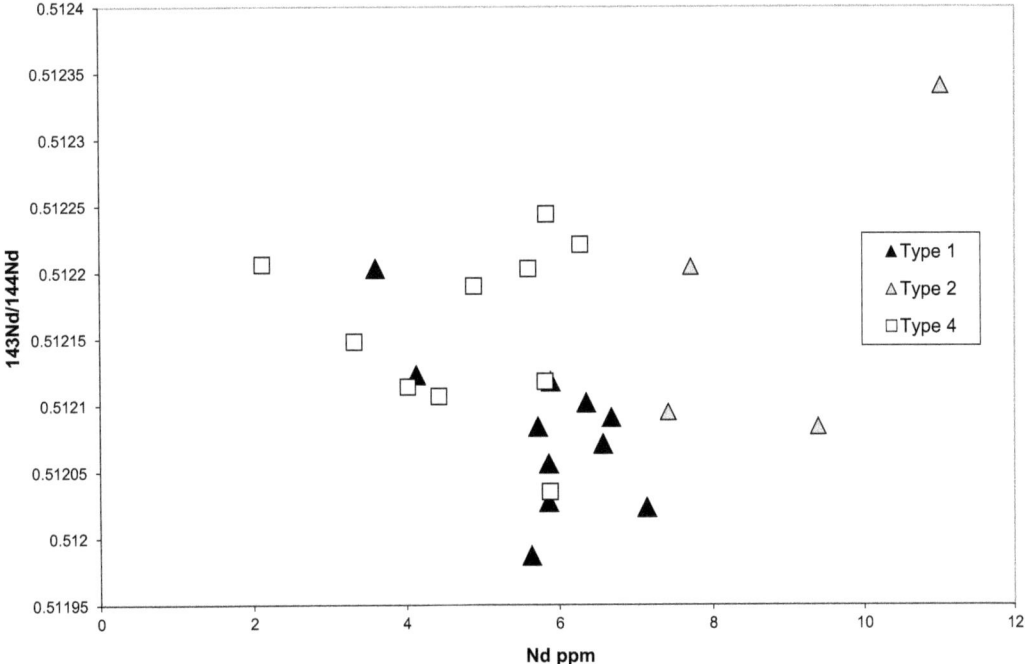

11.10. The concentration of neodymium versus neodymium isotope ratios in glass from al-Raqqa by compositional glass type defined in Henderson *et al.* (2004). Data from Henderson *et al.* 2009.

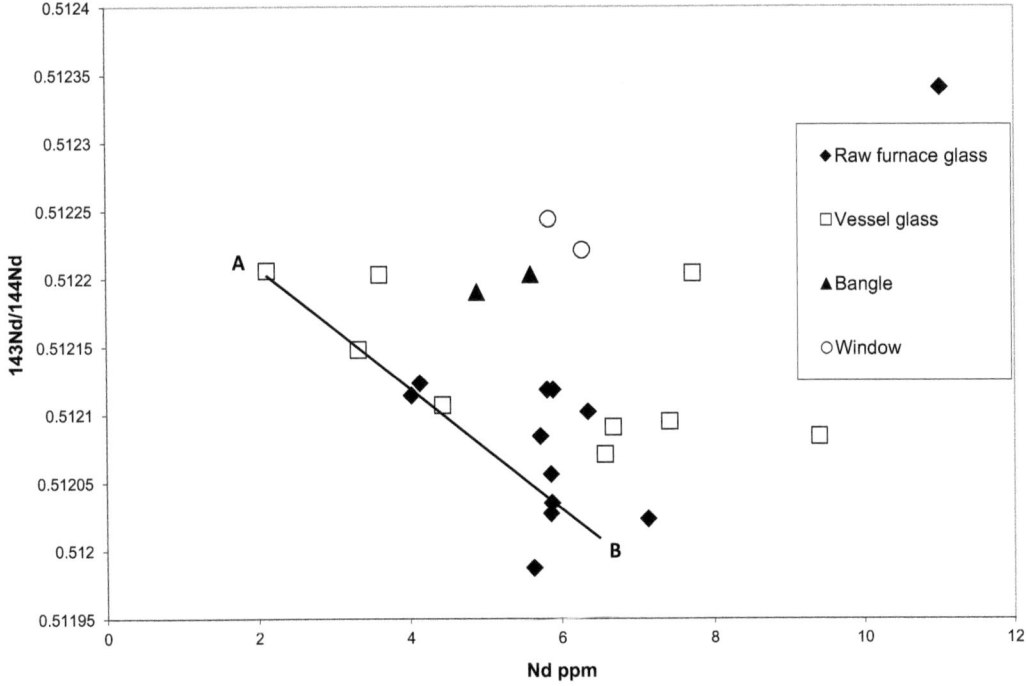

11.11. The concentration of neodymium versus neodymium isotope ratios in glass from al-Raqqa by artefact type showing mixing line A–B. Data from Henderson *et al.* 2009.

The raw glass samples that fall on the mixing line consist of one that has the typical al-Raqqa type 1 signature and three others of type 4. However, the type 1 raw glass is rather distinct isotopically from the main concentration of raw glass, so although it may be compositionally of type 1, the Nd isotope results provide an interesting additional piece of information – that the glass was made with an admixture of younger silica.

Using the results of chemical analyses, we suggested that glass mixing occurred in al-Raqqa with imported raw glass being made with imported raw glass made from chemically distinct raw materials (Henderson *et al.* 2004, 460, types 2 and 4); these isotope results have confirmed this. Glass production at al-Raqqa formed part of a ninth-century experimental phase that occurred there (Chapter 9). Such mixing would probably have occurred using raw (and vessel) glass placed in crucibles in beehive-shaped furnaces before vessel blowing.

11.6.3 Local Production and Imports

Apart from one, all the neodymium results for raw glass samples fall between 0.511987 and 0.51234, and a cluster of eight have tightly grouped Nd concentration values of between 5.639 and 6.348, providing a diagnostic impurity level for the silica source used to make most type 1 glasses at al-Raqqa. Two others have Nd concentrations of 4.02 and 4.132, a third (al-Raqqa 49) was 7.1348. A sample of raw glass (al-Raqqa 40) was clearly made using silica from an entirely different source with a Nd impurity of 11.016 and a ^{143}Nd/^{144}Nd value of 0.512340.

Samples of twisted and plain bangles (al-Raqqa 268 and 269) from eleventh-century Tell Fukhar in al-Raqqa, three fragments of flasks (al-Raqqa 50, 58 and 81) from ninth-century Tell Zujaj and cobalt blue and emerald green window fragments (al-Raqqa 66 and 67) form a group that is united by high neodymium isotope ratios of between 0.512190 and 0512244 (see Fig. 11.10). They were therefore manufactured using a more radiogenic (younger) silica source than that used for the manufacture of typical raw al-Raqqa glasses and were imported to the site. In Figure 11.10 it can be seen that these samples are of chemical types 1, 2 and 4. This underlines a crucial point: different isotopic and chemical types of glasses are not always correlated.

Figure 11.12 shows the neodymium isotope results plotted against oxygen isotope ($\delta^{18}O_{vsmow}$) results for each sample. There is little correlation between the two sets of isotope results, even though both are related primarily to silica. One explanation for this is that oxygen associated with calcium and sodium was present in the melt (see earlier). These oxides make up to between c. 16% and 22% of the glass composition. Most oxygen isotope results for al-Raqqa glass fall within the compositional range produced for Euphrates pebbles (Henderson

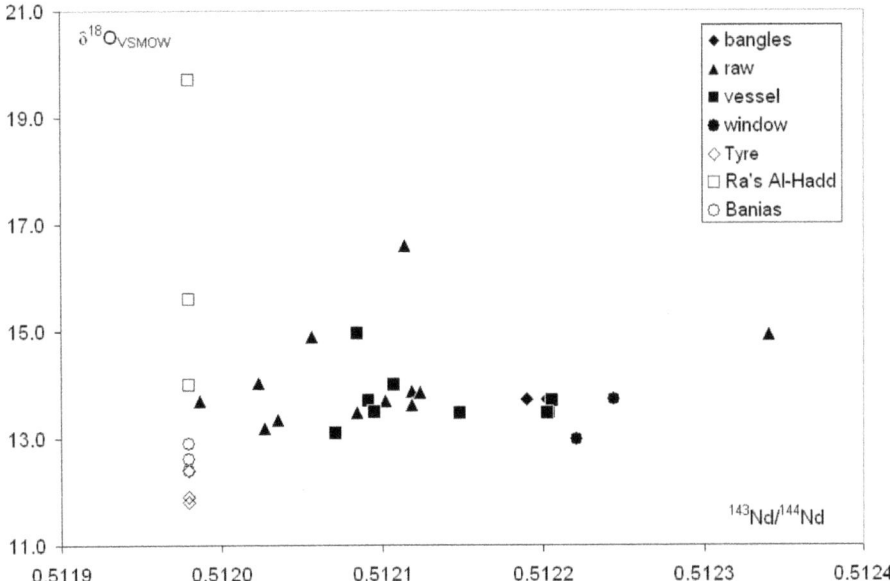

11.12. $^{143}Nd/^{144}Nd$ versus $\delta^{18}O$ vsmow in glasses from al-Raqqa (by artefact type), Tyre, Ra's al-Hadd and Banias. Data from Henderson *et al.* 2009 and Leslie *et al.* 2006.

et al. 2005, fig. 2). Despite this, Figure 11.12 shows that we are able to distinguish between al-Raqqa and Tyre glass. However, there is some overlap between al-Raqqa data and that for Banias and Ra's al-Hadd, showing that the use of oxygen isotopes has its limitations.

Figures 11.13 and 11.14 combine results for neodymium and strontium isotopes. Al-Raqqa 40 (that plots in the top left-hand corner of Fig. 11.13), a piece of raw deep translucent emerald green glass, is interesting because it falls in an area of the diagram that is diagnostic of basaltic rocks for both its strontium (south of the Damascus area) and neodymium composition. This suggests that the glass was produced in a basic, geologically relatively young (basalt) terrain from which the sand was derived directly and that the plants used to make it grew on the same lithology, with the isotope signature of the parent rock being transferred into the glass. As can be seen in Figure 11.13 it was probably made in the same location as Banias glasses, the colourant probably being added. The import of this raw glass from a site where the Banias glass was made to a major glassmaking centre can nevertheless be considered to be unexpected and unusual. Both plots show that the windows and bangles were imported to al-Raqqa. They would also probably have been coloured at a secondary glassworking (and modification) site: the addition of low concentrations of colourants would not distort the isotopic signature. There is no archaeological or scientific evidence for the production of highly coloured (raw) glass at al-Raqqa or on other sites in the Islamic world. There is, however, historical evidence for artisans that specialised in the production of different vessel types

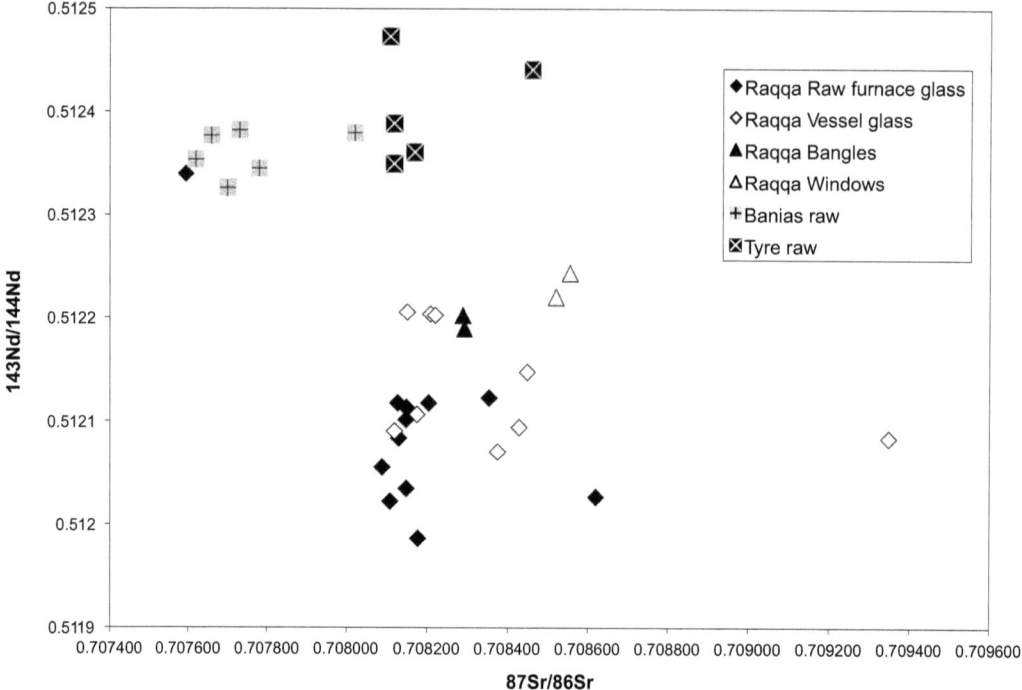

11.13. ^{143}Nd/^{144}Nd versus ^{87}Sr/^{86}Sr in al-Raqqa, Banias and Tyre glasses by artefact type. Data from Henderson *et al.* 2009 and Degryse *et al.* 2010a.

such as lamps, drinking glasses, bottles and flasks (Shatzmiller 1994, 201). There is no particular reason why the primary and secondary production of glass was separated spatially, as has been suggested (see Section 11.2). The presence of primary and secondary production in close proximity, or at least in the same city, would have facilitated communication between the different artisan groups involved.

Two undecorated bowl fragments (al-Raqqa 47 and 54) group tightly with five raw glasses and must be considered to be al-Raqqa products (Fig. 11.13). Four raw glass samples have isotopic compositions that do not match any of the vessel samples, but further isotopic analyses will undoubtedly produce more matches. As for the evidence of mixing, the line plotted in Figure 11.11 shows that five of the seven samples that form the neodymium mixing line are retained in Figure 11.14. The samples consist of two raw type 4 glasses, one type 4 vessel glass, one type 1 raw glass and one type 1 vessel glass. This indicates that the two vessel glasses that fall on the line were produced by mixing raw glass with different isotope signatures. The type 4 samples, in particular, fall close together on a straight line. These results support the model that raw glasses of different compositional types remained separate and were not mixed in the tank furnaces. However, it can be suggested that raw glasses were mixed in the crucibles prior to being extracted for glassblowing. Colourants may have been added to the

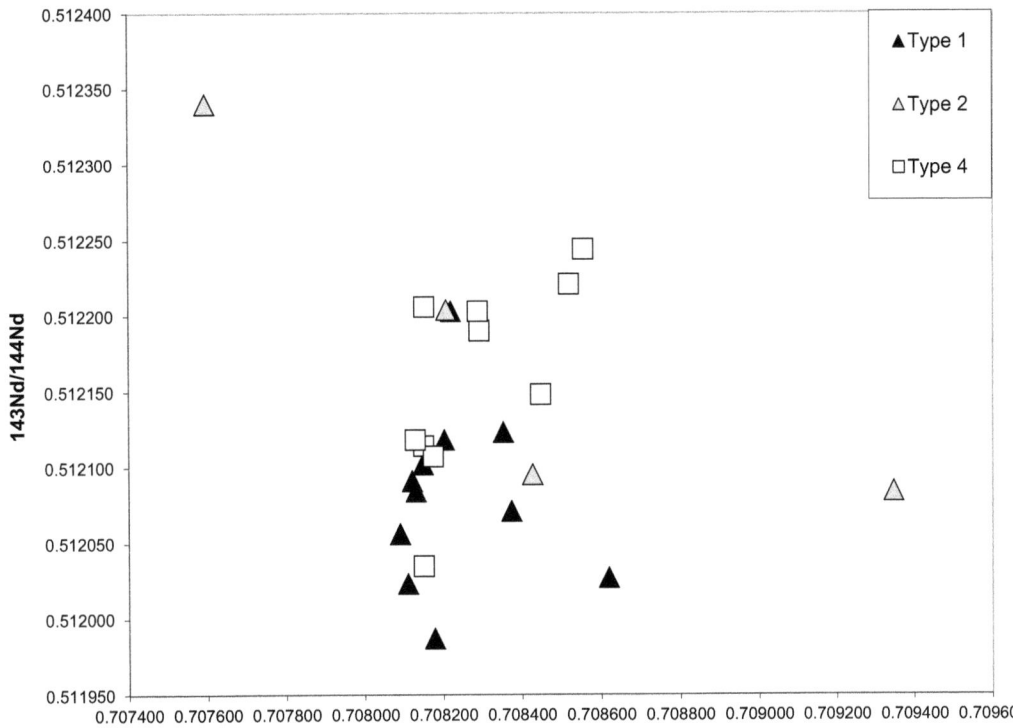

11.14. ^{143}Nd/^{144}Nd versus ^{87}Sr/^{86}Sr in al-Raqqa glasses by compositional type. Data from Henderson *et al.* 2009.

melt at this point, especially if small quantities of opaque glasses were being made that were used for enamel decoration.

Figure 11.13 shows that there is a clear isotopic distinction between raw glasses from three different production sites: Banias, a secondary production site, that could date to between the tenth and thirteenth centuries, Tyre that dates to between the tenth and twelfth centuries (Degryse *et al.* 2010a) and ninth-century al-Raqqa. The neodymium and strontium results for al-Raqqa and Tyre have a similar range of variation, so these may be typical of such production sites. A sample of Banias glass falls into the Tyre group, and a sample of al-Raqqa glass falls into the Banias group, so they must have been imported. The Banias glass results are consistent with it having been melted in the basalt-dominated area that includes southern Syria. These results provide coherent proof that an isotopic approach to provenancing raw plant ash glasses can be a successful one.

11.7 Lead Isotope Analyses of Bronze Age and Islamic Plant Ash Glasses

Brill has carried out a large number of lead isotope determinations of ancient glasses. These have included Egyptian (Lilyquist *et al.* 1993, table 2) and Chinese glass (Brill *et al.* 1991). More recently, Shortland (2006) published more lead isotope results on Egyptian glasses. Lead isotope determinations have also been carried out, including some for Islamic glass from Syria (Henderson *et al.* 2005). One of the key issues in the interpretation of such data is the extent to which lead ore fields have been adequately characterised; another is the technique that is used to display such data (Pollard 2009). Nevertheless, Shortland has shown that for fourteenth-century B.C. Egyptian glasses, there is a core group of results for lead, containing yellow glasses opacified with lead antimonate that incorporates lead deriving from the Gebel Zeit deposit. He has also suggested that the lead was mixed with Mesopotamian lead, although current Sr and Nd isotope results show that this is unlikely for translucent glasses (Henderson *et al.* 2010). Some Mesopotamian yellow glasses plot close to Gebel Zeit data, but others, including Mesopotamian and fifteenth-century B.C. Egyptian glasses, fall away from this field. Shortland (2006, 665) points out that the Gebel Zeit lead deposit has an average lead isotope value that is 'lower than almost all other ore bodies known or suspected to have been used in the ancient world', being distinct from the Sardinia, Cyprus and Lavrion lead ore fields in the Mediterranean, for example. However, some Chinese glasses actually have both lower and similar signatures to Egyptian ones (Brill *et al.* 1991, figs. 1–3).

Figure 11.15 (Henderson *et al.* 2005, fig. 4) exemplifies how complex lead isotope data in ancient glasses can be. The plot is the conventional one used by researchers. First, the results for plant ash glasses (including those from ninth-century al-Raqqa) and third- to fourth-century natron glass from Germany overlap. Second, the plant ash results fall on a mixing line, and this shows that different proportions of lead from different sources are present. The results for glasses found on the eleventh-century Serçe Limani shipwreck off the west coast of Turkey that carried Islamic glass vessels and raw glass fall onto the same mixing line. The latter mainly cluster at the lower end, in the area that may represent an end member; two other results occupy the opposite end of the mixing line and could represent an end member signature. The routes by which the low levels of lead impurities entered the glass are quite difficult to ascertain. One possible route is via lead-rich particulates taken into plants used for making the glasses (Barkoudah and Henderson 2006). The results could potentially be a reflection of several sources, especially if the plants used grew in an industrial environment (Henderson *et al.* 2005). There is a relatively constant lead impurity level in many of the plant ash glasses, and if its introduction is due to thorium-rich minerals in sands used to make the glasses, then the lead isotope ratio would be expected to exhibit variability, and this would be most obvious along the x axis of Figure 11.15 – and indeed this is the case.

It is clear that much more and wide-ranging research needs to be carried out into the identity of potential sources of the lead impurities found in translucent glasses. It is, however, unlikely that lead isotope determinations will provide a

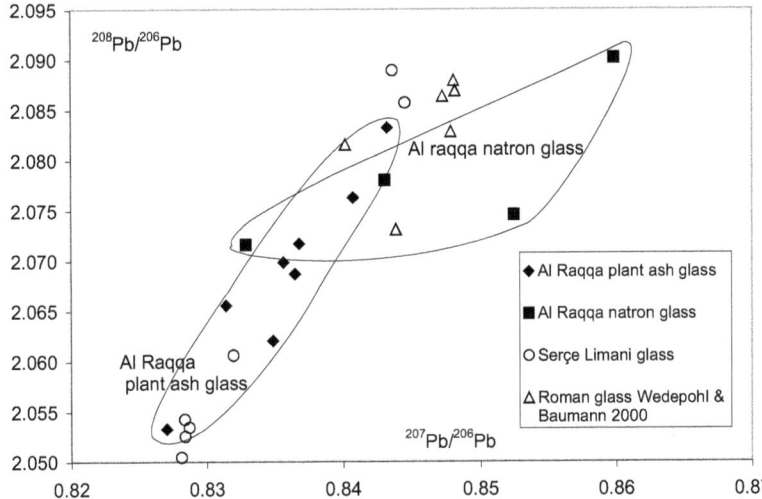

11.15. $^{208}Pb/^{206}Pb$ versus $^{207}Pb/^{206}Pb$ in plant ash and natron glasses from al-Raqqa, plant ash glasses from the Serçe Limani shipwreck and in German fourth-century Roman natron glasses. Data from Henderson *et al.* 2005, Wedepohl and Baumann 2000 and Barnes *et al.* 1986.

provenance for ancient low lead glasses, and it is even debatable whether this is the case for high lead glasses.

11.8 THE FUTURE POTENTIAL OF ISOTOPE STUDIES

The results discussed in the preceding sections include baseline data for a primary plant ash glass production site, two isotopically characterised primary (plant ash) glassmaking sites, evidence for the import of glass artefacts to al-Raqqa and limited movement of raw glass between production sites. Logically, the next step would be to carry out the same exercise for another Islamic glassmaking site and preferably one that is contemporary with the earliest phase of al-Raqqa (ninth-century) glass production, located in a geologically contrasting area. This exercise should include harvesting of plants and collection of silica samples in a wide area around the site, the determination of their $^{87}Sr/^{86}$ and $^{143}Nd/^{144}Nd$ signatures for the raw materials and the chemical and isotopic analysis of raw and artefact glass from the production site. There is a range of reasons for adopting this methodology:

1. It will provide a contrasting database of chemical and isotopic data for a contemporary site that will reflect differences and similarities in the *chaîne opératoire*. The contrasting geology will create differences in plant ash compositions that will be carried through into the glass and the chemical and $^{87}Sr/^{86}$ composition of the raw glass.

2. The $^{87}Sr/^{86}$ and $^{143}Nd/^{144}Nd$ data for plants, silica and glass from a second primary production site will provide a direct comparison with results for the local raw materials used to make glass at al-Raqqa.
3. If the site were in contact with al-Raqqa by trade or exchange, the new results will provide a means of proving an unambiguous link between the two.
4. The discovery of more primary production sites will build on the existing network of chemically and isotopically defined glassmaking sites, providing new and comprehensive information about glass provenance and trade.

When combined with chemical analyses, isotopic studies of plant ash, natron and high alumina glass can contribute in a number of ways. The combination can contribute to evidence for local manufacture, supply systems and trade.

11.8.1 Local Manufacture

11.8.1.1 Bronze Age and Islamic Plant Ash Glasses

Bronze Age plant ash glass was produced on a relatively small scale (see Chapter 5). For Bronze Age Mesopotamia, there is limited evidence for the fusion of glass from primary raw materials. Therefore, any technique that provides an unambiguous provenance for plant ash glasses, associated with evidence of local secondary production, represents a major step forward. Given that the $^{87}Sr/^{86}$ signatures for plants in northern Syria are quite constrained (see Fig. 11.9), this should offer great potential. Therefore, the study of (Islamic) plant ash glass production has provided baseline data for bioavailable $^{87}Sr/^{86}$ data for the calcium source used and $^{143}Nd/^{144}Nd$ signatures for a range of potential silica sources in northern Mesopotamia, probably the area where the first glass was made. It was certainly an important area for the manufacture of some of the earliest glass vessels (see Chapter 5). So the determination of the environmental $^{87}Sr/^{86}$ and $^{143}Nd/^{144}Nd$ signatures in this area is just as relevant to the provenance of Bronze Age glass as it is to that of Islamic glass.

However, obviously Bronze Age and Islamic societies were entirely different (see Chapters 5 and 9). In the Bronze Age, most glass had a high social and religious value, whereas most Islamic glass vessels had utilitarian functions and a relatively low social value. If a sufficient contrast in $^{87}Sr/^{86}$ signatures can be detected between possible production sites farther south in Mesopotamia such as in the area around Eridu (Barag 1970, 133), where some of the earliest glass has been found, then it may be able to establish whether this very early glass was indeed made there – the construction of $^{87}Sr/^{86}$ isotope maps should form part of this study. It would also be possible to investigate whether there were several primary production centres, including for the glass used in the first vessels in the fifteenth century B.C., such as those found at Alalakh (see Chapter 5) and Büklükale, Turkey (*pers. comm.* Dr. K. Matsumura). There may even be supporting archaeological evidence for primary glass production at Alalakh (*pers. comm.* Prof. Dr. A. Yenner). Moreover, data for Bronze Age Mesopotamian glass could be used to investigate the question of whether there was a 'private' or state-sponsored production of glass. From c. 1500 B.C., there were a number of recognisable states, those of Kassite Babylonia, Hittite Anatolia, Egypt, the Mitanni (in northern Syria) and Assyria (see Chapter 5). Each of these states may have controlled its own glass production, and this could be tested using $^{87}Sr/^{86}$ and $^{143}Nd/^{144}Nd$ isotope analyses. The same methodology could be used for the comprehensive evidence of glass production found at Tell el-Amarna (Nicholson 2007; Smirnou and Rehren 2011) and for the evidence of primary glassmaking at Qantir (Rehren and Pusch 2005, 2007). Indeed, a combination of trace element analysis and isotope results is starting to show that it is possible to provenance such glass. Primary production of fourteenth-century B.C. and later glasses occurred in Egypt and Mesopotamia and both areas traded glass with Mycenaean Greece (Henderson *et al.* 2010; Walton *et al.* 2009). As noted in Chapter 6, the results of strontium isotope analysis show that there is sufficient geological contrast in the calcium component that found its way into the glass to demonstrate that there was more than one production centre in Mesopotamia in the fourteenth century B.C.

Between c. second and sixth century A.D., the Sasanians used glass that was either of a plant ash or natron composition (Mirti *et al.* 2008). It is a realistic assumption that the Sasanians made their own plant ash glass and that they imported natron glass from the Levant. The use of isotopic analysis could potentially define more than one production centre for Sasanian plant ash glass.

The next main period of plant ash glass production was from the late eighth century onwards in the Islamic world. Together with the appearance of new characteristically Islamic institutions and objects, there was a phase of experimentation from which new techniques and technologies emerged (Henderson *et al.* 2005a,). One of these was glass production. There is both archaeological and historical evidence for glass production (see Chapter 9). For example, there is historical and/or archaeological evidence for the production of glass in Baghdad and Sāmarrā, Iraq, Tyre, Lebanon, al-Raqqa, Aleppo and Damascus, Syria, Solkhat in the Crimea and at Murcia and al-Andalus in Spain. However, it is not clear in all cases that historical references refer to the fusion of glass from raw materials or to vessel blowing or the manufacture of window glass and jewellery. There is even historical reference to production centres that specialised in the manufacture of thirteen types of glass products including window glass, lamps, drinking glasses, bottles, flasks and beads (Shatzmiller 1994, 201). Whether the production centres that specialised in the manufacture of various objects were also involved in the manufacture of raw glass is an open question. The evidence from ninth-century al-Raqqa shows both that raw glass was fused and that drinking glasses were produced there, but not windows and armlets: window- and bangle-making centres used raw glass that was not fused at al-Raqqa. During the Mamluk caliphate and in the Ottoman period, high alumina glass was used for the manufacture of highly coloured glass armlets found at Khirbat Faris and Tell abu Sarbut, central Jordan; variations of this compositional type were made in India and the Far East (Boulogne and Henderson 2009).

In addition to production centres specialising in different vessel forms, it has been suggested that there were specialised production centres for Islamic glass vessels with distinctive decoration, such as for Ayyūbid and Mamluk enamelled vessels in Damascus, Aleppo and al-Raqqa (Lamm 1929–30), lustre-painted vessels in Damascus and possibly Cairo (Carboni 2001, 52), cameo-decorated vessels in 'Mesopotamia' (ibid., 82) and scratch decorated in Sāmarrā or al-Raqqa (ibid., 80–1).

Carboni (2001) has suggested that scratch-decorated vessels were made with glass fused in al-Raqqa or Sāmarrā and MgO versus CaO plot in Figure 10.11 appears to support a 'Syrian' origin. More scientific research, including radiogenic isotope determinations, is necessary to establish whether glass made in Sāmarrā can be distinguished from glass made in al-Raqqa.

The levels of alumina and calcium oxide detected in ninth-century raw glass fused in al-Raqqa, and the facet-cut vessels and goblets (all plant ash glasses) found there are mainly different from each other (Henderson et al. 2009b, fig. 6). Only a single facet-cut vessel contains similar levels of alumina and calcium oxide to that found al-Raqqa raw glass. Facet-cut vessels dating to between the eighth and tenth centuries have been described as belonging to a transitional period of Islamic glass production, their decoration being typically Sasanian (Carboni 2001, 16–17). Therefore the compositional distinction between facet-cut vessels and al-Raqqa raw glass is likely to be because the vessels were made further east. However, they are also distinctive from the colourless glass, cut in a different way, thought to have been made in Nishapur. The plot of calcium oxide versus magnesia appears to confirm a separate origin for the colourless facet-cut vessels, perhaps in northern Iran (Fig. 10.11). Although almost all of the results for lustre-painted glass vessel samples published by Brill (1999b, VII L) do not derive from archaeological contexts, there are nevertheless interesting correlations between glass body colour and chemical composition. Carboni (2001) has suggested that lustre-painted vessels were produced in either Damascus or Cairo. In Figure 10.11, the lustre vessels with the lowest calcium and magnesium oxide levels and very high alumina levels are all olive green and could be Egyptian products; several colourless bodies contain the highest calcium oxide levels and fall into the 'Levantine' area and cobalt blue samples have a more variable composition, falling into 'Syrian' or 'Levantine' area in the figure. These results suggest that different centres specialised in the manufacture of different glass colours and are in agreement with Carboni's suggestion. Isotopic evidence shows that the raw glasses used as the base glass for coloured glasses were made in separate places. Colourants are likely to have been mixed with raw glass in the same industrial zone in urban contexts. The data in Figures 10.1, 10.10 and 10.11 reflect different traditions in the use of calcium-bearing raw materials (plant ashes and sand) and/or melting conditions, producing different calcium concentrations in the glasses. The results also show that different types of silica (with variable alumina impurities) were used to make plant ash glass.

Therefore, the determination of neodymium and strontium isotopes in samples of decorated vessel types and colours from firmly dated archaeological contexts, together with trace element determinations, would be a means of providing independent locational information. Isotope studies can cut through compositional ambiguities and should reveal local specialisations. Thus, further isotopic determinations of raw glass from primary glass production sites coupled with plants and silica local to those sites and across the landscape will provide a provenance for glass vessel samples for which none currently exists.

11.8.1.2 Phoenician, Hellenistic, Iron Age and Roman Natron Glasses

Since about the 1980s, ample evidence for large-scale primary natron glass production has been discovered at sites such as Apollonia, Dor and Bet Eli'ezer in Israel dating to the second half of the first millennium A.D. (Gorin-Rosen 2000). Evidence for its primary manufacture on the Levantine coast has now been pushed back as far as the first century A.D., and possibly the first century B.C., with the recognition of glass tank furnaces in Beirut (see Chapter 7 and Kowatli et al. 2008). We can assume that natron glass made on the Levantine coast was probably also used to manufacture Phoenician objects. It is also the probable source for the glass that found its way into northern Europe during the Iron Age. The chemical composition of glass that was made on the Levantine coast changed somewhat over time (see Chapter 8). The $^{87}Sr/^{86}$ isotopic signature of natron glass (Freestone et al. 2003; Henderson et al. 2005, fig. 3, Degryse and Schneider 2008) reflects the use of 'modern' seawater taken up in the shell fragments present in the sand used (see Section 11.6). If a more refined provenance for glasses made on the coast of the Levant could be achieved using neodymium isotopes, this would contribute significantly to models of Iron Age, Hellenistic, Roman, Byzantine and Umayyad glass production. For example, it may be possible to distinguish between silica sources referred to by Pliny at the mouth of the river Belus in Israel and that referred to by Strabo at Sidon in the Lebanon. Chemical characteristics of mineral impurities found in sands can sometimes provide a means of distinguishing between glasses made from different sand sources, but contrasting geological ages of silica used in various source areas will provide contrasting ^{143}Nd/^{144}Nd signatures and a more robust means of provenancing glass.

Another potential production area with a different silica source is Alexandria and the North African coast. If its isotopic signature can be distinguished from that in the Levant, this would provide an excellent provenance tool in combination with the chemical distinctions between Levantine and Egyptian natron glasses already observed (Nenna *et al.* 1997, 2000, 2005). The question of whether the glasses can be distinguished is not a simple one to answer, but both glass from Tell el-Ashmumein (Freestone *et al.* 2003) and Tell el-Amarna (Henderson *et al.* 2010) in Egypt have ^{87}Sr/^{86} values of 0.7079 and 0.7080, whereas Levantine sands have values of 0.7090. However, the Egyptian strontium signatures probably relate to glass made at inland locations with a limestone strontium source. What is missing are data for Egyptian coastal sands to make a direct comparison. Nevertheless, Krom *et al.* (1999) found that carbonate-free particulate matter from the Nile delta extending as far as the Bay of Haifa had a value of 0.707.

Some chemical data for Etruscan glass suggests that a production zone other than the Phoenician one in the Levant exists (Towle and Henderson 2008). These glasses would also be expected to produce different ^{143}Nd/^{144}Nd signatures. In the Hellenistic period, it is likely that one or more production centres were located on the Levantine coast, with others on the Alexandrian coast (see Chapter 7). Moreover, it has been suggested that there was an independent primary manufacture of Hellenistic glass on Rhodes (Rehren *et al.* 2005; see Chapter 8), presumably using different sand sources from those used to make glass on the Levantine coast. The ^{143}Nd/^{144}Nd signatures of sand, raw glass and glass artefacts from Rhodes should provide a means of testing this.

There is ample evidence for glassworking in eighth- to first-century B.C. transalpine Iron Age Europe (Gebhardt 1989; Henderson 1989, 1991; Venclová *et al.* 2009). Innovation in colouring glass probably occurred, especially to produce opaque yellow and opaque white glass using tin oxide (Henderson 1985). In the second century B.C., glass technology changed from a base natron glass containing antimony trioxide to one containing manganese oxide. A change in the cobalt sources used to produce blue glasses also occurred then (Henderson 1991). The question remains as to whether Iron Age glass was also made from local sand and imported natron in Iron Age Europe north of the Alps. The proto-urban nature of society in the second century B.C. onwards, with the development of increasingly larger-scale high-temperature industries, is the one time during the Iron Age when primary glass production might be expected to have occurred (Henderson 1991). Chemical analyses of such glasses do not (yet) provide evidence for primary glass production there, and no archaeological evidence for it exists. The determination of ^{143}Nd/^{144}Nd and ^{87}Sr/^{86} signatures, particularly in the raw glass found on proto-urban oppida, could at least provide evidence for the use of different silica sources from those in the Levant and northern Italy, and, by inference, evidence for local production.

In the absence of archaeological evidence for primary glass production in early Roman Italy, an isotopic approach may prove to be the best way of providing evidence for it. As noted in Chapter 4, there is chemical evidence that natron glasses with distinctively different 'European' chemical compositions existed, and this suggests that there was more than one principal production centre for it. Again, the ^{143}Nd/^{144}Nd signatures of a variety of potential sand sources would need to be determined to build up an environmental context against which to compare the results for what are considered to be glass vessels of types made in Italy, such as incised, moulded and polychrome vessels. Such proof would, in turn, provide a means of distinguishing between Italian, Alexandrian and Levantine products. Given that this early Roman glass used for the manufacture of characteristic Italian glass vessels may have been the same (raw) glass that was exported to transalpine Iron Age Europe to manufacture thousands of glass beads and armlets, isotopic investigations could provide a clear connection between the two. Indeed, isotope results for early Roman glass have already provided a distinction between typical Levantine glass and glass apparently made in the western Mediterranean or possibly northwestern Europe (Degryse and Schneider 2008; Degryse et al. 2009). However, it is difficult to know whether this indicates that there was independent glassmaking in Italy or somewhere in northern Europe at the time. The chemical analysis of some Roman glass from Tienen in Belgium falls into the Levantine I compositional type (Freestone et al. 2000), but its isotopic signature indicates that it cannot have been made in the Levant (Degryse et al. 2009, 68). Moreover, the glass of this compositional type is two centuries earlier in date than that defined chemically as Levantine I glass. The neodymium and strontium isotope results have therefore shown how important a combined use of isotopic and compositional analysis is for the study of ancient glass production and provenance.

The production of later Roman, Byzantine and Umayyad natron glass is thought to have involved the same potential supplies of natron and sand, and so precisely the same principles would have been involved, as discussed earlier. As noted in Chapter 8, there were distinctive compositional differences between natron glasses produced in the Hellenistic and later Roman periods. We know that the natron glasses made in the second half of the first millennium A.D. have ^{87}Sr/^{86} signatures that reflect the use of shell fragments in sand as a lime source (Freestone et al. 2003; Henderson et al. 2005). The determination of ^{143}Nd/^{144}Nd signatures in a range of sand samples may be the way forward.

Another aspect of isotope studies that can shed light on local manufacture (discussed earlier in the context of the Islamic plant ash glass) is the mixing of raw glass. Such mixing may provide a means of characterising the *chaîne opératoire* and could characterise specific production sites or production zones. A mixing line for ninth-century natron imported glasses excavated from al-Raqqa, Syria (Henderson et al. 2005, fig. 3), probably reflects variations in the proportion of feldspars present in the Levantine sands, the age of the seashells (the primary lime source) being constant. It also suggests that glass made with sands from different locations on the Levantine coast were mixed.

Both Freestone et al. (2005) and Degryse et al. (2006) have demonstrated that strontium isotope characteristics of technologically distinctive Byzantine glasses can be used to show that they were mixed together. In both cases, high iron, manganese and titanium (HIMT) natron glasses were shown to have been mixed with a more standard natron glass. Freestone et al. provided isotopic evidence for the mixing of raw chunks of HIMT glass (see Chapter 4) and a second type of raw glass. They went as far as suggesting that this mixing may have taken place in Alexandria because the Sr isotope signature relates to one of 0.707 for the carbonate-free particulate matter found on the Nile delta (Krom et al. 1999). Freestone et al. (ibid., 155) have suggested that mixing of the two raw glass types as indicated by the mixed beach and fluviate sands occurred at close proximity to the furnaces that produced each. They have suggested this because the addition of glass with compositional characteristics of a different production centre would produce the chemical and isotopic (lead and strontium) mixing lines they obtained. An alternative explanation is that the raw glass made with beach sand of a contrasting chemical composition was deliberately separated and that the glass was imported from some distance away. The study by Degryse et al. (2006) demonstrated, using lead and strontium isotope analysis, that mixing of natron glass made with a marine sand source and HIMT glass had occurred. However, in this case, the glassworkers were able to identify the glass types, presumably with slightly different working properties, by their colour: the vessel and chunk HIMT glass was dark green and yellow-green; the natron glass was blue. Furthermore, an important implication for glass technology was the discovery that vessel glasses were produced as a result of mixing raw glasses of different isotopic compositions, showing 'a continuous gradation between . . . end members' (Degryse et al. 2006, 497). However, this mixing is unlikely to provide a provenance for the artefacts as suggested by Degryse et al. (2006, 496) because the glasses that lie between the end members could have been made anywhere so the locational information would be lost.

It is clear that radiogenic isotopic analysis of glasses not only can show if glasses were mixed, it can also show that mixing occurred when vessel glasses were produced, presumably therefore on secondary production sites in crucibles.

11.8.2 The Supply System

11.8.2.1 Bronze Age and Islamic Plant Ash Glass

With some exceptions, plant ash glass production must have involved the use of local sources of plant ash and possibly of silica. Bronze Age Mesopotamian glass was manufactured for the social elite often within the context of the palace economy, and it involved specialists (see Chapter 5). Historical references that describe its production refer to the use of what was probably crushed quartz; low levels of neodymium and other trace elements found in late Bronze Age Mesopotamian glasses support this (Henderson *et al.* 2010). If a less pure silica source were used, such as sand, this would also be reflected in trace impurities in glass made from it – and indeed this is the case for late Bronze Age Egyptian glasses (Walton *et al.* 2009; Henderson et al. 2010). Historical documents also describe the use of plant ashes being harvested in the late summer or early autumn for the manufacture of glass. Both principal raw materials would therefore have been gathered from the surrounding environment and supplied to those involved in making glass associated with the palace economy (see Chapter 5). Isotopic research can provide evidence for the zones from which the raw materials originated and whether they were local or non-local.

There is ample evidence that 'raw' ingot glass was used for the manufacture of Bronze Age vessels and jewellery, as mentioned in Chapters 5 and 6 (Bass 1986; Pulak 1997; Henderson 1997; Nicholson 2007; Jackson and Nicholson 2010; Henderson *et al.* 2010). The occurrence of ingots on the thirteenth-century Ulu Burun shipwreck shows that such raw glass was supplied to secondary production centres where glass was formed into objects characteristic to that region – for example, moulded pendants in Mycenaean Greece (Henderson *et al.* 2010). Lead would have been imported, because it was used commonly in combination with antimony as an opacifier (see Chapter 3). This is where lead isotope determinations of (potential) lead sources and of lead-rich glasses could provide evidence for the other component of the supply system (Shortland 2006).

Like Bronze Age glass, Islamic glass production involved not just raw material procurement but also artisans and therefore know-how. Glass artisans would have brought their own traditional techniques, potentially from distant corners of the caliphate. Although the manufacture of Islamic glass on a massive scale may well have involved the import of sand of a known quality, the import of plant ashes from distant (geologically contrasting) areas is considerably less likely given the ubiquity of halophytic plants in and around the production areas. If, for some reason, the 'best' plants for glass production were considered to be located some distance away, this would complicate provenance studies. However, in the case of al-Raqqa, for example, plants with a 'local' signature were used to make the glass there. There is a historical reference to the exploitation of a specific sand source in Syria: Yāqūt (d.1229), who lived in Aleppo, has described the use of fine white sands for glassmaking there (*Buldān* I, 631). The (geological) source of sand mentioned by Yāqūt was Jabal Bishr (Jebel Bishri) on the south side of the Euphrates between al-Rasāfa (to the west of al-Raqqa) and Tadmur (Palmyra; Heidemann 2006, 40–41). Unpublished isotope data confirms that this source was used for glass production in al-Raqqa and that it differs from signatures of silica used in glass from Damascus and Beirut. Naturally, other silica sources could also have been used, but to be able to state that a specific geological source of sand was used to make glass is a new development in glass research. Moreover, during glassmaking at al-Raqqa, it is already evident that two types of sand were largely segregated, as noted earlier. The next step is to attempt to distinguish between potential inland silica sources isotopically.

If an additional calcium-rich raw material was used, it would also contribute a $^{87}Sr/^{86}$ signature to the glass. If from a local calcium-rich mineral formed at the same time as the local limestone, it would simply duplicate the isotope signature introduced with calcium in the plant ashes (so the mixture of calcium sources would not be apparent). If of a different geological age and different proportions of the two calcium types were mixed, an isotopic mixing line would result. Certainly the $^{87}Sr/^{86}$ signatures for plant ash glasses made at al-Raqqa are so constrained that it is impossible to provide supporting evidence for the use of a second calcium-rich raw material suggested using chemical analysis.

11.8.2.2 *Phoenician, Hellenistic, Iron Age and Roman Natron Glasses*

The production of plant ash in the ancient world involved multiple sources of plant ash potentially with their own chemical and isotopic signatures. Far more restricted sources of natron were used for the manufacture of natron glass: chemical compositions of ancient natron glasses show that any compositional distinctions in natron glasses mainly relate to the use of different sand sources. As discussed in Chapter 2, Pliny mentions traders in natron landing on the Levantine coast, the presumed natron source being Wadi el-Natrun in Egypt. Nevertheless, as also noted there are other potential evaporite sources that may have been used in antiquity, such as in Macedonia, Turkey and Syria. However, this raises a number of questions. Can these sources of evaporites be distinguished compositionally or isotopically? If they can, are the isotopic signatures retained in glasses made from different evaporite sources? It is apparent that the $^{87}Sr/^{86}$ contribution from Egyptian natron is negligible (Professor Patrick Degryse, *pers. commun.*); its $^{143}Nd/^{144}Nd$ signature is yet to be determined, but is unlikely to have made a significant contribution. Therefore, isotopic characterisation of the silica (almost inevitably coastal sand) used should allow the sources to be narrowed down, but only a comprehensive survey and isotopic characterisation of coastal sand deposits around the Mediterranean will provide a means of answering the question as to whether sand deposits were exported to production centres.

11.8.3 Trade

11.8.3.1 Trade in Bronze Age and Islamic Plant Ash Glasses

There are historical references to the import of glass (ingots) to Egypt from Mesopotamia, and the Ulu Burun shipwreck off the western coast of Turkey provides the archaeological proof for the trade in glass ingots in the Mediterranean (see Chapter 5). Indeed the goods exchanged between kings formed part of the much larger trade network (Akkermans and Schwartz 2003, 352). Therefore, an isotopic study of Bronze Age raw and artefact glass must take the potential trade in glass into account, and this is why a map of strontium variations across Mesopotamia, Greece and Egypt would form an essential background to tracing the movement of traded glass between these Bronze Age kingdoms. The relatively small-scale production of Bronze Age glass in these zones probably involved only the use of ashed plants and silica local to the palaces, and these would not have been traded. Again, however, the mapping of ^{143}Nd/^{144}Nd signatures of a range of (potential) silica sources is crucial to a study of this kind. An important question to be addressed is the extent to which Egypt was reliant on the import of glass from Mesopotamia. Isotopic results have so far shown that silica and/or calcium sources of contrasting geological ages were used to make fourteenth-century B.C. glass found at Tell el-Amarna, Tell Brak and Nuzi (Chapter 6). Isotopic studies have also shown that glass was exported from both Egypt and Mesopotamia to Greece (Henderson *et al*. 2010; Degryse *et al*. 2010). However, export of glass from Mesopotamia to Egypt, as specified in the Amarna letters, has not (yet) been demonstrated isotopically.

Moreover, it is clear that Ugarit and its port Ras Shamra acted as an important buffer zone between the Mesopotamian, Hittite and Egyptian worlds. The occurrence of an enormous amount of glass there, including ingots, together with probable evidence for the primary production of vitreous materials (see Chapter 5), provides a context in which the determination of radiogenic isotopes in the glass (and faience) could contribute in a significant way in attempting to distinguish between locally produced plant ash glass and glass that was exported to or imported from other centres. With an initial set of isotope data for this thirteenth century glass, it can be demonstrated that two separate sources of silica were used in the glass made at Ugarit (Biron *et al*. 2012).

During the Islamic period of plant ash glass production, there is historical evidence for trade in glass raw materials. In 985, the Arabic geographer al-Muqaddasī mentions that ūshnan (kali) – plant ash – was an export from the province of Aleppo for glass production (Ashtor 1992, 482), but we are not told where it was exported to and what it was exchanged for. With greater definition of the variations in strontium isotope signatures in the Mediterranean, it will become easier to identify the sources of such ashes used in the glasses made from them. Moreover, with the definition of strontium isotope signatures for Syria and the Levant, it should eventually be possible to identify when Levantine and Damascene ashes were used for making Venetian (and northern European) glass between the thirteenth and seventeenth centuries. It would also be possible to define strontium isotope variations in halophytic plants in areas such as southern France, southern Italy, north Africa and the Mediterranean islands. The historical evidence for the export of broken glass vessels, raw glass, plant ash and beach sand from Antioch to Venice in the second half of the thirteenth century (Carboni 1998, 101–2, and n. 12) could similarly be investigated scientifically using a combined chemical and isotopic approach.

In the eleventh century, we have robust archaeological evidence for trade in raw and scrap Islamic glass from east to west in very large amounts. Excavations of the Serçe Limani shipwreck off the coast of Turkey by George Bass revealed two metric tons of raw glass blocks and 11,000 glass lid moils (see Chapter 9; Bass 1984, 65, 67, fig. 4; Bass et al. 2009). It has been suggested that some of the glass exported from the Levantine coast may have been used for the production of eleventh-century (Byzantine) glass mosaic tesserae, to which colourants would have been added as part of a secondary process (Andreescu-Treadgold and Henderson, 2006). The direction of trade in the raw (and vessel) glass could be tested against environmental $^{143}Nd/^{144}Nd$ and $^{87}Sr/^{86}Sr$ signatures, and by doing this the suggestion (*ibid.*) that plant ash glass mosaic tesserae used in Byzantine church mosaics were derived from the Middle East could also be tested. Moreover, the suggestion that (raw) plant ash and natron glasses were mixed in the eleventh century (Henderson 2003) could also be tested by using $^{143}Nd/^{144}Nd$ and $^{87}Sr/^{86}Sr$ determinations. The result of mixing natron and plant ash glass would be to 'dilute' the 'modern' $^{87}Sr/^{86}$ signature derived from shells with the older signature derived from limestone. One interpretation is that the demand for the production of glass mosaic tesserae in the Byzantine world was so great that scrap natron vessel glasses and possibly raw glass were mixed with raw plant ash glasses and then traded (Andreescu-Treadgold and Henderson 2006). It is unlikely that natron glass was still being fused from raw materials after c. 800 A.D.

Given the (apparent) evidence for localised production of specific decorative forms of Islamic glass vessels discussed earlier, the combination of chemical compositions and isotopic work could eventually provide evidence for the origin of Islamic glass vessels which found their way to China along the Silk Raod, where, in the ninth century, there were significant settlements of Muslim Persians and Arabs (Guy 2005, 11–12). An example of such an export is to be found in a ninth-century (Tang dynasty) context in the treasury located in the crypt of the Famen Temple in Shanxi province, China. The finds were placed in the crypt before 874 A.D. They included a highly prized complete deep translucent scratch-decorated plate and a yellow-brown bottle (Gan Fuxi 2009b, figs. 2.31 and 2.32). The chemical analyses of vessel fragments from the treasury (Fenn *et al*. 1991, 62, no. 5886, table 1; Carboni 2001, 80; Gan Fuxi 2009b, table 2.8) provide evidence that two yellow glass fragments (with elevated magnesia levels) derived from Mesopotamia (Iran/Iraq in Fig. 10.11). One of these was from a straight-sided cup. The composition of the blue fragment is suggestive of a possible 'Syrian' provenance.

By determining trace element and isotope signatures in these vessel types, it should be possible to determine with greater confidence whether the glass used to make them was fused in Cairo, the Levant, Damascus, al-Raqqa, Sāmarrā or Nishapur or other centres on the Silk Road. Looking to west, the provenance of enamelled Islamic vessel fragments found on Scandinavian and other sites could also eventually be provenanced using an isotopic approach.

11.8.3.2 Trade in Phoenician, Hellenistic, Iron Age and Early Roman Natron Glasses

The Phoenicians were famous for being actively involved in trade. Their characteristic stratified glass eye beads are found all over the Mediterranean world (Spaer 2001, 81 *ff*.). Although the Levantine coast is the most likely source of the glass used to make these glass beads and the early Hellenistic core-formed glass vessels (see Chapter 7), another possible production centre, is the island of Rhodes, especially from c. sixth century B.C. If indeed Rhodes proves to have been a primary glassmaking centre, then the ^{143}Nd/^{144}Nd and ^{87}Sr/^{86}Sr results for Hellenistic vessels of a Rhodian type would provide the basis for assembling trade patterns for the vessels around the Mediterranean and beyond. If during the Hellenistic period raw glass was fused from raw materials in Alexandria (as well as vessels being formed there; see Chapter 7) then, as noted earlier, distinctive ^{143}Nd/^{144}Nd and perhaps ^{87}Sr/^{86}Sr determinations could provide a fingerprint for their origin and for glass traded to other Mediterranean sites and regions.

The Iron Age peoples of Europe probably imported Hellenistic raw glass made in the Middle East via intermediaries. The third-century B.C. shipwreck of the *Sanguinaire* with 550 kilograms of glass on board, including highly coloured raw glass, is testament to the scale of production and trade in glass at the time. However, there is still a possibility that imported raw glass was modified with the addition of colourants or opacifiers in European Iron Age centres, especially after the second century B.C. when proto-urban oppida developed. Trade in Roman glass occurred on a massive scale. Large-scale manufacture of raw glass occurred in the Middle East via for which we have an increasingly sophisticated scientific characterisation. Therefore there is the potential to demonstrate that trade has occurred. The area we are concerned with, stretching from Western

Europe to northern India, is enormous, and this reminds us that the scientific provenancing of Roman glass is at an early stage: most research has focused on the Mediterranean and Europe. By determining stable and radiogenic isotopes in Roman glass vessel fragments, there is a prospect of detecting the mixing of different glasses. Nevertheless, for good reasons (Degryse *et al.* 2009), most determinations have been carried out on raw glass. Shipwrecks provide evidence of the large-scale distribution and trade of Roman glass, including ingots (Foy *et al.* 2000). Examples of these are the late second–first half of the third century *Julia Felix*, which sank in the Adriatic off the coast of Italy on which 11,000 glass fragments were found (Toniolo 2005), and a first-century B.C. ship that sank near Antikythera, southwestern Greece (Weinberg *et al.* 1965; Weinberg 1992) and many others (Foy and Nenna 2001, 100–12)

11.9 Conclusions

1. Glass raw materials bring with them compositional characteristics that relate to the geological context from which they are derived. Apart from compositionally different plant genera, such as those that are potassium-rich as opposed to sodium-rich, the ratios of oxide concentrations in plant ashes appear to be largely dominated by the geology of the soil in which they grow. Nevertheless the physiology of the plants can override the geological environment and determine the ratio of sodium to potassium. The levels of these alkalis vary inversely to other elements in plant ashes of the genus *Salsola*. The chemical composition of plant ashes alone may eventually be relatable to the glasses made from them, but a number of factors need to be incorporated in any interpretation. These include production processes such as ashing techniques, purification of ashes and the fusion of raw materials in small crucibles – all of which can modify the original composition of the ash. More work to identify the variation in the mineralogical and chemical compositions of ashed halophytic plants from around primary glassmaking sites and a comparison with glasses made on the sites needs to be carried out. It is, however, difficult to know precisely why particular plants and plant colonies were selected and which techniques were used to prepare plants for glass production. By chemically analysing the raw furnace glasses from a primary glassmaking site and comparing them with the chemical compositions of halophytic plant ashes taken from the local area, any similarities and differences will become apparent. For elements that are unaffected by the melting conditions, such data could eventually be used as a means of provenancing glass.

   For large-scale glass production, the levels of many impurities associated with silica would remain unaffected by the production techniques. However, two possible modifications might occur for smaller-scale production. The first is the introduction of impurities through the use of a silica-rich stone implement for crushing quartz pebbles. The second is when glass is enriched with elements as a result of interacting with the inner face of a crucible or furnace. For Bronze Age glass from Mesopotamia, for example, the neodymium level is so low, at c. 1 to 2 ppm, it is likely that the quartz alone – and not the inner face of the crucible, the crushing tool or the plant ash – contributed any neodymium. Therefore, other trace elements associated with silica in these glasses are most likely to have derived from quartz alone.

2. It has been known since Sayre and Smith published their seminal paper in 1961 that the chemical compositions of ancient glasses changed over time. The chemical compositions of ancient glasses (see Chapter 4) can be linked to the geographical zones in which they are found. This is a *geographical provenance*, but it does not independently provide a *geological provenance* for the glasses made there. These chemical compositions are, in fact, the summed result of (a) using raw materials which introduce impurities, (b) the ways in which those raw materials have been prepared, (c) the compositional modifications that may have occurred when the raw materials were fritted and then fully fused into glass and (d) any mixture with scrap glass and colourant materials. Furthermore, the decisions taken about how to select, prepare and fuse raw materials, and about the ways in which the glass is worked, could be determined by the ethnicity of the various groups involved.

3. The determination of radiogenic isotope signatures in glass provides a *geological* provenance for it. This new approach has an enormous potential to revolutionise ancient glass studies. First, it provides an independent means of determining the relative geological age of the silica used to make glass. Second, in the case of plant ash glass, the approach provides a relative geological age for the calcium component in the soil in which the plants used to make glass grew. If these results are set within an environmental context, in which the variations of the strontium and neodymium isotope signatures are determined across the landscape, it has great archaeological potential. In turn, this can show how glass made from raw materials deriving from contrasting geological zones and can be used to build up trade patterns and perhaps even evidence of diplomatic gifts. Moreover, the extent of glass recycling can be demonstrated.

Determination of lead isotope signatures in translucent natron glasses, also thought to reflect the origins of sands through the association of lead-rich minerals in the sands, has substantiated the evidence of raw glass mixing. The isotopic study of natron glass has relied largely on the determination of strontium isotope signatures. However, contrasts in strontium isotopes in different plant ash glassmaking zones provide a far greater potential for identifying geological provenance with a greater geological resolution, than for natron glass.

As noted earlier, for late Bronze Age (plant ash) glasses, changes in raw material compositions as the glass is fused can potentially create problems with attempting to provenance it using chemical analysis. There is nevertheless a trace element distinction between Egyptian and Mesopotamian glass; however, this can be clarified further through the determination of neodymium and strontium isotope signatures associated with the different silica and plant ash sources. This has produced a clear distinction between glasses made in Mesopotamian and Egyptian zones and evidence of separate production zones in Mesopotamia.

TWELVE

CONCLUSIONS

The variety of approaches that can be taken to the study of ancient materials should be mutually enriching. The challenge is not to see the different approaches as separate bodies of information but to combine them to investigate key aspects of the production and use of materials in ancient societies to create overlapping and intersecting results. In this volume, ancient glass was explored in broad environmental, social, archaeological, historical, political and technological contexts.

The transparent, translucent, brilliant, refractive and colourful properties of glass clearly set it apart from other materials. A lack of crystals in transparent and translucent glass allows wavelengths of light to be transmitted, absorbed and reflected, its amorphous character underpinning its unique nature. From the earliest production and use of small glass vessels, pendant discs and perhaps beads in elite political and ritual spheres via the manufacture of highly coloured beads and armlets, as well as moulded vessels forming part of dining rituals, to the mass production of containers following the introduction of glassblowing and the production of cathedral and church windows, glass has been used for a variety of functions for at least 4,500 years. Glass objects have been used in various ways, and their values have changed at different times according to the contexts in which they were manufactured and used by different sectors of society.

Various groups of people were involved in the gathering of raw materials (including fuel) and their preparation, the building of furnaces, the provision of crucibles and glass moulds, the manufacture of blowing irons, and the blowing of glass vessels into moulds with complex decorations. Each of these groups would have had their own traditions, experiences and ethnicities that in combination would have had an impact on the processes of glass production. Each group contributed to the *chaîne opératoire* in slightly different and overlapping ways, and each needed to coordinate and interact with members of the other groups involved. Underlying the production processes were the implicit or explicit rituals associated with other aspects of life, such as the changing seasons, agricultural practice, religious beliefs and associated rituals that would all also have articulated with glass production processes and their significance to society. For example, the (potential) management of halophytic plants and trees and their ashing would have been articulated with the agricultural cycle. Indeed farmers may well have been the social group involved – as they are today in Syria – in ashing the (same) plants for the manufacture of both glass and soap. Groups involved in the provision of high-temperature refractory clay and bricks for the manufacture of crucibles, furnaces and fritting kilns may well have interacted with those making crucibles and furnaces for metal smelting and working, and with potters selecting ashes for glazes, clays for different pottery fabrics and for saggars, pottery kiln construction and kiln furniture (Henderson 2000a). If production of glass and pottery occurred together in the same industrial area, then communication and collaboration between those selecting raw materials, preparing them and constructing kilns and furnaces would potentially have been maximised, with the same groups potentially providing materials to a range of industries.

Glass was the first truly synthetic material, and as a result its manufacture occupied an unusual and innovative place in the history of ancient technology. Production of the first glass must have been preceded by an experimental phase involving raw materials (plant ash and silica) that had no visual connections with the final product. This innovation required a step change and, it has been argued here, a paradigm shift in ancient technology that was in many ways a far greater leap forward than the (first) production of pottery from clay or metal from minerals. Even though elements of glass production must have been influenced by faience, pottery and metal production, the invention of glass in the Middle East c. 2500 B.C. evidently came at a time when the socio-economic and socio-political context supported this kind of innovation, even though glass was apparently used only for the manufacture of small beads for c. 900 years after its invention. The first glass wasn't introduced until c. 800 B.C. in China, and in the Han Dynasty the first high barium glasses were invented there. It can be argued that the semi-translucent material, jade, fulfilled part of the role in society that glass would have had. Elsewhere there appear to have been 'independent' inventions of glass in late Bronze Age southern Europe (of a mixed alkali composition) c. 1200 B.C. and high-lime, high-alumina glass in Africa c. 900 A.D., for example (see chapter 4).

Unlike pottery bodies and metal, the colour and translucency of glass could be radically modified. Even one of the earliest pieces of glass, from Eridu, Iraq, dating to c. 2300 B.C., is of a rich cobalt-blue colour achieved by adding very low levels of cobalt-rich minerals, forming part of what must have been a highly controlled process. Even at this date, what little evidence there is suggests that glass technology was 'fully formed' in terms of the ability to achieve the correct balance of raw materials as well as the correct furnace temperatures and an awareness of the need to anneal the glass. However quickly glass technology initially emerged, whether over one or two generations or over a much longer period, it involved an experimental stage. The evidence for this would include semi-fused and partially formed glass.

Although increasing amounts of research are being published on early glass, especially for (sixth–eleventh centuries B.C.) late Bronze Age glass from the Middle East and southern Europe (see Chapters 5 and 6), little work has been done on the earliest glass. Of course this is because of the restricted amount of material that is available to study. However, what is available has not been subjected to comprehensive scientific study to test what is still hypothetical to some degree, that first glass was made in Iraq or northern Syria. Such work would need to be fully informed by the isotopic and chemical/mineralogical characterisation of plants, sand and quartz sampled from likely production centres and environments. In principle, it should be possible, within the limits of probability, to confirm or disprove where the earliest glass was made. However, despite a relatively recent publication on faience technology (Tite and Shortland 2008), there is much potential research that might be aimed at investigating the relationships between faience and glass technologies in an environmental context, as discussed here.

Environmental aspects of glass raw materials and glass production are discussed in Chapters 2, 3 and 11. It is certainly a truism to state that there is not a simple relationship among the geological, chemical and/or structural characteristics of sand and plants/plant ashes and the glasses made from them. Although some chemical characteristics of raw materials are transferred directly into the glass, many are modified and transformed by the processes of glass production (Rehren 2008; Paynter 2008). A mineral source of alkali such as natron is far purer than ashed plants (see Chapter 2). Therefore, when natron was used in glass production, the glass compositions and associated impurity levels, as well as the all-important working properties of the glass made from it, were far more predictable. Glass made from plant ashes has a composition dominated by the geological characteristics of the soils in which the plants grew, and this could potentially vary widely. Because minerals such as natron are relatively pure, the chemical (and isotopic) characteristics that they introduce have not (currently) provided any provenance information. However, the minerals in the sands used to make natron glass can provide a provenance (Freestone *et al.* 2000). If other sources of natron were exploited in antiquity, such as in Lake Pikrolimni,

20 kilometres northwest of Thessaloniki, it would be important to distinguish them from the deposit at Wadi el Natrun in Egypt, which certainly was (see Chapter 2).

Research into the occurrence and environmental context of glass raw materials has only been (re)initiated relatively recently, especially for glasses made in the West and the Middle East. This is an area of glass research that was (inevitably) initiated by W.E.S. Turner. He published some of the earliest analyses of (Nile) sand and of plant ashes in the context of Bronze Age glass technology (see Chapters 2 and 3). Subsequent research by Sanderson and Hunter (1981), Jackson *et al.* (2005) and Barkoudah and Henderson (2006), for example, has highlighted the complexities and challenges as well as the need for further research. Part of the reason for the resurgence in this research area is due to the need to determine isotopic variations in sands and plants from contrasting geological origins as an essential prerequisite to glass provenance. Sampling and characterisation strategies need to be clearly structured to address specific questions about provenancing and trade involving contrasting geological zones. The same isotopic results can even sometimes be used for glass of different dates. For example, late Bronze Age, Sasanian and Islamic plant ash glasses may have been produced in the same or similar areas and have involved the exploitation of halophytic plants and silica from the same geological zones.

Investigation of glasses made from Indian alkali sources such as reh and oos is at an early stage, and further detailed research into their occurrence, distribution and chemical, isotopic and mineralogical characteristics is necessary. When combined with an isotopic characterisation of silica sources and compared with the chemical and isotopic characteristics of Indian glasses, this could provide provenance information for glasses in addition to that already provided by trace element analyses (see Chapters 2 and 4). Because reh is an alkali source of lesser purity than natron, with higher alumina levels and potentially higher levels of potassium and magnesium, there is a greater potential for provenancing the glass made from it. Of equal significance is the increasing amount of scholarly work that has been published on Russian, central Asian, African and Southeast Asian glasses. In all of these areas, there is further scope to characterise and provenance silica sources, the barium sources used in China, and the lead sources used in the Middle East and Russia, as well as China, where some of the earliest lead-rich glasses were manufactured (see Chapter 4). Brill and coworkers have provided some evidence using lead isotope analysis for regional variations (see Chapter 11). However, the use of several different raw materials that contain lead (including minerals in sand), the fractionation of lead isotopes and their mixing makes it difficult to assess how significant such work will be in the long term. Currently, the use of oxygen isotope analysis of ancient glass provides broad geographical contrasts.

Therefore, the environmental approach to studying glass production and provenance is at a relatively early stage. We are becoming aware of its shortcomings, such as the use of glass raw materials of the same geological age (with the same isotopic signature) in different production zones. Nevertheless, even if a combined chemical and isotopic approach is unable to provide direct evidence for provenance, it will still be possible to exclude some areas of origin from consideration and will help to cut through the complexities of characterising raw materials and glass by using only chemical analysis and will go beyond the reliance solely on archaeological evidence for production (see Chapter 2), providing a *geological provenance*.

As noted in Chapter 4, glasses of a range of ages found in different parts of the world have different chemical compositions, providing a somewhat broad *geographical provenance*. Some of these differences were established half a century ago by Sayre and Smith (1961). However, as increasing numbers of scholars have become involved, there is increasing evidence for more complex variations and for technological innovations. One example comes from recent research in European natron glasses. First, the date for the first tin-opacified glass has been pushed back from the second century B.C. (Werner and Bimson 1967; Henderson and Warren 1983; Henderson 1989, 50–2) to the sixth and seventh centuries B.C. Its use was detected in glass used to decorate Polish glass beads (Purowski *et al*. 2012). This supports the suggestion that tin opacification was invented in prehistoric (eastern?) Europe. Another fascinating characteristic of some of the glasses that Purowski *et al*. describe is that some pale blue samples are not characterised by the virtually universal correlation between magnesium and potassium oxides found in western natron glasses: these Polish glasses contain low magnesium and medium potassium levels (see Chapter 4). Of equal significance is the presence of crystalline silica in some of these Polish natron glasses, providing a possible technological link with mixed alkali (low magnesium-high potassium) European glass produced in the late Bronze Age (Henderson 1988a; Brill 1992; Venclová *et al*. 2011). These three characteristics provide evidence that could eventually show that primary glass production occurred in the seventh and sixth centuries B.C. in Europe. Second, in an analytical survey of prehistoric glass in France, Gratuze (2009) identified some blue natron glasses characterised by high magnesia, low potassium and high alumina levels (of between 4% and 8%). It is difficult to suggest an origin for such early examples of high alumina glasses. Third, from early Byzantine Bulgarian contexts, Rehren and Cholakova (2010) have identified an HIMT (natron) variant that may have been manufactured locally.

Publication of data that reflect the complexities of southern and central Asian, African and Southeast Asian glass technologies (Gan Fuxi 2009b; Li Qinghui *et al*. 2009a; Dussubieux *et al*. 2008; Robertshaw *et al*. 2010) is starting to provide a balance for what has largely been a focus on Western and Middle Eastern glasses. Contributions to tracing the movement of glasses along the Silk Roads at different periods have also been published recently (Gan Fuxi *et al.* 2009; Zorn and Hilgner 2010). As noted in Chapter 4, there are, for example, both similarities and significant differences between 'western' and southeastern glasses.

These exciting discoveries exemplify how our knowledge of ancient glass technology can be quickly modified. The social, ritual, political and economic factors that have led to these changes are key factors that also need to be considered in detail.

In this book, three case studies (in Chapters 5–10) have been used to exemplify how a range of approaches can be used to illuminate ancient glass studies. It is highly significant that even some of the earliest glass was brightly coloured. Glass colour and its modification must be considered to be one of the primary driving forces for its invention (Robson 2001). There is a clear connection between the belief that lapis lazuli, a cobalt-coloured semi-precious stone, had apotropaic properties and the need to imitate it using deeply coloured cobalt blue glass. Some 1,000 years after glass was invented, a reference by the Egyptians was made to 'lapis lazuli from the kiln' and 'lapis lazuli from the mountain' (see Chapter 5), so by this time there was a clear connection between the mineral and glass. The ritual value of early glass almost certainly led to recording its manufacture and the associated rituals in great detail in cuneiform texts. The instructions in these cuneiform tablets are written in Sumarian and may date back to as early as 2000 B.C.

The cuneiform texts mention the elite context of glass production, the complex rituals involved in its production reinforcing its value and the imitation of brilliantly coloured semiprecious stones such as lapis lazuli and turquoise. The Mycenaean Linear B script referred to a 'blue stone', *kyanos* (Dinsmore 1973, 20) and moulded glass pendants in these colours formed part of the Mycenaean palace economy (Ostanso 1999, 328; Nikita 2003, Nightingale 2008). The chemical and isotopic analyses of late Bronze Age glass in the Mediterranean reveal the careful use of colourants and evidence for regional production and trade within and between Mesopotamia, Egypt and Greece. Scientific analysis of raw and artefact glass confirms the specialised manufacture of different glass colours: evidence for the manufacture and trade in brightly coloured blue, purple and opaque turquoise glass this. Opaque turquoise glass had a clear ritual value and role, as the disc pendants found in temples for the goddess Ishtar (Stern and Schlick-Nolte 1994, 119) clearly demonstrate. Thus, the study of Bronze Age glass colour allows historical, archaeological, ritual, social and scientific aspects to feed into a mutually integrated whole (Duckworth 2011; Duckworth 2012). As already noted, a key aspect of scientific research that is now able to provide a provenance for late Bronze Age glass is the determination of radiogenic isotopes. It underpins an exciting way forward for tracing trade routes in glass, which should eventually lead to insights about how glass of the same origin was used in different social, ritual and political contexts in the countries to which it was traded. Such material, whether in ingot or other artefact form, would have had the added values and meanings of having been imported from and manufactured by Mesopotamian or Egyptian artisans, potentially under the control of palace or temple officials. We are therefore at the beginning of a fascinating research area with enormous potential. It can be argued that the chemical compositions of ancient plant ash glasses are partly a reflection of the ways in which the raw materials are prepared and the ways in which they interacted with the buffering material that was used to line the crucible used to make them. However, neodymium and strontium isotopic signatures are dominated by the geological age of the raw materials used and are therefore apparently largely independent of such factors (Henderson *et al.* 2010), offering an objective way through the potential complexities of interpreting chemical analyses. For the relatively small-scale manufacture of Bronze Age glass, isotopes are currently providing some clear answers. For later periods, the technique is important, but in different ways.

Late Hellenistic and early Roman glass production was carried out within the context of fully formed sophisticated city states. Glass objects were designed to be used in a range of urban and rural social contexts. This was an important period for glass production partly because glassblowing was invented in the first century B.C. Glass of the late Hellenistic and early Roman periods is also interesting because it involves a continuation of the use of highly coloured glass from prehistory and the early classical world. Augustus, who reigned from 27 B.C. to A.D. 14, a period when highly coloured glass was used, imported Syrian and Judean craftsmen with glass expertise so as to learn from them and reproduce their skills (Fleming 1997, 3). This occurred before the introduction and use of mass-produced pale-green glass containers, some of which were used as vessels in the burgeoning sea-borne trade system for agricultural and other products such as olive oil. If glassblowing had not been invented, mass-produced highly coloured moulded blue, brown, emerald green and purple open vessel forms would have continued to have been made, each colour undoubtedly having its own significance to the user.

Unlike Bronze Age glass manufactured in crucibles, the natron glass made between the second century B.C. and first century A.D. was produced on a large scale in enormous tank furnaces (see Chapter 7), mainly on the Levantine coast. The tablewares such as bowls and plates, some of which were copies of *terra sigillata* angular moulded ceramic wares, evidently performed key social roles, as part of table services and rituals associated with dining (Cool 2006); the rituals of eating and drinking playing important parts of social life and which played out social stratification.

At the top end of the social scale, there appears to be a clear historical reference to the use of diachroic properties of late Roman glass vessels during prestigious banquets. Dichroism is a rarely seen visual effect in ancient glass: the same glass appears a different colour in reflected and transmitted light (Freestone et al. 2007). A remarkable historical reference published by Whitehouse (1989) could potentially even refer to a vessel such as the late Roman cage cup, the Lycurgus cup. In the early fourth A.D. century Vopiscus wrote a life of the third-century pretender Saturninus. In it he mentions a letter supposedly written by Hadrian to his brother-in-law Severianus in Rome which included the following:

> *I have sent you particoloured cups that change colour, presented to me by the priest of a temple. They are specially dedicated to you and my sister. I would like you to use them at banquets on feast days.*

Such glass vessels were therefore amongst the most socially prestigious to have been used on important social occasions. Whitehouse has suggested that the change in colour from green to red symbolises the ripening of the grape, and that the cup, which had scenes on it that included the god Dyonysus, the god of wine, was intended to be used for banquets dedicated to Dyonysus.

A dip in the scale of production, the quality of glass and the range of vessel forms occurred in the early Byzantine period. Perhaps this occurred because the climate changed as a result of a volcanic explosion and a pandemic (Arjava 2005; Little 2007). The reintroduction of plant ash glass by the Muslims by the ninth century in some respects provided a springboard for an increase in the scale of production of glass in inland locations using locally occurring raw materials. Nevertheless, in the seventh and early eighth centuries, there were interesting interdependences of the Muslims on the Byzantines, with, for example, the provision to the Umayyad caliph of glass mosaic tesserae to decorate the early eighth-century 'imperial' mosque in Damascus. There were early men of science in both the Byzantine and Muslim worlds at the time and the Muslim early scientists became especially prominent from the later eighth and early ninth centuries (Mathews 1998, 91; Al-Hassan 2009). From this time, Islamic glass production appears to have been located in urban or semi-urban environments just as it was in classical contexts. Vessels, window panes, wall decoration, beads and armlets were mass produced. Just as in any other period, they symbolised the society in which they were manufactured, including belief systems, rituals, political control over artisans and other groups associated with production, economic enhancement or restriction affecting the production processes and the sharing of technological and other knowledge. By the ninth century, some vessel types showed Byzantine/Roman influence, and others were more clearly Islamic forms and decorations. Although some excellent work has recently been published in which the relationship between chemical composition and vessel type has been investigated (Kato *et al.* 2010), the precise relationship between chemical compositions and vessels of 'Byzantine' type used in Islamic contexts is yet to be fully investigated.

Conclusions

Ethnographic studies that have a bearing on ancient glass technology can provide insights into social relations between groups involved in glass production as well as the decisions taken by them. Examples are the video directed by Dr. Robert Brill of the *Glassmakers of Herat* (which can be compared to ancient Islamic glass production), the research carried out by Lamb (1965, 1966) into the manufacture of 'Mutisalah' beads in Malaysia and the DVD made by Dr. Marie-Dominique Nenna into the manufacture of 'eye' beads in Turkey (providing clues about the reasons for making beads and the technology involved) and study of glass melting in tank furnaces in Uttar Pradesh, northern India (Sode and Kock 2001), which has helped to interpret excavated remains of glass tank furnaces. An exceptional investigation drawing on ethnographic research amongst local glassblowers is the study by Fischer (2008). Although these studies all contribute to our overall interpretation of glass production, there is not the same rich seam of research in the study of glass as is to be found in traditional potting and metal smelting that can illuminate ancient practices.

Research into ancient glass has already given us fascinating insights into a beautiful, complex and intriguing material. Future advances in methodologies and techniques applied to clearly structured archaeological research projects are bound to 'illuminate' the material even more.

APPENDIX

TECHNIQUES OF SCIENTIFIC ANALYSIS

The techniques of analysis summarised here are the main ones that are currently used in the investigation of ancient glasses. The appropriateness of each technique is determined by the research aims, the availability of the technique, cost and the sample size available/required. For quantitative analysis, standard materials of known compositions are required for all techniques. Brill (1969) summarised the techniques that could be used in the analysis of ancient glasses, which was then a rather young field of research. Since then, new techniques have been introduced and existing ones developed. Two volumes that provide a broad coverage of scientific techniques used to analyse archaeological materials are Ciliberto and Spoto (2000) and Pollard *et al*. (2007).

Inductively Coupled Plasma-Emission Spectroscopy (ICPS)

This is a destructive technique involving the dissolution of the sample that can be used to analyse a wide range of elements, simultaneously. The sample is dispersed in a plasma torch in which argon is combusted at temperatures above 8,000°C. The sample is injected into the flame and breaks up into its constituent parts. It is a sensitive technique that is largely used for trace-element analysis.

Mass Spectrometry

Thermal ionisation mass spectrometry is used to produce isotope ratios in glasses. The aim is to highlight contrasts in the geological ages of the raw materials used to make glasses (as reflected in part by e.g. strontium and neodymium isotope ratios) and, by comparing the results with environmental samples, provide information about where the glasses were fused from raw materials, and thereby produce a geological provenance for the glass.

Results from the use of this technique are discussed at some length in this book, especially in Chapter 6 and 11, and so the technique will be described in some detail here. The method used at the British Geological Survey by Professor Jane Evans is as follows. The glass samples must be cleaned scrupulously prior to analysis to remove any weathering. The samples are broken into fragments, cleaned by lacing in 2.5 molar hydrochloric acid (HCl) in an ultrasonic bath for 5 minutes and rinsed thrice in deionised water for the same length of time. The fragments, once dry, are milled to a fine powder in an agate micro-ball mill. Standard silicate dissolution is then undertaken. The strontium and rare earth element fractions are collected from Dowex resin columns, and the neodymium separated from the rare earth element fraction using Ln (Lanthanide) resin. Neodymium and strontium concentrations were determined by isotope dilution using enriched tracer solutions of ^{150}Nd and ^{84}Sr respectively.

Neodymium concentrations and isotope ratios are measured using a double filament assemblage with a Thermo Triton multi-collector. Strontium is measured using a rhenium single filament with TaF activator after the method of Birck (1986). The reproducibility of the data is based on standard data runs during sample analysis using the NBS987 international strontium standard and the international La Jolla standard for Nd. Typical results are: NBS987 = 0.710282 ± 0.000005 (2σ, n = 9), La Jolla = 0.511848 ± 0.000005, (2σ, n = 3). ^{87}Sr/^{86}Sr data are normalized to NBS987 = 0.710250. Samples are run to a precision of better than ±0.00001 (2SE) for strontium and better than 0.00002 (2SE) for neodymium. The lower neodymium precision can result from very small samples and low concentrations in some of the material. It can be improved on with larger samples and higher concentration of neodymium. All data are corrected to a ^{146}Nd/^{144}Nd ratio of 0.7219 and a ^{86}Sr/^{88}Sr ratio of 0.1194. Blank values were c. 100 pg Nd, and c. 50 pg Sr.

Inductively coupled plasma mass spectrometry (ICP-MS) can provide both (trace) chemical and isotopic analysis. It has been used for the determination of lead isotope ratios in glasses but is insufficiently sensitive for the analysis of strontium and neodymium isotopes. Laser ablation inductively coupled plasma mass spectrometry (LAICPMS) is an ideal very sensitive technique with detection limits of parts per million, or parts per billion for some elements. The technique uses a small solid sample, so the same sample as analysed using electron probe microanalysis can be used. The presence of trace element impurities in ancient glass can help to suggest the types and purities of the raw materials used to make it.

At the British Geological Survey, Dr. Simon Chenery uses the following methodology to determine trace element concentrations in glasses. The LAICPMS system consists of a NewWave UP193FX excimer (193 nm) laser system with built-in microscope imaging coupled to an Agilent 7500 series ICP-MS. Data is collected in a time-resolved analysis mode, with a glass blank being measured before a series of glass ablations, including on the standards. Three replicate ablations are performed on each sample. This highlights any compositional heterogeneity. Each ablation peak is individually integrated. Calibration is performed using the NIST SRM612 glass standard (nominal 500 mg/kg concentrations) with quality control being provided by the NIST SRM610 glass (nominal 500 mg/kg concentrations). Element concentrations are taken from the GeoRem website preferred values compilation (http://georem.mpch-mainz.gwdg.de). The results are normalised to the silica content as determined by electron microprobe analysis to account for differences in laser ablation efficiency. There is a generally good agreement between the elements determined using electron microprobe and the LAICPMS.

Using a magnetic sector machine, ratios of other isotopes can be determined, such as oxygen, samarium, hydrogen, carbon and nitrogen. More recently multiple collector plasma mass spectrometry systems (MC-ICPMS) have also been used to produce data of acceptable levels of accuracy. Quadrupole mass spectrometers can scan the mass spectrum rapidly, although a sacrifice may be the mass resolution.

Mass spectrometry relies on the principle that it is possible to separate electrically charged atoms of an element according to their atomic masses (e.g. ^{87}Sr and ^{86}Sr). This separation is achieved by using magnetic or electrical fields. Simple mass spectrometers have magnetic or electrostatic deflector systems.

Neutron Activation Analysis (NAA)

Small samples such as beads can be analysed in their entirety, or samples are powdered with sizes of between 50 mg and 200 mg. If entire artefacts are analysed, their geometry may make it difficult to produce quantitative results. It is a sensitive technique and is used for trace-element determinations. However, it involves access to an atomic pile, and the samples are radioactive once analysed. The samples are irradiated for a defined period of time. The radioactive samples then decay according to the half-lives of each element in the sample, those with shortest half-lives decaying fastest. The number of counts detected is directly relatable to the concentration of the element in the material being analysed. The gamma radiation produced is directly relatable to the activity of the atomic flux at the time of the radiation. The inclusion of standard materials is vital so that a measure of the reaction of standard and sample to the flux can be made. Normally the standard will be of a similar material to the sample being analysed. The peak wavelengths are characteristic of the elements present; it may be necessary to strip out interference peaks.

NAA can be a sensitive technique with potential detection limits down to parts per million for thirty to thirty-five elements detected simultaneously. It is relatively inaccurate for the detection of some major components.

X-ray Fluorescence Spectrometry (XRF)

This is a basic surface analytical technique that, under the right conditions, can be used for the quantitative determination of major and minor components in ancient glasses. The physical principles of the technique are very similar to the microanalysis in *electron microprobes* and spectrometers attached to *scanning electron microscopes*. X-ray fluorescence analysis can be a totally nondestructive technique. The interaction of primary X-rays (normally produced with an X-ray tube) with the sample generates, among other particles, a spectrum of secondary X-rays that have energies characteristic of each of the elements in the sample. The primary X-rays interact with each of the elements in the sample surface and secondary X-rays are emitted. A variety of energy transitions occur between inner atomic electron shells in each atom of the elements. These lead to the generation of the secondary X-rays. The detected secondary X-rays are displayed as a spectrum of X-ray energies. Wavelength-dispersive spectrometers used in electron probe microanalysis are more sensitive than the energy-dispersive ones used in XRF. The successful quantification of XRF and electron probe microanalysis depends on a number of factors. For quantification to be successful the sample must be homogeneous, completely flat and must lack any surface alteration (which itself would lead to heterogeneity).

The electron beam can be used in scanning or focused modes in *scanning electron microscopes*; the X-ray beam can be collimated in XRF systems. Mobile XRF systems are starting to be used quite commonly, but if no sample preparation is carried out, the results are unlikely to be fully quantitative. Grazing angle XRF systems can be more sensitive than conventional XRF systems.

Electron Probe Microanalysis (EPMA)

The source of exciting energy in *electron probe microanalysis* is a focused beam of electrons. It is one of the most accurate techniques of analysis for ancient glass. Backscattered and secondary electron detectors can be attached to both electron probes and scanning electron microscopes to produce micrographs that reflect compositional or morphological features respectively. The technique has been refined for the analysis of ancient glasses (Henderson 1988a). It is described in some detail here because many of the results discussed in this book were produced using the technique. In the archaeology department of Nottingham University, chemical analyses are carried out using a Jeol JSM8200 Superprobe equipped with 4 wavelength–dispersive spectrometers, one energy–dispersive spectrometer, and secondary and backscattered detectors. The system is calibrated using a series of geological and multi-element glass standards. The glass samples are mounted in epoxy resin discs and polished flat using a series of increasingly fine grades of polishing powders, finishing with 0.25-µm grade diamond paste. The samples are then coated with a thin layer of carbon so as to prevent distortion and deflection of the electron beam. Analyses are performed at 20 KV accelerating voltage and 40 nA beam current using a defocused electron beam of 40 µm in diameter so as to prevent the volatilisation of low-Z components, such as sodium. Multi-element glass standards are analysed on a regular basis so as to detect any change in the performance of the system and so as to establish its accuracy and precision. The following relative analytical accuracies have been obtained using a Corning B standard as unknown: 4% for Na_2O, 1% for SiO_2, 3% for K_2O, 5% for CaO and 2% for PbO. For minor elements the accuracy was 2% for MgO, 1% for Al_2O_3, 3% for P_2O_5, 10% for Fe_2O_3 and CuO, and up to 20% for Cl. The levels of detection varied from 170 ppm for CaO to 1,200 ppm for CuO. A ZAF program is used to correct and quantify the results. The system allows opacifying crystals in the glasses to be analysed separately and photographed under high magnification.

Particle-Induced X-ray Emission (PIXE)

This technique requires a tandem van der Graaf accelerator to generate particles that are accelerated at high speeds towards the sample, where they collide and penetrate it. The sample can either be analysed in an open geometry or in a sample chamber. With an open geometry system, it is also possible to analyse materials containing light elements by bathing the sample in helium to prevent the light secondary X-rays from being absorbed in the air before being detected. The backgrounds produced by PIXE are a factor of ten lower than XRF. The particles bombarding the sample enter at such speed that less scattering of the particles occurs. This makes the technique more sensitive than electron probe microanalysis or XRF so that major, minor and trace levels (down to parts per million) of elements can be detected. Some PIXE systems are fitted with scanning coils so that elemental distribution can be mapped.

X-ray Diffraction Spectrometry (XRD)

XRD can be used to identify unambiguously the type of crystals present in glasses, enamels and glazes, whether deliberately added or present as relic raw materials. It requires a small sample. The technique involves firing radiation of a particular monochromatic wavelength at the crystalline sample, which is mounted at a specific angle to the incoming beam. The interaction of the radiation with the crystal(s) produces an X-ray pattern that is characteristic of the structure of the crystal. Crystals are composed of lattices built up in a regular pattern; their size and spacing is characteristic of the crystal species. Fully automated XRD systems are available: when these are used, a thin slurry of the sample is deposited on a slide. The technique detects d-spacings in crystalline materials (e.g. opaque glasses). To identify the crystals present, those in the sample are compared with libraries of d-spacings for known materials until a match is found. Different species of crystals are formed at different temperatures, and so X-ray diffraction can help to determine what temperature regimes were involved in production processes (e.g. where high temperature crystals form at the interface between the brick furnace floors and the glass that forms on the floor).

Synchrotron-Induced (SR) Radiation

This radiation is produced by an instrument known as a synchrotron. The system consists of an accelerator, which produces high-energy electrons, a booster accelerator and a storage ring. The technique produces intense radiation of all frequencies in the electromagnetic spectrum. The high intensity is accompanied by fine beam dimensions (as low as 1μ), and this allows it to be used for a range of applications associated with a range of analytical techniques, such as SR-induced X-ray micro-diffraction, SR-induced X-ray fluorescence and SR-induced neutron diffraction. Compositional depth profiles can be produced with some instruments. The data produced can therefore either be trace elements or relate to crystal structure.

X-ray Photoelectron Spectroscopy (XPS)

This technique provides information about the chemical environment of an elemental species. The information is produced by a suitable radiation such as Al Kα or a monochromatized synchrotron radiation. It can therefore be used to detect the presence of different bonding states of the same element. This is especially relevant to the study of glass colouration. The technique provides information about surfaces. An X-ray photoelectron spectrum is a plot of binding energies versus the number of electrons detected.

Auger Electron Spectroscopy (AES)

This can be used to examine the chemical state of the atom from which the Auger electrons derive. These particles are produced when a material is bombarded with electrons. The particles have very low kinetic energies and derive from the two or three surface atomic layers. Auger electrons are emitted from the outer orbitals of atoms. The outer orbitals are often involved in chemical bonding, and so the Auger electrons characterise the state of the atom.

Appendix

Particle-Induced Gamma Ray Emission (PIGE)

This is typically carried out at the same time at PIXE (described earlier) by bombarding the sample with protons. Although not commonly used (partly because accelerators are expensive), the technique, which is especially sensitive to the detection of light elements such as lithium and fluorine, can be used quantitatively.

Time-of-Flight Secondary Ion Mass Spectrometry (ToF-SIMS)

This is a technique that has only recently been used to examine ancient glass. SIMS involves sputtering a material with a beam of positively charged ions in an ultra high vacuum chamber. The fragments of the surface that are sputtered range from atomic species to clusters of atoms. Both positively and negatively charged fragments are sputtered, yielding both positive and negative ion mass spectra. Secondary ions are extracted into a mass spectrometer for mass analysis. In the case of ToF-SIMS, a short pulse (of about 1 ns) of primary ions hits the surface, and the resulting secondary ions are accelerated to a constant energy before entering a field-free drift tube. Ions of different masses have different velocities, so that they can be separated by flight time in the drift tube (heavy ions travel more slowly). Their arrival time at a detector is registered, and this produces a spectrum of intensity versus time for each pulse. The sputtered fragments, typically, can escape from only a shallow depth (about 95% of the signal comes from the top two monolayers of atoms). The technique is especially useful for detecting light elements such as boron and lithium, and in principle, it can be used in a quantitative mode. It is the only technique that can be used for both mapping the distribution of components in crystallites and analysing their impurities, down to parts per billion. Isotopic information is also produced.

Transmission Electron Microscopy (TEM)

This is especially useful for the detection, identification and photography of submicron crystallites in (e.g. diachroic) glasses. A thin section of the sample is prepared, and the electrons travel through it.

BIBLIOGRAPHY

Abdurazskov, A.A. 1972. Étude chimique des verres d'Asie Centrale, datant du Moyen Âge, *IXe Congrès International du Verre, Communications artistiques et historiques*, Versailles, 27 September–2 October, Paris, pp. 161–78.

Abdurazskov, A.A. 2009. 'Central Asian glassmaking during the ancient and medieval periods', in eds. Gan Fuxi *et al.*, pp. 201–20.

Aerts, A., Janssens, K. and Adams, F. 1999. 'Trace-level microanalysis of Roman glass from Khirbet Qumrân', Israel, *Journal of Archaeological Science*, 26: 883–91.

Aerts, A., Janssens, K., Velde, B., Adams, F. and Wouters, H. 2000. 'Analyses of the composition of glass objects from Qumrân, Israel', in M.-D. Nenna (ed.), pp. 113–21.

Aerts, A., Velde, B., Janssens, K. and Dijkman, W. 2003. 'Change in silica sources in Roman and post-Roman glass', *Spectrochimica Acta* B58: 659–67.

Akkermans, P.M.M.G. and Schwartz, G.M. 2003. *The archaeology of Syria from Complex hunter-gatherers to early urban societies (ca. 16.000–300 BC)*, Cambridge: Cambridge University Press.

Al-Hassan, A.Y. 2009. 'An eighth-century Arabic treatise on the colouring of glass Kitāb al-durra al-maknūna (the book of the hidden pearl) of Jābir ibn Ḥayyān (c. 721–c. 815)', *Arabic Sciences and Philosophy* 19: 121–56.

Al-Qalqashandī, A. 1913–20. *Subh al-ashā*, 14 volumes, Cairo.

Aldsworth, F., Haggerty, G., Jennings, S. and Whitehouse, D. 2002. 'Medieval Glassmaking at Tyre, Lebanon', *Journal of Glass Studies* 44: 49–66.

Allan, J.P., 1984. *Medieval and post-medieval finds from Exeter 1971–1980*, Exeter Archaeology Reports, 3, Exeter, UK: Exeter University Press.

Allan, J.W. 1973. 'Abdūl Q'āsim's treatise on ceramics', *Iran* 11: 111–20.

Allègre, C.J. and Michard, G. 1974. *Introduction to geochemistry*, Dordrecht and Boston: D. Reidel.

Amouric, H. and Foy, D. 1991. 'De la salicorne aux soudes factices; mutations techniques et variation de la demande', in *L'évolution des Techniques est-elle autonome? (Aix-en-Provence 1989)*, Cahier d'histoire des techniques, Aix-en-Provence: Publications de l'Université de Provence Marseille-Aix 1, pp. 39–75.

Amrein, H. 2001. *L'atelier de verriers d'Avenches l'artisanat du verre au milieu du 1er siècle après J.-C.*, Cahiers d'archéologie romande de la Bibliothèque historique vaudoise No. 87, Aventicum XI, Lausanne: Cahiers d'archeologie romande.

Bibliography

Amrein, H. 2006. 'Quelques réflexions sur l'implantation et l'organisation des ateliers des verriers dans les provinces romaines au nord des Alpes', in eds. G. Creemers, B. Demarsin and P. Cosyns, pp. 58–63.

Andrae, W. 1935. 'Die jüngeren Ischtar-Tempel in Assur', *Wissenschaftliche Veröffentlichungen der Deutschen Orient-Gesellschaft*, 58, Leipzig: Hinrichs.

Andreescu-Treadgold, I. and Henderson, J. with M. Roe, 2006. 'Glass from the Mosaics on the West Wall of Torcello's Basilica', *Arte Medievale*, 5, no. 2: 87–140.

Andreotti, A. and Zanini, A. 1995–6. 'L'insedimento di Fossa Nera di Porcari (Lucca)', *Rivista de Scienze Preistoriche*, 47, 291–330.

Angelini, I., Artioli, G., Bellintani, P., Diella, V., Gemmi, M., Polla, A. and Rossi, A. 2004. "Chemical Analyses of Bronze Age Glasses from Frattesina di Rovigo, Northern Italy," *Journal of Archaeological Science* 31: 1175–84.

Angelini, I., Artioli, G., Bellintani, P., Diella, V., Polla, A. and Residori, G. 2002. 'Project "Glass materials in the protohistory of North Italy": a first summary', in *Atti del II Congresso Nazionale di Archeometica*, Bologna, 29 gennaio–1 febbraio 2002, Bologna: Pàtron editore, pp. 581–95.

Angelini, I., Artioli, G., Bellintani, P. and Polla, A. 2005. Protohistoric vitreous materials of Italy: from early faience to final Bronze Age glasses, *16e Congrès de l'Association Internationale pour l'Histoire du Verre*, London 2003, Nottingham: AIHV, pp. 32–6.

Angelini, I., Polla, A., Giussani, B., Bellintani, P. and Artioli, G. 2009. 'Final Bronze-Age glass in northern and central Italy: is Frattesina the only glass production centre?' Proceedings of the 36th International Symposium on Archaeometry, Cahiers d'archéologie du CELAT no. 25 (eds. Moreau, J.-F., Auger, R., Chabot, J. and Herzog A.), Québec: Université Laval, pp. 329–337.

Aperghis, G.G. 2004. *The Seleukid Royal Economy: The Finances And Financial Administration of the Seleukid Empire*, Cambridge: Cambridge University Press.

Applebaum, S. 1989. *Judaea in Hellenistic and Roman Times: Historical and Archaeological Essays*, Leiden: Brill.

Arenoso Callipo, C.M.S. and Bellintani, P. 1994. 'Dati archeologici e paleoambientali del territorio di Frattesina di Fratta Polesine (RO) tra la tarda età del Bronzo e la prima età del Ferro', *Padusa* 30: 7–65.

Arjava, A. 2005. 'The mystery cloud of 536 CE in the Mediterranean sources', *Dumbarton Oaks Papers* 59: 73–94.

Arletti, R., Vezzalini, G., Biaggio Simona, S. and Maselli Scotti, F. 2008. 'Archaemetrical studies of Roman Imperial age glass from canton Ticino', *Archaeometry* 50: 606–26.

Artioli, G., Angelini, I. and Polla, A. 2008a. 'Crystals and phase transitions in prehistoric glass materials', *Phase Transitions* 81: 233–52.

Artioli, G., Nimis, P., Recchia, S., Marelli, M. and Giussani, B. 2008b. 'Geochemical links between copper mines and ancient metallurgy: the Agordo case study', *Rendiconti Società Geologica Italiana* 4: 15–18.

Asch, K., 2005. *IGME 5000: 1:5 Million International Geological Map of Europe and Adjacent Areas – final version for the Internet*. Hannover: BRG.

Ashtor, E. 1976. 'Il commercio levantino di Ancona', *Rivista Storica Italiana* 88: 227–35.

Ashtor, E. 1992. 'Levantine alkali ashes and European industries', in *Technology, Industry and Trade* (ed. B.Z. Kedar), Brookfield, VT: Variorum, pp. 475–522.

Ashtor, E. and Cevidalli, G. 1983. 'Levantine alkali ashes and European industries', *Journal of European Economic History* 12: 487–91.

Atasoy, N. and Raby, J. 1989. *Iznik: The Pottery of Ottoman Turkey*, London: Alexandria Press.

Azemar, R., Billaud, Y., Bories, G., Costantini, G. and Gratuze, B. 2000. 'Les perles protohistoriques en verre de l'Aveyron', *Cahiers d'archéologie aveyronnaise* 14: 75–88.

BIBLIOGRAPHY

Bachhuber, C. 2006. 'Aegean interest in the Uluburun ship', *American Journal of Archaeology* 110: 345–63.

Bagheri, H., Moore, F. and Alderton, D.H.M. 2007. Cu-Ni-Co-As (U) mineralization in the Anarak area of central Iran, *Journal of Asian Earth Sciences* 29: 651–65.

Bailey, J. 1987. 'Viking glassworking: the evidence from York', *Annales du 10e Congrès de l'Association Internationale pour l'Histoire du Verre*, Madrid 1985, Amsterdam: AIHV, pp. 245–54.

Bailey, J. 2000. 'Glassworking in early medieval England', in *Glass in Britain and Ireland AD 350–1100* (ed. J. Price), British Museum Occasional Paper no. 127, London: British Museum Press, pp. 137–42.

Baker, G. 1968. 'Micro-forms of hay-silica and of volcanic glass', *Mineralogical Magazine* 36: 1012–23.

Baker, P.L. 2004. 'Glass in Early Islamic Society – towards a greater understanding', *Orient* 39: 3–17.

Balista, C. and De Guio, A. 1997. 'Ambiente ed insediamenti dell'età del bronzo nelle Valli Grandi Veronesi', in eds. M. Bernabò Brea, A. Cardarelli and M. Cremaschi, pp. 137–60.

Ball, W. 2001. *Rome in the East: the transformations of an empire*, London: Routledge.

Bamford, C.R. 1977. *Colour generation and control in glass*. Amsterdam: Elsevier Scientific Publishing.

Bar-Yosef, O., Vandermeersch, B., Arensburg, B., Belfer-Cohen, A., Goldberg, P., Laville, H., Meignen, L., Rak, Y., Speth, J.D., Tchernov, E., Tillier, A.-M. and Weiner, S. 1992. 'The excavations in Kebara Cave, Mt. Carmel', *Current Anthropology* 33: 497–550.

Barag, D. 1970. 'Mesopotamian Core-Formed Vessels 1500–500 BC', in Oppenheim *et al.*, pp. 131–99.

Barag, D. 1985. *Catalogue of the western Asiatic glass in the British Museum*, Volume 1, London: British Museum Publications.

Barber, D. and Freestone, I.C. 1990. 'An investigation of the origin of the colour of the Lycurgus Cup by analytical transmission electron microscopy, *Archaeometry* 32: 33–45.

Barfield, L.H. 1978. 'North Italian faience buttons', *Antiquity* 52: 150–3.

Barker, G. 1975. 'Prehistoric territories and economics in central Italy', in *Palaeoeconomy* (ed. E.S. Higgs), Cambridge: Cambridge University Press.

Barkoudah, Y. and Henderson, J. 2006. 'Plant ashes from Syria and the Manufacture of Ancient Glass: Ethnographic and Scientific Aspects', *Journal of Glass Studies* 48: 297–321.

Barnes, I.L., Brill, R.H., Deal, E.C. and Piercy, J. 1986. 'Lead isotope studies of the finds from the Serçe Limani shipwreck', *Proceedings of the 24th International Archaeometry Symposium* (eds. J.S. Olin and J.M. Blackman), Washington, D.C.: Smithsonian Institution Press, pp. 1–12.

Barrera, J. and Velde, B. 1989. 'A study of French Medieval glass composition', *Archéologie Médiévale* 19: 81–130.

Basa, K.K., Glover, I. and Henderson, J. 1991. 'The relationship between early southeast Asian and Indian glass', in *Indo-Pacific Prehistory* (ed. P. Bellwood), Volume 1, Bulletin of the Indo-Pacific Prehistory Association, 10, Canberra and Jakarta: The Indo-Pacific Prehistory Association, pp. 366–85.

Bass, G.F. 1984. 'The nature of the Serçe Limani glass', *Journal of Glass Studies* 26: 64–9.

Bass, G.F. 1986. 'A bronze age shipwreck at Uluburun (Kaş) 1984 campaign', *American Journal of Archaeology* 90: 269–96.

BIBLIOGRAPHY

Bass, G.F. 1997. 'Prolegomena to a study of maritime traffic in raw materials to the Aegean during the fourteenth and thirteenth centuries B.C.', in *Techne. Craftsmen, craftswomen and craftsmanship in the Aegean Bronze Age* (eds. R. Laffinauer and Ph. Betancourt), Liège: Université de Liège, Histoire de l'art et archéologie de la Grèce antique; Austin: University of Texas at Austin, Program in Aegean Scripts and Prehistory, pp. 153–70.

Bass, G.F., Lledo, B., Matthews, S. and Brill, R.H. 2009. *Serçe Limani: Glass of an eleventh-century shipwreck*, Volume 2, Rachel Foundation Nautical Archaeology Series, College Station, Texas: Texas A & M University Press.

Bass, G.F. and Van Doorninck, F.H. 1978. 'An eleventh century shipwreck at Serç Limani, Turkey', *International Journal of Nautical Archaeology and Underwater Exploration* 7: 119–32.

Basso, E., Messiga, B. and Riccardi, M.P. 2008. 'Stones from medieval glassmaking: a suitable waste product for reconstructing an early stage of the melting process in the Mt. Lecco glass factory', *Archaeometry* 50: 822–34.

Batanouny, K.H. 2001. *Plants in the deserts of the Middle East*, Berlin: Springer-Verlag.

Bateson, H.M. and Turner, W.E.S. 1939. 'A note on the solubility of sodium chloride in a soda-lime-silica glass', *Journal of the Society of Glass Technology* 23: 265–67.

Baxter, M.J., Cool, H.E.M., Heyworth, M.P. and Jackson, C.M. 1995. 'Compositional variability in colourless Roman vessel glass', *Archaeometry* 37: 129–41.

Baxter, M.J., Cool, H.E.M. and Jackson, C.M. 2005. 'Further studies in the compositional variability of colourless Romano-British vessel glass', *Archaeometry* 47: 47–68.

Beck, H.C. 1934. 'Glass before 1500 BC', *Ancient Egypt and the East*, 19: 7–21.

Beck, H.C. and Seligman, C.G. 1934. 'Barium in ancient glass', *Nature* 133, no. 6: 982.

Bellintani, P. 1997. 'Frattesina, l'ambre e la prudozione vitrea nel contesto delle relazioni transalpine; in *Ori delle Alpi* (ed. L. Endrizzi and F. Marzatico), Quaderni della Archeologica Castello del Buonconsiglio, Provincia autonoma di Trento: Servizio Beni Culturali, pp. 117–33.

Bellintani, P. 2000. 'I bottoni conici ed altri materiali vetrosi delle fasi non avanzate della media età del Bronzo dell'Italia settenrionale e centrale', *Padusa* 34: 95–110.

Bellintani, P. and Biavati, A. 1997. 'Gli ornamenti in materiale vetroso', in ed. M. Bernabò Brea, A. Cardarelli and M. Cremaschi, pp. 610–12.

Bellintani, P., Biavati, A. and Veritá, M. 1998. 'Alcune considerazione su materiali vetrosi da contesti dell'età del bronzo media e recente dell'Italia settentrionale, in Il vetro dall'antichità all'età contemporanea: aspetti tecnologici, funzionali e commerciali', *Atti 2e giornate nazionali di studio AIHV- comitato nazionale Itliano*, 14–15 Dicembre 1996, Milano: AIHV, pp. 15–24.

Beltsios, K.G., Oikonomou, A., Zacharias, N. and Triantafyllidis, P. 2012 'Characterisation and provenance of archaeological glass artifacts from Mainland and Aegean Greece', in *The dating and provenance of natural and manufactured glasses* (eds. Y. Liritsis and C.M. Stevenson), Albuquerque: University of New Mexico Press, 166–84.

Bernabò Brea, M., Cardarelli, A. and Cremaschi, M., eds. 1997. *Le terramare–la più antica civiltà Padana*, Milano: Electa.

Besborodov, M.A. 1957. 'A chemical and technological study of ancient Russian glasses and refactories', *Journal of the Society of Glass Technology* 41: 168T–84T.

Besborodov, M.A. 1975. *Chemie und Technologie der antiken und mittelälterlichen Gläser*, Mainz: P. Von Zabern.

Besborodov, M.A. and Abdurazakov, A.A. 1964. Newly excavated glassworks in the USSR 3rd–14th centuries AD, *Journal of Glass Studies* 6: 64–69.

Beydoun, Z.R. 1977. 'The Levantine countries: The geology of Syria and Lebanon (maritime regions)', in *The ocean basins and margins, vol. 4A, The eastern Mediterranean* (eds. Nairn, A.E.M., Kanes, W.H. and Stehli, F.G.), New York and London: Plenum Press, pp. 319–53.

Bhardwaj, H.C. 1979. *Aspects of ancient Indian technology*, Delhi: Motilal Banarsidass.

Bhardwaj, H.C., ed. 1987a. Archaeometry of glass, in *Proceedings of the Archaeometry Session of the XIV International Congress on Glass, 1986*, New Dehli and Calcutta: The Indian Ceramic Society.

Bhardwaj, H.J. 1987b. 'A review of archaeometric studies of Indian glasses' in H.J. Bhardwaj (1987a), pp. 64–75.

Bhatia, N.P., Orlic, I., Siegele, R., Ashwath, N., Baker, A.J.M. and Walsh, K.B. 2003. 'Elemental mapping using PIXE shows the main pathways of nickel movement is principally symplastic within the fruit of the hyperaccumulator *Stackhousia tryonii*', *New Phytologist*, 160: 479–88.

Bianchin, S., Brianese, N., Casellato, U., Fenzi, F., Guerriero, P., Vigato, P.A., Nodari, L., Russo, U., Glagani, M. and Mendera, M. 2005. 'Medieval and Renaissance glass technology in Valdelsa (Florence). Part 2: vitreous finds and sands', *Journal of Cultural Heritage*, 6: 39–54.

Biavati, A. 1983. 'L'arte Vetraria nella civiltà protovillanoviana di Frattesina, Fratta Polesine, Rovigo: analisi chimica dei reperti archeologici', *Padusa* 5: 59–63.

Biek, L. and Bayley, J. 1979. 'Glass and other vitreous materials', *World Archaeology* 11, 1: 1–25.

Bietti Sestieri, A.M. 1988. 'The 'Mycenaean connection' and its impact on the central Mediterranean societies', *Dialoghi di Archeologia* 6, 1: 23–51.

Bietti Sestieri, A.M. 1996. *Protostoria, teoria e pratica*, Rome: Nuova Italia Scientifica.

Bietti Sestieri, A.M. 1997. 'Italy in Europe in the Early Iron Age', *Proceedings of the Prehistoric Society* 63: 371–402.

Bietti Sestieri, A. M. 2008. 'l'Età del Bronzo finale nella penisola italiana', *Padusa*, 44: 7–54.

Bietti Sestieri, A.M. and Mazzorin, J. 1995. 'Importazione di materiale prime organische di origine esotica nell'abitato protostorico di Frattesina (Rovigo)', *Atti del Primo Convegno Nazionale di Archeozoologia, Quaderni di Padusa* 1: 367–70.

Bimson, M. and Freestone, I.C. 1983. 'An analytical study of the relationship between the Portland vase and other Roman cameo glasses', *Journal of Glass Studies* 25: 55–64.

Bimson, M. AND Freestone, I.C. 1985. Appendix 3, 'Scientific examination of opaque red glass of the second and first millennia BC' in Barag (1985), pp. 119–22.

Birck, J.L. 1986. 'Precision of K-Rb-Sr Isotopic Analysis – Application to Rb-Sr Chronology', *Chemical Geology* 56: 73–83.

Biron, I., Matoïan, V., Henderson, J. and Evans, J. 2012. Scientific analysis of glass from Ras Shamra – Ugarit (Syria), *Proceedings of the 18th International Congress on the History of Glass*, Thessalonika, September 2009, Thessalonika: AIHV, pp. 29–34.

Blom-Böer, I. 1994. 'Zusammensetzung altägyptischer Farbpigmente und ihre Herkunftslagerstätten', in *Zeit und Rau*, Oudheidkundige mededeelingen van het Rijksmuseum van Oudheden, Ledien, 74: 55–107.

Boffe, M. and Letocart, G. 1962. 'The influence of the size of raw materials on the rate of melting of glass', *Glass Technology* 3 (4): 117–23.

Bonner, M. ed. 2004. *Arab-Byzantine relations in early Islamic Times. The formation of the classical Islamic world*, vol. 8, Aldershot, England: Ashgate Variorum.

Bopearachchi, O. 2002. 'Les relations commerciales et culturelles entre Sri Lanka et Inde de sud: nouvelles données archéologiques et épigraphiques', *Cahier du cercle d'études et de recherches Sri Lankaises* 4: 1–16.

Borgna, E. 1992. *Il ripostiglio di Madriolo presso Cividale e i pani a piccone del Friuli-Venezia Giulia*, Studi e ricerche di Protostoria mediterranea 1, Rome.

Bottéro, J. 1992. *Mesopotamia: Writing, reasoning and gods* (trans. Z. Bahrani and M. Van De Mieroop), Chicago: University of Chicago Press.

Boulogne, S. and Henderson, J. 2009. 'Indian Glass in the Middle East? Medieval and Ottoman Glass Bangles from Central Jordan', *Journal of Glass Studies* 51: 53–75.

Bouquillon, A. and Matoïan, V., 2007. 'Les emplois du bleu égyptien à Ougarit', in *Faïences et matières vitreuses de l'Orient ancien* (ed. A. Caubet), Gand, Belgium, and Paris: Éditions Snoeck and Musée du Louvre, pp. 39–40.

Braghin, C. 2002. 'Polychrome and monochrome glass of the Warring States and Han periods', in *Chinese glass, Archaeological studies on the uses and social context of glass artefacts from the Warring States to the Northern Song period (fifth century BC to twelfth century AD)* (ed. C. Braghin), Orientalia Venetiana 14, Firenze: Leo S. Olschki, pp. 3–43.

Braidwood, J. 1960. *Excavations in the plain of Antioch*, Oriental Institute Publications, Chicago: University of Chicago.

Brain, C. 2002. 'The technology of 17th century flint glass', *Glass Technology* 43C: 357–60.

Braun, Ch. 1983. 'Analysen von Gläsern aus der Hallstattzeit mit einem Exkurs über römische Fenstergläser', in O.-H. Frey, *Glasperlen der Vorrömischen eisenzeit*, Marburger studien zur vor- und frühgeschichte, band 5, Mainz am Rhein: Phillip von Zabern, pp. 129–15.

Brill, R.H. 1962. 'A note on the scientist's definition of glass', *Journal of Glass Studies* 4: 127–38.

Brill, R.H. 1965. 'The chemistry of the Lycurgus cup', *Proceedings of the 7th International Congress on Glass*, Brussels 1965, paper no. 223.

Brill, R.H. 1968. 'The scientific investigation of ancient glass', in *Proceedings of the 8th International Congress on Glass*, Sheffield: Society of Glass Technology, pp. 47–68.

Brill, R.H. 1970. 'The chemical interpretation of the texts', in eds. A. Leo Oppenheim *et al.*, pp. 105–28.

Brill, R.H. 1970a. 'Chemical studies of Islamic lustre glass', Chapter 16 in *Scientific methods in medieval archaeology* (ed. R. Berger), Berkeley: University of California Press.

Brill, R.H. 1972. 'Report on the chemical analysis of some glasses from Afghanistan', *Transactions of the American Philosophical Society* 62: 51–53.

Brill, R.H. 1973. 'Chemical considerations', Appendix in *A history of glass in Japan* (ed D. Blair), Japan: Kodansha International and the Corning Museum of Glass, pp. 448–57.

Brill, R.H. 1986. Note in G. Bass, 'A Bronze Age' shipwreck at Ulu Burun (Kaş), Turkey: 1984 Campaign, *American Journal of Archaeology*: 90, 269–96.

Brill, R.H. 1987. 'Chemical analyses of some early Indian glasses', in *Archaeometry of Glass, Proceedings of the Archaeometry Session of the XIV International Congress on Glass* (ed. H.C. Bhardwaj), 1986, New Delhi, India, part 1, Kolkata (Calcutta): Indian Ceramic Society, section 1, pp. 1–25.

Brill, R.H. 1988. 'Scientific investigations of the Jalame glass and related finds' in ed. G.D. Weinberg, pp. 257–94.

Brill, R.H. 1989. 'Thoughts on the glass of central Asia with analyses of some glasses from Afghanistan', in *Proceedings of the XV International Congress on Glass, Leningrad, 1989, Archaeometry*, Moscow: Rotoprint VNIIZSM, pp. 19–24.

Brill, R.H. 1991. 'Scientific investigations of some glasses from Sedeinga', *Journal of Glass Studies* 33: 11–29.

Bibliography

Brill, R.H. 1991a. 'Chemical analysis of some early Chinese glasses', in *Scientific Research in Early Chinese Glass* (eds. R.H. Brill and J.H. Martin), Proceedings of the Archaeometry of Glass Sessions of the 1984 International Symposium on Glass, Beijing, September 7, 1984, Corning, NY: The Corning Museum of Glass, pp. 31–58.

Brill, R.H. 1991b. 'Scientific investigations of some glasses from Sedeinga', *Journal of Glass Studies* 33: 11–28.

Brill, R.H. 1992. 'Chemical analyses of some glasses from Frattesina', *Journal of Glass Studies* 34: 11–22.

Brill, R.H. 1995. 'Chemical analyses of some glass fragments from Nishapur in the Corning Museum of Glass' in J. Kröger, *Nishapur: Glass of the Early Islamic Period*, Appendix 3, New York: The Metropolitan Museum of Art, pp. 211–33.

Brill, R.H. 1999a. *Chemical analyses of early glasses, volume 1, catalogue of samples*, Corning, NY: The Corning Museum of Glass.

Brill, R.H. 1999b. *Chemical analyses of early glasses, volume 2, tables of analyses*, Corning, NY: The Corning Museum of Glass.

Brill, R.H. 2000. 'Chemical analyses of some glasses in the collection of Mr. Simon Kwan in S. Kwan', *Early Chinese glass*, The Art Museum, The Chinese University of Hong Kong, pp. 448–71.

Brill, R.H. 2001. 'The glassmakers of Firozabad and the glassmakers of Kapadwanj: Two pilot video projects', *Annales du 15e Congrès de l'Association Internationale pour l'Histoire du Verre*, Corning, NY: AIHV, pp. 267–68.

Brill, R.H. 2001a. Some thoughts on the Chemistry and Technology of Islamic Glass', in *Glass of the Sultans*, New York: The Metropolitan Museum of Art in association with The Corning Museum of Glass, Benaki Museum, and Yale University Press, pp. 25–45.

Brill, R.H. 2009. 'Chemical analyses' in G.F. Bass, R.H. Brill, B. Lledó and S.D. Matthews, *Serçe Limani, volume II: The glass of an eleventh-century shipwreck*, College Station: Texas A&M University Press, pp. 459–96.

Brill, R.H. 2009a. 'Opening remarks and setting the stage: lecture at the 2005 Shanghai International Workshop on the archaeology of glass along the Silk Road', in eds. Gan Fuxi *et al.*, pp. 109–47.

Brill, R.H., Barnes. I.L. and Adams, B. 1974. 'Lead isotopes in some ancient Egyptian objects', *Recent Advances in the Science and Technology of Materials*, 3, New York and London: Plenum Press, pp. 9–27.

Brill, R.H., Barnes, I.L. and Jeol, E.C. 1991. 'Lead isotope studies of early Chinese glasses, Scientific Studies in Early Chinese Glass', Proceedings of the Archaeometry of Glass Sessions of the 1984 International Symposium on Glass, Beijing, September 7, 1984, (eds. R.H. Brill and J.H. Martin), pp. 65–83.

Brill, R.H. and Fenn, P.M. 1992. 'Glasswares in Famen temple', *Selected papers from the First International Symposium on the History and Culture of the Famen Temple* (ed. Zhang Qizhi), Shaanxi People's Education Press, pp. 254–58.

Brill, R.H., Fenn, P.M. and Lange, D.E. 1995. 'Chemical analyses of some Asian glasses', Proceedings of the 17th International Congress on Glass, Beijing 1995, volume 1, Beijing: International Academic Publishers, pp. 270–79.

Brill, R.H. and Fullagar, P.D. 2009.' Isotope studies of historical glasses and related materials', *Annales of the 17th Congress of the International Association for the History of Glass*, Antwerp, Belgium, 4–8 September 2006, Antwerp: University Press, pp. 552–57.

Brill, R.H. and Martin, J.H. (eds.) 1991. 'Scientific Research in Early Chinese Glass', *Proceedings of the Archaeometry of Glass Sessions of the 1984 International Symposium on Glass*, Beijing, September 7, 1984, Corning, NY: The Corning Museum of Glass.

Brill, R.H. and Pongracz, P. 2004. 'Stained glass from Saint-Jean-des-Vignes (Soissons) and comparisons with glass from other medieval sites', *Journal of Glass Studies* 46: 115–44.

Brill, R.H. and Shirahata, H. 1997. 'Laboratory Analyses of Some Glasses and Metals from Tell Brak', in Oates, J. *et al.*, pp. 89–94.

Brill, R.H., Tong, S.S.C. and Dohrenwend, D. 1991. 'Chemical analyses of some early Chinese glasses' in eds. R.H. Brill and J.H. Martin, pp. 31–58.

Brill, R.H. and Wosinski, J.F. 1965. 'A huge slab of glass in the ancient necropolis of Beth She'arim', *Proceedings of the 7th International Congress on Glass*, Brussels, paper no. 219.

Brill, R.H. and Wosinski, J.F. 1988. 'Glass manufacture' in ed. G.D. Weinberg, pp. 283–8.

Brill, R.H., Yamazaki, K., Barnes, I.L., Rosman, K.J.R. and Diaz, M., 1979. 'Lead isotopes in some Japanese and Chinese glasses', *Ars Orientalis* 11: 87–109.

Brill, R.H., Barnes, L. and Jeol, E. 1991. 'Lead isotope studies of early Chinese glasses' in *Scientific research in early Chinese glass* (ed. R.H. Brill and J.H. Martin), Corning, NY: The Corning Museum of Glass, pp. 65–83.

Brumfield, E. and Earle, T. 1987. 'Specialization, exchange, and complex societies: An introduction', *Specialization, Exchange, and Complex Societies* (eds. E. Brumfield and T. Earle), Cambridge: Cambridge University Press, pp. 1–10.

Brun, N., Mazerolle, L. and Pernot, M. 1991. 'Microstructure of opaque red glass containing copper', *Journal of Materials Science Letters* 10: 1418–20.

Brun, N. and Pernot, M. 1992. 'The opaque red glass of Celtic enamels from continental Europe', *Archaeometry* 34: 235–52.

Butcher, K. 2003. *Roman Syria and the Near East*, London: British Museum Press.

Cabart, H. 2004. 'Une activité verrière á la fin du 1er siècle á Reims', *Coeur de verre. Production et diffusion du verre antique* (ed. D. Foy), Gollion, Switzerland: Infolio.

Cable, M. 1958. 'An investigation of the effect of sand grain size on the refining of a pure soda-lime-silica glass in a laboratory furnace', *Journal of the Society of Glass Technology* 42: 20–31.

Cable, M. 1998. 'The operation of wood-fired glass melting furnaces', *The Prehistory and History of Glassmaking Technology* (ed. P. McCray), Proceedings of the Prehistory and History of Glassmaking Technology, Proceedings of the Symposium held at the 99th Annual Meeting of the American Ceramics Society in Cincinnati, Ohio, May 4-7, 1997. Ceramics and Civilisation 8, volume 378, pp. 315–29.

Carboni, S. 1998. 'Gregorio's tale; or, Of enameled glass production in Venice', in ed. R. Ward, pp. 101–6.

Carboni, S. 2001. *Glass from Islamic lands, the al-Sabah collection*, London: Thames and Hudson.

Carboni, S., Giancarlo, L. and Whitehouse, D. 2003. 'Glassmaking in medieval Tyre: The written evidence', *Journal of Glass Studies* 45: 139–49.

Carboni, S. and Whitehouse, D. 2001. *Glass of the Sultans*, New York: Metropolitan Museum of Art.

Carter, S.W., Wiegand, B., Mahood, G.A., Dudas, O.F., Wooden, J.L., Sullivan, A.P. and Bowring, S.A. 2011. 'Strontium isotope evidence for prehistoric transport of gray ware ceramic materials in the eastern Grand Canyon region, USA', *Geoarchaeology* 26(2): 189–218.

Casellato, U., Fenzi, F., Guerriero, P., Sitran, S., Vigato, P.A., Russo, U., Galgani, M., Mendera, M. and Manasse, A. 2003. 'Medieval and Renaissance glass technology in medieval Valdelsa (Florence). Part 1: raw materials, sands and non-vitreous materials', *Journal of Cultural Heritage* 4(4): 337–53.

Castillo, F. and Martínez, R. 2000. 'Un taller de vidrio en Bayyana-Pechina (Almería)', in ed. P. Cressier, pp. 83–101.

Catalogue (du) Musée National de Damas, 1976, Damascus.

Caubet, A. 1995. 'Art and architecture in Canaan and ancient Israel' in ed. J.M. Sasson, pp. 2671–91.

Cavanagh, W.G. and Mee, C. 1998. *A private place: Death in prehistoric Greece*, Studies in Mediterranean Archaeology, no. 85, Göteborg, Sweden: Åström.

Charleston, R.J. 1978. 'Glass furnaces through the ages', *Journal of Glass Studies* 20: 9–34.

Charleston, R.J. 1984. *English glass*, London: Allen and Unwin.

Chebab, M. 1979. 'Tyre a l'époque des Croisades', *Bulletin du Musée de Beyrouth* 31, Beirut.

Chen Xianqiu, Huang Ruifu, Jiang Lingzhang and Yu Ling. 1986. 'The structural nature of Jianyang Hare's Fur Temmoku Imitations', *Scientific and technological insights on chinese pottery and porcelain* (ed. Shanghai Institute of Ceramics), Beijing: Science Press, pp. 236–40.

Christensen, A.M. 2000. The Orientalizing glass finds from Poggio Civitale (Murlo), *Annales du 14ᵉ Congrès de l'association internationale pour l'histoire du verre*, Venezia and Milano, Italy 1998, Lochem: AIHV, pp. 16–19.

Ciliberto, E. and Spoto, G. 2000. *Modern analytical methods in art and archaeology*, New York: John Wiley.

Coles, R.A. 1983. *The oxyrhynchus papyri* 50, London: Egypt Exploration Society.

Collis, J. 1984. *Prehistoric Europe*, London: Batsford.

Constanze, C., Höpken, C. and Schäfer, F.F. 2006. 'Glasverarbeitung und Glaswerkstätten in Köln', in eds. G. Creemers, B. Demarsin and P. Cosyns, pp. 74–85.

Cool, H.E.M. 2003. 'Local production and trade in glass vessels in the British Isles in the first to seventh centuries AD', in eds. Foy and Nenna, pp.139–45.

Cool, H.E.M. 2006. *Eating and drinking in the Roman world*, Cambridge: Cambridge University Press.

Cool, H.E.M. and Baxter, M.J. 1999. 'Peeling the onion: an approach to comparing vessel glass assemblages', *Journal of Roman Archaeology* 12: 72–100.

Cool, H.E.M. and Baxter, M.J. 2005. 'Cups for gentleman', in *Annales du 16e Congrès de l'association Internationale pour l'Histoire du verre*, London 2003, Nottingham: International Association for the History of Glass, pp. 127–30.

Cooperson, M. 2005. *Al Ma'mum*, Oxford: Oneworld Publications.

Cosyns, P., Janssens, K., Van der Linden, V. and Schalm, O. 2006. 'Black glass in the Roman Empire: A work in progress', in eds. G. Creemers, B. Demarsin and P. Cosyns, pp. 30–41.

Creemers, G., Demarsin B. and Cosyns P., eds. 2006. *Roman glass in Germania Inferior. Interregional comparisons and recent results*, Proceedings of the International Conference, held in the Gallo-Roman Museum in Tongeren, 13 May 2005, Tongeren, Belgium: Provincial Gallo-Roman Museum.

Cressier, P., ed. 2000. *El vidrio en al-Andalus*, Madrid: Coeditions de la Casa de Velázquez.

Crossley, D.W. and Åberg, F.A. 1972. 'Sixteenth-century glass-making in Yorkshire: excavations of furnaces at Hutton and Rosedale, North Riding, 1968–1971', *Post-medieval Archaeology* 6: 107–59.

Dakoronia, P. and Deger-Jalkotzy, S. 1988. 'Elateia: Mykenaiko Nekrotafeio', *Archaiologikon Deltion*, 43, Chronika B1, 229–32.

Dakoronia, P., Deger-Jalkotzy, S. and Sakellariou, A. 1996. 'Die Siegel aus der Nekropole von Elatia-Alonaki, Kleinere griechische Sammlungen', *Corpus der minoischen und mykenischen Siegel* (ed. I. Pini), Akademie der Wissenschaften und der Literatur, v. 5, supp. 2, Mainz: F. Matz, v–xxx and indexes I and II.

David, A.R. 2000. 'Mummification', in *Ancient Egyptian materials and technology* (eds. P. T. Nicholson and I. Shaw), Cambridge: Cambridge University Press, 372–89.

Davison, C.C. 1972. *Glass beads in African archaeology, results of neutron activation analysis supplemented by results of X-ray fluorescence analysis*, unpublished PhD thesis, University of California, Berkeley.

Day, P.M. 2004. 'Marriage and mobility: Tradition and the dynamics of pottery production in twentieth century East Crete' (eds. P.P. Betancourt, C. Davaras and R. Hope Simpson), *Pseira VIII. The archaeological survey of Pseira Island Part I*, Prehistory Monographs 11, Philadelphia: INSTAP Academic Press, pp. 105–42.

Day, P.M., Relaki, M. and Todaro, S. 2010. Living from pots? Ceramic perspectives in the economies of prepalatial Crete, political economies of the Bronze Age Aegean (ed. D. Pullen), *Proceedings of the Langford Conference, Tallahassee, Florida, 22–24 February 2007*, Oxford: Oxbow Books, 205–29.

De Francesco, A.M., Scarpelli, R., Barca, D., Ciarallo, A. and Buffone, L. 2010. 'Preliminary chemical characterization of Roman glass from Pompeii', *Periodico di Mineralogia* 79: 11–19.

De Gayangos, P. 1984. *The history of the Mohammadan dynasties in Spain*, 2 volumes, Delhi: Idarah-i-Adabiyat-i-Delli.

De Guio, A. 1991. 'Alla ricerca del potere: alcune prospettive italiane', *Papers of the Fourth Conference of Italian Archaeology*, 1 (eds. E. Herring, R. Whitehouse and J. Wilkins), *The Archaeology of Power*, part 1, London: Accordia Research Centre, pp. 153–92.

De Marinis, R. 1975. 'L'età del bronzo', in, *Preistoria e protostoria nel Reggiano, ricerche e scavi 1940–1975* (ed. M. Cremaschi), Reggio Emelia: Civici Musei, pp. 31–53.

De Min, M. 1987. 'La necropolis protovillanoviana di Frattesina di Fratta Polesine', *Prospettive storico-antropologiche in archeologia preistorica*, Quaderni di Dialoghi di Archeologia 3, pp. 277–82.

De Raedt, I.K., Janssens, J., Veeckman, L., Vincze, B., Vekermans, B. and Jeffries, T.E. 2001. 'Trace analysis for distinguishing between Venetian and façon-de-Venise glass vessels of the 16th and 17th century', *Journal of Analytical Atomic Spectrometry* 4: 1012–17.

Degryse, P., Henderson, J. and Hodgins, G., eds., 2009. *Isotopes in vitreous materials*, Leuven, Belgium: Leuven University Press.

Degryse, P. and Poblome, J. 2002. 'The implications of the use of nepheline for the glass of Roman Sagalassos', *Hyalos vitrum glass, history, technology and conservation of glass and vitreous materials in the Hellenic world* (ed. G. Kordas), Athens: Glasnet Publications, pp. 349–52.

Degryse, P., Schneider, J., Haack, U., Lauwers, V., Poblome, J., Waelkens, M. and Muchez, Ph., 2006. 'Evidence for glass 'recycling' using Pb and Sr isotopic ratios and Sr-mixing lines: the case of Early Byzantine Sagalassos', *Journal of Archaeological Science* 33, 4: 494–501.

Degryse, P. and Schneider, J. 2008. 'Pliny the Elder and Sr-Nd isotopes: tracing the provenance of raw materials for Roman glass production', *Journal of Archaeological Science* 35: 1993–2000.

Degryse, P., Schneider, J., Lauwers, V., Henderson, J., Van Daele, B., Martens, M., Huisman, H.D.J., De Muynck, D. and Muchez, P. 2009. 'Neodymium and strontium isotopes in the provenance determinations of primary natron glass production', in eds. P. Degryse, J. Henderson and G. Hodgins, pp. 53–72.

Degryse, P., Boyce, A., Erb-Satullo, N., Eremin, K., Kirk, S., Scott, R., Shortland, A.J., Schneider, J. and Walton, M. 2010. 'Isotopic discriminants between late bronze Age glasses from Egypt and the Near East' *Archaeometry* 52: 380–88.

Degryse, P, Freestone, I. Schneider, J. and Jennings, S. 2010a. 'Technology and provenance study of Levantine plant ash glass using Sr-Nd isotope analysis. (eds. Drauke, J. and Keller, D.) *Glass in Byzantium – Production, Usage, Analyses*. Mainz: Romisch-Germanischen Zentralmuseums, pp. 83–91.

Dekówna, M. 1978. 'Les verres de Haithabu', *Annales du 7e Congrès de l'association internationale pour l'histoire du verre*, 1977, Liège: AIHV, pp. 167–88.

Dekówna, M. 1979. 'Remarques sur la chronologie de l'introduction dans la verrerie européenne médiévale de la technologie potassique et de celle au plomb non-alcaline', *Annales di 8e Congrès de l'Association Internationale pour l'Histoire du Verre*, London-Liverpool, 18–25 February 1979, pp. 145–60.

Dekowná, M. 1980. Methods of examining ancient glasses (ed. J. Schild), *Unconventional Archaeology*, Wroclaw: Ossolineium, Wydawnictwo Polskiej Akademii Nauk, 213–33.

Dillon, E. 1907. *Glass*, London: Methuen and Co. and New York: G.P. Putnam's Sons.

Dimbleby, G. 1978. *Plants and archaeology*, London: John Baker.

Dinsmore, W.B. 1973. *The architecture of ancient Greece*, revised and enlarged by William J. Anderson and R. Phené Spiers, New York: Biblo and Tannen.

Dodwell, C.R. 1961. *Theophilus, De Diversis artibus. Theophilus, the various arts*, Oxford: Clarendon Press.

Doppelfeld, O. 1966. *Römisches und Fränkisches Glas in Köln*, Köln: Greven Verlag.

Doremus, R.H. 1994. *Glass science*, New York: John Wiley.

Dotsika, E., Poutoukis, D., Tzavidopoulos, I., Maniatis, Y., Ignatiadou, D. and Raco, B. 2009. 'A natron source at Pikrolimni Lake in Greece? Geochemical evidence', *Journal of Geochemical Exploration* 103: 133–43.

Drower, M.S. 1973. 'Syria c. 1550–1400 B.C.' in I.E.S. Edwards *et al.*, *The Cambridge Ancient History*, Vol. 3, pt. 1 and Vol. 1, pt. 2.

Duckworth, C.N. 2011. *The created stone: Chemical and archaeological perspectives on the colour and material properties of early Egyptian glass, 1500–1200 BC*. Unpublished Ph.D. thesis, University of Nottingham, England.

Duckworth, C.N. 2012. 'Imitation, artificiality and creation: the colour and perception of the earliest glass in New Kingdom Egypt', *Cambridge Archaeological Journal* 22, 3: 309–27.

Duckworth, C.N., Henderson, J., Rutten, F.J. and Nikita, K. 2012. 'Opacifiers in Late Bronze Age glasses: The use of ToF-SIMS to identify raw ingredients and production techniques', *Journal of Archaeological Science*. 39, 2143–52.

Dungworth, D. 2005. 'The scientific study of late seventeenth-century glassworking at Silkstone, England', in *Annales de l'Association Internationale pour l'Histoire du Verre*, Vol. 16, London, 2003, Nottingham: AIHV, pp. 254–57.

Dungworth, D., Degryse, P. and Schneider, J. 2009. Kelp in historic glass: the application of strontium isotope analysis', in eds. P. Degryse, J. Henderson and G. Hodgins, pp. 113–30.

Dussart, O. 1995. 'Les verres de Jordanie et de Syrie du sud du IVe au VIIe sicle, nouvelles données chronologiques', *Le verre de l'antiquite tardive et du haut moyen age* (ed. D. Foy), Guiry-en-Vexin: Musee archaeologique departmental du val d'Oise, pp. 343–59.

Dussart, O. 1998. *Le verre en Jordanie du Sud*, Bibliotèque Archéologique et Historique, Beirut: Institut français d'archéologie du Proche-Orient, 152.

Dussart, O. 2000. 'Quelques indices d'ateliers de verriers en Jordanie et en Syrie du sud de la fin de l'époque Hellénistique á l'époque Islamique', in ed. M.-D. Nenna, pp. 91–96.

Dussart, O., Velde, B., Blanc, P.-M. and Sodini, J.-P., 2004. 'Glass from Qal'at Se'man (northern Syria): The reworking of glass during the transition from the Roman to the Islamic compositions', *Journal of Glass Studies* 46: 67–83.

Dussubieux, L. 2001. *L'apport de l'ablation laser couplée á l'ICP-MS á la caractérisation des verres: application á l'étude du verre archéologique de l'Océan Indien*, Unpublished thesis presented to the University of Orleans to obtain the grade of Doctor of the University of Orleans, Chemistry.

Dussubieux, L. and Gratuze, B. 2003. 'Origine et diffusion du verre dans le monde Indien et en Asie du sud-est: L'importance du dosage des elements-traces', *Revue d'Archéometrie* 27: 67–73.

Dussubieux, L., Kusimba, C.M., Gogte, V., Kusimba, S.B., Gratuze, B. and Oka, R. 2008. 'The trading of ancient glass beads: new analytical data from south Asian and east African soda-alumina glass beads', *Archaeometry* 50: 797–821.

Eisen, G. 1916. 'The origin of glass blowing', *American Journal of Archaeology* 20: 134–43.

El Cheikh, N.M. 2004. *Byzantium viewed by the Arabs*, Harvard Middle Eastern Monographs XXXVI, Cambridge, MA: Harvard Center for Middle Eastern Studies.

Emery, K.O. and Neev, D. 1960. 'Mediterranean beaches of Israel', *Geological Survey of Israel Bulletin* 26: 1–16.

Evans, J.A., Montgomery, J., Wildman, G. and Boulton, N. 2010. 'Spatial variations in biosphere $87Sr/86Sr$ in Britain', *Journal of the Geological Society* 167: 1–4.

Evans, J.A. and Tatham, S. 2004. 'Defining 'local signature' in terms of Sr isotope composition using a tenth–twelfth century Anglo-Saxon population living on a Jurassic clay-carbonate terrain, Rutland, UK', *Forensic Geoscience: Principles, techniques and applications* (eds. K. Pye and D.J. Croft), Geological Society of London Special Publication, 232, pp. 237–48.

Eyre, J.C. 1987. 'Work and organisation of work in the New Kingdom', *Labor in the Ancient Near East*, Newhaven (ed. M.A. Powell), New Haven, CT: Yale University Press, pp. 167–221.

Farnsworth, M. and Ritchie, P.D. 1938. 'Spectrographic studies on ancient glass. Egyptian glass mainly of the 18th Dynasty, with special reference to its cobalt content', *Technical Studies in the Field of the Fine Arts* 6(3): 155–73.

Faure, G. 1986. *Principles of isotope geology*, New York: John Wiley.

Fenn, P.M., Brill, R.H. and Shi Meiguang. 1991. Appendix to chapter 4 in eds. R.H. Brill and J.H. Martin, pp. 59–64.

Feugère, M. 1992. 'Le verre pré-romain en Gaule méridionale: acquis récents et questions ouvertes', *Revue archéologique de Narbonnaise* 25: 151–76.

Feugère, M. and Leyge, F. 1989. 'La cargaison de verrerie augustéenne de l'épave de la Tradelière (Iles de Lérins)', *Le verre préromain en Europe occidentale* (ed. M. Feugère), Montagnac: Éditions Monique Mergoil, pp. 169–76.

Figueiral, I. 1992. 'The charcoals', in 'Iron-making at the Chesters villa, Woolaston, Gloucestershire: Survey and Excavation 1987–91', M.G. Fulford and J.R.L. Allen *Britannia* 23, 1: 188–91.

Finley, M.I. 1985. *The ancient economy*, Berkeley and Los Angeles: University of California Press.

Fischer, A. 2008. *Hot pursuit: Integrating anthropology in search of ancient glass-blowers*, Lanham, MD: Lexington Books.

Fischer, A. and McCray, W.P. 1999. 'Glass production activities as practised at Sepphoris, Israel (37 BC–AD 1516)', *Proceedings of the International Symposium on Archaeometry* (eds. J. Henderson, H. Neff and T. Rehren), University of Illinois at Urbana Champaign (UIUC), 20–24 May 1996, *Journal of Archaeological Science* 26, 8: 893–906.

Fleming, S.J. 1997. *Roman Glass: Reflections on Cultural Change*, Pennsylvania: University of Pennsylvania Museum of Archaeology.

Fleury, M., Brut, C. and Velde, B. 2002. '13th century drinking glasses from the Cour Carree, Louvre', *Journal of Glass Studies* 44: 95–110.

Folk, R.L. and Hoops G.K. 1982. 'An early Iron Age layer of glass made from plants at Tel Yin'am, Israel', *Journal of Field Archaeology* 9: 455–66.

Follman-Schultz, A.B. 1991. 'Fours de verriers romains dans la province de Germaine inférieure', in eds. D. Foy and G. Sensequier, pp. 35–40.

Fossing, P. 1940. *Glass vessels before glass blowing*, Copenhagen: E. Munksgaard.

Foster, H.E. and Jackson, C.M. 2009. 'The composition of 'naturally coloured' late Roman vessel glass from Britain and the implications for models of glass production and supply', *Journal of Archaeological Science* 36: 189–204.

Foy, D. 1989. *Le verre mediéval et son artisanat en France mediterranéenne*, Paris: CNRS.

Foy, D. 1991. 'Ateliers de verre de l'antiquité et du Haut Moyen Âge en France. Méthodologie et results – un état de la question', in eds. D. Foy and G. Senequier, pp. 55–69.

Foy, D. 2003. 'Le verre en Tunisie: L'apport des fouilles récentes tuniso-françaises', *Journal of Glass Studies* 45, 59–89.

Foy, D. 2005. 'Une production de bols moulés à Beyrouth à la fin de l'époque hellénistique et le commerce de ces verres en Méditerranée occidentale', *Journal of Glass Studies* 47: 11–36.

Foy, D. and Jézégou, M.-P. 1998. 'Commerce et technologie du verre antique: Le témoignage de l'épave 'Ouest Embiez 1', *Méditerranée antique. Pêche, navigation, commerce* (ed. E. Rieth), 121e Congrès national des sociétés historiques et scientifiques, Nice 1996, pp. 121–34.

Foy, D., Jézégou, M.-P. and Fontaine, S. 2005. 'La circulation du verre en Méditerranée au début du IIIe siècle: le témoignage de l'épave Ouest Emblez 1 dans le sud de la France (fouilles 2001–2003), *Annales du 16e Congrès de l'Association Internationale pour l'Histoire du Verre*, London 2003, Nottingham: AIHV, pp. 122–26.

Foy, D. and Nenna, M.-D. 2001. *Tout feu tout sable: mille ans de verre antique le Midi de la France*, Aix-en-Provence, France: Musées de Marseille, Éditions Édisud.

Foy, D. and Nenna, M.-D., eds. 2003. *Échanges et commerce du verre dans le monde antique*, Actes du colloque de l'Association Francaise pour l'Archéologie du Verre, Aix-en-Provence et Marseille, 7–9 juin 2001, Montagnac: Éditions Monique Mergoil.

Foy, D., Picon, M. and Vichy, M. 2000. Les matières premières du verre et la question des produits semi-fins. Antiquité et Moyen-Âge, in *Arts du feu et productions artisanales*, XXth Rencontres Internationales d'Archéologie et d'Histoire d'Antibes, Antibes: Éditions APDCA, pp. 420–32.

Foy, D., Picon, M. and Vichy, M. 2000a. 'L'ingots de verre en Méditerranée occidenatale (IIIe siècle av. J.-C. – VIIe siècle après J-C: approvisionnement et mise en oevre, donées archéologiques et donées de laboratoire', *Annales du Verre*, Venice-Milan 1998, Lochem: AIHV, pp. 51–57.

Foy, D., Picon, M., Vichy, M. and Thirion-Merle, V. 2003. 'Caractérisation des verres de la fin de l'Antiquité en Méditerranée occidentale: l'émergence de nouveaux courants commerciaux', in eds. D. Foy and M.-D. Nenna, pp. 41–85.

Foy, D. and Sennequier, G., eds. 1991. *À travers le verre, du Moyen Âge à la Renaissance*, Rouen. France: Musées departementaux de la Seine-Maritime.

Foy, D., Thirion-Merle, V. and Vichy, M. 2004. 'Contribution á l'étude es verres antiques décolorés á l'antimoine', *Revue d'Archéometrie* 28: 169–77.

Frank, S. 1982. *Glass and archaeology*, London: Academic Press.
Freestone, I.C. 1987. 'Composition and microstructure of early opaque red glass', *Early vitreous materials* (eds. M. Bimson and I.C. Freestone), British Museum Occasional Paper no. 56, London: British Museum Press, 1987, pp. 173–91.
Freestone, I.C., 1991. 'Looking into glass', *Science and the past* (ed. S. Bowman), London: British Museum Press, pp. 37–56.
Freestone, I.C. 1994. 'Chemical analysis of "raw" glass fragments', in *Excavations at Carthage, vol. II, 1 The Circular Harbour, North Side 290* (ed. H.R. Hurst), Oxford: Oxford University Press for the British Academy.
Freestone, I.C. 2001. 'Post-depositional changes in archaeological ceramics and glasses' *Handbook of archaeological science* (eds. D.R. Brothwell and A.M. Pollard), Chichester: John Wiley and sons, pp. 615–26.
Freestone, I.C. 2002. 'Composition and affinities of glass from the furnaces on the island site, Tyre', *Journal of Glass Studies* 44: 67–77.
Freestone, I., Meeks, N., Sax, M. and Higgitt, M. 2008. 'The Lycurgus Cup – A Roman Nanotechnology', *The Gold Bulletin* 40: 1–8.
Freestone, I.C., Bimson, M. and Buckton, D. 1990. 'Compositional categories of Byzantine glass tesserae', *Proceedings of the 11th Congress of the International Association for the History of Glass*, Basle, 1988, Amsterdam: AIHV, pp. 271–80.
Freestone, I.C and Gorin-Rosen, Y. 1999. 'The great glass slab at Bet She'arim, Israel: an Early Islamic Glassmaking Experiment?', *Journal of Glass Studies* 41: 105–16.
Freestone, I.C., Gorin-Rosen, Y. and Hughes, M.J. 2000. 'Primary glass from Israel and the production of glass in late antiquity and the early Islamic period', in ed. M.-D. Nenna 2000, pp. 65–84.
Freestone, I.C., Greenwood, R. and Gorin-Rosen, Y. 2002. 'Byzantine and early Islamic glassmaking in the eastern Mediterranean: Production and distribution of primary glass', *Hyalos-Vitrum-Glass: History, technology and conservation of glass in the Hellenistic World*, First International Conference, Rhodes, 1–4 April, Athens: G. Kordas, Glasnet Publications, pp. 167–74.
Freestone, I.C., Leslie, K.A., Thirlwell, M. and Gorin-Rosen, Y. 2003. 'Strontium isotopes in the investigation of early glass production: Byzantine and early Islamic glass from the Near East', *Archaeometry* 45(1): 19–32.
Freestone, I.C. and Stapleton, C.P., 1998. 'Composition and technology of Islamic enamelled glass of the thirteenth and fourteenth centuries' in ed. R. Ward, pp. 122–8.
Freestone, I., Meeks, N., Sax, M., Higgitt, C. 2008. 'The Lycurgus Cup – a Roman nanotechnology', *Gold Bulletin* 40: 1–8.
Freestone, I.C., Wolf, S. and Thirlwell, M. 2005. 'The production of HIMT glass: elemental and isotopic evidence', *Annales of the 16th Congress of the International Association for the History of Glass* 2003, London, Nottingham: AIHV, pp. 153–57.
Fremersdorf, F. 1965–1966. Die Anfänge der römishen Glashütten Kölns, Kölner *Jahrbuch Jahrbuch für Vor- und Frühgeschichte* 8: 24–43.
Freydier, R., Michard, A., De Lange, G. and Thomson, J. 2001. 'Nd isotopic compositions of Eastern Mediterranean sediments: tracers of the Nile influence during sapropel S1 formation?' *Marine Geology* 177: 45–62.
Gagarin, M. and Fantham, E. 2009. *The Oxford Encyclopedia of Ancient Greece and Rome*, volume 1, Oxford: Oxford University Press.
Gale, N.H. and Stos-Gale, Z. 2000. 'Lead isotope analysis applied to provenance studies', chapter 17 in Ciliberto, E. and Spoto, G. eds. pp.503–84.
Gan Fuxi 2005. *Development of Chinese ancient glass*, Shanghai: Fudan University.
Gan Fuxi 2009a. 'Origin and evolution of ancient Chinese glasses', in eds. Gan Fuxi *et al.*, pp. 1–40.

Gan Fuxi 2009b. 'The silk road and ancient Chinese glass', in eds. Gan Fuxi, R.H. Brill and T. Shouyun, pp. 41–108.

Gan Fuxi, Chen Huansheng and Li Qinghui. 2006. 'Origin of ancient glasses – study on the earliest Chinese ancient glasses', *Science and China* 49, 6: 701–13.

Gan F.X., Huang, Z.E. and Xiao, B.Y. 1978. 'The origin of ancient Chinese glass', *Journal of Chinese Ceramics* 12: 99–104.

Gan Fuxi, Brill, R.H. and Shouyun, T., eds. 2009. *Ancient glass research along the Silk Road*, Singapore: World Scientific Publishing Co.

Gan Fuxi, Cheng, H.S., Hu Y.Q., Ma Bo and Gu D.H. 2009. 'Study on the most early glass eye-beads in China unearthed from Xu Jialing tomb in Xichuan of Henan province, China', *Science in China series E: Technological sciences* 52: 922–27.

Gandolfi, G. and Paganelli, L. 1984.' Petrografia della sabbie del litorale tirrenico fra i Monti dell'Uccellina e Monte di Procida', *Mineralogica et Petrographica Acta*, 38: 173–91.

Garner, H. 1956a. 'An early piece of glass from Eridu', *Iraq* 18: 147–49.

Garner, H. 1956b. 'The use of imported and native cobalt in Chinese blue and white', *Oriental Art* 2, 3: 48–50.

Gates, M.-H. 1987. 'Alalakh and Chronology again', *High, Middle or Low?* (ed. P. Åström), *Acts of an International Colloquium on Absolute Chronology* held at the University of Gothenburg, August 1987, part 2, Gothenburg, Sweden: Åström, 60–82.

Gates, M.-H. 1989. 'Discussion following M.H. Gates' paper', in *High, Middle or Low? Acts of an International Colloquium on Absolute Chronology* held at the University of Gothenburg, August 1987, part 3 (ed. P. Åström), Gothenburg, Sweden: Åström, 67–73.

Gebhardt, R. 1989. *Der Glasschmuck aus dem oppidum von Manching*. Die Augrabungen in Manching, vol. 11, Stuttgart: Franz Steiner Verlag.

Gebhardt, R. 2010. 'Celtic glass' in eds. B. Zorn and A. Hilgner, pp. 3–14.

Geilmann, W. 1962. 'Beitrage zur Kenntnis alter Gläser. VII Kobalt als färbungsmittel', *Glastechnische Berichte* 35, 4: 186–92.

Geilmann, W. and Brückbauer, T. 1964. 'Beiträge zur Kenntnis alter Gläser II. Der Mangangehalt alter Gläser', *Glastechnische Berichte* 37, 456–59.

Genequand, D. 2005. 'From 'desert castle' to medieval town: Qasr al-Hary al-Sharki (Syria)', *Antiquity* 79(304): 350–61.

Ghirshman, R. 1966. *Tchoga-Zanbil I: La Ziggurat*, Mémoires de la delegation en Parse 39, Paris.

Gibb, H.A.R. 2004. 'Arab-Byzantine relations under the Umayyad caliphate' in ed. M. Bonner, pp. 65–79 [originally published in *Dumbarton Oaks Papers* 12, 1958: 219–33].

Giardino, C. 1995. *Il Meditarraneo Occidentale fra XIV e VIII sec. a.c.*, Oxford: British Archaeological Reports, S612.

Gimeno, D., Garcia-Valles, M., Fernandez-Turiel, J. L., Bazzocchi, F., Aulinas, M., Pugès, Tarozzi, C., Pia Riccardi, M., Basso, E., Fortina, C., Mendera, M. and Messiga, B. 2008. 'From Siena to Barcelona: deciphering colour recipes of Na-rich Mediterranean stained glass windows at the XIII–XIV century transition', *Journal of Cultural Heritage*, 9: e10–e15.

Glover, I.C. 1990. 'Ban Don Ta Phet: The 1984–85 excavation' *Southeast Asian Archaeology 1986* (eds. I. Glover and E. Glover), Oxford: British Archaeological Reports, International Series, S-561, pp. 139–83.

Glover, I. and Henderson, J. 1995. 'Early glass in south and south-east Asia and China', *China and Southeast Asia: Art, commerce and interaction* (eds. Rosemary Scott and John Guy), London: Percival David Foundation of Chinese Art, pp. 141–70.

Godfrey, E.S. 1975. *The development of English glassmaking*, Oxford: Clarendon Press.

Goerk, I.H. 1977. 'Glass materials and batch preparation', *Proceedings of the 9th International Congress on Glass*, Prague 4–8 July, vol. 2, pp. 38–111.

Goitein, S.D. 1968–93. *A Mediterranean society: The Jewish communities of the Arab world as portrayed in the documents of the Cairo Geniza*, 6 vols., Berkeley: University of California.

Goldstein, S.M. 1979. *Pre-Roman and early Roman glass in the Corning Museum of Glass*, Corning, NY: The Corning Museum of Glass.

Goldstein, S.L., O'Nions, R.K. and Hamilton, P.J. 1984. 'A Sm-Nd isotopic study of atmospheric dusts and particulates from major river systems', *Earth and Planetary Science Letters* 70: 221–36.

Goldstein, S.M. 2005. *Glass, from Sasanian antecedents to European imitations*, New York: Palgrave Macmillan.

Gorin-Rosen, Y. 1995. 'Hadera, Bet Eli'ezer', *Excavations and Surveys in Israel* 13: 42–43.

Gorin-Rosen, Y. 2000. 'The ancient glass industry in Israel: Summary of the finds and new discoveries', in ed. M.-D. Nenna 2000a, pp. 49–64.

Graber, O. 2004. Islamic art and Byzantium in ed. E.M. Bonner, pp. 263–93. [Originally published in *Dumbarton Oaks Papers* 18, 1964, 69–88].

Graber, O., Holod, R., Knudstad, J. and Trousdale, W. 1978. *City in the desert: Qasr al-Hayr East*, 2 vols., Cambridge, MA: Harvard University Press.

Gratuze, B. 2009. 'Les premiers verres au natron retrouvés en Europe occidentale: composition chimique et chrono-typologie', *Annales of the 17th Congress of the International Association of the History of Glass*, Antwerp: University of Antwerp Press, pp. 8–14.

Gratuze, B. and Barrandon, J.-N. 1990. 'Islamic glass weights and stamps: analysis using nuclear techniques', *Archaeometry* 32, 2: 155–62.

Gratuze, B., Dussubieux, L. and Bopearachchi, O. 2000. 'Etude de perles de verre trouvées au Sri-Lanka, IIIe siècle v. J.-C. – Iie siècle ap. J.-C'. *Annales de l'Association Internationale pour l'Histoire du Verre*, Venice and Milan, Lochem: AIHV, pp. 46–50.

Gratuze, B., Foy, D., Lancelot, J. and Téreygeol, F. 2003. 'Les "lissoirs" carolingiens en verre au plomb: mise en evidence de la valorisation des scories issues du traitement des galènes argentifères de Melle (Deux-Sèvres)', in *Échanges et commerce du verre dans le monde antique* (eds. D. Foy and M.-D. Nenna), Actes du colloque de l'AFAV, Aix-en-Povenence et Marseille, 7–9 June 2001, Montagnac, France: Éditions Monique Mergoil, pp. 101–8.

Gratuze, B., Louboutin, C. and Billaud, Y. 1998. 'Les perles protohistoriques en verre du Musées des Antiquités nationales', *Antiquités Nationales* 30: 11–24.

Gratuze, B., Soulier, I., Barrandon, J.N. and Foy, D. 1995. 'The origin of the cobalt blue pigment in French glass from the thirteenth to the eighteenth centuries', *Trade and discovery: The scientific study of artefacts from post-medieval Europe and beyond* (eds. D.R. Hook and D.R.M. Gaimster), British Museum Occasional Paper No. 109, London: British Museum Press, pp. 123–32.

Gratuze, B., Soulier, I., Blei, I. and Vallauri, L. 1992. 'De l'origine du cobalt dans les verres', *Revue d'Archéometrie*, 16: 97–108.

Greaves, G.N., Gurman, S.J., Catlow, C.R.A., Cahrdwick, A.V., Howde-Walter, S., Henderson, C.M.B. and Dobson, B.R. 1991. 'A structural basis for ionic diffusion in oxide glasses', *Philosophical Magazine*, A 64: 1059–72.

Grose, D.F. 1979. 'The Syro-Palestinian glass industry in the later Hellenistic period', *Muse* 13: 54–67.

Grose, D.F. 1981. 'The Hellenistic glass industry reconsidered', *Annales du 8e Congrès de l'association Internationale pour l'histoire du verre*, London-Liverpool, 18–25 September 1979, Liège: AIHV, pp. 61–72.

Grose, D.F. 1989. *The Toledo Museum of Art: Early Ancient Glass: Core-formed, rod-formed, and cast vessels and objects form the late Bronze Age to the early Roman Empire, 1600 B.C. to A.D. 50*, New York: Hudson Hills Press in association with the Toledo Museum of Art.

Grousset, F.E., Biscaye, P.E., Zindler, A., Prospero, J. and Chester, R. 1988. 'Neodymium isotope as tracers in marine sediments and aerosols: North Atlantic', *Earth and Planetary Science Letters* 87: 367–78.

Guido, M., Henderson, J., Cable, Bayley J. and Biek L. 1984. 'A Bronze Age glass bead from Wilsford, Wiltshire: Barrow G. 42 in the Lake group', *Proceedings of the Prehistoric Society* 50: 245–54.

Guilaine, J., Gratuze, B. and Barrandon J.-N. 1991. 'Les perles de verre du Chalcolithique et de l'âge du bronze: Analyses d'exemplaires trouvés en France', in *L'âge du Bronze Atlantique*, Actes du 1er Colloque du parc archéologique du Beynac, 10–14 Septembre 1990, Beynac et Cazenac: Association des Musées du Sarladais, pp. 255–66.

Guinier, A. 1984. *The structure of matter, from the blue sky to liquid crystal*, London: Edward Arnold.

Gutas, D. 1998. *Greek thought, Arabic culture: The Graeco-Arabic translation movement in Baghdad and early Abbasid society (second–fourth/eighth–tenth centuries)*, London: Routledge.

Guy, J. 2005. Early ninth-century Chinese export ceramics and Persian Gulf connection: The Belitung shipwreck evidence, *Chine-Méditerranée, Routes et échanges de la céramique avant le XVIè siècle*, Taoci 4, Editions Findakly, pp. 9–20.

Haevernick, T.E. 1959. Beiträge zur Geschichte des antiken Glases II: Stachelflächen, *Jahrbuch des Römisch-germanischen Zentralmuseums Mainz* 6: 63–65.

Haevernick, T.E. 1967. 'Die Verbreitung der "Zarte Rippenschalen"', *Jahrbuch des Römisch-germanischen Zentralmuseums Mainz* 14: 153–66.

Haevernick, T.E. 1978. 'Urnenfelderzeitliche Glasperlen', *Zeitschrift für schweizerische Archäologie und Kunstgeschichte* 35: 145–57.

Hahn-Weinheimer, P. 1955. 'Spektrochemische und Physikalische Untersuchungen an latènezeitlichen Glasfunden aus dem oppidium von Manching', *Beilage zum Sammelblatt des Historischen Vereins Ingolstadt*, 65, Frankfurt-am-Main.

Hall, H.R. 1930. *A season's work at Ur: Al-'Ubaid, Abu Sharain (Eridu), and elsewhere*. London: Methuen and Co.

Haller, A. 1954. *Die Gräber und Grüfte von Assur*, Wissenschaftliche Veröffentlichungen der Deutschen Orient-Gesellschaft 65, Berlin.

Hamidullah, M. 1960. 'Nouveaux documents sur les rapports de l'Europe avec l'orient musulman au Moyen âge', *Arabica* 7: 281–300.

Hannon, A.C. and Parker, J.M. 2000. 'The structure of aluminate glasses by neutron diffraction', *Journal of Non-crystalline Solids* 274: 102–9.

Harden, D.B. 1968. 'Ancient glass I: pre-Roman', *The Archaeological Journal* 125: 46–72.

Harden, D.B. 1968a. 'The Canosa Group of Hellenistic Glasses in the British Museum', *Journal of Glass Studies* 10: 21–47.

Harden, D.B. 1981. *Catalogue of the Greek and Roman glass in the British Museum*, vol. 1, London: The British Museum Press.

Harden, D.B. 1987. *Glass of the Caesars*, Milan: Olivetti.

Harding, A. 1971. 'The earliest glass in Europe', *Archeolohicke Rozhledy* 23: 188–200.

Harding, A.F. and Warren, S.E. 1973. 'Early bronze age faience beads from central Europe', *Antiquity* XLVII: 64–66.

Harris, J.R. 1961. *Lexicographical studies in ancient Egyptian minerals*, Veröffentlichung 54.1, Berlin: Institut für Orientforschung.

Harrison, J.V. 1968. 'Minerals', *The Cambridge History of Iran*, vol. I (ed. W.B. Fisher), pp. 489–516.

Hartmann, G. 1994. 'Late Medieval glass manufacture in the Eichsfeld region (Thuringia, Germany)', *Chemie der Erde* 54: 103–28.

Hartmann, G., Kappel, I., Grote, K. and Arndt, B. 1997. 'Chemistry and Technology of Prehistoric Glass from Lower Saxony and Hesse', *Journal of Archaeological Science* 24: 547–59.

Hartmann, S. and Grünewald, M. 2010. 'The late antique glass from Mayen (Germany): first results of chemical and archaeological studies', in eds. B. Zorn and A. Hilgner, 15–28.

Hasson, R. 1979. *Early Islamic glass*, Jerusalem: L.A. Mayer Memorial Institute for Islamic Art.

Hatcher, H., Henderson, J., Kazmarzyck, A., Lakovara, P. and Vandiver, P. In preparation. 'Archaeological and technical studies of faience and faience making from Kerma, Sudan', Boston: Boston Museum of Fine Arts.

Hawthorne, J.G. and Smith C.S. 1963. *On Divers Arts. The Treatise of Theophilus*, Chicago: Chicago University Press.

Hedges, R.E.M. 1976. 'Pre-Islamic glazes in Mesopotamia-Nippur', *Archaeometry* 18, 2: 209–13.

Hedges, R.E.M. 1982. 'Early glazed pottery and faience in Mesopotamia', in *Early pyrotechnology, the evolution of the first fire-using industries* (eds. Th.A. Wertime) and S.F. Wertime, Washington, DC: The Smithsonian Institution Press, pp. 93–103.

Hedges, R.E.M. and Moorey, P.R.S. 1975. Pre-Islamic ceramic glazes at Kish and Nineveh in Iraq, *Archaeometry* 17: 25–43.

Heide, K. and Heide, G. 2011. 'Vitreous state in nature-origin and properties', *Chemie der Erde* 71: 305–35.

Heidemann, S. 2006. 'The history of the industrial and commercial area of 'Abbasid Al-Raqqa, called Al-Raqqa Al-Muḥtariqa', *Bulletin of SOAS* 69, 1: 33–52.

Heltzer, M. 1978. *Goods, prices and the organisation of trade in Ugarit*, Wiesbaden: Dr. Ludwig Reichert Verlag.

Henderson, J. 1983. *X-ray fluorescence analysis of iron age glass beads*, Unpublished Ph.D. thesis, University of Bradford.

Henderson, J. 1985. 'The raw materials of early glass production', *Oxford Journal of Archaeology* 4(3): 267–91.

Henderson, J. 1986. 'Beads and rings', in D. Tweddle: *Finds from Parliament Street and other sites in the city centre*, C.B.A. Research Report, 17/4, The Archaeology of York, York: CBA, pp. 209–26.

Henderson, J. 1987. 'Glass and glass working', in B. Cunliffe: *Hengistbury Head, Dorset, Volume 1: The Prehistoric and Roman settlement 3500 B.C.–A.D. 500*. Oxford: University of Oxford Committee on Archaeology Monograph 13, pp. 60–3 and pp. 80–5.

Henderson, J. 1988a. 'Electron probe microanalysis of mixed-alkali glasses', *Archaeometry* 30, 1: 77–91.

Henderson, J. 1988b. 'Glass production and Bronze Age Europe', *Antiquity* 62: 435–51.

Henderson, J. 1989. 'The scientific analysis of ancient glass and its archaeological interpretation', *Scientific analysis in archaeology and its interpretation* (ed. J. Henderson), Oxford University Committee on Archaeology Monograph no. 19, UCLA Institute of Archaeology Research Tools 5, Oxford: Oxbow Books, pp. 30–62.

Henderson, J. 1991. 'Industrial specialization in late Iron Age Britain and Europe', *Archaeological Journal* 148: 104–48.

Henderson, J. 1991a. 'Chemical and structural analysis of Roman enamels from Britain, *Archaeometry '90*, Basel: Birkhäuser Verlag, pp. 285–94.

Henderson, J. 1991b. 'Chemical characterisation of Roman glass vessels, enamels and tesserae', in *Materials issues in art and archaeology II* (eds. P.B. Vandiver, J. Druzik and G.S. Wheeler), Research Society Symposium Proceedings, vol. 185, Pittsburgh, PA: Materials Research Society, pp. 601–7.

Henderson, J. 1991c. 'The glass', in *Excavations at Lurk Lane Beverley 1979–82* (eds. P. Armstrong, D. Tomlinson and D.H. Evans), Sheffield Excavation Reports 1, Sheffield, England: J.R. Collis, pp.124–30 and fiche B6–C1.

Henderson, J. 1992. 'The scientific analysis of vitreous materials from Kentria and Theologos-Tsiganadika tombs', *Protoistoiki Thasos: Ta Nekrotafeia tou Oikismou Kastri* (ed. Chaidos Koukouli-Chrysanthaki), Publications of Archaiologikon Deltion, no. 45, Athens: Archaeological Receipts Fund, pp. 804–6.

Henderson, J. 1993. 'Aspects of Early Medieval glass production in Britain', *Annales du 12e Congrès de l'Association Internationale pour l'Histoire du verre*, Vienna 26–31 August 1991, Amsterdam: AIHV, pp. 247–59.

Henderson, J. 1993a. 'Chemical analysis of the glass and faience from Hauterive-Champréveyres, Switzerland', in A.-M. Rychner-Faraggi, *Hauterive-Champréveyres, 9: Métal et parure au Bronze final*, Archéologie Neuchâteloise 17, Neuchâtel: Musée cantonal d'archéologie, pp. 111–17.

Henderson, J. 1995. 'An investigation of early Islamic glass production at Raqqa, Syria'. *Proceedings of the Materials Research Society Conference*, Issues in Art and Archaeology IV (ed. P. Vandiver *et al.*). Columbus, Ohio: The Materials Research Society, pp. 433–44.

Henderson, J. 1995a. 'Investigations into marvered glass II', *Islamic Art in the Ashmolean Museum* (ed. J. Allan), Oxford Studies in Islamic Art, vol. 10, part 1, Oxford: Oxford University Press, pp. 31–50.

Henderson, J. 1996. 'Scientific analysis of selected Fishbourne vessel glass and its archaeological interpretation', in *Excavations at Fishbourne 1969–1988*, Chichester Excavations 9 (eds. B. Cunliffe, A. Down and D. Rudkin), Chichester, England: Chichester District Council, pp. 189–92.

Henderson, J. 1997. 'Scientific analysis of glass and glaze from Tell Brak and its archaeological significance', in D. and J. Oates *Excavations at Tell Brak*, vol. 1 Cambridge: The Macdonald Institute, University of Cambridge, pp. 94–100.

Henderson, J. 1998. 'Islamic glass production and the possible evidence for trade in a cobalt-rich colorant', in ed. R. Ward, pp. 116–21.

Henderson, J. 1999a. Una nueva caracterización? La investigación científica de las cuentas de fayenza encontradas en la Cova des Càrritx (Menorca, Sa Cometa des Morts I (Mallorca), Son Maimó (Mallorca) y Este (Véneto, Italia), in *La Cova des Càrritx y la Cova des Mussol* (eds. V. Lull, R. Micó, C.R. Herrada and R. Risch), Ideología y sociedad en la prehistoria de Menorca, Barcelona: Universitat Autònoma de Barcelona, pp. 631–41.

Henderson, J. 1999b. 'Archaeological and scientific evidence for the production of early Islamic glass in al-Raqqa, Syria', *Levant* 31: 225–40.

Henderson, J. 2000a. *The science and archaeology of materials*, Routledge: London and New York.

Henderson, J. 2000b. 'Chemical analysis of ancient Egyptian Glass and its interpretation', in *Ancient Egyptian materials and technology* (eds. P.T. Nicholson and J. Shaw), Cambridge: Cambridge University Press, pp. 206–24.

Henderson, J. 2002a. 'An archaeological and scientific study of Viking-age glass', in D. Freke, *Excavations on St. Patrick's Isle, Peel, Isle of Man 1982–88*, Centre for Manx Studies Monographs 2, Liverpool: Liverpool University Press, pp. 349–62.

Henderson, J. 2002b. 'Tradition and experiment in 1st millennium AD glass production – the emergence of early Islamic glass technology in late antiquity', *Accounts of Chemical Research*, 35: 594–602.

Henderson, J. 2003. 'Localised production or trade? Advances in the study of Cobalt blue and Islamic glasses in the Levant and Europe', *Patterns and Process, A festschrift in honor of Dr. Edward V. Sayre* (ed. L. van Zelst), Suitland, MD: Smithsonian Center for Materials Research and Education, pp. 227–47.

Henderson, J. 2003a. 'Glass trade and chemical analysis: A possible model for Islamic glass production', in eds. D. Foy and M.-D. Nenna, pp. 109–23.

Henderson, J. 2003b. 'Technological analysis of Mamluk glass', in *Mamluk enamelled and gilded glass in the Museum of Islamic Art, Qatar*, Doha, Qatar: The Museum of Islamic Art, pp. 28–31.

Henderson, J. 2006. 'Medieval and post-medieval glass finewares from Lincoln: An investigation of the relationships between technology, chemical compositions, typology and value', *The Archaeological Journal* 162: 256–322.

Henderson, J. 2009. 'The provenance of archaeological plant ash glasses', *From mine to microscope, advances in the study of ancient technology* (eds. A.J. Shortland, I.C. Freestone and T. Rehren), Oxford: Oxbow Books, pp. 129–38.

Henderson, J. 2012. 'Scientific analysis of the beads', in K.D. Politis, *Sanctuary of Lot at Deir 'Ain 'Abata in Jordan. Excavations 1988–2003*, Amman: Jordan Distribution Agency, pp. 86–92.

Henderson J. In press b. 'Scientific and archaeological analysis of glass from Mantai', in J. Carswell, *Excavations at Mantai, Sri Lanka*.

Henderson, J. and Allan, J.A. 1990. 'Enamels on Ayyubid and Mamluk glass fragments', *Archaeomaterials* 4: 167–83.

Henderson, J., Challis, K., O'Hara, S., McLoughlin, S., Gardner, A. and Priestnall, G. 2005. 'Experiment and innovation: Early Islamic industry at al-Raqqa, Syria', *Antiquity*, 79: 130–45.

Henderson, J., Evans, J.A., Sloane, H.J., Leng, M.J. and Doherty, C. 2005a. 'The use of oxygen, strontium and lead isotopes to provenance ancient glasses in the Middle East', *Journal of Archaeological Science* 32: 665–73.

Henderson, J., Evans, J. and Barkoudah, Y. 2009. 'The roots of provenance: glass, plants and isotopes in the Islamic Middle East', *Antiquity* 83: 414–29.

Henderson, J., Evans, J. and Barkoudah, Y. 2009a. 'The provenance of Syrian plant ash glass: an isotopic approach' in eds. P. Degryse, J. Henderson and G. Hodgins, pp. 73–98.

Henderson, J., Evans, J., Barkoudah, Y. and Bertier, S. 2009b. 'The roots of provenance: Radiogenic isotopes and glass production in the Islamic Middle East', a lecture presented at the 18th Congress of the International Association for the History of Glass, Thessalonika, 21–25 September.

Henderson, J., Evans, J. and Nikita, K. 2010. 'Isotopic evidence for the primary production, provenance and trade of late Bronze Age glass in the Mediterranean', *Mediterranean Archaeology and Archaeometry* 10, 1: 1–24.

Henderson, J. and Faber, E. 2008. 'Report on the scientific analysis of glass and glass-bearing crucibles from Sigtuna, Sweden', in *På väg mot Paradiset* (ed. A. Wilksröm), Sigtuna: Sigtuna Museum.

Henderson, J. and Holand I. 1992. 'The glass from Borg, an early medieval chieftain's farm in northern Norway', *Medieval Archaeology* 36: 29–58.

Henderson, J., Janaway, R.C. and Richards, J. 1987. 'A curious clinker', *Journal of Archaeological Science* 14: 353–65.

Henderson, J., McLoughlin, S. and McPhail, D. 2004. 'Radical changes in Islamic glass technology: evidence for conservatism and experimentation with new glass recipes from early and middle Islamic Raqqa, Syria', *Archaeometry* 46: 439–68.

Henderson, J., Tregear, M. and Wood, N. 1989. 'The technology of 16th and 17th century Chinese cloisonné enamels', *Archaeometry* 31: 133–46.

Henderson, J. and Warren, S.E. 1983. 'Prehistoric lead glass analyses', in *Proceedings of the 22nd International Symposium on Archaeometry* (eds. A. Aspinall and S.E. Warren), Bradford, UK, Bradford: The University of Bradford, 62–8.

Henkes, E.H. and Henderson, J. 1998. 'The spun-stem *Roemer*, a hitherto overlooked *Roemer* type: typology, technology, and distribution', *Journal of Glass Studies* 40: 89–103.

Hennig, H. 1995. 'Zur Frage der Datierung des Grabhügels 8 'Hexembergh' von Wehringen, Lkr. Augsburg, Bayerisch-Schwaben, in *Trans Europa*, Bonn: Rudolph Habelt.

Highley, D.E. 1977. *Silica*, Mineral dossier no. 18, London: HMSO.

Hobsbawm, E. and Ranger, T. 1994. *The invention of tradition*, Cambridge: Cambridge University Press.

Hodges, H. 1970. *Technology in the ancient world*, New York: Barnes and Noble.

Hodges, H. 1992. *Technology in the ancient world*, New York: Barnes and Noble, 2nd edition.

Hodges, R. and Whitehouse, D. 1983. *Mohammad, Charlemagne and the origins of Europe*, London: Duckworth.

Holmes, L.L. and Harbottle, G. 1994. 'Compositional characterization of French limestone: a new tool for art historians', *Archaeometry* 36: 25–39.

Höpken, C. 2005. Die römische Keramikproduktion in Köln, *Kölner Forschungen* 8, Mainz.

Horat, H. 1991. *Der Glasschmelzofen des Piesters Theophilus*, Bern and Stuttgart: Paul Haupt.

Hoskin, P.W.O. and Ireland, T.R. 2000. 'Rare earth element chemistry of zircon and its use as a provenance indicator', *Geology* 28: 627–30.

Hourani, A. 2002. *A history of the Arab peoples*, London: Faber and Faber.

Hsueh-Man, S. 2002. 'Liao yu Bei Song she li ta nei cang jing zhi yan jiu (Scripture deposits in northern Song and Liao pagodas)', *Taida Journal of Art History* 12: 169–212.

Hsueh-Man, S. 2002a. 'Luxury or necessity: Glassware in *Sarīra* relic pagodas of the Tang and northern Song periods', *Chinese glass, archaeological studies on the uses and social context of glass artefacts from the Warring States to the Northern Song period (fifth century BC to twelfth century AD)* (ed. C. Braghin), Orientalia Venetiana 14, Firenze: Leo S. Olschki, pp. 71–110.

Hsiao-Chun, H., Yoshiyuki, I., Bellwood, P., Nguyen, K.D., Bellina, B., Silapanth, P., Dizon, E., Santiago, R., Datan, I. and Manton, J.H. 2007. 'Ancient jades map 3,000 years of prehistoric exchange in Southeast Asia', *Proceedings of the National Academy of Sciences* 104, no. 50: 19745–50.

Hughes, M.J. 1972. 'A technical study of opaque red glass of the Iron Age in Britain', *Proceedings of the Prehistoric Society* 38: 98–107.

Humphrey, J.W., Oleson, J.P. and Sherwood, A.N. 1998. *Greek and Roman Technology: A Sourcebook : Annotated Translations of Greek and Latin Texts and Documents*, London: Routledge.

Ige, O.A. and Swanson, S.E. 2008. 'Provenance studies of Esie sculptural soapstone from southwestern Nigeria', *Journal of Archaeological Science* 35: 1553–85.

Ignatiadou, D., Dotsika, E., Kouras, A. and Maniatis, Y. 2005. 'Nitrum chalestricum: the natron of Macedonia', *Annales de 16ᵉ Congrès de l'association internationale pour l'histoire du verre*, London, Nottingham: AIHV, pp. 64–67.

Ilan, D., Vandiver, P. and Spaer, M. 1993. 'An early glass bead from Tell Dan', *Israel Exploration Journal* 43: 230–34.

Ingram, R.S. 2005. *Faience and glass beads from the late Bronze Age shipwreck at Uluburun*, unpublished MA thesis, Texas A&M University, College Station, TX.

Insoll, T., Polya, D.A., Bhan, K., Irving, D. and Jarvis, K. 2004. 'Towards an understanding of the carnelian bead trade from western India to sub-Saharan Africa: the application of UV-LA-ICP-MS to carnelian from Gujarat, India, and West Africa', *Journal of Archaeological Science* 31: 1161–73.

In-Sook Lee. 1993. 'Chemical analyses of some ancient glasses from Korea', *Annales du 12e Congrès de l'association internationale pour l'histoire du verre*, Amsterdam: AIHV, pp. 163–75.

In-Sook Lee. 2004. 'Characteristics of early glasses in ancient Korea, with respect to Asia's maritime bead trade', *Proceedings of the XXth International Congress on Glass*, Kyoto, Japan, 26 September–1 October.

In-Sook Lee. 2009. 'Characteristics of early glasses in ancient Korea with respect to Asia's maritime bead trade', in eds. Gan Fuxi, R.H. Brill and T. Shouyun, pp. 183–90.

Invernizzi, R. and Vecchi, L. 1998. 'Una tomba femminile protoimperiale nel territorio di *Placentia*', in ed. C. Maccabrini, pp. 45–52.

Irwin, R. 1998. 'A note on textual sources for the history of glass', in ed. R. Ward, pp. 24–26.

Israeli, Y. 1991. 'The invention of blowing', *Roman glass: Two centuries of art and invention* (eds. M. Newby and K. Painter), Society of Antiquaries Occasional Paper vol. 13, London: Society of Antiquaries, pp. 46–55.

Israeli, Y. 2005. 'What did Jerusalem's first-century BCE glass workshop produce?' *Annales du 16e Congrès de l'association Internationale pour l'Histoire du verre*, London 2003, Nottingham: AIHV, pp. 54–58.

Jackson, C.M. 1992. *A compositional analysis of Roman and early post-Roman glass and glass-working waste from selected British sites*, unpublished Ph.D. thesis, University of Bradford, England.

Jackson, C.M. 2005. 'Making colourless glass in the Roman world', *Archaeometry* 47: 763–80.

Jackson, C.M., Baxter, M.J. and Cool, H.E.M. 2003. 'Identifying group and meaning: An investigation of Roman colourless glass', in eds. D. Foy and M.-D. Nenna, pp. 33–39.

Jackson, C.M., Booth, C.A. and Smedley, J.W. 2005. 'Glass by design? Raw materials, recipes and compositional data', *Archaeometry* 47: 781–95.

Jackson, C.M., Nicholson, P.T. and Gneisinger, W., 1998. 'Glassmaking at Tell el-Amarna: an integrated approach' *Journal of Glass Studies*, 40: 11–23.

Jackson, C.M. and Nicholson, P. 2007. 'Compositional analysis of the vitreous materials found at Amarna', in P. Nicholson, *Brilliant things for Akhenaten, the production of glass, vitreous materials and pottery at Amarna site 04,125*, London: The Egypt Exploration Society, pp. 101–16.

Jackson, C.M. and Nicholson, P.T. 2010. 'The provenance of some glass ingots from the Uluburun shipwreck', *Journal of Archaeological Science* 37: 295–301.

Jackson, C.M. and Smedley, J.W. 2004. 'Medieval and post-medieval glass technology: Melting characteristics of some glasses melted from vegetable ash and sand mixtures', *Glass Technology* 45: 36–42.

Jackson-Tal, R.E. 2004. 'The late Hellenistic glass industry in Syro-Palestine: A reappraisal', *Journal of Glass Studies* 46: 11–32.

Jaggi, O.P. 1977. *Science and technology of medieval India*, vol. 7, New Delhi: Atma Ram.

Jaksch, H., Seipel, W., Weiner, K.L. and El Goresy, A. 1983. 'Egyptian blue-cuprorivaite: A window to ancient Egyptian technology', *Naturwissenschaften* 70: 525–35.

Jaques of Vitry. 1611. Historia orientalis seu Hierosolymitana, in rivo qui est inter Valeniam, sub castro Margath, et Maracleam in *Gesta Dei per Francos I* (ed J. Bongars), Hannau.

Jennings, S. 2000. 'Late Hellenistic and early Roman cast glass from the Souks Excavations (BEY 006), Beirut, Lebanon', *Journal of Glass Studies* 42: 41–60.

Jennings, S. 2006. *Vessel glass from Beirut BEY 006, 007 and 045*, Berytus Archaeological Studies volumes XLVIII–XLIX 2004–2005, Archaeology of the Beirut Souks 2, AUB and ACRE Excavations in Beirut, 1994–1996, Beirut: The Faculty of Arts and Sciences, the American University of Beirut.

Jiménez Castillo, P. 2000. *El vidrio andalusí en Murcia*, in ed. P. Cressier, pp. 117–48.

Jiménez Castillo, P. 2006. 'Talleres técnicas y producciones de vidrio en al-Andalus', in *Vidrio islámico en al-Andalus* (eds. E. Rontomé and P. Pastor), La Granja de San Ildefonso: Fundacíon Centro Nacional del Vidrio and Junta de Castilla y León, pp. 51–73.

Jiménez Castillo, P., Navarro, J. and Thiriot, J. 2005. 'Taller de vidrio y casas andalusíes en Murcia. La excavación arqueológica del cáson de Puxmarina', *Memorias de Arqueología*, 13: 419–58.

Jones, G.O. 1956. *Glass*, London: Methuen.

Jones, R.E., Vagnetti, L., Levi, S.T., Williams, J., Jenkins, D. and de Guio, A. 2002. 'Mycenaean pottery from northern Italy. Archaeological and archaeometric studies', *Studi Micenei ed Egeo-Anatolici* 44: 221–61.

Jordan, J. 1912. 'Aus den Grabungsberichten aus Assur. November bis April 1912', *Mitteilungen der Deutschen Orient-Gesellschaft zu Berlin* 49. 25–8.

Karwowski, M. 2004. *Latènezeitlicher Glasringsschmuck aus Ostösterreich*, Österreichische Akademie der Weissenschaften Philosophisch-historische Klasse. Mitteilungen der prähistorischen Kommission 55, Vienna: Verlag der Österreichischen Akademie der Wissenschaften.

Kazmarczyck, A. 1986. 'The source of cobalt in ancient Egyptian pigments', *Proceedings of the 24th International Symposium on Archaeometry* (eds. J.S. Olin and J. Blackman), Washington, DC: Smithsonian Institution Press, pp. 369–76.

Kaczmarczyk, A. and Hedges, R.E.M. 1983. *Ancient Egyptian faience: An analytical survey of Egyptian faience from predynastic to Roman times*, Warminster, England: Aris and Phillips.

Kanungo, A. K. and Brill, R. H. 2009. 'Kopia, India's First Glassmaking Site: Dating and Chemical Analysis', *Journal of Glass Studies* 51: 11–25.

Kato, N., Nakai, I. and Shindo, Y. 2009. 'Change in chemical composition of early Islamic glass excavated in Raya, Sinai Peninsula, Egypt: on-site analyses using a portable X-ray fluorescence spectrometer', *Journal of Archaeological Science* 36: 1698–707.

Kato, N., Nakai, I. and Shindo, Y. 2010. Transitions in Islamic plant ash glass vessels: On-site chemical analyses conducted at the Raya/al Tur area on the Sinai Peninsula, Egypt', *Journal of Archaeological Science* 37: 1381–95.

Keller, C.A. 1983. 'Problems of dating glass industries of the Egyptian New Kingdom: Examples from Malkata and Lisht', *Journal of Glass Studies* 25: 19–28.

Keller, D. 2007. Paper presented at the Conference *Roman Finds: Context and theory: Proceedings of a conference held at the University of Durham* (eds. R. Hingley and S.Willis), Oxord: Oxbow Books.

Kennedy, D.L. 20007. *Gerasa and the Decapolis: A 'virtual island' in northwest Jordan*, London: Duckworth.

Kennedy, H. 2004. 'Byzantine-Arab diplomacy in the Near East from the Islamic conquests to the mid eleventh century', in ed. M. Bonner, pp. 81–91.

Kerr, R. and Wood, N., 2004. *Ceramic technology, Science and Civilization in China*, vol. 5, pt. 12, Cambridge: Cambridge University Press.

Khalil, I. and Henderson, J. 2011. 'An interim report on new evidence for early Islamic glass production at al-Raqqa, northern Syria', *Journal of Glass Studies* 53: 237–42.

Kingery, W.D., Bowan, H.K. and Uhlmann, D.R. 1976. *Introduction to ceramics*, 2nd ed., New York: John Wiley.

Kleinmann, B. 1991. 'Cobalt-pigments on the early Islamic blue glazes and the reconstruction of the way of their manufacture', *Archaeometry '90* (eds. E. Pernicka and G.A. Wagner), Heidelberg and Basel: Birkhauser Verlag, pp. 327–36.

Kock, J. and Sode, T. 1995. *Glass, glass beads and glassmakers in Northern India*, Vanlose, Denmark: THOT.

Kowatli, I., Curvers, H.H., Stuart, B., Sablerolles, Y., Henderson, J. and Reynolds, P. 2008. 'A pottery and glassmaking site in Beirut (015)', *Bulletin de Archéologie et d'Architecture Libanaises*, 10: 103–20.

Kramarovsky, M. 1998. 'The import and manufacture of glass in the territories of the Golden Horde', in ed. R. Ward, 96–100.

Kröger, J. 1995. *Nishapur: Glass of the early Islamic period*, New York: Metropolitan Museum of Art.

Kröger, J. 2010. 'Tradition und Innovation bei Glas aus islamischen Länderen', M. *Vorsicht Glas! Zerbrechliche Kunst 700–2010* (eds. A. Becker, J. Kröger, M. Kühn, K. Müller and Wajsberg, J., Museum für Islamische Kunst, Munchen: Minerva GmbH and Staatliche Museen zu Berlin, pp. 26–31.

Krom, M.D., Cliff, R.A., Eijsink, L.M., Herut, S. and Chester, R. 1999. 'The characterisation of Saharan dusts and Nile particulate matter in surface sediments from the Levantine basin using Sr isotopes', *Marine Geology* 155: 319–30.

Kuisma-Kursula, P., Räisänen, J. and Matiskainen, H. 1997. 'Chemical analyses of European forest glass', *Journal of Glass Studies* 39: 57–68.

Kunicki-Goldfinger, J., Kierzek, J., Dnierżanowski, P. and Kasprzak, A.J. 2005. 'Central European crystal glass of the first half of the eighteenth century', in *Annales of the 16e Congrès de l'Association Internationale pour l'Histoire du Verre*, London 2003, Nottingham: AIHV, pp. 258–62.

Laffineur, R. and Greco, E. 2005. 'Emporia. Aegeans in the central and eastern Mediterranean', (eds. R. Laffineur and E. Greco), Proceedings of the 10th International Aegean Conference, Athens, Italian School of Archaeology, 14–18 April 2004, 10e Rencontre Egéenne Internatiuonale, Liège: Université de Liège, Histoire de l'art et archéologie de la Grèce antique, Aegaeum 25, and Austin (TX), University of Texas at Austin, program in Aegean Scripts and Prehistory.

Lal, B.B. 1987. 'Glass technology in early India', in *Archaeometry of Glass, Proceedings of the Archaeometry Session of the XIV International Congress on Glass*, 1986 (ed. H.C. Bhardwaj), New Delhi, India, part 1, Kolkata (Calcutta): Indian Ceramic Society, section 1, pp. 44–56.

Lamb, A. 1965. 'Some glass beads from the Malay Peninsula', *Man* 65: 36–38.

Lamb, A. 1966. 'A note on the glass beads from the Malay Peninsula', *Journal of Glass Studies* 15: 86–94.

Lambert, J.B. 1997. *Traces of the past: Unravelling the secrets of archaeology through chemistry*. Reading, MA: Perseus Books.

Lambert, N. 1972. 'La Seube: Témoin de l'art du verre en France méridionale di Bas-Empire à la fin du Moyen Age,' *Journal of Glass Studies* 14: 77–116.

Bibliography

Lamm, J.C. 1928. *Das Glas von Samarra*, Forschungen zur Islamischen Kunst, 2. Die Ausgrabungen von Samarra, 4, Berlin.

Lamm, J.C. 1929–30. *Mittelalterliche Gläser und Steinschnittarbeiten aus dem Nahen Osten*, 2 vols., Forschungen zur islamischen Kunst, 5, Berlin.

Lamm, J.C. 1941. *Oriental glass of medieval date found in Sweden and the early history of lustre-painting*, Stockholm: Kungliga Vitterhetsakademien Historic och Anrikvitets Akademiens Handlingar 50:1.

Lankton, J.W. and Dussubieux, L. 2006. 'Early glass in Asian maritime trade: A review and an interpretation of compositional analyses', *Journal of Glass Studies*, 48: 121–44.

Lappe, U. and Möbes, G. 1984. 'Glashütren im Eichsfeld', *Alt Thüringen* 20: 207–32.

Le Fèvre-Lehöerff, A. 1992. 'Les moules de l'age du bronze dans la plaine orientale du Po: vestiges de mise en forme des alliages base cuivre', *Padusa* 28: 131–243.

Leach, B.H. 1940. *A potter's book*, London: Faber and Faber.

Lee, L. and Quirke, S. 2000. 'Painting materials', in eds. Paul T. Nicholson and Ian Shaw 2000, pp. 104–20.

Lemche, P.I. 1995. 'The history of ancient Syria and Palestine: An overview' in ed. J.M. Sasson 1995, pp. 1195–218.

Lemke, C. 1998. 'Reflections of the Roman Empire: The first-century glass industry as seen through traditions of manufacture', The prehistory and history of glassmaking technology (ed. P. McCray), *Ceramics and Civilization*, vol. VIII, Westerville, OH: The American Ceramic Society, pp. 269–92.

Leslie, K.A., Freestone, I.C., Lowry, D. and Thirlwell, M. 2006. 'The provenance and technology of Near Eastern glasses: Oxygen isotopes by laser fluorination as a complement to strontium', *Archaeometry* 48(4): 253–70.

Lester, E., Hilal, N. and Henderson, J. 2004. 'Porosity in ancient glass from Syria (c. 800 AD) using gas adsorption and atomic force microscopy', *Surface and Interface Analysis* 36: 1323–9.

Levi-Provençal, E. 1963. *Histoire de l'Espagne musulmane*, vol. III, Paris.

Li Qinghui, Xu Yongchun, Zhang Ping, Gan Fuxi and Cheng Huansheng. 2009a. 'Chemical composition analyses of early glasses of different historical periods found in Xinjiang, China', in eds. Gan Fuxi, R.H. Brill and T. Shouyun, pp. 331–57.

Li Quinghui, Wang Weizhao, Xiong Zhaoming, Gan Fuxi and Cheng Huansheng. 2009b. 'PIXE study on the ancient glasses of the Han Dynasty unearthed in Hepu county, Guangxi', in eds. Gan Fuxi, R.H. Brill and T. Shouyun, pp. 397–411.

Lilyquist, C. 1993. 'Granulation and glass; chronological and stylistic investigations at selected sites, ca. 2500–1400 B.C.E.', *Bulletin of the American Schools of Oriental Research*, 290/291: 29–94.

Lilyquist, C. and Brill, R.H., eds. 1993. *Studies in early Egyptian glass*. New York: Metropolitan Museum of Art.

Lilyquist, C., Brill, R.H., Wypyski, M. and Koestler, R. 1993. Part 2. Glass, in eds. C. Lilyquist and R. H. Brill, pp. 23–58.

Little, L.K., ed. 2007. *Plague and the end of antiquity: The pandemic of 541–750*, Cambridge: Cambridge University Press.

Lledó, B. 1997. 'Mold siblings in the 11th century cullet from Serçe Limani', *Journal of Glass Studies* 39: 43–55.

Long, P.O. 2003. *Technology and society in the medieval centuries: Byzantium, Islam, and the West, 500–1300*, Washington, D.C.: American Historical Association.

Loveland, W. 1989. 'Environmental sciences'. *Lanthanide probes in life, chemical, and earth sciences* (eds. J. Bunzli and G. Choppin), New York: Elsevier, pp. 391–411.

Lucas, A. 1934 [1921]. *Ancient Egyptian materials and industries*, London: Edward Arnold.

Lucas, A. 1948. *Ancient Egyptian materials and industries*, 3rd edition, London: Edward Arnold

Lucas, A. and Harris, J.R. 1989 [1962]. *Ancient Egyptian materials and industries*, 4th ed., London: Histories and Mysteries of Man.

Maccabruni, C., ed. 1998. *Vetro e vetri, preziose iridescenze*, Museo Archeologico, Milano, 1 November 1998–18 April 1999, Milan: Electra.

Magou, H. 1992. 'Scientific analysis' in Gladys Davidson Weinberg and M. C. McClellan, *Glass Vessels in Ancient Greece: Their History Illustrated from the Collection of the National Museum, Athens*, Publications of the Archaiologikon Deltion, no. 47, Athens: Archaeological Receipts Fund.

Magou, H. 2001. 'Chemical analyses of three opaque glass beads and a sword from the Pylona cemetery', *The Mycenaean Cemetery at Pylona on Rhodes* (ed. Efi Karantzali), British Archaeological Reports International Series 988, Oxford: Archaeopress, 2001, pp. 117–18.

Malleret, L. 1962. *L'Archéologie du delta du Mékong*, vol. 3, Paris: Publications de l'Ecole Française d'Extrême-Orient.

Mallowan, M.E.L. 1947. 'Excavations at Tell Brak and Chagar Bazar', *Iraq* 9: 1–259.

Marii, F. 2007. *Glass, glass cakes and tesserae from the Petra church in Petra, Jordan*, Unpublished Ph.D. thesis, University of London.

Martinón-Torres, M. and Rehren, Th. (eds.) 2008. *Archaeology, History and Science: Integrating approaches to ancient materials*, London: Institute of Archaeology Publications, California: Left Coast Press.

Marzatico, F. 1997. 'L'industria metallurgica nel Trentino durante l'età del bronzo', in eds. M. Bernabò Brea, A. Cardarelli and M. Cremaschi, pp. 570–6.

Mass, J.L., Stone, R.E. and Wypyski, M.T. 1998. 'The mineralogical and metallurgical origins of Roman opaque colored glasses', *The prehistory and history of glassmaking technology* (eds. W.D. Kingery and P. McCray), Ceramics and Civilization, Westerville, OH: The American Ceramic Society, 8: pp. 121–44.

Mass, J.L., Wypyski, M.T. AND Stone, R.E. 2002. 'Malkata and Lisht glassmaking technologies: towards a specific link between second millennium B.C. metallurgists and glassmakers', *Archaeometry* 44(1): 67–82.

Mathews, T.F. 1998. *The art of Byzantium: Between antiquity and the Renaissance*, London: Weidenfeld and Nicholson.

Matoïan, V. 2000. 'Données nouvelles sur le verre en Syrie au IIe millénaire av. J.-C.: le cas de Ras Shamra-Ougarit', in ed. M.-D. Nenna, pp. 23–48.

Matoïan, V. 2002–2003. 'Matières vitreuses au royaume d'Ougarit', *Annales Archéologiques Arabes Syriennes*, XLV–XLVI: 153–62.

Matoïan, V. and Bouquillon, A., 1999. 'La céramique argileuse à glaçure du site de Ras Shamra-Ougarit (Syrie)', *Syria*, 76: 57–82.

Matoïan, V. and Bouquillon, A. 2000.' Le "bleu égyptien" à Ras Shamra-Ougarit (Syrie)', in *Proceedings of the First International Congress on the Archaeology of the Ancient Near East* (eds. P. Matthiae, A. Enea, L. Peyronel and F. Pinnock), Rome, 18–23 May 1998, Rome, pp. 983–1000.

Matoïan, V. and Bouquillan A. 2003. 'Vitreous materials in Ugarit: New data', *Culture through objects. Ancient Near Eastern studies in Honour of P.R.S. Moorey* (eds. T. Potts, M. Roaf and D. Stein), Oxford: Griffith Institute, pp. 333–46.

Matoïan, V. and A. Bouquillan 2007. 'Matériaux vitreux découverts à Tell Ashara – Terqa (chantier E, quinzième campagne), II. Un lingot de matière vitreuse (TQ 15.411)

de l'époque du royaume de Hana', *Les rives de l'Euphrate* (eds. J.-C. Margueron, O. Rouault and P. Lombard), Akh Puritum, vol. 2, Mémoires d'Archéologie et d'Histoire Régionales Interdisciplinaires, Lyon: Maison de l'Orient et de la Méditerranée, pp. 187–197.

Matson. F.R. 1948. 'The manufacture of eighth-century Egyptian glass weights and stamps', *Early Arabic glass weights and stamps* (ed. C. Miles), Numismatic Notes and Monographs, no. 111, New York: The American Numismatic Society, pp. 31–69.

Matson, F.R. 1951. 'The composition and working properties of ancient glass', *The Journal of Chemical Education* 28: 82–87.

Mayell, H. 2005. 'Asteroid rained glass over entire earth, scientists say', *National Geographic News* (April 15). Available at http://news.nationalgeographic.com/news/2005/04/0415_050418_chicxulub.html.

McArthur, J.M., Howareth, R.J. and Bailey, T.R. 2001. 'Strontium isotope stratigraphy: LOWESS version 3: Best fit to the marine Sr-isotope curve for 0-509 Ma and accompanying look-up table for deriving numerical age', *Journal of Geology* 109: 155–170.

McClellan, T.L. 1989. 'The chronology and ceramic assemblages of Alalakh', *Essays in ancient civilization presented to Helen J. Kantor* (eds. A. Leonard, Jr. and B.B. Williams), Chicago: The Oriental Institute, University of Chicago, pp. 181–212.

McClelland, M. 1984. *Core-formed glass from dated contexts*, unpublished Ph.D. dissertation, University of Philadelphia.

McClelland, M. 1992. 'Core-formed glass vessels, 525 BC to 10 AD.', *Glass vessels in ancient Greece: Their history illustrated from the collection of the National Archaeological Museum, Athens*, Publications of the Archaiologikon Deltion (ed. G.D. Weinberg), Athens: Archaeological Receipts Fund.

McCray, P. 1999. *Glassmaking in Renaissance Venice: The fragile craft*, Brookfield, Vermont and Aldershot, England: Ashgate.

McCray P., Osborne, Z. and Kingery, W.D. 1995. 'The history and technology of Renaissance Venetian chalcedony glass,' *Rivista Della Stazione Sperimentale del Vetro* 5: 259–78.

McKinnon, E.E. and Brill, R.H. 1987. 'Chemical analyses of some glasses from Sumatra', *Archaeometry of Glass, Proceedings of the Archaeometry session of the XIV International Congress on Glass, 1986* (ed. H. C. Bhardwaj), New Delhi, India, part 1, Kolkata (Calcutta): Indian Ceramic Society, section 2, pp. 1–14.

McLoughlin, S.D. 2003. *The characterisation of archaeological glasses using advanced analytical techniques*, unpublished D.Phil. thesis, Imperial College of Science, Technology and Medicine, London.

McVay, G.L. and Day, D.E. 1970. 'Diffusion and internal friction in in Na-Rb silicate glasses,' *Journal of the American Ceramic Society* 53: 508–13.

Mecquenem, R. de and Michalon, J. 1953. *Recherches a Tchoga Zanbil*, Mémoires de la mission archéologique en Iran, 33, Paris.

Meek, A.S. 2005. *Tracing continuity: The scientific investigation of glass compositions from al-Raqqa, Syria by EPMA/WDX and LA-ICP-MS*, unpublished M.Sc. dissertation, University of Nottingham, England.

Meek, A.S. 2011. *The chemical and isotopic analysis of English forest glass*, unpublished Ph.D. thesis, University of Nottingham, England.

Meek, A.S., Henderson, J. and Evans, J.A. 2009. 'North-western European forest glass: Working towards an independent means of provenance', *Proceedings of the 37th International Symposium on Archaeometry* (ed. I. Turbanti-Memmi), Berlin and Heidelberg: Springer-Verlag, pp. 105–11.

Meek, A., Henderson, J. and Evans, J.A. 2012. ' "As for glassmakers, they be scant in this land". The isotopic analysis of English forest glass from the Weald and Staffordshire', *Journal of Analytical Atomic Spectroscopy.* 27, 786–95.

Meinecke, M. 1996. 'Forced labor in early Islamic architecture: The case of ar-Raqqa/ar-Rāfiqa on the Euphrates', in *Patterns of stylistic changes in Islamic architecture. Local traditions versus migrating artists* (ed. M. Meinecke), New York and London: Hagop Kevorkian Series on Near Eastern Art and Civilization, pp. 5–30.

Meiwes, K.L. and Beese, F. 1988. 'Ergebnisse der Untersuchung des Stoffhaushaltes eines Buchenwaldökosystems auf Kalkgestein', *Berichte des Forsuchingszentrum Walldökosysteme,* 9: 1–142.

Mendera, M. 1991. 'Produrre vetro in Valselsa: l'officina vetraria di Germagnana (Gambassi-Fi) (secc. XIII-XIV)', *Archeologia e storia della produzione del vetro preindustriale* (ed. M. Mendera), Florence: Edizioni all'insegna del Giglio, pp. 15–20.

Merchant, I., Henderson, J., Crossley, D. and Cable, M. 1997. 'An examination of the chemical interaction of glass and ceramics in medieval glass technology', *Archaeological Sciences 1995, Proceedings of a conference on the application of scientific techniques to the study of archaeology*, (eds. A. Sinclair, E. Slater and J. Gowlett), Liverpool July 1995, Oxbow Monograph 64, Oxford: Oxbow Books, pp. 31–7.

Millett, M. 1990. *The Romanisation of Britain*, Cambridge: Cambridge University Press.

Mirti, P., Casoli, A. and Appolonia, L. 1993. 'Scientific analysis of Roman glass from Augusta Pretoria', *Archaeometry* 35: 225–40.

Mirti, P., Pace, M., Negro Ponzi, M.M. and Aceto, M. 2008. 'ICP-MS analysis of glass fragments of Parthian and Sasanian epoch from Seleucia and Veh Ardašīr (central Iraq)', *Archaeometry* 50: 429–50.

Misra, M., Ragland, K. and Baker, A. 1993. 'Wood ash composition as a function of furnace temperature', *Biomass and Bioenergy* 4: 103–16.

Montgomery, J., Evans, J.A. and Cooper, R.E. 2007. 'Resolving archaeological populations with Sr-isotope mixing models', *Applied Geology* 22: 1502–14.

Montgomery, J., Evans, J.A. and Wildman, G., 2006. 'Sr-87/Sr-86 isotope composition of bottled British mineral waters for environmental and forensic purposes', *Applied Geochemistry,* 21: 1626–34.

Moorey, P.R.S. 1985. *Materials and manufacture in ancient Mesopotamia: The evidence of archaeology and art*, Oxford: British Archaeological Reports, International Series, 237.

Moorey, P.R.S. 1986. 'The emergence of the light, horsedrawn chariot in the Near East c. 2000–1500 B.C.', *World Archaeology* 18: 196–215.

Moorey, P.R.S. 1989. 'The Hurrians, the Mittani and technological innovation', *Archaeologica Iranica et Orientalis Miscellanea in Honorem Louis Vanden Berghe*, vol. I (eds. L. de Meyer and E. Haerinick), Ghent: University of Ghent, pp. 273–86.

Moorey, P. R. S. 1994. *Ancient Mesopotamian materials and industries. The archaeological evidence*, Oxford: The Clarendon Press.

Moorey, P.R.S. 2001. 'The mobility of artisans and opportunities for technology transfer between Western Asia and Egypt in the Late Bronze Age', in *The Social Context of Technological Change: Egypt and the Near East, 1650–1550 B.C.* (ed. A.J. Shortland), Oxford: Oxbow Books, pp. 1–14.

Moran, W.L. 1992. *The Amarna Letters*, Baltimore: The Johns Hopkins Press.

Morel, J., Amrein, H., Meylan, M.F. and Chavalley, C. 1991. 'Un atelier de verrier du milieu du 1er siècle après J.-C. á Avenches', *Archäologie der Schweiz* 15: 2–17.

Moretti, C. and Hreglich, S. 1984. 'Opacification and colouring of glass by the use of 'anime', *Glass Technology* 25: 277–82.

Moretti, C. and Hreglich, S. 2005. 'Opaque glass manufacturing techniques used by Venetian glassmakers between the 15th and 20th centuries', *Rivista della Stazione Sperimentale del Vetro* 5: 15–32.

Morey, G.W. 1964. *Phase-equilibrium relations of the common rock-forming oxides except water*, U.S. Geological Survey Professional Paper No. 440-L, Chapter L.

Mortimer, C. 1995. 'Analysis of post-medieval glass from Old Broad Street, London with reference to other contemporary glasses from London and Italy', *Trade and discovery: The scientific study of artifacts from post-medieval Europe and beyond*, (eds. D.R. Hook and R.M. Gaimster), British Museum Occasional Paper 109, London: The British Museum Press, pp. 135–44.

Muhly, J. 1988. 'The beginnings of metallurgy in the Old World', *The beginnings of metals and alloys* (ed. R. Maddin), Cambridge, MA: MIT Press, pp. 9–27.

Na'aman, N. 1981. 'Economic aspects of the Egyptian occupation of Canaan', *Israel Exploration Journal* 31: 172–85.

Näsman, U. (1979) 'Die Herstellung von Glasperlen', in M. Bencard 'Wikingerzeitliches Handwerk in Ribe', *Acta Archaeologica (København)* 49: 113–138.

Nenna, M.-D. 1995. Les plaques d'incrustation: une industrie égyptienne du verre, Alexandria and the Hellenistic-Roman world, Rome: l'esmadi Bretschneider, 377–84.

Nenna, M.-D. 1999. *Les verres*, Exploration archéologique de Délos 37, Paris: Ecole Française d'Athènes.

Nenna, M.-D., ed. 2000, *La Route du verre: Ateliers primaires et secondaires du second millénaire av. J.-C. au Moyen Age*, Lyon: Maison de l'Orient Méditeraneéan-Jean Pouilloux.

Nenna, M.-D. and Gratuze, B. 2009. 'Étude diachronique des compositions de verres eomployés dans les vases mosaïqués antiques: résultants préliminaries', *Annales of the 17th Congress of the International Association for the History of Glass*, Antwerp, 2006, Antwerp: University Press Antwerp, pp. 199–205.

Nenna, M.-D., Picon, M., Thirion-Merle, V. and Vichy, M. 2005. 'Ateliers primaires du Wadi Natrun: nouvelles découvertes', *Annales of the 16ᵉ Congrès de l'Association Internationale pour l'Histoire du Verre*, London 2003, Nottingham: AIHV, pp. 59–63.

Nenna, M.-D., Picon, M. and Vichy, M. 2000. 'Ateliers primaries et secondaires en Égypte á l'époque gréco-romaine', in ed. M.-D. Nenna 2000a, pp. 97–112.

Nenna, M.-D., Vichy, M. and Picon, M. 1997. 'L'atelier de verrier de Lyon, du 1er siècle après J.-C., et l'origine des verres "Romains"', *Revue d'Archéométrie* 21: 81–87.

Neri, A. 1612. *L'arte vetraria* (The art of glassmaking), Firenze: nella Stampa de' Giunto. Translation and commentary by Christopher Merrett 1662.

Newton, R. and Davison, S. 1989. *Conservation of glasses*, London: Butterworths.

Newton, R.G. 1978. 'Colouring agents used by medieval glassmakers', *Glass Technology* 19: 59–60.

Newton, R.G. and Renfrew, C. 1970. 'British faience beads reconsidered', *Antiquity* 54: 199–206.

Nicholson, P.T. 1993. *Ancient Egyptian faience and glass*, London: Shire Publications.

Nicholson, P.T. 1995. 'Glass-making and glass-working at Amarna: some new work', *Glass Technology* 37: 11–19.

Nicholson, P.T. 2003. 'New excavations at a Ptolemaic-Roman faience factory at Memphis, Egypt', *Proceedings of the 15th International Congress of the Association for the History of Glass*, Corning, NY and Nottingham: AIHV, pp. 49–52.

Nicholson, P.T. 2007. *Brilliant things for Akhenaten. The production of glass, vitreous materials and pottery at Amarna site 04, 125*, London: Egypt Exploration Society.

Nicholson, P.T., Jackson, C.M. and Trott, K.M. 1997. 'The Uluburun Glass Ingots, Cylindrical Vessels and Egyptian Glass', *Journal of Glass Studies* 39: 143–53.

Nicholson, P.T. and Shaw, I., eds. 2000. *Ancient Egyptian materials and industries*, Cambridge: Cambridge University Press.

Nightingale, G. 2008. 'Tiny, fragile, common, precious. Mycenaean glass and faience beads and other objects, in Vitreous materials in the Late Bronze Age Aegean (eds. C.M. Jackson and E.C. Wager), Sheffield Studies in Aegean Archaeology, Oxford: Oxbow Books, pp. 64–104.

Nikita, K. 2003. 'Mycenaean glass beads: technology, forms and function', in *Ornaments of the past* (eds. I.C. Glover, H. Hughes-Brock and J. Henderson), London and Bangkok: Bead Study Trust, 23–37.

Nikita, K. and Henderson, J. 2006. 'Glass analyses from Mycenae, Thebes and Elateia: Compositional evidence for a Mycenaean glass industry', *Journal of Glass Studies* 48: 71–120.

Noble, J.V. 1969. 'The technique of Egyptian faïence', *American Journal of Archaeology* 73: 435–9.

Noll, W. 1981. 'Mineralogy and technology of the painted ceramics of ancient Egypt', in *Scientific Studies of Ancient Ceramics* (ed. M. Hughes), British Museum Occasional Paper 19, London: British Museum Press, pp. 143–54.

Nolte, B. 1968. 'Die Glasgefässe im alten Ägypten', Berlin: Münchner Ägyptologische, Studien 14.

Northedge, A. and Faulkner, R. 1987. 'The 1986 survey season at Samarra', *Iraq* 49: 143–73.

Oates, D. 1968. 'The excavations at Tell el Rimah', *Iraq* 30, 115–38.

Oates, J., Oates, D. and MacDonald, H. 1997. *Excavations at Tell Brak*, vol. 1. Cambridge: The Macdonald Institute of Archaeological Research.

O'Hara, S., Jones, A., Henderson, J. and Challis, K. In preparation.'Environmental research at al-Raqqa, Syria'.

O'Hea, M. 2005. 'Late Hellenistic glass from some military and civilian sites in the Levant: Jebel Khalid, Pella and Jerusalem', *Annales du 16e Congrès de l'association Internationale pour l'Histoire du verre*, London 2003, Nottingham, England: AIHV, pp. 44–48.

Oliver, A. Jr. 1968. *The reconstruction of two Apulian tomb groups* (Antike Kunst Beiheft 50), Bern: Francke.

Oppenheim, A.L. 1973. 'Towards a History of Glass in the Ancient Near East', *Journal of the American Oriental Society* 93: 259–66.

Oppenheim, A.L., Brill, R.H., Barag, D.P. and Von Saldern, A. 1970. *Glass and glassmaking in ancient Mesopotamia*, Corning, NY: The Corning Museum of Glass [reprint 1988].

Painter, K. and Whitehouse, D. 1990. 'Style, date and place of manufacture' [of the Portland vase], *Journal of Glass Studies* 32: 122–25.

Panagiotaki, M., Papazoglou-Manioudaki, G., Chatzi-Spiliopoulou, E., Andreopoulou-Mangou, E., Maniatis, Y., Tite, M.S. and Shortland, A. 2005. 'A glass workshop at the Mycenaean citadel of Tiryns in Greece', *Annales de l'Association Internationale pour l'Histoire du Verre*, London, 2003, Nottingham, England: AIHV, pp. 14–18.

Partington, J.R. 1935. *Origins and development of applied chemistry*, London.

Patyk-Karar, N.G., Bardeeva, E.G. and Sjevelev, A.G. 1999. 'Titano-zirconium placers in the sedimentary cover platforms', *Episodes* 22: 89–97.

Paul, A. 1990. *Chemistry of glasses*, London: Chapman and Hall.

Paynter, S. 2006. 'Analyses of colourless Roman glass from Binchester, County Durham', *Journal of Archaeological Science* 33: 1037–57.

Paynter, S. 2008. 'Experiments in the reconstruction of Roman wood-fired glassworking furnaces: waste products and their formation processes', *Journal of Glass Studies* 50: 271–90.

Paynter, S. and Dungworth, D. 2011. 'Recognising frit: Experiments reproducing Post-medieval plant ash glass', *Proceedings of the 37th International Symposium on Archaeometry* (ed. I. Turbani-Memmi), Heidelberg and Berlin: Springer-Verlag, 133–38.

Peacock, D.P.S. and Williams, D.F. 1986. *Amphorae and the Roman economy*, London: Longman.

Pearce, M. 2000. 'Metals make the world go round: the copper supply for Frattesina', *Metals make the world go round: The supply and circulation of metals in Bronze Age Europe* (ed. C. F.C. Pare), Proceedings of a conference held at the University of Birmingham in June 1997, Oxford: Oxbow Books, pp. 108–15.

Pearce, M. 2007. *Bright blades and red metals*, essays on north Italian metalwork, volume 14, Accordia Specialist studies on Italy, London: Accordia Research Institute, University of London.

Pearce, M. and De Guio, A. 1999. 'Between the mountains and the plain: an integrated metals production system in later Bronze Age north-eastern Italy', *Prehistoric Alpine environment, society and economy*, Papers of the International Colloquium PAESE, 1997, held in Zurich, Universitätsforschungen zur prähistorischen Archäologie 55, Bonn: Rudolf Habelt, pp. 289–93.

Pellegrini, E. 1992. 'Nuovi dati su due ripostigli dell'età del bronzo finale del Grossetano: Piano di Tallone e tra "Manciano e Samprugnano", *Bullettino di Paletnologia italiana* 83: 341–60.

Peltenburg, E.J. 1969. 'Al Mina glazed pottery and its relations', *Levant* 1: 73–96.

Peltenburg, E.J. 1987. 'Early faience: Recent studies, origins and relations with glass', in *Early vitreous materials* (eds. M. Bimson and I.C. Freestone), British Museum Occasional Paper no. 56, London: British Museum Press, pp. 5–29.

Peroni, R. 1997. 'Le terramare nel quadro dell'età del bronzo europea', in eds. M. Bernabò Brea, A. Cardarelli and M. Cremaschi, pp. 30–6.

Peters, J.P. 1898. *Nippur, or Explorations and Adventures on the Euphrates vol. II*, New York and London: G. P. Putnam's sons.

Phenyo, T., Wilmsen, E.N., Killick, D. and Rosenstein, D.D. 2009. 'Mmopi le mmpoa: potter and clay. Making pottery today and c. 1000 years ago', *Botswana Notes and Records* 41: 25–38

Phillips, K.M., Jr. 1993. *In the hills of Tuscany: Recent excavations at the Etruscan site of Poggio Civitale (Murlo, Siena)*, Philadelphia: University of Pennsylvania Museum of Archaeology and Anthropology.

Photos-Jones, E, Ballin-Smith, B., Hall, A.J. and Jones, R.E. 2007. 'On the intent to make cramp: an interpretation of vitreous seaweed cremation 'waste' from prehistoric burial sites in Orkney, Scotland', *Oxford Journal of Archaeology*, 26: 1–23.

Pollard, A.M. 2009. 'What a long strange trip it's been: Lead isotopes in archaeology', *Mine to microscope – advances in the study of ancient technology* (eds. A. Shortland, I.C. Freestone and Th. Rehren), Oxford: Oxbow Books, pp. 181–189.

Pollard, A.M. and Heron, C. 1996. *Archaeological chemistry*, London: Royal Society of Chemistry.

Pollard, A.M. and Højlund, F. 1983. 'High magnesium glazed sherds from Bronze Age tells on Failaka, Kuwait', *Archaeometry* 25: 196–200.

Pollard, A.M. and Moorey, P.R.S. 1982. 'Some analyses of middle Assyrian faience and related materials', *Archaeometry* 24: 45–50.

Pollard, A.M., Batt, C., Stern, B. and Young, S.M.M. 2007. *Analytical chemistry in archaeology*, Cambridge: Cambridge University Press.

Pomerancblum, M. 1966. 'The distribution of heavy minerals and their hydraulic equivalents in sediments of the Mediterranean continental shelf of Israel', *Journal of Sedimentary Petrology* 36: 162–74.

Porada, E., Hansen, D., Dunham, S. and Babcock, S. 1992. 'The chronology of Mesopotamia, ca. 7000 to 1600 B.C.', Chapter 5 in *Chronologies in Old World archaeology*, Chicago: Chicago University Press, pp. 77–121.

Price, J. 1985. 'Late Hellenistic and early imperial vessel glass at Berenice: A survey of imported tablewares found during excavations at Sidi Khreish, Benghazi', *Cyrenaica in antiquity*, Society for Libyan Studies Occasional Papers 1, Oxford, pp. 287–96.

Price, J. 1991. 'Decorated mould-blown glass tablewares in the first century AD', *Roman glass, two centuries of art and invention* (M. Newby and K. Painter eds.), Occasional Paper 8, London: The Society of Antiquaries.

Price, J. 2000. 'Late Roman vessels in Britain, from AD 350 to 410 and beyond', *Glass in Britain and Ireland AD 350–1100* (ed. J. Price), British Museum Occasional Paper No. 127. London: British Museum, 1–32.

Price, J. 2005. 'Glass working and glassworkers in cities and towns', *Roman working lives and urban living* (eds. A. Mac Mahon and J. Price), Oxford: Oxbow Books, pp. 167–90.

Price, T.D. and Gestsdóttir, H. 2006. 'The first settlers of Iceland: An isotopic approach to colonisation', *Antiquity* 80: 130–44.

Pulak, C. 1988. 'A Bronze Age shipwreck at Ulu Burun, Turkey: 1985 campaign', *American Journal of Archaeology* 92: 1–37.

Pulak, C. 1997. 'The Uluburun shipwreck', *Res Maritimae, Cyprus and the eastern Mediterranean from Prehistory to Late Antiquity* (eds. S. Swiny, R.L. Hohlfelder and H.W. Swiny), Nicosia, Cyprus 18–21 October 1994, American Schools of Oriental Research Archaeological Reports 4, Atlanta: ASOR, pp. 233–62.

Pulak, C. 2001. 'The cargo of the Uluburun ship and evidence for trade with the Aegean and beyond, in Italy and Cyprus in Antiquity, 1500–450 B.C.', Proceedings of an international symposium held at the Italian academy for advanced studies in America at Columbia University (eds. L. Bonfante and V. Karageorghis), November 16–18, 2000. The Costakis and Leto Severis Foundation, Nikosia, pp. 13–60.

Pulak, C. 2005. 'Discovering a Royal Ship from the age of King Tut: Uluburun, Turkey' *Beneath the Seven Seas Adventures with the Institute of Nautical Archaeology* (ed. George F. Bass), London: Thames and Hudson.

Purowski, T., Dzierżanowski, P., Bulska, E., Wagner, B. and Nowak, A. 2012. 'A study of glass beads from the Hallstatt C-D from southwestern Poland: implications for glass technology and provenance', *Archaeometry* 54: 144–66.

Pusch, E. 2000. 'Doors – statues – musical instruments? Large-scale bronze production and casting in the residence of Ramses' (eds. E. Hickmann, I. Laufs and R. Eichmann), *Studien zur Musikarchaeologie* II, 219–32.

Pusch, E.B. and Rehren, Th. 2007. 'Hochtemperatur-Technologie in der Ramses-Stadt'. *Rubinglas für den Pharao*, Forschungen in der Ramses-Stadt 6, Gerstenberg:Verlag, Hildesheim.

Quartieri, S., Riccardi, M.P., Messiga, B. and Boscherini, F. 2005. 'The ancient glass production of the medieval Val Gargassa glasshouse: Fe and Mn XANES study', *Journal of Non-Crystalline Solids* 351: 3013–22.

Raftery, B. 1976. 'Rathgall and Irish hillfort problems', in *Hillforts, later prehistoric earthworks of Britain and Ireland* (ed. D.W. Harding), London: Academic Press, pp. 339–57.

Raftery, B. 1987. 'Some glass beads of the later Bronze Age in Ireland in Glasperlen der Verromischen Eisenziet II', Marburger Sudien zur Vor- und Frühgeschichte vol. 9, Mainz: Phillip von Zabern, pp. 39–48.

Reade, W., Freestone, I.C. and Simpson, S.J. 2005. 'Innovation or continuity? Early First millennium BCE glass in the Near East: the cobalt blue glasses from Assyrian Nimrud', in *Annales of the 16e Congrès de l'Association Internationale pour l'Histoire du Verre*, London 2003, Nottingham, England: AIHV: 23–7.

Rebourg, A. 1989a. 'Les ateliers antiques, les fours d'Autun', in *À travers le verre, du Moyen Âge à la Renaissance* (eds. D. Foy and G. Sennequier), Rouen: Musées departementaux de la Seine-Maritime, pp. 49–52.

Rebourg, A. 1989b. 'Un atelier de Verrier Gallo-Romain à Autun (Saône-et-Loire)', *Revue Archéologique de l'Est et du Centre-Est* 40: 249–58.

Rechinger, K.H. 1973. *Flora Aegaea, flora der inseln und halbinseln des Ägäischen meeres*, Vienna: Springer Verlag, 1943, repr. Koenigstein: Otto Koeltz Antiquariat.

Redford, D.B. 1990. *Egypt and Canaan in the New Kingdom* (ed. Shmuel Ahituv and Beer Sheva). Beersheva, Israel: Ben Gurion University.

Redford, D.B. 1992. *Egypt, Canaan and Israel in ancient times*, Princeton, NJ: Princeton University Press.

Redman, C.L., Anzalone, R.D. and Rubertone, P.E. 1979. 'Medieval archaeology at Qsar es-Seghir, Morocco', *Journal of Field Archaeology* 6: 1–16.

Rehren, Th. 1997. 'Rammeside glass colouring crucibles', *Archaeometry* 39: 355–68.

Rehren, Th. 2000. 'New aspects of ancient Egyptian glassmaking', *Journal of Glass Studies* 42: 13–24.

Rehren, Th. 2001. 'Aspects of the production of cobalt-blue glass in Egypt', *Archaeometry* 43, no. 4: 483–89.

Rehren, Th. 2008. 'A review of factors affecting the composition of early Egyptian glasses and faience: alkali and alkali earth oxides', *Journal of Archaeological Science* 35: 1345–54.

Rehren, Th. and Cholakova, A. 2010. 'The early Byzantine HIMT glass from Dichin, Northern Bulgaria', *Interdisciplinary Studies* 22/23: 81–96.

Rehren, Th. and Pusch, E.B. 2005. 'Late Bronze Age glass production at Qantir-Piramesses, Egypt', *Science* 308: 1756–58.

Rehren, Th., Spenser, L. and Triantafyllidis, P. 2005. 'The primary production of glass at Hellenistic Rhodes', in *Annales du 16e Congrès de l'association Internationale pour l'Histoire du verre*, London 2003, Nottingham: International Association for the History of Glass, pp. 39–43.

Renfrew, C. 1975. 'Trade as action at a distance: Questions of Integration and Communication', in *Ancient Civilisation and Trade* (eds. J.A. Sabloff and C.C. Lamberg-Karlovsky), Albuquerque: University of New Mexico Press, pp. 3–59. Republished in Renfrew, C. 1984. *Approaches to Social Archaeology* 86–134. Edinburgh University Press, Edinburgh.

Renfrew, J.E., Dixon, J.E. and Cann, J.R. 1966. 'Obsidian and early cultural context in the Near East', *Proceedings of the Prehistoric Society* 32: 30–72.

Riis, P.J. and Poulson, V. Hama 1931–1938. *Les verreries et poteries médiévales*, Fouilles et recherches de la foundation Carlesberg, vol. 4, no. 2, Copenhagen.

Riva, C. 2007. 'The archaeology of Picenum: The last decade', in *Ancient Italy: Regions without boundaries* (eds. G. Bradley, E. Isayev and C. Riva), Exeter: Exeter University Press, pp. 96–100.

Robertshaw, P., Benco, N., Wood, M., Dussubieux, L., Melchiorre, E. and Ettahiri, A. 2010. 'Chemical analysis of glass beads from medieval al-Basra (Morocco)', *Archaeometry* 52: 355–79.

Robertshaw, P., Glascock, M.D., Wood, M. and Popelka, R.S. 2003. 'Chemical analysis of ancient African glass beads: A very preliminary report', *African Journal of Archaeology* 1, 1: 139–46.

Robertshaw, P., Rasoarifetra, B., Wood, M., Melchiorre, E., Popelka-Filcoff, R.S. and Glascock, M.D. 2006. 'Chemical analysis of glass beads from Madagascar', *Journal of African Archaeology* 4: 91–109.

Robertshaw, P., Wood, M., Melchiorre, E., Popelka-Filcoff, R.S. and Glascock, M. 2010. 'Southern African glass beads: Chemistry, glass sources and patterns of trade', *Journal of Archaeological Science* 37: 1898–912.

Robinson, C., Baczyńska, B. and Polańska, M. 2004. 'The origins of faience in Poland', *Sprawozdania Archeologiczne* 56: 79–121.

Robson, E. 2001. 'Technology in society: Three textual case studies from Late Bronze Age Mesopotamia', in *The social context of technological change: Egypt and the Near East 1650–1150 BCE* (ed. A. Shortland), Oxford: Oxbow Books 2001, 39–57.

Rohl, B. and Needham, S. 1998. *The circulation of metal in the British Bronze Age: The application of lead isotope analysis*, British Museum Occasional Paper No. 102, London: The British Museum.

Rooksby, H.P. 1962. 'Opacifiers in opal glasses', *G.E.C. Journal of Science and Technology* 29, 1: 20–6.

Rosen, S.A., Tykot, R.H. and Gottesman, M. 2005. 'Long distance trinket trade: Early Bronze Age obsidian from the Negev', *Journal of Archaeological Science* 32: 775–84.

Rossi, F. and Chiaravalle, M. 1998. 'Due corredi funerary della prima età imperial dalla pianura bresciana', in *Vetro e vetri* (ed. C. Maccabruni), pp. 25–44.

Rouault, O. 1977. Mukannišum: I .'administration et l'économie palatiales á Mari, *Archives royales de Mari*, XVIII, Paris: Librairie orinetaliste Paul Geuthner

Rouault, O. 1993. 'Les fouilles de Terqa (Syrie). Quinzième saison (1994)', *Orient-Express* 1993/2, 11–12.

Rouault, O. 1998. 'Recherches récente à Tell Ashara-Terqa (1991–1995)', À propos de Subartu, Études consacrées à la Haute Mésopatamie, I, Paysage, Archéologie, Peuplement (ed. M. Lebeau), Brepolis, pp. 313–23.

Roy, P. and Varshney, Y.P. 1953. Ancient Kopia glass, *Glass Industry* 34: 366–392.

Royce-Roll, D. 1994. 'The colors of Romanesque stained glass', *Journal of Glass Studies* 36: 71–80.

Rutten, F., Briggs, D., Henderson, J. and Roe, M.J. 2009. 'The application of time-of-flight secondary ion mass spectrometry (ToF-SIMS) to the characterisation of opaque ancient glasses', *Archaeometry* 91: 66–86.

Rye, O.S. 1976. 'Khār: Sintered plant ash in Pakistan', Appendix 3 in *Traditional pottery techniques of Pakistan* (eds. O.S. Rye and C. Evans), Smithsonian Contributions to Anthropology No. 21, Washington, DC: Smithsonian Institution Press, pp. 180–5.

Sablerolles, Y. n.d. The glass vessels from Beirut 015, unpublished report for the Dutch excavations in Beirut.

Sablerolles, Y and Henderson J. 2012. 'The glass' in *Het domein van de boer en de ambachtsman*, ed. J. Dijkstra, ADC monografie 12.

Saggs, H.W.F. 1962. *The greatness that was Babylon*, London: Sedgwick and Jackson.

Saitowitz, S.J., Reid, D.L. and van der Merwe, N.J. 1996. 'Glass bead trade from Islamic Egypt to South Africa, c. AD 900–1250', *South African Journal of Science* 92: 101–4.

Salibi, N. 1954–5. 'Rapport préliminaire sur la deuxième campagne de fouilles à Raqqa', *Annales Archéologiques de Syrie* 4–5: 69–76, 205–12.

Saliou, C. 1996. *Le traité d'urbanisme de Julien d'Ascalon: Droit et architecture en Palestine au Vie Siècle*, Paris: De Boccard.

Salzini, L. 1986. 'Abitati preistorici e protostorici dell'alto e del medio Polsesine', *L'Antico Polesine: Testimonianze archeologiche e paleoambientali* (eds. M. De Min and R. Peretto), Cat. Mostra Adria e Rovigo, pp. 103–16.

Salzini, L. 1989. 'Necropoli dell'età del bronzo finale alle Narde di Fratta Polenise. Prima nota', *Padusa* 25: 5–42.

Salzini, L. 1990–91. 'Necropoli dell'età del bronzo finale alle Narde di Fratta Polenise'. Seconda nota, *Padusa* 26–27: 125–206.

Sanderson, D.C.W. and Hunter, J.R. 1981. 'Compositional variability in vegetable ash', *Science and Archaeology* 23: 27–30.

Sandys, George. 1670. *Sandys Travells*, 6th ed., London: printed by Rob. Clavel and others,

Santagostino Barbone, A., Gliozzo, E., d'Acapito, F., Memmi Turbanti, I., Turchiano, M. and Volpe, G. 2008. 'The *Sectilia* panels of Faragola (Ascoli-Satriono, southern Italy): A multi-analytical study of the red, orange and yellow glass slabs', *Archaeometry* 50: 451–73.

Santopadre, P. and Verità, M. 2000. 'Analyses of the production technologies of Italian vitreous materials of the Bronze Age', *Journal of Glass Studies* 42: 25–40.

Sarton, G. 1927. *Introduction to the history of science*, vol. 1, Baltimore: Williams and Wilkins.

Sasson, J.M., ed. 1995. *Civilizations of the Near East*, 2 vols., Peabody Museum, Peabody, MA: Hendrickson Publishers.

Sartre, M. 2005. *The Middle East under Rome*, Cambridge, Mass.: Harvard University Press.

Sayre, E.V. 1964. 'Some ancient glass specimens with compositions of particular archaeological significance', in *Brookhaven National Laboratory, Atomic Energy Commission USA Reports*, Upton, NY: The Brookhaven National Laboratory, pp. 1–25.

Sayre, E.V. 1967. Summary of the Brookhaven Program of analysis of ancient glass, *Applications of Science in Examinations of Works of Art* (ed. W.J. Young), Proceedings of the seminar held in the Boston Museum of Fine Arts, Boston, MA, 7–16 September 1965, Boston: Museum of Fine Arts, Boston, pp. 145–54.

Sayre, E.V. and Smith, R.W. 1961. 'Compositional categories of ancient glass', *Science* 133: 1824–26.

Sayre, E.V. and Smith R.W. 1963. 'The intentional use of antimony and manganese in ancient glasses', in *Proceedings of the VI International Congress on Glass. Advances in Glass Technology*, pt. 2 (eds. F.R. Matson and G.E. Rindone), History papers and discussion of the technical papers, New York: Plenum Press, pp. 263–82.

Sayre, E.V. and Smith, R.W. 1967. 'Some materials of glass manufacturing in antiquity', *Archaeological chemistry, a symposium* (ed. M. Levey), Third Symposium of Archaeological Chemistry, Atlantic City, NJ, Philadelphia: University of Pennsylvania Press, pp. 279–312.

Sayre, E.V. and Smith, R.W. 1974. 'Analytical studies of ancient Egyptian glass', in *Recent advances in the science and technology of materials* (ed. A. Bishay), vol. 3, New York: Plenum Press, pp. 47–70.

Scanlon, G.T. and Pinder-Wilson, R. 2001. *Fustat glass of the early Islamic period. Finds excavated by the American Research Center in Egypt 1964–1980*, London: Altajir World of Islam Trust.

Schaeffer, C.F.A. 1962. *Ugaritica IV*, Paris: Geuthner.

Schalm, O., Caluwé, D., Wouters, H., Janssens, K., Verhaeghe, F. and Pieters, M. 2004. 'Chemical composition and deterioration of glass excavated in the 15th–16th century fishermen town of Raversijde (Belgium)', *Spectrochimica Acta*, Part B, 59: 1647–56.

Schäfer, F. 2005. *Das Praetorium in Köln und weitere Statthalterpaläste im Imperium Romanum*, Unveröffentliche Doktorarbeit, Köln: Universität zu Köln.

Schlick-Nolte, B. and Werthmann, R. 2003. 'Glass vessels from the burial of Nesikhons', *Journal of Glass Studies* 45: 11–34.

Schreurs, J.W.H. and Brill, R.H. 1984. 'Iron and sufur related colors in ancient glasses', *Archaeometry* 26: 199–209.

Scrivener, A.E., Vance, D. and Rohling, E.J. 2004. 'New neodymium isotope data quantify Nile involvement in Mediterranean anoxic episodes', *Geology* 32: 565–8.

Segnit, E.R. 1987. 'Evaporite minerals from the Dakhleh Oasis,' in *Ceramics from the Dakhleh Oasis; Preliminary studies* (eds. W.I. Edwards, C.A. Hope and E.R. Segnit), Victoria College Archaeology Research Unit, Occasional Paper 1, Burwood, Victoria, pp. 97–102.

Seligman, C.G., Ritchie, P.D. and Beck, H.C. 1936. 'Early Chinese glass from pre-Han to Tang times', *Nature* 138: 721.

Selimkhanov, I.R. 1975. 'Sur l'étude du fragment de vase de Tello . . . et le problème de l'utilisation de l'antimone dans l'antiquité', *Annales du Laboratoire de Recherche des Musées de France*, 44. 45–52.

Sellner, C., Oel, H.J. and Camera, B. 1979. 'Untersuchung alter Gläser (Waldglas) auf Zusammenhang von Zusammensetzung, Farbe und Schmelzatmosphäre mit der Elektronenspektroskopie und der Electronenspinresonanz (ESR)', *Glastechniche Berichte* 52: 255–64.

Shahid, K. and Glasser, F. 1972. 'Phase equilibria in the glass-forming region of the system Na_2O-CaO-MgO-SiO_2', *Physics and Chemistry of Glasses* 13: 27–42.

Shatzmiller, M. 1994. *Labour in the medieval Islamic world*, Leiden: Brill.

Shepherd, J.D. and Heyworth, M. 1991. 'Le travail du verre dans Londres romain (Londinium): Un état de la question', in *Ateliers de Verriers de l'antiquité á la période pré-industrielle*, Association Française pour l'archéologie du verre, Actes de 4 èmes Rencontres, Rouen, 24–25 Novembre 1989, Rouen: AFAV, pp. 13–22.

Sherratt, A. and Sherratt, S. 2001. 'Technological change in the East Mediterranean Bronze Age', in *The social context of technological change* (ed. A. Shortland), Oxford: Oxbow Books, pp. 15–38.

Shi Meiguang, He Ouli and Zhou Fuzheng. 1986. 'An investigation of some Chinese potash glasses excavated in Han dynasty tombs', *Journal of the Chinese Silicate Society* 14(3), 307–11.

Shi Meiguang, He Ouli and Zhou Fuzheng 1987. 'Investigation of some Chinese potash glasses excavated in Han Dynasty tombs', (ed. Bhardwaj 1987a), section II, pp. 15–20.

Shortland, A.J. 2000. *Vitreous materials at Amarna: The production of glass and faience in 18th Dynasty Egypt*, British Archaeological Reports International Series 827, Oxford: Archaeopress.

Shortland, A.J. 2001. 'Social influences on the development and spread of glass technology', in ed. A. Shortland, 2001a, pp. 211–22.

Shortland, A.J., ed. 2001a. *The social context of technological change, Egypt and the Near East, 1650–1550 BC*, Oxford: Oxbow Books.

Shortland, A.J. 2002. 'The use and origin of antimonate colorants in early Egyptian glass', *Archaeometry* 44: 517–30.

Shortland, A.J. 2004. 'Evaporites of the Wadi Natrun: Seasonal and annual variation and its implication for ancient exploitation', *Archaeometry* 46, 4: 497–517.
Shortland, A.J. 2005. 'The raw materials of early glasses: The implications of new LA-ICPMS analyses', in *Annales de l'Association Internationale pour l'Histoire du Verre*, 16, London, 2003, Nottingham, England: AIHV, pp. 1–5.
Shortland, A.J. 2006. 'Application of lead isotope analysis to a wide range of late Bronze Age Egyptian materials', *Archaeometry* 48: 657–69.
Shortland, A.J. and Eremin, K. 2006. 'The analysis of second millennium BC glass from Egypt and Mesopotamia, part 1: new WDS analyses', *Archaeometry* 48,1: 581–605.
Shortland, A.J., Nicholson, P. and Jackson, C. 2001. 'Glass and faience at Amarna: different methods of both supply for production, and subsequent distribution', in ed. A.J. Shortland, 2001a, pp. 147–60.
Shortland, A.J., Rogers, N. and Eremin, K. 2007. 'Trace element discriminants between Egyptian and Mesopotamian Late Bronze Age glasses', *Journal of Archaeological Science* 34: 781–9.
Shortland, A.J., Tite, M.S. and Ewart, I. 2006. 'Ancient exploitation and use of cobalt alums from the western oases of Egypt', *Archaeometry* 48: 153–68.
Shortland, A.J. and Tite, M.S. 1998. 'The interdependence of glass and vitreous faience production at Amarna', *The prehistory and history of glassmaking technology* (ed. P. McCray), *Ceramics and Civilization*, vol. VIII, Westerville, OH: The American Ceramic Society, pp. 251–68.
Shortland, A.J. and Tite, M.S. 2000. 'Raw materials of glass from Amarna and implications for the origins of Egyptian glass', *Archaeometry* 42, 1: 141–52.
Shortland, A.J., Tite, M.S. and Ewart, I. 2006. 'Ancient exploitation and use of cobalt alums from the western oases of Egypt', *Archaeometry*, 48, 1: 153–68.
Shugar, A. 2000. 'Byzantine opaque red glass tesserae from Beit Shean, Israel', *Archaeometry* 42: 375–84.
Shugar, A. and Rehren, T. 2002. 'Formation and composition of glass as a function of firing temperature', *Glass Technology* 43C: 145–50.
Sillen, A., Hall, G. and Armstrong, R. 1998. '$^{87}Sr/^{86}Sr$ ratios in modern and fossil food webs of the Sterkfontein valley: implications for early hominid habitat preference', *Geochimica et Cosmochimica Acta* 62: 2463–78.
Silvestri, A., Molin, G. and Salviulo, G. 2008. 'The colourless glass of *Julia Felix*', *Journal of Archaeological Science* 35: 331–41.
Silvestri, A., Molin, G., Salviulo, G. and Schievenin, G. 2006. 'Sand for Roman glass production: An experimental and philological study on source of supply', *Archaeometry* 48, 3: 415–32.
Silvestri, A., Longinelli, A. and Molin, G. 2010. '$\delta^{18}O$ measurements of archaeological glass (Roman to Modern age) and raw materials: possible interpretation', *Journal of Archaeological Science* 37: 549–60.
Singh, N.T. 2005. *Irrigation and soil salinity in the Indian subcontinent: Past and present*, Lehigh: Lehigh University Press.
Singh, R.N. 1989. *Ancient Indian glass, archaeology and technology*, Delhi: Parimal Publications.
Smedley, J.W. and Jackson, C.M. 2002. 'Medieval and post-medieval glass technology: batch measuring practices', *Glass technology* 43: 22–27.
Smedley, J.W., Jackson, C.M. and Booth, C.A. 1998. 'Back to the roots: The raw materials, glass recipes and glassmaking practices of Theophilus', in *Ceramics and civilization, volume VIII. The prehistory and history of glassmaking technology* (ed. P. McCray), Westerville, OH: The American Ceramic Society, pp. 145–66.

Smirniou, M. and Rehren, Th. 2011. 'Direct evidence of primary glass production in late Bronze Age Amarna, Egypt', *Archaeometry* 53: 58–80.

Šmit, Ž., Janssens, J., Schalm, O. and Kos, M. 2004. 'Spread of façon de Venise glass-making through central and western Europe', *Nuclear Instruments and Methods* B 213: 717–22.

Smith, C.S. 1981. *A search for structure*, Cambridge, MA: M.I.T Press.

Sode, T. and J. Kock 2001. 'Traditional raw glass production in northern India: The final stage of an ancient technology', *Journal of Glass Studies* 43: 155–69.

Sosman, R.B. 1965. *The phases of silica*, New Brunswick, NJ: Rutgers University Press.

Soucek, J. 1960. *Encyclopedia of Islam*, 2nd ed., Leiden: E.J. Brill.

Spaer, M. 2001. *Ancient glass in the Israel Museum. Beads and other small objects*, Jerusalem: The Israel Museum.

Stapleton, C.P. 2003. 'Composition and technology of mosaic vessel glass from the early Roman period', *Annales de 15e Congrès de l'association internationale pour l'histoire du verre*, Corning, NY, and Nottingham: AIHV, pp. 29–32.

Starinsky, A., Bielski, M. and Lazar, B. 1980. 'Marine $^{87}Sr/^{86}Sr$ ratios from the Jurassic to Pleistocene: evidence from groundwaters in Israel' *Earth and Planetary Science Letters*, 47: 75–80.

Starr, R.F.S. 1939. *Nuzi, report on the excavations at Yorgan Tepa near Kirkuk 1927–1931*, Cambridge, MA.

Stein, D.L. 1989. 'A reappraisal of the "Saustatar Letter" from Nuzi', *Zeitschrift für Assyriologie und vorderasiatische Archäologie* (Berlin) 79: 36–60.

Stein, G.J. and Blackman, J. 1993. 'The organizational context of specialized craft production in early Mesopotamian states', *Research in Economic Anthropology* 14: 29–59.

Stern, E.M. 1987. 'On the glass industry of Arikamedu (ancient Poduke)', *Archaeometry of Glass, Proceedings of the Archaeometry session of the XIV International Congress on Glass* (ed. H.C. Bhardwaj), 1986, New Delhi, India, part 1, Kolkata (Calcutta): Indian Ceramic Society, section 1, pp. 26–36.

Stern, E.M. 1991a. 'Early Roman export glass in India', in *Rome and India: The Ancient Sea Trade*, Madison: The University of Wisconsin Press.

Stern, E.M. 1991b. 'Early exports beyond the empire', *Roman glass: Two centuries of Art and Invention*, London: British Museum Press, pp. 149–52.

Stern, E.M. and Schlick-Nolt, B. 1994. *Early glass of the ancient world, 1600 BC–AD 50*, Ostfildern: Verlag Gerd Hatje.

Stern, W.B. and Gerber, Y. 2004. 'Potassium-calcium glass: New data and experiments', *Archaeometry*, 46: 137–56.

Stern, W.B. and Gerber, Y. 2009. 'Ancient potassium-calcium glass and its raw materials (wood-ash, fern-ash, potash) in central Europe', *Mitteilungen der Naturforschenden Gesellschaften beider Basel* 11: 107–22.

Stos-Gale, Z.A., Maliotis, A.G., Gale, N.H. and Annetts, N. 1997. 'Lead isotope characteristics of the Cyprus copper ore deposits applied to provenance studies of copper oxhide ingots', *Archaeometry* 39: 83–123.

Tachikawa, K., Roy-Barman, M., Michard, A., Thouron, D., Yeghicheyan, D. and Jeandel, C. 2004. 'Neodymium isotopes in the Mediterranean Sea: Comparison between seawater and sediment signals', *Geochimica et Cosmochimica Acta* 68: 3095–106.

Tal, O., Jackson-Tal, R.E. and Freestone, I.C. 2004. 'New evidence of the production of raw glass at late Byzantine Apollonia-Arsuf, Israel', *Journal of Glass Studies* 46: 51–66.

Tanimoto, S. 2007. *Experimental study of late Bronze Age glass-making practice*, unpublished Ph.D. thesis, University of London.

Tarkian, M., Bock, W.D. and Neumann, M. 1984. 'Geology and mineralogy of the Cu-Ni-Co-U deposits at Talmessi and Meeskani, central Iran', *Mineralogy and Petrology* 32, nos. 2–3: 111–33.

Theophilus Presbiter 1979. *On Divers Arts* (trans. J.G. Hawthorne and C.S. Smith), New York: Dover Books.

Thirion-Merle, V. 2005. 'Les verres de Beyrouth et les verres du Haut Empire dans le monde occidental: étude archéométrique', *Journal of Glass Studies* 47: 37–53.

Tite, M.S. and Bimson, M. 1986. 'Faience: An investigation of the microstructures associated with different methods of glazing', *Archaeometry* 28: 69–78.

Tite, M.S. and Bimson, M. 1989. 'Glazed steatite: An investigation of the methods of glazing used in ancient Egypt', *World Archaeology* 21: 87–100.

Tite, M.S., Bimson, M. and Cowell, M. 1987. 'The technology of Egyptian blue', *Early Vitreous Materials* (eds. M. Bimson and I.C. Freestone), British Museum Occasional Paper No. 56, London: British Museum Press, pp. 69–78.

Tite, M.S., Freestone, I.C. and Bimson, M. 1983. 'Egyptian faience: An investigation of the methods of production', *Archaeometry* 25: 17–27.

Tite, M.S., Hatton, G.D., Shortland, A.J., Maniatis, Y., Kavoussanaki, D. and Panagiotaki, M. 2005. 'Raw materials used to produce Aegean Bronze Age glass and related vitreous materials', in *Annales de l'Association Internationale pour l'Histoire du Verre*, London 2003, Nottingham, England: AIHV, pp .10–13.

Tite, M.S., Manti, P. and Shortland, A. 2007. 'A technological study of ancient faience from Egypt', *Journal of Archaeological Science*, 34: 1568–83.

Tite, M. S., Pradell, T. and Shortland, A. 2008. 'Discovery, production and use of tin-based opacifiers in glasses, enamels and glazes from the Late Iron Age onwards: a reassessment', *Archaeometry* 50: 67–84.

Tite, M.S. and Shortland, A. 2003. 'Production technology for copper- and cobalt-blue vitreous materials from the New Kingdom site of Amarna: A reappraisal', *Archaeometry* 45(2): 285–309.

Tite, M.S. and Shortland, A.J. (eds.) 2008. *Production Technology of Faience and Related Early Vitreous Materials*, Oxford: Oxford University School of Archaeology.

Tite, M.S., Shortland, A., Maniatis, Y., Kavoussanaki, D. and Harris, S.A. 2006. 'The composition of the soda-rich and mixed alkali plant ashes used in the production of glass', *Journal of Archaeological Science* 33: 1284–92.

Tite, M., Shortland, A. and Paynter, S. 2002. 'The beginnings of vitreous materials in the Near East and Egypt', *Accounts of Chemical Research* 35: 585–93.

Tonghini, C. and Henderson, J. 1998. 'An eleventh century pottery production workshop at Al-Raqqa, preliminary report', *Levant*, 30: 113–27.

Toniolo, A. 2005. 'Musealizzazione del relitto di epoca imperial romana *Julia Felix* di Grado (Italia) nel nuovo Museo di Archeologia Subacquea di Grado, Il relotto e il suo carico', in *Proceedings of the III Congreso International sobre Musealizaciòn de yacimientos arqueològicos*, Zaragoza, November 15–18 2004, pp. 161–64.

Towle, A. 2002. *A scientific and archaeological investigation of prehistoric glasses from Italy*, unpublished Ph.D. thesis, University of Nottingham, England.

Towle, A. and Henderson, J. 2008. 'The glass bead game: Archaeometric evidence for the existence of an Etruscan glass industry', *Etruscan Studies* 10: 47–66.

Towle, A., Henderson, J., Bellintani, P. and Gambacurta, G. 2001. 'Frattesina and Adria: report of the scientific analysis and early glass from Veneto', *Padusa*, 37: 7–68.

Triantafyllidis, P. 2000. 'New evidence of the glass manufacture in classical and Hellenistic Rhodes', *Annales du 14e Congrès de l'association Internationale pour l'Histoire du verre*, Venice and Milan, 1998, Lochem, the Netherlands: AIHV, pp. 30–34.

Triantafyllidis, P. 2000a. *Rhodian Glassware I. The luxury hot-formed transparent vessels of the classical and early Hellenistic periods*. Athens: Ministry of the Aegean Sea.

Triantafyllidis, P. 2003. 'Classical and Hellenistic glass from Rhodes', in eds. D. Foy and M.-D. Nenna, pp. 131–38.

Triantafyllidis, P. 2009. 'Early core-formed glass from a tomb at Ialysos, Rhodes', *Journal of Glass Studies*, 51: 26–39.

Trowbridge, M.L. 1930. *Philological studies in ancient glass*, University of Illinois Studies in Language and Literature XIII, nos. 3–4, Urbana: University of Illinois.

Tsaferis, V. and Israeli, S. 1994. 'Banias-1992', *Excavations and Surveys in Israel*, 14: 1–3.

Tufnell, O. 1958. *Lachish IV, The Bronze Age*. London: Arcaeological Research Expedition to the Near East Publication.

Turgeon, L. 2004. Beads, bodies and regimes of value: From France to North America c. 1500–c. 1650 (ed. T. Murray), *The archaeology of contact in settler societies*. Cambridge: Cambridge University Press, pp. 19–47.

Turner, W.E.S. 1954. 'Studies of ancient glass and glassmaking processes. Part I: Crucibles and melting temperatures employed in Ancient Egypt at about 1370 B.C.', *Journal of the Society of Glass Technology* 38: 436T–44T.

Turner, W.E.S. 1956a. 'Studies of ancient glass and glassmaking processes. Part III: The chronology of glassmaking constituents', *Journal of the Society of Glass Technology* 40: 39–52.

Turner, W.E.S. 1956b. 'Studies of ancient glass and glassmaking processes. Part V: Raw materials and melting processes', *Journal of the Society of Glass Technology*, 40: 277–300.

Turner, W.E.S. and Rooksby, H.P. 1959. 'A study of opalising agents in ancient opal glasses throughout three thousand four hundred years', *Glastechnische Berichte*, 32K, VII: 17–28.

Turner, W.E.S. and Rooksby, H.P. 1961. 'Further historical studies based on X-ray diffraction methods of the reagents employed in making opal and opaque glasses', *Jahrbuch des Römisch-Germanischen Zentralmuseums* 8: 1–16.

Turner, W.E.S. and Rooksby, H.P. 1963. 'A study of the opalising agents in ancient glasses throughout 3400 years, part II', *Proceedings of the 6th International Congress on Glass: Advances in glass technology* (eds. F.R. Matson and G.E. Rindone), New York: Plenum Press, pp. 306–7.

Tykot, R.H. 2002. 'Chemical fingerprinting a source tracing of obsidian: the central Mediterranean trade in black gold', *Accounts of Chemical Research* 35: 618–27.

Van der Meiroop, M. 2004. *A history of the ancient Near East ca. 3000–323 B.C.*, Oxford: Blackwell.

Van Soldt, W.H. 1995. 'Ugarit, a second-millennium kingdom on the Mediterranean coast', in ed. M. Sasson, pp. 1255–66.

Vandiver, P.B. 1983a. 'Egyptian faience technology'. Appendix A in A. Kaczmarczyk and R.E.M. Hedges, *Ancient Egyptian faience: An analytical survey of Egyptian faience from predynastic to Roman times*, Warminster, England: Aris and Phillips, pp. A1–144.

Vandiver, P.B. 1983b. 'Glass technology at the mid-second millennium BC Hurrian site of Nuzi', *Journal of Glass Studies* 25: 239–47.

Vandiver, P.B. 1998. 'A review and proposal of new criteria for production technologies of Egyptian faience', *La couleur dans la peinture et l'émaillage de l'Égypte ancienne* (eds. S. Colinart and M. Menu) Bari: Edipuglia, pp. 121–39.

Vandiver, P.B. and Kingery, W.D. 1986. 'Egyptian faience: The first high-tech ceramic', *High technology ceramics past, present and future, Ceramics and Civilization* vol. III (ed. W.D. Kingery), Westerville, OH: The American Ceramic Society, pp. 19–34.

Velde, B. and Barrera, J. 1986. 'Composition of medieval blown glass fragments found at Orleans', *Archéologie Médiévale*,16: 93–103.

Venclová, N., Hulínský, V., Frána, J. and Fikrle, M. 2009. 'Němčice and glass-working in La Tène Europe', *Archeologické rozhledy* LXI: 383–426.

Venclová, N., Hulínský, V., Henderson, J, Chenery, S., Šulová, L. and Hložek, J. 2011. 'Late Bronze Age mixed-alkali glasses from Bohemia', *Archeologické rozhledy* LXIII: 559–85.

Verità, M. 1985. 'L'invenzione del cristallo muranese: Una verifica analitica delle fonti storiche', *Rivista della Stazione Sperimentale del Vetro* 15: 17–29.

Verità, M. 1995. 'Analytical investigation of European enameled beakers of the 13th and 14th centuries', *Journal of Glass Studies* 37: 83–98.

Verità, M. 2000. 'Tecniche di fabbricazione dei materiali musivi vitrei: indagini chimiche e mineralogiche', in *Medieval Mosaics: Light, Color, Materials*, (eds. E. Borsook, F. Gioffredi Superbi and G. Pagliarulo), Florence: Silvana Editorale, pp. 47–64.

Verità, M. 2005. 'Comments on W.B. Stern and Y. Gerber, "Potassium-calcium glass: New data and experiments", *Archaeometry* 46(1) (2004), 137–56', *Archaeometry* 47: 667–69.

Verità, M. and Biavati, A. 1989. 'The glass from Frattesina a glassmaking centre in the Late Bronze Age', *Rivista Stazione Sperimentale del Vetro* 19, 4: 295–99.

Verità, M. and Tontinato, T. 1990. 'A comparative analytical investigation on the origins of the Venetian glassmaking', *Rivista della Stazione Sperimentale del Vetro*, 20: 169–75.

Verità, M., Renier, A. and, Zecchin, S. 2002. 'Chemical analyses of ancient glass findings excavated from the Venetian lagoon', *Journal of Cultural Heritage* 3: 261–71.

Vickers, M. 1996. 'Rock crystal: the key to cut glass and diatreta in Persia and Rome' *Journal of Roman Archaeology* 9: 48–65.

Von Saldern, A. 1970. 'Other Mesopotamian glass vessels (1500–600 BC)', in *Glass and Glassmaking in Ancient Mesopotamia* (eds. A.L. Oppenheim, R.H. Brill, D.P. Barag and A. von Saldern), Corning: The Corning Museum of Glass, pp. 203–28.

Wadia, D.N. 1975. *Geology of India*, 4th ed., New Delhi: Tata McGraw-Hill.

Walton, M.S., Shortland, A., Kirk, S. and Degryse, P. 2009. 'Evidence for the trade of Mesopotamian and Egyptian glass to Mycenaean Greece', *Journal of Archaeological Science* 36: 1496–503.

Ward, R., ed. 1998. *Gilded and enameled glass from the Middle East*, London: The British Museum Press.

Watanabe, T., Broadley, M.R., Jansen, S., White, P.J., Takada, J., Satake, K., Takamatsu, T., Tuah, S.J. and Osaki M. 2007. 'Evolutionary control of leaf element composition in plants', *New Phytologist* 174: 516–23.

Watson, O. 1985. *Persian lustre ware*, London: Faber and Faber.

Watson, O. 2004. *Ceramics from Islamic lands*. London: Thames and Hudson.

Wedepohl, K.H. 2000. 'The change in composition of medieval glass types occurring in excavated fragments from Germany', *Proceedings of the 14th Congress of the International Association for the History of Glass*, Venice and Milan, Lochem, the Netherlands: AIHV, pp. 253 57.

Wedepohl K.H. 2003. *Glas in Antike und Mittelalter*, Geschichte eines Werkstoffs. Stuttgart: Schweizerbartsche Verlagbuchhandlung.

Wedepohl, K.H. 2008. 'Mittelalterliches Holzache-Glas', in *Glashüttenlandschaft Europopa* (eds. H. Flachenecker, G. Himmelsbach and P. Steppuhn, P.), Regensburg: Schnell and Steiner.

Wedepohl, K.H. and Baumann, A. 2000. 'The use of marine molluskan shells for Roman glass and local raw glass production in the Eifel area (western Germany)' *Naturwissenschaften* 87:129–32.

Wedepohl, K.H., Krueger, I. and Hartmann, G. 1995. 'Medieval lead glass from North western Europe', *Journal of Glass Studies* 37: 65–82.

Wedepohl, K.H. and Simon, K. 2010. 'The chemical composition of medieval wood ash glass from Central Europe', *Chemie der Erde* 70: 89–97.

Wedepohl, K.H., Winkelmann, W. and Hartmann, G. 1997. 'Glasfunde aus der karolingischen Pfalz in Paderborn und die frühe Holzasche-Glasherstellung', *Ausgrabungen und Funde in Westfalen Lippe* 9A: 41–53.

Weigand, P.C., Harbottle, G. and Sayre, E.V. 1977. 'Turquoise source and source analysis: Mesoamerica and the southwestern USA', *Exchange Systems in Prehistory* (eds. T.K. Earle and J.E. Erikson), New York: Academic Press, pp. 15–34.

Weinberg, G.D. 1969. 'Glass manufacture in Hellenistic Rhodes', *Archaeologikon Deltion* 24: 143–51.

Weinberg, G.D. 1970. 'Hellenistic glass from Tel Anafa in Upper Galilee', *Journal of Glass Studies* 12: 17–27.

Weinberg, G.D. 1983. 'A Hellenistic glass factory on Rhodes: Progress report', *Journal of Glass Studies* 15: 37.

Weinberg, G.D., ed. 1988. *Excavations at Jalame site of a glass factory in late Roman Palestine*, Columbia: University of Missouri Press.

Weinberg, G.D. 1992. *Glass vessels in ancient Greece*, Athens: Archaeological Receipts Fund.

Weinberg, G.D. Grace, V.R. and Edwards, G.R. 1965. *The Antikythera shipwreck reconsidered*, Philadelphia: Transactions of the American Philosophical Society.

Weldeab, S., Kay-Christian, E., Hemleben, C. and Siebel, W. 2002. 'Provenance of lithogenic surface sediments and pathways of riverine suspended matter in the Eastern Mediterranean sea: Evidence from $^{143}Nd/^{144}Nd$ and $^{87}Sr/^{86}Sr$ ratios', *Chemical Geology* 186: 139–49.

Werner, A.E. and Bimson, N. 1963. 'Some opacifying agents in oriental glass', in *Proceedings of the 6th International Congress on Glass: Advances in Glass Technology* (eds. F.R. Matson and G.E. Rindone), New York: Plenum Press, pp. 303–5.

Werner, A.E. and Bimson, N. 1967. 'Technical report on the Glass Gaming-pieces' in I.M. Stead, 'A La Tène III burial at Welwyn Garden City', *Archaeologia* 101: 16–17.

Weyl, W.A. 1937. The chemistry of coloured glass: II, *The Glass Industry*: 117–20.

Weyl, W.A. 1951/1992. *Coloured glasses*, Sheffield: Society of Glass Technology.

Whitehouse D.B. 1989, 'Roman Dichroic Glass: Two Contemporary Descriptions?', *Journal of Glass Studies* 31: 119–21.

Whitehouse, D.B., Pilosi, L. and Wypyski, M.T. 2000. 'Byzantine silver stain', *Journal of Glass Studies* 42: 85–96.

Wiedemann, H. and Bayer, G. 1982. 'The bust of Nefertiti: the analytical approach', *Analytical Chemistry* 54: 619A–28A.

Wilhelm, G. 1995. 'The Kingdom of Mitanni in Second-Millennium Upper Mesopotamia', in ed. M. Sasson, pp. 1243–54.

Willaume, M. 1987. 'Les objects de la vie quodidienne', *Archéologique d'Entremont au muse Granet*, exhibition catalogue, Aix-en-Provence, pp. 107–42.

BIBLIOGRAPHY

Willmott, H. 2002. *Early post-medieval vessel glass in England c. 1500–1670*. Council for British Archaeology, Research Report no. 132, York: Council for British Archaeology.

Wolf, S., Stos, S., Mason, R. and Tite, M.S. 2003. 'Lead isotope analysis of Islamic pottery glazes from Fustat, Egypt', *Archaeometry* 45, 3: 405–20.

Wood, N. 2009. 'Some implications of the use of wood ash in Chinese stoneware glazes of the 9th–12th centuries', in *From mine to microscope* (eds. A.J. Shortland, I.C. Freestone and Th. Rehren), Oxford: Oxbow Books, pp. 51–60.

Wood, N., Tite, M.S., Doherty, C. and Gilmore, B. 2007. 'A technological examination of ninth-century AD Abbasid blue-and-white ware from Iraq, and its comparison with eighth century AD Chinese blue-and-white *sancai* ware', *Archaeometry* 49, 4: 665–84.

Woolley, L. 1927 The Excavations at Ur, 1926–7, *Antiquaries Journal* 7, no. 4: 385–421.

Woolley, L. 1955. *Alalakh, an account of the excavations at Atchana in the Hatay, 1937–49*, Oxford: Oxford University Press.

Woolley, L. 1965. *Ur Excavations*, vol. VIII, London.

Wright, T. 1848. *Early travels in Palestine*, London.

Wypyski, M.T. 2002. 'Renaissance enameled jewelry and 19th century Renaissance revival: Characterization of enamel compositions', *Materials Research Society Symposium Proceedings*, vol. 712, pp. 223–33.

Wypyski, M.T. 2006. 'Technical analysis of glass mosaic tesserae from Amorium, *Dumbarton Oaks Papers* 59: 183–92.

Wypyski, M.T. and Richter, R.W. 1997. 'Preliminary compositional study of 14th and 15th C. European enamels', *Techne* 6: 48–57.

Yi Jialiang and Tu Shujin. 1991. 'Chinese glass technology in Boshan around the 14th century', in eds. R.H. Brill and J.H. Martin, pp. 99–102.

Young, S. 1956. 'The analysis of Chinese Blue-and-White', *Oriental Art*, 11: 43.

Youngblood, E., Fredrikson, B.J., Kraut, F. and Fredrikson, K. 1978. 'Celtic vitrified forts: Implications of a chemical-petrological study of glasses and source rocks', *Journal of Archaeological Science* 5: 99–121.

Zang Fukang. 1987. 'Origin and development of early Chinese glasses' in ed. H.J. Bhardwaj, 1987a, pp. 25–28.

Zang Fukang and Cowell, M. 1987. 'Report on the examination of some ceramics and glass employing cobalt as a colorant', British Museum Department of scientific research, project no. 5460, unpublished.

Zang Weiyong. 2005. 'Ancient glass technology in the Yuan, Ming and Qing dynasties', in *Development of Chinese ancient glass* (ed. Gan Fuxi), Shanghai: CIP, pp. 141–65.

Zecchin, L. 1987. *Vetro e vetrai di Murano*, vol. 1. Venice: Arsenale Editrice.

Zecchin, L. 1990. *Vetro e vetrai di Murano*, vol. 3. Venice: Arsenale Editrice.

Zorn, B. and Hilgner, A. (eds.) 2010. Glass along the Silk Road from 200 BC to AD 1000 Römisch-Germanisches Zentralmuseum – Tagungen volume 9, Mainz: Römisch-Germanisches Zentralmuseum.

INDEX

Abbāsid 57, 74
'Abbāsid caliphate 253, 303
Abila, Jordan 206
absolane 70, 73
absorption band 76
Abu Shahrein (Eridu) 134, 145
Abu Tahir family of potters 262
Abū'l Qāsim 24, 73, 262, 274
Achaemanid 205, 226
Aci Gölü, Turkey 28
Acre/Akko, Israel 61, 210, 256
Acre Bay, Israel 28, 29
Afghanistan 4, 112, 158, 296, 298–300
Agra, India 275
Ahhotep 147, 168
Aigues-Mortes, France 23
Aix-en-Provence, France 246
Akda' Madeni, Turkey 74
A'Ke-Si-Pi-Li castle, China 121
Akhenaten, pharaoh, 'king' 130
Akkad, Iraq 127, 128, 134
Akkadian 127, 132, 133, 134
Al-Biruni 263–4
Al-Hakim II mosque 255
al-Idrīsī 268
al-Jābir 265, 282–4, 289
al-Marwazī 255
al-Muqaddasī 85, 254, 256, 262–3, 267, 269
al-Qazwīnī 267
al-Rāfiqa, Syria 256, 336, 338
al-Raqqa, Syria 31, 33, 44, 47, 65, 69, 85, 87, 93, 97–100, 110, 112, 117, 119, 162, 253, 257, 261, 263, 266–69, 271–2, 274–8, 280, 282–97, 299–302, 335–350, 354–6, 359,
311–13, 321–2, 324, 327–8, 335–50, 354–6, 359
al-Tabarī 255–6
alabastra 209, 223, 225–6
Alalakh (Tell Atchana), Turkey 16, 128, 135–6, 139, 143–4, 349
Alashiya 130
albite 60, 65
alchemy 254, 262, 282, 303
Aldrvandin type beakers 81
Aleppo, Syria 24, 31, 33, 42, 92, 128, 130, 253, 262–3, 266–9, 278, 292, 304, 335, 337, 349–50, 356–7
Alexander the Great 205, 208–9, 224
Alexandria, Egypt 30, 63, 64, 263, 330–1, 352–4, 359
Alicante, Spain 23
Allerona-type sword 153
allunogen 71
Al mass'oudi 264
Almeria, Spain 253, 269
alum 72, 324,
alumina 3, 14, 29, 42, 43, 46, 47, 53, 64, 66, 71, 163, 172–6, 187–8, 199, 236–9, 241–7, 249, 279–87, 291–4, 297–300, 312, 314–5, 318, 324–5, 327, 332, 337, 340, 348, 350–1, 364, 366, 367
aluminium 159, 163, 188
aluminium sulphate 71
Amarna letters 130, 168
amber 131, 138, 150, 153, 155
Amenhotep II 169
Amenhotep III 170
Amorites 128, 136
amorphous 2, 3, 19, 20, 363
amphibole 329
amphoriskoi 209, 223, 225
Anabasis 23, 39, 41, 262
Anabasis articulatum 30, 32
Anabasis syriaca 31, 32, 35, 36
Anarak, Iran 72, 148
andesite 117
Angel Gabriel 252
Anglo-Saxon 94
anneal 18, 20, 223, 229, 269
anorthite 60
Anthony 208–9
Antigonids 205, 209
Antikythera shipwreck, Greece 360
antimony 74, 83, 147–8, 181, 188, 210, 212, 232
antimony trioxide 189, 210, 246
Antioch, Syria 205, 206, 253–4, 277
Antiochus I 205–6
Antiochus XIII Asiaticus 205–6
antler 150
Antonio Neri 24
aolian sand 57
Aosta: see Augusta pretoria
Apamea 205, 208
Apkallu 132
Apollonia, Israel 97, 99, 246, 270, 351
Appenine 63
Aqar Quf (Dur Kurigalzu), Iraq 136
Aradus, Syria 206
aragonite 46, 329
Arikamedu 112–3
Aristophanes 204
armlet/bangle/bracelet 162, 285, 296–300, 302, 336, 338, 341–4, 350, 353, 363, 370
Armorium, Turkey 78
arsenic 70, 72, 73, 159, 172, 188
arsenic oxide 159

Index

Arthrocnemum 23, 35, 39, 40
Arthrocnemum strobilaceum 31, 32, 36, 46
artifex vitri 204
Artiplex leucoclada 31, 35
aryballoi 209
Ascalon/Ashkelon 168, 208
Ashur, Iraq 134, 144
Assur, Iraq 127–8, 135–6, 142
Assyria 349
Astarte, goddess 140
asteroid 5, 6
astronomy 254
Athens, Greece 181
atomic absorption spectrometry 26, 160, 166, 189, 247, 249
atomic number 66
atomic weight 66
Attalids 205
Augusta Pretoria (Aosta), Italy 88, 235, 247, 249
Augustus, emperor 369
Autun, France 230–1
Avenches, France 228–9, 235
Avignon, France 236
Ayyūbid caliphate 253, 257–8, 267–8, 277, 280, 294, 350

Bab el Hawa, Syria 337
Babylon 128–30, 132, 143, 349
Babylonia 127–8, 130–2, 143–4, 146
Bactria 114
Baghdad, Iraq 253–4, 257, 266–7, 289, 301, 303, 349
Bahrain 294–5
Balikh valley 33, 41, 128, 311, 313, 337
Ban Don Ta Phet, Thailand 124, 125
Bangle: See armlet
Banias, Israel 276, 290, 338–9, 343–5
barium 48, 60, 62, 96, 102, 104, 115–16, 118–23, 239–40, 243, 249–50, 312–3, 332, 364
barium disilicate 123
barium-lead glass 119
basaltic rocks 60
Basilicata, Italy 150
Basra, Iraq 253, 254
Basra, Morocco 302–3
Baycheh Bagh mine, Iran 73
beach sand 46, 72, 294, 309, 329, 330–2, 354, 358
bead 2, 84, 131, 133–6, 138–40, 144, 146, 148, 153–5, 160, 162–8, 170, 172–3, 180–1, 183–4, 186, 188–9, 193–5, 197–8, 200, 363–4, 367, 370–1
beaker 204, 226
Bedouin 262
bee-hive shaped furnace 100, 228–9, 272, 274

Beirut, Lebanon 93–4, 97–9, 204, 206, 209, 211, 215, 217–22, 230–1, 235–47, 250, 290, 294–5, 302, 332, 338, 351, 356
Bekaa Valley, Lebanon 206
Bel, France 195–8
Belus river, Israel 28, 63, 61, 64, 203, 240, 247, 331, 333, 351
Beni Salama, Egypt 231, 248
Benjamin of Tudela 267
Bet Eli'ezer, Israel 93, 238, 241–3, 253, 270–1, 274, 332, 351
Bet She'an, Israel 144, 206, 332
Beth She'arim 280
Beverley, UK 103
Bihar, India 112
Bir Hooker, Egypt 248
Biringuccio 4, 75
Birjand, Iran 148
bismuth 72, 169
Black Forest, Germany 74, 172, 174, 188
blowing 2, 18, 100, 104, 204, 211, 213, 227–32, 235–40, 243–5, 249–51, 260, 265–7, 270, 363, 369, 371
Boğazköy/Hattusa, Turkey 128, 144
Bohemond of Antioch 277
Bologna, Italy 151–2
bone 65, 191
bone ash 6, 46, 198
Bonn, Germany 228
bornite 72
boron 48, 50
Boshan glass factory 123
Bosra, Syria 208, 222
bracelet: See armlet
Bracken 28, 106
Bringairet, France 192–3, 198
Bukhara, Uzbekistan 112, 296
Büklükale, Turkey 349
burkeite 52, 53
Bursa, Turkey 253, 256
Bury Ball, Devon, UK 78
button, conical 190, 195
button-based goblet 136, 143, 147
Byblos, Lebanon 128, 130, 144
Byzantine 81, 88, 94, 96
Byzantium 254–6

Caesaraea, Israel 28, 61, 208
cage cup 250, 370
Cairo (Fustat), Egypt 28, 253, 260, 268, 270, 277, 279, 281, 292–3, 295, 304, 334
Calabria 149, 150
calcite 60, 168
calcium 2, 5, 13, 16, 22, 27, 29, 36, 37, 41, 46, 48, 50, 53, 54, 60, 61, 65, 160, 179, 186, 188, 193, 199, 201
calcium antimonate 19, 77, 78, 79, 147

calcium carbonate 279, 315, 328, 333
calcium fluoride 79, 80
calcium oxide 42, 45, 46, 47, 48, 49, 53, 90–2, 94–5, 97, 101, 103–4, 108, 112–3, 117, 119, 122–6, 161–3, 170, 172, 183, 186, 190–2, 195–6, 199, 237–8, 241–8, 279, 281, 284–7, 291, 293–7, 300–2, 312, 315, 317, 322–3, 326, 329, 333, 350
calcium phosphate 28, 65
calcium stannate 79
Caliph 252
Caliph al-Ma'mun 254, 256, 303
Caliph al-Mutasim 254, 257, 266
Caliph al-Walīd 254–5, 257
Caliph Hārūn al-Rashid 253, 256–7, 266, 282, 303–4
Camargue, France 24
cameo 280, 350
Canal Bianco, Italy 183
Canatha (Qanawat) 205, 207
Canosa, Italy 211, 224–5, 233
Capitolias (Beit Ras) 206
carbon 81
carbon dioxide 29
carbon monoxide 81
Carchemish, Syria 131
carnelian 138, 306
Carpentras, France 236
Carthage, Tunisia 23, 62, 245
carving 13
cassiterite 13
cast vessel 204, 209–213, 223–4, 226, 228, 233, 235–6, 238–9, 242, 244–5, 249–50
Castello del Tartaro 153, 155
cedar 131
Cem Sultan tomb, Turkey 256
Cenezoic 337
cerium 58, 80, 117
cerium anomaly 117
Chagar Bazar, Syria 134
chaîne opératoire 12, 14, 21, 84, 187, 259–60, 276, 332, 340, 347, 354, 364
chalcedony 56
Chalcis, Syria 205
chalcocite 72
chalcopyrite 72
Chalestra, Macedonia, Greece 52
Chaoyang, China 119
charcoal 69, 89
chemical environment 68
Chenopodium album 31, 35
Chenopodium murale 31, 35
chert 56
Chiaromonte, Italy 183, 185, 188
China 197
chinane plant 263
chlorine 14, 29, 30
chlorite 13
Chotin, Slovakia 120

INDEX

Christian 254, 259–60, 266, 269, 282–3, 291, 303–4
Christopher Merrett 107, 108
chromite 60, 309
chromium 61, 62, 66, 67, 180, 182–3
Claudius 208
Cleopatra 208–9
cloisonné enamels 124
cobalt 8, 10, 66, 67, 69, 70, 71, 72, 73, 75, 131, 134, 142, 147–8, 157–9, 161–8, 170–5, 177, 181, 188–9, 191, 200–1, 210, 230–2, 365, 368
cobalt alum 159, 163
cobalt sulphide 70
cobaltite 70, 73, 74, 78, 159
Codova, Spain 255
colloid 76
Cologne, Germany 228
Colonia Julia Augusta Felix 215
conchoidal 4
Constantinus Harmenopulus, legal manual 208
copiapite 71
copper 7, 13, 15, 17, 42, 44, 49, 66, 67, 73, 75, 76, 80, 81, 128, 89, 96, 104, 163–6, 168, 172, 174, 188, 197, 210, 215, 232, 312–3
copper ruby glass 76
copper sulphide 11
coquimbite 71
core-formed vessel 20, 133, 135–6, 139, 142–5, 147, 166, 209–10, 223–6, 232–3
craftsmen 147
cretaceous 337
cretaceous limestone 34
criolite 122
Cristallo glass 46, 68, 76, 102, 107
cristobalite 56
Crottes à Roaix, Vaucluse, France 197
crucible 84, 85, 145, 148, 154, 156, 169, 183, 191, 198, 201, 214–5, 223, 229, 316–318, 320, 324–6, 324, 344, 354, 360–1, 363–4, 369
cullet 69, 142, 214, 232, 249, 275–7, 307, 319, 335
Cuma, Italy 63
cuneiform text 17, 20, 28, 156–7
cupellation 110
cuprous oxide 77, 79, 80
cuprovarite 13
Cyprus 130–1, 138, 153

Dagan 127
Dakhla oasis, Egypt 71, 159, 172
Damascus, Syria 31, 40, 138, 205, 207, 252–6, 259, 263, 266–69, 277–8, 290, 292, 294–5, 304, 335, 337–8, 343, 349–50, 356, 358–9
Darius III Codomanus 205

De Diversis Artibus 67
decapolis 206–7
decoloriser 76, 83
Deir 'Ain 'Abata, Jordan 160
Delos, Greece 211, 253
Demon mask 136, 142
dichroism 76
digenite 72
Dingcun, Vietnam 125
dinner service 224, 233–4
diorite 117
Dium (Tell Ashari?) 206
Dog Star 52
dolomitic limestone 28, 66
Doñana, Spain 29
Dor, Israel 93, 211, 270, 332, 351
Dorestad, the Netherlands 103
Dusu-coloured glass 21
Dyonysus 370
Dzhambul, Kazakhstan 112, 296
Dzharkutan, Uzbekistan 114

Eastern Han dynasty 123
Eastern Zhou dynasty 121
Ebla, Syria 127–8, 139, 144
ebony 131
Edirne, Turkey 253, 256
efflorescence technique 15
Egyptian blue 13, 215
eḫlipakku 168
Elam 130
Elamite 128, 144
Elateia, Greece 86, 89, 91, 161, 167, 170–6, 181, 188–9, 192–3
Electron probe microanalysis 160, 163, 166–7, 190, 283
Elkeb, Egypt 88
Emperor Constantine VI 256
Emperor Nicephorus I 256
Emporia 255
Empress Irene 256
enamel 80, 81, 88, 122, 124, 126, 257, 259, 264, 267–70, 277
energy quanta 68
Entremont, France 226, 246
eocene 337
epidote 60
erdkobalt 172
Eridu (Tell abu Shahrain), Iraq 133–4, 145, 158–9, 348, 364
erythrite 70, 159
Erzgebirge, Germany 74
Ethiopian highlands 331
Ethnographic studies *371*
Etruria 149–53, 155
Etruscan 94, 155, 352
Etymologiarum sive originum 52, 63
Euphrates River 33, 182, 246, 264, 266–7, 301, 332, 342, 356
Europe 85
europium 60

evaporite 52, 91, 92, 98, 100, 114
Exeter, UK 108
experiment 12, 284, 286–92, 301, 303–4, 314, 316, 342, 349
experimental archaeology 19

Fabbricia dei Soci, Italy 153–5
facet cut 249–50
Façon de Venise 102, 107–9, 111
Fagus syvatica (Beech) 48, 49, 68, 102, 105, 106
Faience 7, 9, 12, 13, 14, 16, 17, 59, 71, 75, 131, 133, 136, 138–9, 156, 163–4, 166–7, 183–92, 195–9, 201, 364–5
Failaka, Kuwait 16
Famen temple, Shanxi, China 358
Fars, Iran 73
Fatimid caliphate 253
Fayum, Egypt 169
feldspar 28, 54, 57, 60, 238, 240, 284, 286, 309, 324–5, 329–30, 332, 339, 354
Fengmen-ling, China 123
Fergana, Uzbekistan 112, 296
ferric oxide 42, 287–9, 325–7
filtration 186
Firozabbad, India 114
Fishbourne, UK 86 88, 120, 235
flint 4, 56, 264
fluorine 122
fluorite 122
Fondo Paviani, Italy 153–4
Fondo Zanotto, Italy 153
forest glass 48, 67
fractionation 169,
Fratta Polesine, Italy 152
Frattesina, Italy 89–90, 91, 125, 148, 150–5, 174–6, 183–5, 187–196, 198–9, 200–1
Freiburg, Germany 74
frit, 'frit' 18 20, 21, 24, 50, 63, 65, 91, 99, 115, 133, 136, 142, 148, 163–4, 166, 169, 198, 215, 231, 246, 274–6, 285, 316, 320, 361, 364
Friuli plain, Italy 148–9
fuel 98, 254, 259, 263, 273, 275
furnace 2, 84, 209–10, 212–23, 228–32, 363–5, 369, 371
Fustat: *See* Cairo

gabbroic rocks 60
Gadara (Umm Qais), Jordan 206
galena 74, 169
Gambassi, Italy 64, 65
garum (fish sauce) 208
Gaul 64
Gebel Zeit, Egypt 346
geographical provenance 326, 332, 361, 367
Geography 52

427

Index

geological provenance 95, 163, 179, 182, 199, 314, 328, 332, 361, 367
George Sandys 263
Georgius Agricola 50
Gerasa (Jerash) 206–7
Germagnana, Italy 64, 106
Gerterode, Germany 107
Girgensonia 31, 35, 312
Giribawa, Sri Lanka 115
glass batch 10, 11, 12, 36, 162, 187, 193, 198, 314–6, 319–20, 326–7
glass ceramic 18, 19, 156
glass colour 1, 66, 67
glass melt 311, 315–20, 327, 333, 342–345
glass weights 280, 289
glassy-faience 183–91, 195–6, 198–9, 201
Glastonbury, UK 105
glaze 194, 254, 256, 262–4, 271, 334, 364
glazed steatite 13
glazed vessel 133, 161–3
glebas nitri/glebeas nitri 52
goblet 290–1, 350
gold 10, 75, 76, 131–2, 153, 186, 194
gold ruby glass 76
gold-glass 209, 214
Golfe de Fos, France 235–40, 243–4
Gord, France 186, 195–8
granite 117, 331, 333
granitic rocks 60, 66
Greater Iraq 253
Greater Syria 253, 277
Grotte Vittorio Vecchi, Italy 183, 185
guild 132
gypsiferous limestone 33, 40, 47

Hadera, Israel 93, 238, 241–3, 253, 270–1, 274, 290
Hadrian 370
hafnium 88
Hagoshrim, Israel 211
Haifa, Israel 61, 62, 352
halite 52
Hallstatt C 94, 96
Hallstatt D 94, 96
Halocnemum 23
Halopeplis 23, 31, 32, 35, 36, 39
halophytic plant 23, 24, 28, 46, 54, 85
Haloxylon articulatum 31, 262
Haloxylon recurvum 29
Haloxyylon salicornium 28
Hama, Syria 144, 268
Hammada 23, 31
Hammada scoparia 24, 28, 262
Hammurabi 128
Han dynasty 5, 123, 364

harvest season 310, 313, 335
Hasanlu, Iran 94
Hasmonaeans 206
Hastinapur, India 113, 125
Hatshepsut 165, 168
Hatti 130, 138
Hattusili I 128
Hauterive Champrevèyres, Switzerland 9, 89, 189, 192–3, 195–7
Hazar-tam, China 121
Hazor, Israel 144
heirloom 194, 200
Hellenistic 4, 76, 83, 94, 95, 203–215, 217, 221, 222–33, 310, 351
hematite 81
Hengistbury Head, UK 88
Heracla, Syria 33
Herat, Afghanistan 28, 371
Herod 206, 208, 213
High lime-high alumina glass 117, 118, 364
High iron, managanese, titanium glass (HIMT) 95, 96, 354
Hippopotamus 131
Hippus (Qalaat el-Husn) 206
Hittite 92, 128–30, 136, 144, 163, 168, 349, 357
horse 130–31
horse riding equipment 80
horsemanship 130
Hurrian kingdom of Mitanni 162
Hurrian language 128–30
Hutton, UK 104
Hüttplatz, Germany 107
hyalos 204
hydriae 209

Ialysos, Rhodes 165
Iberian Peninsula 64
Ibn Jubair 256
Igbo Ukwa, Nigeria 118
igneous rocks 319
Ile-Ife, Nigeria 117
ilmenite 58, 60
incised decoration 291, 353
inductively coupled plasma atomic emission spectrometry 32, 44, 45
inductively coupled plasma emission spectrometry 26, 65, 114
inductively coupled plasma optical emission spectrometry 32
ingot copper 131,
ingot glass 131, 135–7, 139–42, 145, 151, 154, 162–4, 166, 170, 179, 181, 191, 202, 249, 334, 355, 357, 360, 369
ingot tin 131, 188–90, 192–6, 198
ingot, pick 151–3
innovation 84, 235, 243, 364, 367

iron 3, 42, 44, 57, 64, 65, 66, 67, 68, 70, 80, 159, 165, 172, 188, 210, 264, 267
Iron Age 6, 64, 70, 73, 74, 351–3, 356, 359
iron oxide 58, 59, 81, 159, 175, 188
iron titanite 58,
Ishtar, goddess 128, 140, 142
Isidore of Seville 51, 63
Islamic glass 83, 204, 252, 279–93, 295–7, 299–305, 309, 311, 316, 320–1, 323–4, 326, 330, 334–5, 338–40, 343, 345–50, 353, 355, 357–9, 369–70
ivory 131, 153, 155

Jabal Bishri, Syria 267, 356
Jaboul, Syria 33, 41
jade 119, 306, 364
Jalame, Israel 97–99, 231–2, 242, 247, 251
James of Vitry 268
Jebel Khalid, Syria 212, 233
Jerusalem, Israel 130, 144, 208, 211–13, 233, 250, 267
Jewish 260, 262, 266–69, 271, 279, 303
Jezaziyat, Iraq 28
John Greene 108
John the Grammarian 256
Josephus 62, 203, 211
Judaea 206, 208
juglet 226
Julia Felix, shipwreck, Italy 360
jurassic limestone 34

Kakovatos, tholos tomb, Greece 144, 165
Ka-La-Ke'er, China 120
Kaloula property, Rhodes 214
Kandahar, Afghanistan 28
Kar-Tukulti-Ninurta 135
Kashan, Iran 24, 73, 159, 262
Kassite 129, 135, 143, 349
Kausambi, India 112–13
Keli/Ūshnan 26–8, 53, 261–3, 357
Kemsar, Iram 159
Kerala coast, India 112
Kerma, Sudan 15, 184, 187, 195–7
Kharga oasis, Egypt 71, 159
Khirbat Faris, Jordan 116, 297–9, 350
kiln 215–8, 254, 271, 273–4, 364, 368
Kish, Iran 16
Kition, Cyprus 16
Kiziltur, Xinjiang, China 88, 122
Kizzuwatna, Anatolia 130
Knightons, Surrey 28
Kom Helil, Memphis, Egypt 16

Index

Kopia, Uttar Predesh, India 115
Kota Cina, Sumatra 122, 126
Kufa, Iraq 254

La Seube, France 75
La Tène 64
Lachish 144, 168
Laser ablation inductively coupled plasma mass spectrometry 60, 241
lajvard 73
lanthanum 180, 183, 309, 318
Laodice 205
lapis lazuli 1, 4, 10, 70, 73, 138, 143, 148, 156, 168, 202, 368
Latakia, Syria 253, 262
Late Helladic period 167
lattice 3
Lavrion, Greece 169, 369
Le Narde, France 153
leaching 50
lead 22, 43, 44, 57, 70, 73, 74, 75, 80, 83, 84, 96, 99, 109–12, 117, 119–26, 169–70, 188, 215
lead antimonate 77, 147, 160, 215, 334, 346
lead arsenate 79
lead isotope 74, 77, 131, 169–70, 333–5, 345–6, 355, 361, 366
lead ore 334, 346
lead oxide 267
lead oxide-silica glass 119, 126, 264, 302–3
lead-tin oxide 79, 267
Leicester, UK 248
Levant 203–4, 206, 210, 212–13, 215, 222–3, 225, 228–33, 235–43, 246–8, 250–1
ligand field theory 68
Limlinggerode, Germany 107
Limoges enamel 74
Lincoln, UK 87 105, 108–9, 111
linear absorption coefficient 66
linear-cut bowl 236–7, 239–40, 244
linen smoothers 109–10
linneite 73
Lisht, Egypt 165
Literno, Italy 63
lithification 60
Little Birches, U.K 104
London, UK 93 108, 204, 229
lost-wax technique 128
Lou Lan, Xinjiang, China 112
low magnesium-high potassium glass (LMHK) 167, 173–4, 183, 196
Lübeck, Germany 111
Luco/Laugen culture, Italy 152
Lustre decoration 280–2, 291–2, 296–7, 303, 350
Luxor, Egypt 61
Lycurgus cup 76, 370

Macedon 209
Madagascar 115–7
Magdalensburg, Austria 88
Maghreb 254, 267
magma 5
magnesium 13, 14, 16, 26, 27, 29, 36, 37, 40, 41, 53, 64, 72, 159–60, 167, 186
magnesium carbonate 279, 315
magnesium oxide (magnesia) 48, 49, 53, 76, 81, 83, 85, 88–91, 94, 96, 99, 101–2, 104, 109, 116, 122, 158, 160, 163, 170–2, 183, 186–7, 190–2, 195–6, 199, 235, 237, 245, 264–5, 279–80, 282–5, 290–7, 295, 301, 312, 315–8, 321–6, 337, 350, 359, 367
Mahilaka, Madagascar 116
Maidstone, UK 108
Mai'e Pur, China 93
Maktalta, al-Raqqa, Syria 271
malachite 168
Malaga, Spain 23, 253, 269
Malkata, Egypt 160, 165, 169, 170, 179–181, 200
Mamluk 70, 116
Mamluk caliphate 253, 269–70, 280, 292, 294–7, 299, 302, 304, 350
Mancetter, UK 248
manganese 36, 42, 50, 67, 68, 73, 74, 75, 76, 159, 163, 165, 172, 188, 210, 264–5
manganese oxide 42, 43, 48, 158, 166, 172–5, 189, 265, 287–9, 353
Mantai, Sri Lanka 59, 115, 299
Manutention, Lyon, France 228–9, 236, 246–7
Marco Polo 116
Maresha, Israel 211–2
Mari, Syria 127–8, 130, 132
Mariconda, Italy 89, 90, 125, 175–6, 183, 188–193, 199
Mark Anthony 208
Marlik, Iran 134–5
Marseilles, France 228, 246
marvered decoration 280
marvering 5, 20
Mayen, Germany 96
Mecca, Saudi Arabia 252
Medina 252, 255, 269
Megiddo, Israel 143
meku 139, 168
Melle, France 111
Meroitic Sudan 89
Meskani, Iran 72
Meskene, Syria 16
Mesopotamia 7, 10, 85, 88, 92, 309, 317–8, 323–5, 330, 346, 348–50, 355, 357, 359, 361–2
Metal hoard 150, 153
metamorphic rocks 57, 319, 331, 333

mica 60, 61
Midi, France 197–8
millefiori 257, 261, 265, 277
Minet el-Beida 138–9
Mishlab, al-Raqqa, Syria 271
mishna 211
Misr, Iraq 254
Mitanni 349
mixed-alkali glass 90–2, 94, 123–6, 167, 172–6, 183–201, 315, 323–5
mixing of glass 282, 284, 288, 291–2, 298–9, 303, 315, 319, 327, 334–5, 337, 341–2, 344, 346, 353–4, 358, 360–1
Mohammad 252
moil 229, 270, 358
monazite 58, 309
Mongol 262
Montagnana, Italy 151, 153
Monte Lecco, Italy 65
Moorgate, London, UK 229
Morgantina, Sicily 246
mosaic glass vessel 134–6, 139, 142–3, 147–8, 156–7, 209–10, 231, 233, 235, 242–3, 320
Moscoti di Cingol, Italy 150
mosque lamp 257, 269–70, 277, 280, 293–6, 304
mould-blown vessel 204
moulding 2, 13, 18, 236–40, 243–5, 249, 251, 260, 266, 363
muff window glass 105
mummification 51
Murad II, mosque 256
Murano, Italy 1
Murcia, Spain 253, 269, 349
Murex shell 131
Mursili 128–9
Muslim 94, 98, 109, 252, 254–5, 259, 263, 266–7, 269, 271, 277
Mutisalah beads 371
Mycenaean Greece 7, 130, 144, 150, 153–4, 158, 160, 164–7, 170–3, 176–8, 181, 199, 349, 355
Mycenaean glass 318, 323–5, 349, 355

Nabataeans 207
naga plant 156
Nahr Na'man 203
Nāir Muhammad ibn Qulāūn 270
Narbonne, France 23
Narim-Sin 128
Nasīr al-Dīn al-Tūsī 267
natrofairchildite 45
natron 13, 16, 51, 52, 66, 365–6
natron glass 46, 51, 64, 80, 81, 83, 88–103, 110, 112, 115–6, 118, 121–2, 184–5, 187, 194, 203–4, 212, 222–3, 228, 231, 235, 237, 239, 241–3, 245–6, 248, 280–4, 286–92,

429

297, 300–4, 307–10, 315, 329–331, 333, 346–9, 351–4, 356, 358–9, 361, 365, 367, 369
naturalis historia 51, 52, 250
Nawar 128
Nazimaruttash, king 143
Nefertiti 132
Němčice, Moravia 227
Neo-Assyrian 212
neodymium 43, 176–9, 201, 309, 318–9, 331, 340
neodymium isotopes 65, 98, 115, 163, 175, 180, 199, 330–1, 333, 335–6, 362, 369
neogene 33, 337
nephaline 54, 310
nephaline syanite 54
Nesikhons 92
network former 2
network modifier 2
network stabiliser 3, 27
neutron activation analysis 116
New Kingdom (Egypt) 160, 163, 166, 199
New Mexico 5
niccolite 72
nickel 49, 70, 159, 163, 166, 172, 188
nickel oxide 174–5
Nile river 61, 331–2, 352, 354, 366
Nimrud, Iraq 93–4
Nineveh, Iraq 16, 134, 143–4
Nippur, Iraq 16, 143–4
Nishapur, Iran 74, 253, 268, 277, 279, 281, 291, 293–5, 301, 304, 321–2, 350, 359
Noaea micronata 31, 35
Nuzi, Iraq 42, 132–3, 135, 140–4, 160–2, 164–5, 170–1, 180–1, 202, 323–5, 357
nyerereite 45

obsidian 2, 4, 306
ochre 197
oinochoai 155, 209, 223
olivine 60
Olympia, Greece 246
Oman 330, 338
Omar b. Abd al-Aziz 255
oos 53
opacity 10, 77, 82, 83, 84
opal 56
opalisers 77
oppida 156, 226, 353, 359
optical emission spectroscopy 164
Orontes River 205
orthoclase feldspar 65
ostrich egg shell 153
Ottoman 116, 256, 297–9, 350
Ouest Embiez I shipwreck 77
oxidising atmosphere 67
Oxus River 226

oxygen 2, 56, 69
oxygen isotope 166, 330, 333, 342–3, 366

palace of Niqme-pa, Alalakh, Turkey 137
Palafitte 149, 154
paleogene 33
Palestine/Palestinian 128, 130, 137, 144, 147, 205, 208, 210, 212, 225–7
Palmyra (Tadmur) 208, 267
Pandemic 370
paradigm shift 12, 17–19, 156, 364
pattern-moulded 270
Pavia, Italy 263
Peel castle, Isle of Man, UK 102
Pella (Tabaqat el-Fahil) 206, 212, 233
pendant 136, 138, 140, 162, 167, 170, 200, 363
Pendjikent, Tajikistan 112
Pergamon 205
Periplous Maris Erythraei 112
Persepolis 226
Petra, Jordan 78
Petronius's *Satyricon* 251
Peyrestortes, France 236
Pfahlbautonnchen mit Spirale bead 188
Philadelphia (Amman) 206–7
Phoenician 94, 203, 205–6, 223, 351–2, 356, 359
phosphorus 36, 46, 64, 186, 191, 302–3, 311, 315
phosphorus pentoxide 42, 49, 187, 191, 196, 285–6, 297, 312
pickeringite 71
Pikrolimni lake, Greece 366
pillar moulded bowl 234
Pingba, China 119
Pinus sp. (pine) 48
Piriform bottle 139, 140, 142–3
Pirotechnica 75
plagioclase feldspar 329
plant ash 1, 8, 10, 15, 18, 25, 27, 28, 51, 66, 71, 80, 81, 91, 134, 156, 158, 160, 163, 171, 174–5, 177–9, 182–4, 186, 201, 204, 212, 279, 282–4, 286, 310–12, 314, 315, 364–6, 369–70
plant ash glass 83, 85–6, 88–9, 91–2, 94, 96, 98–102, 108–9, 112–3, 116–7, 134, 159, 162–3, 165, 167, 170, 172–3, 185–89, 191, 197, 199, 201, 204, 235, 237, 239–40, 261–6, 271, 274, 276–7, 279, 280, 282–4, 286–8, 307–8, 311, 314, 315
plant ash purification 25, 27, 50, 279, 284, 306, 315, 319–20, 326, 360
plant taste 320
plant transpiration 328
plaque 161–2, 166–7, 172–3

Pliny the Elder 51, 62, 65, 92, 97, 203–6, 211, 222–3, 231, 308, 330, 351, 356
Po River, Italy 149, 150, 152, 155, 332
Poggio Civitale, Italy 155
Poland 184, 195–7
polis 205, 216
Pompeii 41
pontil 18
porcelain 6
Portland vase 230
potassium 5, 16, 26, 27, 30, 35, 36, 37, 40, 45, 50, 60, 64, 65, 71, 85, 167, 183–4, 186–8, 197–8, 200, 311–12, 314, 316–8, 322, 324, 360
potassium carbonate 315
potassium chloride 23, 35, 45, 315
potassium feldspar 46, 66, 71, 90, 94, 98, 199
potassium glass 183–4, 186–7, 189, 196–8
potassium oxide 36, 42, 45, 48, 49, 53, 66, 80, 81, 85, 88–92, 94, 96–9, 101–104, 107–8, 111–3, 117, 122, 124–6, 158–60, 164, 170–2, 182–3, 189, 191, 193–8, 235, 249, 282, 296–300, 312, 315, 321, 367
potassium sulphate 28
potassium-lime glass 124
Poviglio, Italy 183, 185, 188, 198
Prato di Frabulino, Italy 183, 188
priest 128, 132
priestess 128
primary glass production (glassmaking) 132–4, 139, 141, 146–8, 155, 158, 163, 165, 167–9, 172–3, 180–2, 199, 200–1, 204, 209, 210, 222, 226, 228–32, 235, 238, 242, 251, 262–3, 266–8, 274–7, 279–80, 282–4, 286, 289–90, 292, 294, 296, 300, 302, 304–305, 307–11, 313–4, 316–7, 327–32, 335–6, 338, 343–5, 337, 348–9, 351–4, 356–7, 359–61
production centre 163–4, 184, 191, 194, 197–202
proto-geometric period 167, 173
proton-induced X-ray emission (PIXE) 13, 163
Proto-villanovan 151, 152, 154, 155
provenance 165, 170, 177–8, 306–7, 309–14, 320–1, 326–8, 333–5, 338, 340, 345, 347, 349–55, 359–60
Pteridium aquilinium 48
Ptolemy I 205–6, 209
Ptolemy IV 222
Ptolemy V 222
Puglia, Italy 150–51
Pylona cemetery, Rhodes 166
pyrite 10

Index

pyrolucite 75
pyroxene 61, 329

Qadisiyya quarter, Baghdad, Iraq 267
Qal'at Sem'an 282,
Qamasar, Iran 73
Qantir-Piramesses, Egypt 18, 165, 169, 200, 317, 349
Qasr al-Banāt, al-Raqqa, Syria 267
Qasr al-Hayr al-Sharqi, Syria 28, 280, 282–3
Qasr al-Seghir, Morocco 280–1
Qatna 129
Qom process 14, 15
Qom, Turkey 28
quartz 13, 14, 15, 43, 54, 56, 57, 61, 90, 134, 142–3, 145, 156, 181–2, 184–5, 187, 198–9, 263–4, 276, 284, 309, 330, 332–3, 339, 355, 360–1, 365
quaternary 33
quaternary basalt 34
Queen Nefertiti 13
Quercus sp. (Oak) 48, 106
Qumrân, Israel 248

Ra's al-Hadd 338–9, 343
ramelsbergite 72
Ramla, Israel 238, 262
Raphana 206
rare earth anomaly 117
Ras Shamra, Syria 357
Rasāfa, Syria 356
Rathgall, Ireland 89, 91, 175–6, 186, 192–4, 199
Ravenscroft 111
Raya, Egypt 96, 295, 297, 300–2
recrystalisation 186
recycling 44, 248–50, 307, 320, 334, 361
reducing atmosphere 67, 68
reh 53, 54, 66, 115, 310
Reims, France 228
Rhanterium epapposum 28
Rhodes 88, 138, 209–10, 214–5, 223, 225–6, 246, 352, 359
rhyolitic rocks 60
rhyton 131
Rightly Guided Caliphs 252
ritual 1, 364
Rochetta 24
rod glass 143–4
Roman 4, 76, 77, 78, 81, 83, 88, 94–6, 120, 351
Rome, Italy 204, 206, 229, 247, 249–50, 370
Rosedale, UK 104
Rosetta, Egypt 263
rubidium 48, 104, 328
Rūm (Byzantines) 255

Rupar, India 112
Ruscino, France 235–240, 243–4

Sa Huyna culture, Vietnam 125
safflorite 72
Sagalassos, Turkey 54, 96
saggar 16
sagging 211–3, 226–7, 245
Şah Malek Paşa Camii, Turkey 256
Sahara desert 6
Sainte, France 247
Sākinu 132
Salicornia 23, 24, 28, 29
Salicornia kali 30
Salsola 30, 35, 37, 38, 39, 40, 41, 43, 47, 160, 182, 311–4, 321–5, 360
Salsola europaea 173
Salsola fruticosa 172
Salsola jordanicola 31, 32, 32, 36, 44, 46, 311–2
Salsola kali 24, 32, 35, 40, 42, 173
Salsola soda 24, 172, 262
Salsola sueda 32, 262
Salsola tragus 24
Salsola vermiculata 23, 25, 32, 35, 36, 42, 44, 46, 173
saltpetre 186
Samaria 206, 208
Samarium 330
Samarkand, Uzbekistan 112, 296
Sāmārra, Iraq 74, 253, 256, 359
Samudera-Pasai, Indonesia 115
Sandrakatsky, Madagascar 116
sandstone 60
Sanguinaire shipwreck, Corsica 227, 359
Santa Maria de Pedralbes church, Barcelona, Spain 101
Saqal-tam, China 121
Sar Dheri, India 112
Sargon 127
Sarmin, Syria 262
Sasanian 76, 88, 204, 212, 260, 265–6, 282, 284–6, 289, 291, 293–5, 303, 308–9, 321–4, 326, 349–50
scanning electron microscopy 13, 163, 190, 245
scarab 132
scaraboïde 140
sceptre 139
scratch decorated 281, 292, 296–7, 300–1, 304, 350, 359
Scythopolis: *See* Beth She'an
seal 167
seashell 28, 46, 329
seaweed 6, 106–107
secondary geological provenance 332
secondary glass production (glassworking) 139, 141, 143, 145–7, 153, 155, 163, 201, 242, 245,

247, 266–8, 269, 272, 274, 292, 343, 349, 352
Seleucia (Tell Umar), Iraq 87, 88, 205
Seleucid 205–6, 209, 215–6
Seleucos I (Nicator) 205
Seljuqs 253, 258
semiprecious stone 146, 148, 156–7, 368
Sepphoris, Israel 208, 231
Serçe Limanı shipwreck 74, 253, 270, 276–7, 290, 293–5, 301–2, 335, 346–7, 358
Severianus 370
Severus Alexander 208
Shahr-i-Banu, Afghanistan 112, 299
Shamshi-Adad 128
Shechem, Israel 208
Shikohabad, India 53
short glass 100
Sidon, Lebanon 51, 63, 204–5, 210, 240
Siena, Italy 101
Sigtuna, Sweden 110
silica 1, 2, 5, 6, 8, 10, 15, 17, 18, 22, 44, 54, 56, 57, 59, 60, 62, 75, 77, 84, 88, 92, 94–5, 102, 104, 109–13, 119–121, 123–6, 364, 366–7
silica tetrahedra 56, 68
silicon 2, 27
silk road 88, 91, 109, 112, 120, 122, 126, 197, 254, 268, 358–9, 368
silver 70, 72, 76, 111, 168, 172
sinter 21, 142
Siraf, Iran 281
skutterudite 70, 72, 188
slag 5, 6, 8, 188
Sâone River, France 247
soap 85, 262–3, 265, 268–9, 364
soapstone 306,
soda-alumina glass 113–15, 117, 121
sodium 2, 5, 27, 41, 54, 60, 186, 189
sodium carbonate 44, 52, 53, 72, 262, 313, 315, 333
sodium chloride 23, 24, 29, 45, 52, 53, 92
sodium fluorosilicate 122
sodium oxalate 44
sodium oxide 30, 36, 42, 45, 66, 100, 161, 170, 194, 198
sodium phosphate 79
sodium sesquicarbonate 52, 187
sodium sulphate 23, 53
Solkhat, Crimea 269, 349
spacer bead 144, 147
specialisation 84, 307, 351, 355
sphalerite 169
spinel 73
Sri Lanka 114, 298–300
St. Jean de Garguier, France 236
St. Maria Assunta, Torcello, Italy 11, 101

431

stamnoi 209
Staré Hradisco, Moravia 79
Stato's tower 208
stibnite 77, 78
stoke hole 217
stoneware 6
Strabo 52, 63, 203–5, 210–11, 222, 229–30
striking glass 77, 80, 147, 157
strontium 14, 42, 49, 61, 64, 104, 115–6, 238–41, 249
strontium isotope 26, 72, 103, 115, 163, 175–6, 178–9, 180, 199, 200–1, 294, 301–2, 313, 328–30, 332–9, 343, 349, 351, 353–4, 358, 361–2, 369
strontium oxide 300
Sueda 32
Sulaiman 73
sulaimani 73
sulphur 50, 70, 74, 159, 172
sulphur trioxide 30, 48, 78
Sultan Mehmed I mosque complex, Turkey 256
Sumer 127–8
Sumerian 132–3
sūq al-marzuqīn, Aleppo, Syria 267
Susa, Iran 202
Syriac texts 304

Tabriz, Iran 73, 253, 256
Tacitus 63
talc 7, 13
Talmasi, Iran 72
Tamarisk 28
Tamarix meyeri 28
Tamarix tetrandra 28
Tang dynasty 122, 126, 358
tank furnace 18, 69, 95, 97, 100, 114, 209, 214–222, 229–31, 269–76, 371
Taposiris Magna, Egypt 231, 247–8
Tārīkh 255
Tashkent, Uzbekistan 296
Tchoga Zanbil, Iran 143, 160, 172
Tebtynis, Egypt 242
Tel Mevorakh, Israel 144
Tel Yin'am, Israel
Tell abu Ali, al-Raqqa, Syria 271, 273–4
Tell abu Sarbut, Jordan 116, 297–9, 350
Tell al-Fakhar, Iraq 134
Tell al-Fakhariya, Syria 130
Tell Anafa, Israel 210–11, 233, 246
Tell Ashara-Terqa (Syria) 164
Tell Asmar (ancient Eshnunna), Iraq 134
Tell Aswad, al-Raqqa, Syria 273
Tell Atchana, Turkey: *See* Alalakh
Tell Bellor, al-Raqqa, Syria 271, 275, 325

Tell Brak (Nagar), Syria 8, 16, 42, 85–6, 89, 127, 129, 134–5, 137, 139–40, 143, 156, 159–62, 164, 170–1, 176, 179–81, 323–5, 357
Tell Dan, Israel 160
Tell el Amarna, Egypt 18, 61, 85, 86, 89, 140, 161, 164–5, 167–9, 171, 176–7, 179–181, 200, 323–5, 349, 352, 357
Tell el-Ashmunein, Egypt 241, 352
Tell el-Rimah, Iraq 16, 160, 172
Tell Fukhar, al-Raqqa, Syria 271, 294–5, 325, 336, 342
Tell Judeideh, Syria 133–4
Tell Rimah, Iraq 134–6, 142, 160, 172
Tell Zujaj, al-Raqqa, Syria 271, 274, 289, 325, 335–6, 342
Terqa (Tell Ashara), Syria 135
Terra sigillata shaped vessels 236–240, 244, 246, 250, 369
Terramare 150, 155, 198
tertiary 33
tertiary limestone 241
tesserae 78, 80, 81, 229, 231–2, 254–5, 259, 277, 280, 293–5, 302, 304, 345, 350, 359, 370
Tetrapolis 205, 212
textile 131
Thasos, Greece 91, 144
Thebes, Greece 161, 167, 171–3, 175, 181
Theologos, Greece 192–3, 196
Theophilus presbiter 21, 67, 255
thermal ion mass spectrometry (TIMS) 26
Thessaloniki, Greece 366
Thisbe, Greece 166
thorium 60
Tiberius 208, 233
Ticino, Italy 204
Tigranes 207
Tigris River, Italy 182, 266, 268, 301, 332
tile 256
time of flight-secondary ion mass spectrometry (ToF-SIMS) 169, 200
tin 43, 44, 70, 77, 188, 367
Tiryns 166
titanium 61, 96, 180, 183
titanium oxide 172, 249, 300
titanomagnetite 13
Torcello, Italy 59, 93, 100–1
trace element 83, 85, 88, 102–3, 159, 165, 171, 173, 178–9, 181–3, 185, 199–201, 279, 302, 304, 312, 318, 330–1, 349, 351, 355, 359, 361
trade 209, 211–12, 227–233, 246, 249–50, 306–8, 320, 330, 332, 348–9, 356–61
trade wind beads 114

Tradelière shipwreck, France 227, 249
transition temperature 5, 17, 19
transmission electron microscopy (TEM) 77
Trentino, Italy 149, 151
trianite 70
tridymite 56, 184–5, 189, 198
Trinitapoli, Italy 183, 198
tripod 161
Tripoli, Syria 262
trona 13, 51, 52, 53, 187, 310
tube 172, 213–5
tube-blowing 250
Tulunids 253
turquoise 1, 10, 156, 306, 368
Tuthmosis II 147
Tuthmosis III 147 165, 168–9
Tuthmosis IV 165 169
Tuttul (Raqqa), Syria 127–9
Tyre, Lebanon 61, 168, 210, 256, 267–8, 274, 280, 290, 292, 295, 301, 335, 338–9, 343–5, 349

Üç Serefeli mosque, Turkey 256
Ugarit, Syria 16, 128, 135–6, 160, 163, 163, 164, 338, 357
Ukhadir, Syria 256
Ulu Burun shipwreck, Turkey 131, 161, 166, 170–1, 181, 355, 357
Umayyad caliphate 94, 252–4, 256, 259–261, 266, 280–2, 292, 304, 330, 351, 353
Umbria, Italy 151
ummânu 132
Untashgal, king 144
Ur, Iraq 127–8, 133, 143
uranium 72, 115–6, 122
Urkesh 128
Urshu 128
Ūshnan: *See* Keli
Uttar Pradesh, India 114, 122

Val d'Adige, Italy 149
Valle d'Aosta, Italy 149
Valli Grande Veronesi, Italy 154–55
vanadium 62
varec (kelp) 186
Veh Ardašīr, Iraq 88
Veii, Italy 152
Venetians 24, 27, 40, 76, 77, 101–2, 107–9, 262–3, 277, 316, 321, 358
Veneto 90, 101, 149, 151, 153
Venice, Italy 253, 262, 277
Vergina, Greece 246
Verucchio, Italy 152
vessel 2, 14, 15, 16, 20
Vicofertile, Italy 183, 185
Vienne, France 246
Viking 94, 102, 110, 255

Index

Villamarzana, Italy 151, 153–4
Villanovan 94, 151–152, 154
viscosity 5, 244–5
Vitrum blancum 102, 107–9, 111
vizier 132
Volturno River, Italy 62
Vopiscus 370

Wadi el Tur, Egypt 300–2
Wadi el Natrun, Egypt 51, 92, 231, 247–8, 300, 310, 331, 356, 366
Wady Qirud 147, 168
Warring States dynasty 94, 123
Wasshukkani 130
Weald, United Kingdom 104
Western Zhou dynasty 122
Wijnaldum, the Netherlands 81
William of Tyre 268
winding 2, 20

window glass 106, 257, 267, 270, 291, 302–3, 307, 336, 338, 341–4, 349–50, 370
wine 264
Wingerode, Germany 107
wollastonite 13
wood ash 50, 51, 74, 102–7, 109, 111, 309
Woolwich (UK) sands 108
workshop 334

X-ray diffraction spectroscopy (XRD) 26, 30, 32, 44, 45, 54, 77, 315
X-ray fluorescence spectroscopy (XRF) 30, 42, 65, 115, 162, 163, 241, 300

Yamkhad (kingdom of) 128–9
Ya'qubi 254
Yāqūt 267

yttrium 58
Yuan dynasty 123

Zagros mountains, Iran 129
Zakik, Egypt 231, 248
Zanjan, Iran 73
zatte rippenschale 235, 242
Zengids 253
Zhalainuer tomb, Inner Mongolia 121
Zhongyang dong, Korea 121
ziggurat 144
Zimrilim, king 132
zinc 42, 43, 44, 48, 57, 70, 71, 73, 74, 104, 159, 166, 169, 172, 188
zinc oxide 159
zircon 58, 59, 60, 64, 238, 309
zirconium 61, 62, 88, 96, 115, 180, 183, 238–9, 241–3, 248–9, 318

For EU product safety concerns, contact us at Calle de José Abascal, 56–1°, 28003 Madrid, Spain or eugpsr@cambridge.org.

www.ingramcontent.com/pod-product-compliance
Ingram Content Group UK Ltd.
Pitfield, Milton Keynes, MK11 3LW, UK
UKHW050110230326
469255UK00020B/476